# Intelligent Systems Reference Library 39

**Editors-in-Chief**

Prof. Janusz Kacprzyk
Systems Research Institute
Polish Academy of Sciences
ul. Newelska 6
01-447 Warsaw
Poland
E-mail: kacprzyk@ibspan.waw.pl

Prof. Lakhmi C. Jain
School of Electrical and Information
Engineering
University of South Australia
Adelaide
South Australia SA 5095
Australia
E-mail: Lakhmi.jain@unisa.edu.au

T0137362

For further volumes:
http://www.springer.com/series/8578

George A. Anastassiou and Iuliana F. Iatan

# Intelligent Routines

Solving Mathematical Analysis with Matlab,
Mathcad, Mathematica and Maple

*Authors*
George A. Anastassiou
Department of Mathematical Sciences
University of Memphis
Memphis
USA

Iuliana F. Iatan
Department of Mathematics and
Computer Science
Technical University of Civil Engineering
Bucharest
Romania

ISSN 1868-4394
ISBN 978-3-642-43073-2
DOI 10.1007/978-3-642-28475-5
Springer Heidelberg New York Dordrecht London

e-ISSN 1868-4408
ISBN 978-3-642-28475-5 (eBook)

Printed on acid-free paper

Springer is part of Springer Science+Business Media (www.springer.com)

Dedicated to our families.

"Homines dum docent discunt."

Seneca, Epistole 7

"Nihil est in intellectu, quod non prius fuerit in sensu."

John Locke

"Les beaux (grands) esprits se rencontrent."

Voltaire

"Men should be what they seem,
Or those that be not, would they might seem none."

Shakespeare, Othello III. 3

"Science needs a man's whole life. And even if you had two lives, they would not be enough. It is great passion and strong effort that science demands to men ..."

I. P. Pavlov

VIII

"Speech is external thought, and thought internal speech."

A. Rivarol

"Nemo dat quod non habet."

Latin expression

"Scientia nihil aliud est quam veritatis imago."

Bacon, Novum Organon

# Preface

Real Analysis is a discipline of intensive study in many institutions of higher education, because it contains useful concepts and fundamental results in the study of mathematics and physics, of the technical disciplines and geometry.

This book is the first one of the kind that solves mathematical analysis problems with all four related main software Matlab, Mathcad, Mathematica and Maple.

Besides the fundamental theoretical notions, the book contains many exercises, solved both mathematically and by computer, using: Matlab 7.9, Mathcad 14, Mathematica 8 or Maple 15 programming languages.

Due to the diversity of the concepts that the book contains, it is addressed not only to the students of the Engineering or Mathematics faculties but also to the students at the master's and PhD levels, which study Real Analysis, Differential Equations and Computer Science.

The book is divided into nine chapters, which illustrate the application of the mathematical concepts using the computer. The introductory section of each chapter presents concisely, the fundamental concepts and the elements required to solve the problems contained in that chapter. Each chapter finishes with some problems left to be solved by the readers of the book and can verified for the correctness of their calculations using a specific software such as Matlab, Mathcad, Mathematica or Maple.

The first chapter presents some basic concepts about the theory of sequences and series of numbers.

The second chapter is dedicated to the power series, which are particular cases of series of functions and that have an important role for some practical applications; for example, using the power series we can find the approximate values of some functions so we can appreciate the precision of a computing method.

In the third chapter are treated some elements of the differentiation theory of functions.

The fourth chapter presents some elements of Vector Analysis with applications to physics and differential geometry.

The fifth chapter presents some notions of implicit functions and extremes of functions of one or more variables.

Chapter six is dedicated to integral calculus, which is useful to solve various geometric problems and to mathematical formulation of some concepts from physics.

Seventh chapter deals with the study of the differential equations and systems of differential equations that model the physical processes.

The chapter eight deals with the line and double integrals. The line integral is a generalization of the simple integral and allows the understanding of some concepts from physics and engineering; the double integral has a meaning analogous to that of the simple integral: like the simple definite integral is the area bordered by a curve, the double integral can be interpreted as the volume bounded by a surface.

The last chapter is dedicated to the triple and surface integral calculus. Although it is not possible a geometric interpretation of the triple integral, mechanically speaking, this integral can be interpreted as a mass, being considered as the distribution of the density in the respective space. The surface integral is a generalization of the double integral in some plane domains, as the line integral generalizes the simple definite integral.

This work was supported by the strategic grant POSDRU/89/1.5/S/ 58852, Project "Postdoctoral programme for training scientific researchers" cofinanced by the European Social Fund within the Sectorial Operational Program Human Resources Development 2007-2013.

The authors would like to thank Professor Razvan Mezei of Lander University, South Carolina, USA for checking the final manuscript of our book.

January 10, 2012

George Anastassiou,
Memphis
USA


Iuliana Iatan
Bucharest
Romania

# Contents

**1  Sequences and Series of Numbers** ........................ 1
1.1  Cauchy Sequences ........................................ 1
1.2  Fundamental Concepts .................................... 3
    1.2.1  Convergent Series ............................... 3
        1.2.1.1  Cauchy's Test .......................... 5
    1.2.2  Divergent Series ................................ 7
    1.2.3  Operations on Convergent Series ................. 11
1.3  Tests for Convergence of Alternating Series .............. 14
1.4  Tests of Convergence and Divergence of Positive Series .... 16
    1.4.1  The Comparison Test I ........................... 16
    1.4.2  The Root Test ................................... 19
    1.4.3  The Ratio Test .................................. 22
    1.4.4  The Raabe's and Duhamel's Test .................. 26
    1.4.5  The Comparison Test II .......................... 29
    1.4.6  The Comparison Test III ......................... 30
1.5  Absolutely Convergent and Semi-convergent Series ........ 31
1.6  Problems ............................................... 34

**2  Power Series** .............................................. 41
2.1  Region of Convergence ................................... 41
2.2  Taylor and Mac Laurin Series ............................ 49
    2.2.1  Expanding a Function in a Power Series ........... 49
2.3  Sum of a Power Series ................................... 60
2.4  Problems ............................................... 65

**3    Differentiation Theory of the Functions** ................. 71
  3.1  Partial Derivatives and Differentiable Functions
     of Several Variables .................................... 71
     3.1.1  Partial Derivatives .............................. 71
     3.1.2  The Total Differential of a Function .............. 83
     3.1.3  Applying the Total Differential of a Function
         to Approximate Calculations ..................... 90
     3.1.4  The Functional Determinant ..................... 93
     3.1.5  Homogeneous Functions ......................... 99
  3.2  Derivation and Differentiation of Composite Functions
     of Several Variables .................................... 102
  3.3  Change of Variables ................................... 119
  3.4  Taylor's Formula for Functions of Two Variables ........ 126
  3.5  Problems ............................................. 143

**4    Fundamentals of Field Theory** ......................... 157
  4.1  Derivative in a Given Direction of a Function ........... 157
  4.2  Differential Operators ................................. 162
  4.3  Problems ............................................. 179

**5    Implicit Functions** ..................................... 187
  5.1  Derivative of Implicit Functions ....................... 187
  5.2  Differentiation of Implicit Functions ................... 193
  5.3  Systems of Implicit Functions ......................... 203
  5.4  Functional Dependence ................................ 209
  5.5  Extreme Value of a Function of Several Variables
     Conditional Extremum ................................ 214
  5.6  Problems ............................................. 229

**6    Terminology about Integral Calculus** ................... 245
  6.1  Indefinite Integrals ................................... 245
     6.1.1  Integrals of Rational Functions ................... 245
     6.1.2  Reducible Integrals to Integrals of Rational
         Functions ...................................... 251
         6.1.2.1  Integrating Trigonometric Functions ....... 251
         6.1.2.2  Integrating Certain Irrational Functions ... 252
  6.2  Some Applications of the Definite Integrals in Geometry
     and Physics ........................................... 260
     6.2.1  The Area under a Curve ........................ 260
     6.2.2  The Area between by Two Curves ................ 265
     6.2.3  Arc Length of a Curve ......................... 269
     6.2.4  Area of a Surface of Revolution ................. 274
     6.2.5  Volumes of Solids ............................. 276
     6.2.6  Centre of Gravity ............................. 277

6.3   Improper Integrals.................................... 280
      6.3.1   Integrals of Unbounded Functions ................ 280
      6.3.2   Integrals with Infinite Limits..................... 288
      6.3.3   The Comparison Criterion for the Integrals ........ 293
6.4   Parameter Integrals.................................. 296
6.5   Problems ........................................... 302

7   **Equations and Systems of Linear Ordinary**
    **Differential Equations** .................................. 317
    7.1   Successive Approximation Method ...................... 317
    7.2   First Order Differential Equations Solvable
          by Quadratures ..................................... 320
          7.2.1   First Order Differential Equations with Separable
                  Variables ..................................... 322
          7.2.2   First Order Homogeneous Differential Equations ... 324
          7.2.3   Equations with Reduce to Homogeneous
                  Equations .................................... 326
          7.2.4   First Order Linear Differential Equations .......... 330
          7.2.5   Exact Differential Equations .................... 332
          7.2.6   Bernoulli's Equation ........................... 339
          7.2.7   Riccati's Equation ............................. 340
          7.2.8   Lagrange's Equation ........................... 342
          7.2.9   Clairaut's Equation ............................ 346
    7.3   Higher Order Differential Equations.................... 349
          7.3.1   Homogeneous Linear Differential Equations
                  with Constant Coefficients....................... 349
          7.3.2   Non-homogeneous Linear Differential Equations
                  with Constant Coefficients....................... 357
                  7.3.2.1   The Method of Variation of Constants..... 357
                  7.3.2.2   The Method of the Undetermined
                            Coefficients .......................... 360
          7.3.3   Euler's Equation ............................... 367
          7.3.4   Homogeneous Systems of Differential Equations
                  with Constant Coefficients....................... 369
          7.3.5   Method of Characteristic Equation .............. 371
          7.3.6   Elimination Method ............................ 373
    7.4   Non-homogeneous Systems of Differential Equations
          with Constant Coefficients............................. 377
    7.5   Problems ........................................... 380

8   **Line and Double Integral Calculus** ...................... 395
    8.1   Line Integrals of the First Type ........................ 395
          8.1.1   Applications of Line Integral of the First Type ..... 396
    8.2   Line Integrals of the Second Type ..................... 406
    8.3   Calculus Way of the Double Integrals ................... 421

8.4   Applications of the Double Integral ..................... 430
    8.4.1  Computing Areas ............................. 430
    8.4.2  Mass of a Plane Plate......................... 431
    8.4.3  Coordinates the Centre of Gravity of a Plane
         Plate ...................................... 433
    8.4.4  Moments of Inertia of a Plane Plate .............. 436
    8.4.5  Computing Volumes ........................... 438
8.5   Change of Variables in Double Integrals ................. 441
    8.5.1  Change of Variables in Polar Coordinates.......... 441
    8.5.2  Change of Variables in Generalized Polar
         Coordinates ................................. 445
8.6   Riemann-Green Formula ............................. 447
8.7   Problems........................................... 454

9   **Triple and Surface Integral Calculus** .................... 475
9.1   Calculus Way of the Triple Integrals .................... 475
9.2   Change of Variables in Triple Integrals ................. 477
    9.2.1  Change of Variables in Spherical Coordinates ...... 477
    9.2.2  Change of Variables in Cylindrical Coordinates..... 480
9.3   Applications of the Triple Integrals .................... 485
    9.3.1  Mass of a Solid............................... 485
    9.3.2  Volume of a Solid ............................ 493
    9.3.3  Centre of Gravity ............................ 499
    9.3.4  Moments of Inertia ........................... 509
9.4   Surface Integral of the First Type ..................... 512
9.5   Surface Integral of the Second Type ................... 520
    9.5.1  Flux of a Vector Field.......................... 524
    9.5.2  Gauss- Ostrogradski Formula ................... 529
    9.5.3  Stokes Formula .............................. 540
9.6   Problems........................................... 551

**References** ............................................... 573

**List of Symbols** .......................................... 577

**Index** .................................................... 579

# 1
# Sequences and Series of Numbers

## 1.1 Cauchy Sequences

**Definition 1.1** (see [41], p.13). A sequence $(x_n)_{n \in \mathbb{N}}$ is **convergent** if $(\exists)$ $a \in \mathbb{R}$ such that $(\forall)\ \varepsilon > 0, (\exists)\ n_\varepsilon \in \mathbb{N}$ such that $|x_n - a| < \varepsilon, (\forall)\ n \geqslant n_\varepsilon$.

**Definition 1.2** (see [41], p.23). A sequence $(x_n)_{n \in \mathbb{N}}$ is called a **Cauchy sequence** if the terms of the sequence, eventually all become arbitrarily close to one another, i.e. if for $(\forall)\ \varepsilon > 0, (\exists)\ n_\varepsilon \in \mathbb{N}$ such that $(\forall)\ n \geqslant n_\varepsilon$ one obtains $|x_{n+p} - x_n| < \varepsilon, (\forall)\ p \in \mathbb{N}$.

**Example 1.3.** Prove that the following sequence is a Cauchy sequence :

a) $x_n = \dfrac{\cos 1!}{1 \cdot 2} + \dfrac{\cos 2!}{2 \cdot 3} + \cdots + \dfrac{\cos n!}{n(n+1)}, \quad (\forall)\ n \in \mathbb{N}^*$

b) $x_n = \dfrac{\cos \alpha}{1^2} + \dfrac{\cos 2\alpha}{2^2} + \cdots + \dfrac{\cos n\alpha}{n^2}, \quad (\forall)\ n \in \mathbb{N}^*,\ (\forall)\ \alpha \in \mathbb{R}.$

**Solutions.**

a) We shall have

G.A. Anastassiou and I.F. Iatan: Intelligent Routines, ISRL 39, pp. 1–40.
springerlink.com © Springer-Verlag Berlin Heidelberg 2013

$$|x_{n+p} - x_n| = \left| \sum_{k=n+1}^{n+p} \frac{\cos k!}{k(k+1)} \right| \leqslant \sum_{k=n+1}^{n+p} \frac{1}{k(k+1)}$$

$$= \sum_{k=n+1}^{n+p} \left( \frac{1}{k} - \frac{1}{k+1} \right) = \frac{1}{n+1} - \frac{1}{n+2}$$

$$+ \frac{1}{n+2} - \frac{1}{n+3} + \cdots + \frac{1}{n+p} - \frac{1}{n+p+1}$$

$$= \frac{1}{n+1} - \frac{1}{n+p+1} < \frac{1}{n+1} \xrightarrow[n \to \infty]{} 0.$$

We shall use the definition of convergence of the sequence $y_n = \frac{1}{n+1}$, $n \in \mathbb{N}^*$ in 0 to determine the rank $n_\varepsilon \in \mathbb{N}$ such that $(\forall)\ n \geqslant n_\varepsilon$ we have $|x_{n+p} - x_n| < \varepsilon, (\forall)\ p \in \mathbb{N}$.

Hence, using the Definition 1.1, $(\forall)\ \varepsilon > 0, (\exists)\ n_\varepsilon \in \mathbb{N}$ such that $\frac{1}{n+1} < \varepsilon$, $(\forall)\ n \geqslant n_\varepsilon$.

It results that $n > \frac{1}{\varepsilon} - 1$; we choose

$$n_\varepsilon = \left[ \frac{1}{\varepsilon} - 1 \right] + 1;$$

we achieve

$$n > \left[ \frac{1}{\varepsilon} - 1 \right] + 1 > \frac{1}{\varepsilon} - 1.$$

As $(\forall)\ \varepsilon > 0, (\exists)\ n_\varepsilon = \left[ \frac{1}{\varepsilon} - 1 \right] + 1 \in \mathbb{N}$, such that $(\forall)\ n \geqslant n_\varepsilon$ we have $|x_{n+p} - x_n| < \varepsilon$; according to the Definition 1.2 it results that

$$x_n = \sum_{k=1}^{n} \frac{\cos k!}{k(k+1)},\ n \in \mathbb{N}^*$$

is a Cauchy sequence.

b) We can write

$$|x_{n+p} - x_n| = \left| \sum_{k=n+1}^{n+p} \frac{\cos k\alpha}{k^2} \right| \leqslant \sum_{k=n+1}^{n+p} \frac{1}{k^2} \leq \sum_{k=n+1}^{n+p} \frac{1}{k(k-1)}$$

$$= \sum_{k=n+1}^{n+p} \left( \frac{1}{k-1} - \frac{1}{k} \right) = \frac{1}{n} - \frac{1}{n+1}$$

$$+ \frac{1}{n+1} - \frac{1}{n+2} \cdots + \frac{1}{n+p-1} - \frac{1}{n+p}$$

$$= \frac{1}{n} - \frac{1}{n+p} < \frac{1}{n} \xrightarrow[n \to \infty]{} 0.$$

We shall use the definition of convergence of the sequence $y_n = \frac{1}{n}$, $n \in \mathbb{N}^*$ in 0 to determine the rank $n_\varepsilon \in \mathbb{N}$ such that $(\forall)$ $n \geqslant n_\varepsilon$ we have $|x_{n+p} - x_n| < \varepsilon$, $(\forall)$ $p \in \mathbb{N}$.

Hence, using the Definition 1.1, $(\forall)$ $\varepsilon > 0$, $(\exists)$ $n_\varepsilon \in \mathbb{N}$ such that $\frac{1}{n} < \varepsilon$, $(\forall)$ $n \geqslant n_\varepsilon$. It results that $n > \frac{1}{\varepsilon}$; we choose

$$n_\varepsilon = \left[\frac{1}{\varepsilon}\right] + 1;$$

we achieve

$$n > \left[\frac{1}{\varepsilon}\right] + 1 > \frac{1}{\varepsilon}.$$

As $(\forall)$ $\varepsilon > 0$, $(\exists)$ $n_\varepsilon = \left[\frac{1}{\varepsilon}\right] + 1 \in \mathbb{N}$, such that $(\forall)$ $n \geqslant n_\varepsilon$ we have $|x_{n+p} - x_n| < \varepsilon$; according to the Definition 1.2 it results that

$$x_n = \sum_{k=1}^{n} \frac{\cos k\alpha}{k^2}, \quad n \in \mathbb{N}^*$$

is a Cauchy sequence.

## 1.2   Fundamental Concepts

### 1.2.1   Convergent Series

**Definition 1.4** (see [41], p. 29). A number series

$$\sum_{n=1}^{\infty} a_n = \sum_{n \geq 1} a_n = a_1 + a_2 + \cdots + a_n + \cdots , \qquad (1.1)$$

is called **convergent** if the sequence of its **partial sums**

$$S_n = \sum_{i=1}^{n} a_i \qquad (1.2)$$

has the finite limit

$$S = \lim_{n \to \infty} S_n. \qquad (1.3)$$

**Definition 1.5** (see [41], p. 30). The quantity from (1.3) is called the **sum of the series** $\sum_{n \geq 1} a_n$.

**Example 1.6.** Prove that the following series are convergent:

a) $\dfrac{1}{5} + \dfrac{1}{45} + \cdots + \dfrac{1}{16n^2 - 8n - 3} + \cdots$

b) $\displaystyle\sum_{n \geq 1} \dfrac{1}{\left(n + \sqrt{2}\right)\left(n + \sqrt{2} + 1\right)}.$

**Solutions.**
a) We can notice that

$$16n^2 - 8n - 3 = 16n^2 + 4n - 12n - 3 = (4n - 3)(4n + 1);$$

therefore

$$S_n = \sum_{k=1}^{n} a_k = \sum_{k=1}^{n} \dfrac{1}{(4k - 3)(4k + 1)} = \dfrac{1}{4} \sum_{k=1}^{n} \left( \dfrac{1}{4k - 3} - \dfrac{1}{4k + 1} \right)$$
$$= \dfrac{1}{4} \left( 1 - \dfrac{1}{4n + 1} \right) \xrightarrow[n \to \infty]{} \dfrac{1}{4},$$

i.e.

$$\sum_{n \geq 1} \dfrac{1}{16n^2 - 8n - 3}$$

is convergent.                                                                □

We can also notice that using the following Matlab 7.9 sequence:
>>**syms n**
>> **symsum(1/(16\*n^2-8\*n-3),n,1,inf)**
**ans =**
**1/4**
or in Mathcad 14:

$$\sum_{n=1}^{\infty} \dfrac{1}{16n^2 - 8n - 3} \to \dfrac{1}{4}$$

or  using Mathematica 8:
 **In[1]:=Sum[1/(16\*n^2 - 8\*n - 3), {n, 1, Infinity}]**
 **Out[1]=$\frac{1}{4}$**
 or with Maple 15:

$$\sum_{n=1}^{\infty} \dfrac{1}{16\,n^2 - 8 \cdot n - 3}$$

$$\dfrac{1}{4}$$

b) We have

$$S_n = \sum_{k=1}^{n} a_k = \sum_{k=1}^{n} \frac{1}{\left(k + \sqrt{2}\right)\left(k + \sqrt{2} + 1\right)}$$

$$= \sum_{k=1}^{n} \left( \frac{1}{k + \sqrt{2}} - \frac{1}{k + \sqrt{2} + 1} \right)$$

$$= \frac{1}{1 + \sqrt{2}} - \frac{1}{n + 1 + \sqrt{2}} \xrightarrow[n \to \infty]{} \frac{1}{1 + \sqrt{2}};$$

hence

$$\sum_{n \geq 1} \frac{1}{\left(n + \sqrt{2}\right)\left(n + \sqrt{2} + 1\right)}$$

is convergent.

In Matlab 7.9 we shall have:

>>**syms n**

>> **symsum(1/((n+sqrt(2))\*(n+sqrt(2)+1)),n,1,inf)**

**ans =**

**1/(1+2^(1/2))**

or in Mathcad 14:

$$\sum_{n=1}^{\infty} \frac{1}{\left(n + \sqrt{2}\right)\left(n + \sqrt{2} + 1\right)} \text{ simplify} \to \sqrt{2} - 1$$

or with Mathematica 8:

**ln[2]:=Sum[1/((n + Sqrt[2])\*(n + Sqrt[2] + 1)), {n, 1, Infinity}]**

**Out[2]=-1 + $\sqrt{2}$**

or with Maple 15:

$$\sum_{n=1}^{\infty} \frac{1}{\left(n + \sqrt{2}\right)\cdot\left(n + \sqrt{2} + 1\right)}$$

$$\frac{1}{\sqrt{2} + 1}$$

## 1.2.1.1    Cauchy's Test

**Proposition 1.7** (see [41], p. 36). The necessary and sufficient condition for the convergence of the series $\sum_{n \geq 1} a_n$ is: $(\forall)\ \varepsilon > 0, (\exists)\ n_\varepsilon \in \mathbb{N}$ such that:

$$|a_{n+1} + \cdots + a_{n+p}| < \varepsilon, (\forall)\ n \geq n_\varepsilon, (\forall)\ p \in \mathbb{N}.$$

**Example 1.8**. Use the Cauchy's test for testing the convergence of the series

$$\text{a) } \sum_{n\geq 1} \frac{\cos nx}{5^n}, \quad (\forall) \ x \in \mathbb{R}$$

$$\text{b) } \sum_{n\geq 1} \frac{\sin n^2 x}{n^3}, \quad (\forall) \ x \in \mathbb{R}.$$

**Solutions.**
a) We have

$$|a_{n+1} + \cdots + a_{n+p}| = \left| \sum_{k=n+1}^{n+p} \frac{\cos kx}{5^k} \right| \leq \sum_{k=n+1}^{n+p} \frac{1}{5^k}$$

$$= \frac{1}{5^{n+1}} \left( 1 + \frac{1}{5} + \cdots + \frac{1}{5^p} \right)$$

$$= \frac{1}{5^{n+1}} \cdot \frac{\frac{1}{5^p} - 1}{\frac{1}{5} - 1} = \frac{1}{4 \cdot 5^n} \cdot \left( 1 - \frac{1}{5^p} \right) < \frac{1}{5^n} \xrightarrow[n\to\infty]{} 0.$$

Therefore, $(\forall) \ \varepsilon > 0, (\exists) \ n_\varepsilon \in \mathbb{N}$ such that $\frac{1}{5^n} < \varepsilon, (\forall) \ n \geqslant n_\varepsilon$, i.e.

$$5^n > \frac{1}{\varepsilon} \implies n > -\frac{\ln \varepsilon}{\ln 5};$$

it results that $(\forall) \ \varepsilon > 0, (\exists) \ n_\varepsilon = \left[ -\frac{\ln \varepsilon}{\ln 5} \right] + 1 \in \mathbb{N}$, such that $(\forall) \ n \geqslant n_\varepsilon$ we have $|a_{n+1} + \cdots + a_{n+p}| < \varepsilon$, i.e.

$$\sum_{n\geq 1} \frac{\cos nx}{5^n}, \quad (\forall) \ x \in \mathbb{R}$$

is convergent.
b) We achieve:

$$|a_{n+1} + \cdots + a_{n+p}| = \left| \sum_{k=n+1}^{n+p} \frac{\sin k^2 x}{k^3} \right| \leq \sum_{k=n+1}^{n+p} \frac{1}{k^3}$$

$$\leq \sum_{k=n+1}^{n+p} \frac{1}{k^2} \leq \sum_{k=n+1}^{n+p} \frac{1}{k(k-1)}$$

$$= \sum_{k=n+1}^{n+p} \left( \frac{1}{k-1} - \frac{1}{k} \right) = \frac{1}{n} - \frac{1}{n+p} < \frac{1}{n} \xrightarrow[n\to\infty]{} 0.$$

Hence, $(\forall)\ \varepsilon > 0,\ (\exists)\ n_\varepsilon \in \mathbb{N}$ such that $\frac{1}{n} < \varepsilon,\ (\forall)\ n \geqslant n_\varepsilon$, i.e.

$$n > \frac{1}{\varepsilon};$$

it results that $(\forall)\ \varepsilon > 0,\ (\exists)\ n_\varepsilon = \left[\frac{1}{\varepsilon}\right] + 1 \in \mathbb{N}$, such that $(\forall)\ n \geqslant n_\varepsilon$ we have $|a_{n+1} + \cdots + a_{n+p}| < \varepsilon$, i.e.

$$\sum_{n \geq 1} \frac{\sin n^2 x}{n^3}, \quad (\forall)\ x \in \mathbb{R}$$

is convergent.

### 1.2.2  Divergent Series

**Definition 1.9** (see [41], p. 30). If the limit $\lim_{n \to \infty} S_n$ does not exist (or it is infinite), the series is then called **divergent**.

**Proposition 1.10** (see [15], p. 93). If the terms of a series sequence is not convergent to 0, the series is divergent, namely if for the series $\sum_{n \geq 1} a_n$ we have

$$\lim_{n \to \infty} a_n = a \neq 0,\text{'}$$

then $\sum_{n \geq 1} a_n$ diverges.

**Example 1.11.** Prove that the following series are divergent:

a) $0,07 + \sqrt{0,07} + \sqrt[3]{0,07} + \cdots + \sqrt[n]{0,07} + \cdots$

b) $\displaystyle\sum_{n \geq 1} \frac{1}{\sqrt{2n+1} - \sqrt{2n-1}}$

c) $\displaystyle\sum_{n \geq 1} \frac{2^n + 3^n}{2^{n+1} + 3^{n+1}}$

d) $\displaystyle\sum_{n \geq 1} \ln \frac{n+1}{n}$

e) $\displaystyle\sum_{n \geq 1} n \sin \frac{a}{n},\ a \neq 0.$

**Solutions.**

a) As

$$\lim_{n \to \infty} \sqrt[n]{0,07} = \lim_{n \to \infty} (0,07)^{\frac{1}{n}} = 1 \neq 0$$

we deduce that $\sum_{n \geq 1} \sqrt[n]{0,07}$ diverges.

We can prove that using Matlab 7.9:

**>> syms n**

**>> limit(0.07^(1/n),inf)**

**ans =**

**1**

or in Mathcad 14:

$$\lim_{n \to \infty} \sqrt[n]{0.07} \to 1.0$$

or with Mathematica 8:

In[2]:= **Limit[(0.07)^(1/n), n → Infinity]**

Out[2]= **1.**

or in Maple 15:

$$\lim_{n \to \infty} \sqrt[n]{0.07}$$

1.

b) Since

$$\lim_{n \to \infty} \frac{1}{\sqrt{2n+1} - \sqrt{2n-1}} = \lim_{n \to \infty} \frac{\sqrt{2n+1} + \sqrt{2n-1}}{2n+1-2n+1} = \infty,$$

using the Definition 1.9 it results that the series diverges.

We can also deduce that using Matlab 7.9:

**>>syms n**

**>> limit(1/(sqrt(2*n+1)-sqrt(2*n-1)),inf)**

**ans =**

**Inf**

or in Mathcad 14:

$$\lim_{n \to \infty} \frac{1}{\sqrt{2 \cdot n + 1} - \sqrt{2 \cdot n - 1}} \to \infty$$

or with Mathematica 8:

In[3]:= `Limit[1 / (Sqrt[2 * n + 1] - Sqrt[2 * n - 1]), n → Infinity]`

Out[3]= ∞

or in Maple 15:

$$\lim_{n \longrightarrow \infty} \frac{1}{\sqrt{2 \cdot n + 1} - \sqrt{2 \cdot n - 1}}$$

∞

c) One obtains

$$\lim_{n \to \infty} \frac{2^n + 3^n}{2^{n+1} + 3^{n+1}} = \lim_{n \to \infty} \frac{3^n \left( \frac{2^n}{3^n} + 1 \right)}{3^{n+1} \left( \frac{2^{n+1}}{3^{n+1}} + 1 \right)} = \frac{1}{3} \neq 0;$$

we have used that

$$\lim_{n \to \infty} q^n = \begin{cases} 0, & \text{if } q \in (-1, 1) \\ 1, & \text{if } q = 1 \\ \infty, & \text{if } q > 1 \\ \text{it doesn't exists,} & \text{if } q \in (-\infty, -1]. \end{cases}$$

Hence, based on the Proposition 1.10 it follows that the series diverges. It will also result that in Matlab 7.9:

>> **syms n**
>> **limit((2^n+3^n)/(2^(n+1)+3^(n+1)),inf)**
ans =
1/3
or in Mathcad 14:

$$\lim_{n \to \infty} \frac{2^n + 3^n}{2^{n+1} + 3^{n+1}} \to \frac{1}{3}$$

or with Mathematica 8:

In[4]:= `Limit[ (2^n + 3^n) / (2^(n + 1) + 3^(n + 1)), n → Infinity]`

Out[4]= $\frac{1}{3}$

or in Maple 15:

$$\lim_{n \to \infty} \frac{2^n + 3^n}{2^{n+1} + 3^{n+1}}$$

$$\frac{1}{3}$$

d) We have

$$\lim_{n \to \infty} \ln \frac{n+1}{n} = 0,$$

namely we can say nothing about the series nature.

As

$$S_n = \sum_{k=1}^{n} \ln \frac{k+1}{k} = \ln \frac{2}{1} + \ln \frac{3}{2} + \cdots + \ln \frac{n+1}{n}$$

$$= \ln \left( \frac{2}{1} \cdot \frac{3}{2} \cdot \cdots \cdot \frac{n+1}{n} \right) = \ln (n+1)$$

and

$$\lim_{n \to \infty} S_n = \infty$$

one deduces that the series is divergent.

The same result can be achieved using Matlab 7.9:
>> **syms n k**
>> **limit(symsum(log((k+1)/k),k,1,n),n,inf)**
**ans =**
**Inf**
or Mathcad 14:

$$\lim_{n \to \infty} \sum_{k=1}^{n} \ln \left( \frac{k+1}{k} \right) \to \infty$$

or with Mathematica 8:

In[5]:= **Limit[Sum[Log[(k + 1) / k], {k, 1, n}], n → Infinity]**

Out[5]= ∞

or in Maple 15:

$$\lim_{n\to\infty}\sum_{k=1}^{n}\ln\left(\frac{k+1}{k}\right)$$

$$\infty$$

e) It will result:

$$\lim_{n\to\infty} n\cdot\sin\frac{a}{n}=\lim_{n\to\infty}a\frac{\sin\frac{a}{n}}{\frac{a}{n}}=a\neq 0,$$

therefore the series is divergent.

We can also obtain that using Matlab 7.9:

>> **syms n a**

>> **limit(n\*sin(a/n),inf)**

**ans =**

**a**

or Mathcad 14:

$$\lim_{n\to\infty}\left(n\cdot\sin\left(\frac{a}{n}\right)\right)\to a$$

or with Mathematica 8:

In[6]:= **Limit[n \* Sin[a / n], n → Infinity]**

Out[6]= **a**

or in Maple 15:

$$\lim_{n\to\infty} n\cdot\sin\left(\frac{a}{n}\right)$$

$$a$$

## 1.2.3  Operations on Convergent Series

**Proposition 1.12** (see [15], p. 95). Let $\sum_{n\geq 1}a_n$ and $\sum_{n\geq 1}b_n$ be two convergent series, which have the sums $A$ and $B$ respectively and let be $\alpha\in\mathbb{R}$. Then:

a) the series $\sum_{n\geq 1}\alpha a_n$ converges and it has the sum $\alpha A$, namely a convergent series may be multiplied term by term by any number $\alpha$;

b) the sum (difference) of the two convergent series is a convergent series $\sum_{n\geq 1}(a_n \pm b_n)$ and it has the sum $A \pm B$.

**Example 1.13.** Use the operations on convergent series to compute the sum of the series:

$$\text{a)} \sum_{n\geq 1} \frac{2n+3}{n(n+1)(n+2)}$$

$$\text{b)} \sum_{n\geq 1} \frac{2^n + 3^{n+1} - 6^{n-1}}{12^n}.$$

**Solutions.**

a) One can notice that

$$a_n = \frac{2n+3}{n(n+1)(n+2)} = \frac{3}{2n} - \frac{1}{n+1} - \frac{1}{2(n+2)}$$

$$= \frac{1}{2n} + \frac{2}{2n} - \frac{1}{n+1} - \frac{1}{2(n+2)}$$

$$= \frac{1}{2n} - \frac{1}{2(n+2)} + \frac{1}{n} - \frac{1}{n+1} = \frac{1}{2}\left(\frac{1}{n} - \frac{1}{n+2}\right) + \left(\frac{1}{n} - \frac{1}{n+1}\right);$$

hence

$$\sum_{n\geq 1} \frac{2n+3}{n(n+1)(n+2)} = \sum_{n\geq 1} \frac{1}{2}\left(\frac{1}{n} - \frac{1}{n+2}\right) + \sum_{n\geq 1}\left(\frac{1}{n} - \frac{1}{n+1}\right).$$

We shall denote

$$\begin{cases} u_n = \alpha a_n = \frac{1}{2}\left(\frac{1}{n} - \frac{1}{n+2}\right) \\ \\ v_n = \frac{1}{n} - \frac{1}{n+1}. \end{cases}$$

One can notice that

$$\sum_{i=1}^{n} a_i = \sum_{i=1}^{n}\left(\frac{1}{i} - \frac{1}{i+2}\right)$$

$$= 1 - \frac{1}{3} + \frac{1}{2} - \frac{1}{5} + \cdots + \frac{1}{n-2} - \frac{1}{n} + \frac{1}{n-1} - \frac{1}{n+1} + \frac{1}{n} - \frac{1}{n+2}$$

$$= 1 + \frac{1}{2} - \frac{1}{n+1} - \frac{1}{n+2}$$

and

$$\lim_{n\to\infty} \left(1 + \frac{1}{2} - \frac{1}{n+1} - \frac{1}{n+2}\right) = \frac{3}{2};$$

hence it results that the series $\sum_{n\geq 1} a_n$ converges.

Applying the Proposition 1.12 it follows the series $\sum_{n\geq 1} \alpha a_n$ will also converges and

$$\sum_{i=1}^{n} \alpha a_i = \frac{1}{2} \cdot \frac{3}{2} = \frac{3}{4}.$$

Similarly,

$$\sum_{i=1}^{n} v_i = \sum_{i=1}^{n} \left(\frac{1}{i} - \frac{1}{i+1}\right) = 1 - \frac{1}{2} + \frac{1}{2} - \frac{1}{3} + \cdots + \frac{1}{n} - \frac{1}{n+1}$$

$$= 1 - \frac{1}{n+1}$$

and

$$\lim_{n\to\infty} \left(1 - \frac{1}{n+1}\right) = 1;$$

hence, it results that the series $\sum_{n\geq 1} v_n$ converges.

Applying the Proposition 1.12 it follows that the series $\sum_{n\geq 1} (u_n + v_n)$ will also converge and it has the sum

$$U + V = \frac{3}{4} + 1 = \frac{7}{4}.$$

b) We shall have

$$\sum_{n\geq 1} \frac{2^n + 3^{n+1} - 6^{n-1}}{12^n} = \underbrace{\sum_{n\geq 1}\left(\frac{1}{6}\right)^n}_{a_n} + 3\underbrace{\sum_{n\geq 1}\left(\frac{1}{4}\right)^n}_{b_n} - \frac{1}{6}\underbrace{\sum_{n\geq 1}\left(\frac{1}{2}\right)^n}_{c_n}.$$

It results that

$$A = \lim_{n \to \infty} \sum_{i=1}^{n} a_i = \frac{1}{6} \cdot \frac{1}{1 - \frac{1}{6}} = \frac{1}{5},$$

$$B = \lim_{n \to \infty} \sum_{i=1}^{n} b_i = \frac{1}{4} \cdot \frac{1}{1 - \frac{1}{4}} = \frac{1}{3},$$

$$C = \lim_{n \to \infty} \sum_{i=1}^{n} c_i = \frac{1}{2} \cdot \frac{1}{1 - \frac{1}{2}} = 1;$$

hence, the series

$$\sum_{n \geq 1} \left(\frac{1}{6}\right)^n, \ \sum_{n \geq 1} \left(\frac{1}{4}\right)^n, \ \sum_{n \geq 1} \left(\frac{1}{2}\right)^n$$

are convergent and applying the Proposition 1.12 it follows that the series

$$\sum_{n \geq 1} \frac{2^n + 3^{n+1} - 6^{n-1}}{12^n}$$

will also converge and it has the sum

$$S = \frac{1}{5} + 1 - \frac{1}{6} = \frac{31}{30}.$$

## 1.3    Tests for Convergence of Alternating Series

**Definition 1.14** (see [15], p. 94). **An alternating series** is an infinite series of the form $\sum_{n \geq 1} (-1)^{n+1} a_n$, where all the $a_n$ are non-negative, namely $a_n > 0$ for all $n \in \mathbb{N}$.

**Proposition 1.15 (Leibnitz's test,** see [15], p. 94). If the sequence $(a_n)_n$ is monotone decreasing and it equals $0$ as $n$ approaches infinity, then the series

$$\sum_{n \geq 1} (-1)^{n+1} a_n$$

converges.

**Example 1.16.** Test the convergence of the series:

a) $\displaystyle\sum_{n \geq 1} (-1)^{n+1} \sin \frac{1}{n}$

b) $\displaystyle\sum_{n \geq 1} (-1)^{n+1} \frac{27n^2 + 36n + 11}{(3n + 1)(3n + 2)(3n + 3)}$

$$\text{c) } \sum_{n \geq 1} (-1)^{n+1} \frac{10^{n-1}a + 10^{n-2}a + \cdots + 10a + a}{10^n}, \ a > 0.$$

## Solutions.

a) With $a_n = \sin \frac{1}{n}$, we shall deduce

$$a_{n+1} - a_n = \sin \frac{1}{n+1} - \sin \frac{1}{n} = 2 \sin \frac{\frac{1}{n+1} - \frac{1}{n}}{2} \cos \frac{\frac{1}{n+1} + \frac{1}{n}}{2}$$

$$= -2 \sin \frac{1}{2n(n+1)} \cos \frac{2n+1}{2n(n+1)} < 0,$$

namely $(a_n)_n$ is monotone decreasing; in addition we have

$$\lim_{n \to \infty} \sin \frac{1}{n} = 0;$$

hence, using the Leibnitz's test it follows that the alternating series is convergent.

b) Since

$$a_n = \frac{27n^2 + 36n + 11}{(3n+1)(3n+2)(3n+3)} = \frac{1}{3n+1} + \frac{1}{3n+2} + \frac{1}{3n+3}$$

it results that

$$a_{n+1} - a_n = \frac{1}{3n+4} + \frac{1}{3n+5} + \frac{1}{3n+6} - \frac{1}{3n+1} - \frac{1}{3n+2} - \frac{1}{3n+3}$$

$$< \frac{1}{3n+1} + \frac{1}{3n+2} + \frac{1}{3n+3} - \frac{1}{3n+1} - \frac{1}{3n+2} - \frac{1}{3n+3} = 0,$$

namely $(a_n)_n$ is monotone decreasing; in addition we have

$$\lim_{n \to \infty} \frac{27n^2 + 36n + 11}{(3n+1)(3n+2)(3n+3)} = 0;$$

therefore, using the Leibnitz's test it follows that the alternating series is convergent.

c) We can notice that

$$a_n = \frac{10^{n-1}a + 10^{n-2}a + \cdots + 10a + a}{10^n} = a \cdot \frac{10^{n-1} + 10^{n-2} + \cdots + 10 + 1}{10^n}$$

$$= a \cdot \frac{10^n - 1}{10^n(10 - 1)} = \frac{a}{9} \cdot \frac{1 - \frac{1}{10^n}}{1} = \frac{a}{9} \cdot \left(1 - \left(\frac{1}{10}\right)^n\right) \xrightarrow[n \to \infty]{} \frac{a}{9} \neq 0;$$

hence, the alternating series is divergent.

## 1.4   Tests of Convergence and Divergence of Positive Series

**Definition 1.17.** (see [15], p. 93). The series $\sum_{n\geq 1} a_n$ is with **positive terms** if $a_n > 0$, $(\forall)\ n \in \mathbb{N}$.

### *1.4.1   The Comparison Test I*

**Proposition 1.18 ( The comparison test I,** see [15], p. 93): Let $\sum_{n\geq 1} a_n$ and $\sum_{n\geq 1} b_n$ be two positive series such that $a_n \leq b_n$. Then if:

a) $\sum_{n\geq 1} b_n$ converges it follows that $\sum_{n\geq 1} a_n$ also converges;

b) $\sum_{n\geq 1} a_n$ diverges it results that $\sum_{n\geq 1} b_n$ also diverges.

**Example 1.19.** Test the convergence of the positive series:

$$\text{a) } \sum_{n\geq 1} \frac{1}{\sqrt{10n}}$$

$$\text{b) } \sum_{n\geq 1} \frac{1}{\sqrt{n^3 + n}}$$

$$\text{c) } \sum_{n\geq 1} \frac{\sqrt[4]{n}}{\sqrt[3]{n^4 + n + 1}}$$

$$\text{d) } \sum_{n\geq 1} \frac{5n}{n^2 - 3}$$

$$\text{e) } \sum_{n\geq 1} \frac{1 + \frac{1}{2} + \cdots + \frac{1}{n}}{n}$$

$$\text{f) } \sum_{n\geq 1} \frac{\sqrt{n+1} - \sqrt{n}}{n}$$

$$\text{g) } \sum_{n\geq 1} \frac{(n+1)^{n-1}}{n^{n+1}}.$$

**Solutions.**

a) We choose

$$a_n = \frac{1}{n\sqrt{10}}$$

and

$$b_n = \frac{1}{\sqrt{10n}}.$$

One knows that the generalized harmonic series $\sum_{n \geq 1} \frac{1}{n^p}$ converges for $p > 1$ and diverges for $p \leq 1$.

As $\sum_{n \geq 1} \frac{1}{n^p}$ diverges, using the comparison test I one deduces that

$$\sum_{n \geq 1} \frac{1}{\sqrt{10n}}$$

diverges.

b) Since

$$\underbrace{\frac{1}{\sqrt{n^3 + n}}}_{a_n} \leq \frac{1}{\sqrt{n^3}} = \underbrace{\frac{1}{n^{\frac{3}{2}}}}_{b_n}$$

and $\sum_{n \geq 1} \frac{1}{n^{3/2}}$ converges, using the comparison test I it results that

$$\sum_{n \geq 1} \frac{1}{\sqrt{n^3 + n}}$$

converges.

c) We shall have

$$\underbrace{\frac{\sqrt[4]{n}}{\sqrt[3]{n^4 + n + 1}}}_{a_n} \leq \frac{n^{1/4}}{(n^4)^{1/3}} = \frac{n^{1/4}}{n^{4/3}} = \underbrace{\frac{1}{n^{\frac{13}{12}}}}_{b_n};$$

therefore, using the comparison test I it results that

$$\sum_{n \geq 1} \frac{\sqrt[4]{n}}{\sqrt[3]{n^4 + n + 1}}$$

converges.

d) We can notice that

$$\underbrace{\frac{5n}{n^2 - 3}}_{b_n} \geq \frac{5n}{n^2} = \underbrace{\frac{5}{n}}_{a_n};$$

hence one deduces that

$$\sum_{n \geq 1} \frac{5n}{n^2 - 3}$$

is divergent.

e) As

$$\underbrace{\frac{1 + \frac{1}{2} + \cdots + \frac{1}{n}}{n}}_{b_n} \geq \frac{\frac{1}{n} + \frac{1}{n} + \cdots + \frac{1}{n}}{n} = \underbrace{\frac{1}{n}}_{a_n}$$

one deduces that

$$\sum_{n \geq 1} \frac{1 + \frac{1}{2} + \cdots + \frac{1}{n}}{n}$$

diverges.

f) We can notice that

$$\frac{\sqrt{n+1} - \sqrt{n}}{n} = \frac{n + 1 - n}{(\sqrt{n+1} + \sqrt{n}) \cdot n} \leq \frac{1}{(\sqrt{n} + \sqrt{n}) \cdot n}$$

$$= \frac{1}{2 \cdot n^{1/2} \cdot n} = \frac{1}{2 \cdot n^{3/2}};$$

hence

$$\sum_{n \geq 1} \frac{\sqrt{n+1} - \sqrt{n}}{n}$$

is convergent.

g) We can notice that

$$\sum_{n \geq 1} \frac{(n+1)^{n-1}}{n^{n+1}} = \sum_{n \geq 1} \left( \frac{n+1}{n} \right)^n \cdot \frac{1}{n(n+1)}.$$

Using the inequality

$$\left( 1 + \frac{1}{n} \right)^n < e < \left( 1 + \frac{1}{n} \right)^{n+1}, \quad (\forall) \; n \in \mathbb{N}^*$$

we deduce

$$\underbrace{\left( 1 + \frac{1}{n} \right)^n \cdot \frac{1}{n(n+1)}}_{a_n} < e \cdot \frac{1}{n(n+1)} \leq \underbrace{e \cdot \frac{1}{n^2}}_{b_n}, \quad (\forall) \; n \in \mathbb{N}^*;$$

therefore, applying the comparison test I it results that

$$\sum_{n \geq 1} \frac{(n+1)^{n-1}}{n^{n+1}}$$

converges.

## 1.4.2   The Root Test

**Proposition 1.20 (The root test,** see [15], p. 94): Let $\sum_{n\geq 1} a_n$ be a positive series. Assume that:

$$\lim_{n\to\infty} \sqrt[n]{a_n} = \lambda.$$

Then we have the following:

A) If $\lambda < 1$, then the series $\sum_{n\geq 1} a_n$ is convergent;
B) If $\lambda > 1$, then the series $\sum_{n\geq 1} a_n$ is divergent;
C) If $\lambda = 1$, then the series $\sum_{n\geq 1} a_n$ may be convergent or it may be divergent, namely we do not have a definite conclusion.

**Example 1.21.** Discuss the convergence of the positive series:

$$\text{a) } \sum_{n\geq 1} \frac{a^n}{n^n}, \ a > 0$$

$$\text{b) } \sum_{n\geq 1} \left( \sqrt[3]{n^3 + n^2 + 1} - \sqrt[3]{n^3 - n^2 + 1} \right)^n$$

$$\text{c) } \sum_{n\geq 1} \frac{n}{\left(1 + \frac{1}{n}\right)^{n^2}}.$$

**Solutions.**
a)  We can notice that

$$\lim_{n\to\infty} \sqrt[n]{a_n} = \lim_{n\to\infty} \frac{a}{n} = 0 < 1;$$

therefore, using the root test it follows that

$$\sum_{n\geq 1} \frac{a^n}{n^n}, a > 0$$

converges.
b)  We can write

$$\sqrt[n]{a_n} = \frac{n^3 + n^2 + 1 - n^3 + n^2 - 1}{\sqrt[3]{(n^3 + n^2 + 1)^2} + \sqrt[3]{(n^3 + n^2 + 1)(n^3 - n^2 + 1)} + \sqrt[3]{(n^3 - n^2 + 1)^2}}$$

$$= \frac{2n^2}{n^2 \left[ \sqrt[3]{\left(1 + \frac{1}{n} + \frac{1}{n^2}\right)^2} + \sqrt[3]{\left(1 + \frac{1}{n} + \frac{1}{n^2}\right)\left(1 - \frac{1}{n} + \frac{1}{n^2}\right)} + \sqrt[3]{\left(1 - \frac{1}{n} + \frac{1}{n^2}\right)^2} \right]};$$

hence

$$\lim_{n \to \infty} \sqrt[n]{a_n} = \frac{2}{3} < 1.$$

Using the root test it follows that

$$\sum_{n \geq 1} \left( \sqrt[3]{n^3 + n^2 + 1} - \sqrt[3]{n^3 - n^2 + 1} \right)^n$$

is convergent.

We shall give a computer solution using Matlab 7.9:

>> **syms n**

>> **limit((((n^3+n^2+1)^(1/3)-(n^3-n^2+1)^(1/3))^n)^(1/n),**
**inf)**

**ans =**

**2/3**

or Mathcad 14:

$$\lim_{n \to \infty} \left[ \left( \sqrt[3]{n^3 + n^2 + 1} - \sqrt[3]{n^3 - n^2 + 1} \right)^n \right]^{\frac{1}{n}} \quad \text{simplify} \; \to \frac{2}{3}$$

or with Mathematica 8:

```
In[5]:= Limit[(((n^3 + n^2 + 1) ^ (1 / 3) - (n^3 - n^2 + 1) ^ (1 / 3)) ^n) ^ (1 / n), n → Infinity]

Out[5]= 2/3
```

or Maple 15:

$$\lim_{n \to \infty} \sqrt[n]{\left( \sqrt[3]{n^3 + n^2 + 1} - \sqrt[3]{n^3 - n^2 + 1} \right)^n}$$

$$\frac{2}{3}$$

c) We shall have

$$\sqrt[n]{a_n} = \sqrt[n]{\frac{n}{\left(1 + \frac{1}{n}\right)^{n^2}}} = \frac{n^{\frac{1}{n}}}{\left(1 + \frac{1}{n}\right)^{\frac{n^2}{n}}} = \frac{n^{\frac{1}{n}}}{\left(1 + \frac{1}{n}\right)^n},$$

where

$$\lim_{n\to\infty} n^{\frac{1}{n}} = e^{\lim_{n\to\infty} \frac{\ln n}{n}} = e^0 = 1,$$

$$\lim_{n\to\infty} \left(1 + \frac{1}{n}\right)^n = e.$$

Whence

$$\lim_{n\to\infty} \sqrt[n]{a_n} = \frac{1}{e} < 1$$

and using the root test it follows that

$$\sum_{n\geq 1} \frac{n}{\left(1 + \frac{1}{n}\right)^{n^2}}$$

converges.

This result can be also obtained using Matlab 7.9:

>> **syms n**
>> **limit((n/((1+1/n)^(n^2)))^(1/n),n,inf)**
ans =
**1/exp(1)**

in Mathcad 14:

$$\lim_{n\to\infty} \left[ \frac{n}{\left(1 + \frac{1}{n}\right)^{n^2}} \right]^{\frac{1}{n}} \to e^{-1}$$

or with Mathematica 8:

In[4]:= **Limit[ (n / ((1 + 1 / n) ^ (n^2))) ^ (1 / n), n → Infinity]**

Out[4]= $\dfrac{1}{e}$

or with Maple 15:

$$\lim_{n\to\infty} \sqrt[n]{\frac{n}{\left(1 + \frac{1}{n}\right)^{n^2}}}$$

$$e^{-1}$$

## 1.4.3    The Ratio Test

**Proposition 1.22 (The ratio test,** see [15], p. 94): Let $\sum_{n \geq 1} a_n$ be a positive series such that $a_n \neq 0$ for any $n \geq 1$. Assume that:

$$\lim_{n \to \infty} \frac{a_{n+1}}{a_n} = \lambda.$$

Then we have the following:

A) If $\lambda < 1$, then the series $\sum_{n \geq 1} a_n$ is convergent;
B) If $\lambda > 1$, then the series $\sum_{n \geq 1} a_n$ is divergent;
C) If $\lambda = 1$, then the series $\sum_{n \geq 1} a_n$ may be convergent or it may be divergent, namely we do not have a definite conclusion.

**Example 1.23.** Discuss the convergence of the positive series:

$$\text{a)} \sum_{n \geq 1} \frac{2 \cdot 7 \cdot 12 \cdot \cdots \cdot (5n - 3)}{5 \cdot 9 \cdot 13 \cdot \cdots \cdot (4n + 1)}$$

$$\text{b)} \sum_{n \geq 1} \frac{(n!)^2}{(2n)!}$$

$$\text{c)} \sum_{n \geq 1} n \tan \frac{\pi}{2^{n+1}}$$

$$\text{d)} \sum_{n \geq 1} n \cdot a^n, \ a > 0.$$

**Solutions.**
a) One obtains

$$\lim_{n \to \infty} \frac{a_{n+1}}{a_n} = \lim_{n \to \infty} \frac{2 \cdot 7 \cdot 12 \cdot \cdots \cdot (5n + 2)}{5 \cdot 9 \cdot 13 \cdot \cdots \cdot (4n + 5)} \cdot \frac{5 \cdot 9 \cdot 13 \cdot \cdots \cdot (4n + 1)}{2 \cdot 7 \cdot 12 \cdot \cdots \cdot (5n - 3)}$$

$$= \lim_{n \to \infty} \frac{5n + 2}{4n + 5} = \frac{5}{4} > 1;$$

therefore, using the ratio test it results that the series

$$\sum_{n \geq 1} \frac{2 \cdot 7 \cdot 12 \cdot \cdots \cdot (5n - 3)}{5 \cdot 9 \cdot 13 \cdot \cdots \cdot (4n + 1)}$$

diverges.
We can see that in Mathcad 14:

$$a(n) := \frac{\prod\limits_{k=1}^{n} (5 \cdot k - 3)}{\prod\limits_{k=1}^{n} (4 \cdot k + 1)}$$

$$\lim_{n \to \infty} \frac{a(n+1)}{a(n)} \to \frac{5}{4}$$

or with Mathematica 8:

```
In[10]:= a[n_] := Product[5*k - 3, {k, 1, n}] / Product[4*k + 1, {k, 1, n}]

In[11]:= Limit[a[n + 1] / a[n], n → Infinity]

Out[11]=  5
          ─
          4
```

or in Maple 15:

$$a := n \to \frac{\prod\limits_{k=1}^{n} (5 \cdot k - 3)}{\prod\limits_{k=1}^{n} (4 \cdot k + 1)}$$

$$n \to \frac{\prod\limits_{k=1}^{n} (5 k - 3)}{\prod\limits_{k=1}^{n} (4 k + 1)}$$

$$\lim_{n \to \infty} \frac{a(n+1)}{a(n)}$$

$$\frac{5}{4}$$

b) We shall notice that

$$\frac{a_{n+1}}{a_n} = \frac{[(n+1)!]^2}{(2n+2)!} \cdot \frac{(2n)!}{(n!)^2} = \frac{(n!)^2 \cdot (n+1)^2 \cdot (2n)!}{(2n+2)(2n+1) \cdot (2n)! \cdot (n!)^2}$$

$$= \frac{(n+1)^2}{(2n+2)(2n+1)} = \frac{1}{4} < 1;$$

hence, using the ratio test one deduces that

$$\sum_{n \geq 1} \frac{(n!)^2}{(2n)!}$$

is convergent.

We can also see that in Matlab 7.9:

```
>> syms n
>> v='n!^2/(2*n)!';
>> v1=subs(v,n,n+1);
>> limit(v1/v,n,inf)
ans =
1/4
```

or in Mathcad 14:

$$a(n) := \frac{(n!)^2}{(2 \cdot n)!}$$

$$\lim_{n \to \infty} \frac{a(n+1)}{a(n)} \to \frac{1}{4}$$

or with Mathematica 8:

```
In[12]:= a[n_] := (n!)^2 / (2*n)!

In[13]:= Limit[a[n+1] / a[n], n -> Infinity]

Out[13]= 1
          ─
          4
```

or in Maple 15:

$$a := n \to \frac{(n!)^2}{(2 \cdot n)!}$$

$$n \to \frac{n!^2}{(2\,n)!}$$

$$\lim_{n \to \infty} \frac{a(n+1)}{a(n)}$$

$$\frac{1}{4}$$

c) We can write:

$$\frac{a_{n+1}}{a_n} = \frac{n+1}{n} \cdot \frac{\tan \frac{\pi}{2^{n+2}}}{\tan \frac{\pi}{2^{n+1}}} = \frac{n+1}{n} \cdot \frac{\sin \frac{\pi}{2^{n+2}}}{\frac{\pi}{2^{n+2}}} \cdot \frac{\frac{\pi}{2^{n+2}}}{\cos \frac{\pi}{2^{n+2}}} \cdot \frac{\frac{\pi}{2^{n+1}}}{\sin \frac{\pi}{2^{n+1}}} \cdot \frac{\cos \frac{\pi}{2^{n+1}}}{\frac{\pi}{2^{n+1}}};$$

therefore

$$\lim_{n \to \infty} \frac{a_{n+1}}{a_n} = \frac{1}{2}$$

and using the ratio test one establishes that

$$\sum_{n \geq 1} n \tan \frac{\pi}{2^{n+1}}$$

is convergent.

We can notice using Matlab 7.9:
```
>>syms n
>> a=@(n) n*tan(pi/(2^(n+1))) ;
>> limit(a(n+1)/a(n),n,inf)
ans =
1/2
```
or Mathcad 14:

$$a(n) := n \cdot \tan\left(\frac{\pi}{2^{n+1}}\right)$$

$$\lim_{n \to \infty} \frac{a(n+1)}{a(n)} \to \frac{1}{2}$$

or with Mathematica 8:

```
In[14]:= a[n_] := n * tan (Pi / 2 ^ (n + 1))
```

```
In[15]:= Limit [a[n + 1] / a[n], n → Infinity]
```

$$Out[15]= \frac{1}{2}$$

or in Maple 15:

$$a := n \rightarrow n \cdot \tan\left(\frac{\pi}{2^{n+1}}\right)$$

$$n \rightarrow n \tan\left(\frac{\pi}{2^{n+1}}\right)$$

$$\lim_{n \rightarrow \infty} \frac{a(n+1)}{a(n)}$$

$$\frac{1}{2}$$

d) We shall have

$$\frac{a_{n+1}}{a_n} = \frac{(n+1) \cdot a^{n+1}}{n \cdot a^n} \underset{n \to \infty}{\to} a;$$

using the ratio test it results that:

- $\sum_{n \geq 1} n \cdot a_n$ is convergent if $a \in (0, 1)$;
- $\sum_{n \geq 1} n \cdot a_n$ is divergent if $a > 1$.

We don' t have a definite conclusion for $a = 1$.

## 1.4.4 The Raabe's and Duhamel's Test

**Proposition 1.24 (The Raabe's and Duhamel's test,** see [15], p. 94):
Let $\sum_{n \geq 1} a_n$ be a positive series. Assume that:

$$\lim_{n \to \infty} n \left(\frac{a_n}{a_{n+1}} - 1\right) = \lambda.$$

Then we have the following:

A) If $\lambda > 1$, then the series $\sum_{n \geq 1} a_n$ is convergent;

B) If $\lambda < 1$, then the series $\sum_{n \geq 1} a_n$ is divergent;

C) If $\lambda = 1$, then the series $\sum_{n \geq 1} a_n$ may be convergent or it may be divergent, namely we do not have a definite conclusion.

**Example 1.25.** Test for convergence the following positive series:

$$\text{a)} \quad \sum_{n \geq 1} \frac{1 \cdot 3 \cdots \cdots (2n-1)}{2 \cdot 4 \cdots \cdots 2n} \cdot \frac{1}{2n+1}$$

$$\text{b)} \quad \sum_{n \geq 1} 7^{\ln n}$$

$$\text{c)} \quad \sum_{n \geq 1} \frac{n!}{\alpha \, (\alpha + 1) \, (\alpha + 2) \cdots \cdots (\alpha + n - 1)}, \quad a > 0.$$

**Solutions.**

a) One obtains

$$
\begin{aligned}
n \left( \frac{a_n}{a_{n+1}} - 1 \right) &= n \left[ \frac{1 \cdot 3 \cdots \cdots (2n-1)}{2 \cdot 4 \cdots \cdots 2n} \cdot \frac{1}{2n+1} \cdot \frac{2 \cdot 4 \cdots \cdots (2n+2)(2n+3)}{1 \cdot 3 \cdots \cdots (2n+1)} - 1 \right] \\
&= n \left[ \frac{(2n+2)(2n+3)}{(2n+1)^2} - 1 \right] \\
&= n \cdot \frac{4n^2 + 6n + 4n + 6 - 4n^2 - 4n - 1}{(2n+1)^2} = n \cdot \frac{6n-1}{(2n+1)^2}.
\end{aligned}
$$

Hence

$$\lim_{n \to \infty} n \left( \frac{a_n}{a_{n+1}} - 1 \right) = \frac{6}{4} > 1$$

and with the Raabe's and Duhamel's test it results that $\sum_{n \geq 1} a_n$ is convergent.

b) One deduces that:

$$n \left( \frac{a_n}{a_{n+1}} - 1 \right) = n \left( \frac{7^{\ln n}}{7^{\ln(n+1)}} - 1 \right) = n \left( 7^{\ln \frac{n}{n+1}} - 1 \right) = \frac{7^{\ln \frac{n}{n+1}} - 1}{\ln \frac{n}{n+1}} \cdot n \cdot \ln \frac{n}{n+1}$$

and

$$\lim_{n \to \infty} n \left( \frac{a_n}{a_{n+1}} - 1 \right) = \lim_{n \to \infty} \frac{7^{\ln \frac{n}{n+1}} - 1}{\ln \frac{n}{n+1}} \cdot \lim_{n \to \infty} n \cdot \ln \frac{n}{n+1}.$$

One knows that

$$\lim_{x \to 0} \frac{a^x - 1}{x} = \ln a, \ a > 0;$$

therefore

$$\lim_{n \to \infty} \frac{7^{\ln \frac{n}{n+1}} - 1}{\ln \frac{n}{n+1}} = \lim_{x \to 0} \frac{7^x - 1}{x} = \ln 7.$$

We shall compute

$$\lim_{n \to \infty} n \cdot \ln \frac{n}{n+1} = \lim_{n \to \infty} \ln \left(\frac{n}{n+1}\right)^n = \ln \lim_{n \to \infty} \frac{1}{\left(1 + \frac{1}{n}\right)^n} = \ln \left(e^{-1}\right) = -1.$$

We have used the fact that

$$\lim_{x \to \infty} \left(1 + \frac{1}{x}\right)^x = e.$$

Finally, one obtains

$$\lim_{n \to \infty} n \left(\frac{a_n}{a_{n+1}} - 1\right) = \ln 7 \cdot \ln \left(e^{-1}\right) = -\ln 7 < 1$$

and using the Raabe's and Duhamel's test it follows that

$$\sum_{n \geq 1} 7^{\ln n}$$

diverges.

We can prove this result in Matlab 7.9:

```
>> syms n
>>a=@(n) 7^(log(n));
>> limit(n*(a(n)/a(n+1)-1),n,inf)
ans =
-log(7)
```

or in Mathcad 14:

$$a(n) := 7^{\ln(n)}$$

$$\lim_{n \to \infty} \left[n \cdot \left(\frac{a(n)}{a(n+1)} - 1\right)\right] \to -\ln(7)$$

or with Mathematica 8:

In[16]:= `a[n_] := 7^(Log[n])`

In[17]:= `Limit[n * (a[n] / a[n + 1] - 1), n → Infinity]`

Out[17]= $-\text{Log}[7]$

or in Maple 15:

$a := n \to 7^{\ln(n)}$

$$n \to 7^{\ln(n)}$$

$$\lim_{n \to \infty} n \cdot \left( \frac{a(n)}{a(n+1)} - 1 \right)$$

$$-\ln(7)$$

c) We shall have

$$n \left( \frac{a_n}{a_{n+1}} - 1 \right) = n \left[ \frac{n!}{\alpha \, (\alpha + 1) \, (\alpha + 2) \cdot \cdots \cdot (\alpha + n - 1)} \right.$$

$$\left. \cdot \frac{\alpha \, (\alpha + 1) \, (\alpha + 2) \cdot \cdots \cdot (\alpha + n - 1) \, (\alpha + n)}{(n+1)!} - 1 \right]$$

$$= n \cdot \frac{\alpha + n - n - 1}{n + 1} \xrightarrow[n \to \infty]{} a - 1;$$

hence, using the Raabe's and Duhamel's test it results that if:

- $a - 1 > 1 \iff \alpha > 2$, then the series is convergent;
- $a - 1 < 1 \iff \alpha < 2$, then the series is divergent.

## 1.4.5   The Comparison Test II

**Proposition 1.26 (The comparison test II,** see [15], p. 94): Let $\sum_{n \geq 1} a_n$ and $\sum_{n \geq 1} b_n$ be two positive series such that:

$$\frac{a_{n+1}}{a_n} \leq \frac{b_{n+1}}{b_n}.$$

Then if:

a) $\sum_{n \geq 1} b_n$ is converges, it follows that $\sum_{n \geq 1} a_n$ also converges;

b) $\sum_{n \geq 1} a_n$ diverges, it results that $\sum_{n \geq 1} b_n$ also diverges.

**Example 1.27.** Discuss the convergence of the positive series:

$$\sum_{n \geq 1} \frac{n^n}{e^n \cdot n!}.$$

**Solution.**
Considering that:

$$b_n = \frac{n^n}{e^n \cdot n!}$$

we shall obtain

$$\frac{b_{n+1}}{b_n} = \frac{(n+1)^{n+1}}{e^{n+1} \cdot (n+1)!} \cdot \frac{e^n \cdot n!}{n^n} = \frac{(n+1)^n \cdot (n+1)}{e \cdot (n+1) \cdot n!} \cdot \frac{n!}{n^n} =$$
$$= \frac{n^n \cdot \left(1 + \frac{1}{n}\right)^n}{e \cdot n^n} = \frac{\left(1 + \frac{1}{n}\right)^n}{e}.$$

Since

$$\left(1 + \frac{1}{n}\right)^n < e < \left(1 + \frac{1}{n}\right)^{n+1}, \quad (\forall) \; n \in \mathbb{N}^*$$

we shall have

$$\frac{b_{n+1}}{b_n} \geq \frac{\left(1 + \frac{1}{n}\right)^n}{\left(1 + \frac{1}{n}\right)^{n+1}} = \frac{\left(\frac{n+1}{n}\right)^n}{\left(\frac{n+1}{n}\right)^{n+1}} = \frac{n}{n+1} = \frac{\frac{1}{n+1}}{\frac{1}{n}} = \frac{a_{n+1}}{a_n};$$

hence using the comparison test II it follows that $\sum_{n \geq 1} b_n$ diverges.

### 1.4.6   The Comparison Test III

**Proposition 1.28 (The comparison test III**, see [15], p. 94): Let $\sum_{n \geq 1} a_n$ and $\sum_{n \geq 1} b_n$ be two positive series such that:

$$\lim_{n \to \infty} \frac{a_n}{b_n} = K.$$

If:

a) $0 < K < \infty$ then the two series have the same nature;

b) $K = 0$ and $\sum_{n \geq 1} b_n$ converges it results that $\sum_{n \geq 1} a_n$ also converges;

c) $K = \infty$ and $\sum_{n \geq 1} b_n$ diverges it results that $\sum_{n \geq 1} a_n$ also diverges.

**Example 1.29.** Discuss the convergence of the positive series:

$$\sum_{n\geq 1} \sin \frac{1}{n}.$$

**Solution.**
As

$$\lim_{n\to\infty} \frac{\sin \frac{1}{n}}{\frac{1}{n}} = 1$$

taking into account that $\sum_{n\geq 1} \frac{1}{n}$ diverges and using the comparison test III, it follows that

$$\sum_{n\geq 1} \sin \frac{1}{n}$$

also diverges.

## 1.5 Absolutely Convergent and Semi-convergent Series

**Definition 1.30** (see [41], p. 41). The series $\sum_{n\geq 1} a_n$ is called **absolutely convergent** if the series of absolute values $\sum_{n\geq 1} |a_n|$ is convergent.
**Remark 1.31** (see [15], p. 95). For investigating the absolute convergence of the series $\sum_{n\geq 1} |a_n|$, one employs the convergence tests of positive series.
**Definition 1.32** (see [41], p. 42). The convergent series $\sum_{n\geq 1} a_n$ is called **conditionally convergent** (or **semi- convergent**) if the series of absolute values $\sum_{n\geq 1} |a_n|$ is divergent.
**Example 1.33.** Determine if each of the following series is absolute convergent or semi- convergent:

a) $\sum_{n\geq 1} \frac{\cos n\alpha}{n^2}, \ \alpha \in \mathbb{R}$

b) $\sum_{n\geq 2} (-1)^n \frac{1}{\sqrt{n(n-1)}}$

c) $\sum_{n\geq 1} (-1)^{n+1} \frac{n+1}{n^2}$

d) $\sum_{n\geq 1} \frac{\sin n\alpha}{3^n}, \ \alpha \in \mathbb{R}$

$$\text{e) } \sum_{n \geq 1} \frac{b_n}{n\left(n + \sqrt{3}\right)},$$

where $(b_n)_{n \geq 1}$ is a bounded sequence.

**Solutions.**

a) Noting that:

$$\left| \frac{\cos n\alpha}{n^2} \right| = \frac{|\cos n\alpha|}{n^2} \leq \frac{1}{n^2}$$

and the series $\sum_{n \geq 1} \frac{1}{n^2}$ is convergent, using the comparison test I one deduces that

$$\sum_{n \geq 1} \left| \frac{\cos n\alpha}{n^2} \right|, \alpha \in \mathbb{R}$$

also converges, i.e.

$$\sum_{n \geq 1} \frac{\cos n\alpha}{n^2}, \alpha \in \mathbb{R}$$

is absolutely convergent.

b) One can notice that

$$\sum_{n \geq 2} (-1)^n \frac{1}{\sqrt{n(n-1)}}$$

is convergent according to the Leibnitz's test.

We shall have

$$\left| (-1)^n \frac{1}{\sqrt{n(n-1)}} \right| = \frac{1}{\sqrt{n(n-1)}} \geq \frac{1}{\sqrt{n \cdot n}} = \frac{1}{n};$$

as $\sum_{n \geq 1} \frac{1}{n}$ is divergent, with the comparison test I one deduces that

$$\sum_{n \geq 2} \left| (-1)^n \frac{1}{\sqrt{n(n-1)}} \right|$$

is divergent; therefore

$$\sum_{n \geq 2} (-1)^n \frac{1}{\sqrt{n(n-1)}}$$

is semi-convergent.

c) As

$$\left| \frac{n+1}{n^2} \right| = \frac{n+1}{n^2} \geq \frac{n+1}{(n+1)^2} = \frac{1}{n+1}$$

and $\sum_{n\geq 1} \frac{1}{n+1}$ is divergent, then using the comparison test I, it results that

$$\sum_{n\geq 1} \left| \frac{n+1}{n^2} \right|$$

is divergent; therefore

$$\sum_{n\geq 1} (-1)^{n+1} \frac{n+1}{n^2}$$

is semi-convergent.

d) We can notice that

$$\left| \frac{\sin n\alpha}{3^n} \right| \leq \frac{1}{3^n}$$

Since the geometric series $\sum_{n\geq 1} q^n$ is:

- convergent for $|q| < 1$;
- divergent for $|q| \geq 1$

one deduces that the series $\sum_{n\geq 1} \frac{1}{3^n}$ is convergent; hence, using the comparison test I, it results that

$$\sum_{n\geq 1} \left| \frac{\sin n\alpha}{3^n} \right|$$

is convergent, i.e.

$$\sum_{n\geq 1} \frac{\sin n\alpha}{3^n}$$

is absolutely convergent.

e) If is $(b_n)_{n\geq 1}$ is a bounded sequence it results that:

$$(\exists)\ M > 0 \text{ such that } |b_n| \leq M,\ (\forall)\ n \in \mathbb{N}.$$

We shall have

$$\left| \frac{b_n}{n\left(n+\sqrt{3}\right)} \right| = \frac{|b_n|}{n\left(n+\sqrt{3}\right)} \leq \frac{M}{n\left(n+\sqrt{3}\right)} \leq \frac{M}{n^2}.$$

As the series $\sum_{n\geq 1} \frac{1}{n^2}$ is convergent, using the comparison test I, it results that

$$\sum_{n\geq 1} \left| \frac{b_n}{n\left(n+\sqrt{3}\right)} \right|$$

is convergent, i.e.

$$\sum_{n \geq 1} \frac{b_n}{n\left(n + \sqrt{3}\right)}$$

is absolutely convergent.

## 1.6   Problems

1. Is the following sequence

$$x_n = \sum_{k=1}^{n} \frac{\cos \frac{k\pi}{4}}{k!}, \quad (\forall)\ n \in \mathbb{N}^*$$

a Cauchy sequence?

2. Prove that the following sequence is a Cauchy sequence:

$$x_n = 1 + \frac{1}{2^2} + \frac{1}{3^2} + \cdots + \frac{1}{n^2}, \quad (\forall)\ n \in \mathbb{N}^*.$$

3. Using the Cauchy's test, test the convergence of the series

$$\sum_{n \geq 1} \cos \frac{x^n}{n^2}, \quad (\forall)\ x \in \mathbb{R}.$$

4. Use the Cauchy's test to test the convergence of the series

$$\sum_{n \geq 1} \frac{\sin n\alpha}{2^n}, \quad (\forall)\ \alpha \in \mathbb{R}.$$

5. Test for convergence the following alternating series:

$$1 - \frac{2}{7} + \frac{3}{13} + \cdots + (-1)^{n+1} \frac{n}{6n - 5} + \cdots$$

**Computer solution.**
We can notice using Matlab 7.9:
```
>>syms n
>>a=@(n) n/(6*n-5);
>>limit(a(n),inf)
ans=
1/6
```
or Mathcad 14:

$$a(n) := \frac{n}{6 \cdot n - 5}$$

$$\lim_{n \to \infty} a(n) \to \frac{1}{6}$$

or with Mathematica 8:

In[23]:= `a[n_] := n / (6 * n - 5)`

In[24]:= `Limit[a[n], n → Infinity]`

Out[24]= $\frac{1}{6}$

or in Maple 15:

$$a := n \to \frac{n}{6 \cdot n - 5}$$

$$n \to \frac{n}{6n - 5}$$

$$\lim_{n \to \infty} a(n)$$

$$\frac{1}{6}$$

i.e. the alternating series is divergent.

6. Is the following alternating series

$$\sum_{n \geq 1} (-1)^{n+1} \tan \frac{1}{n\sqrt{n}}$$

convergent?
**Computer solution.**
Using Matlab 7.9:
`>> syms n`
`>> a=@(n)tan(1/(n*sqrt(n)));`
`>> limit(a(n),inf)`
`ans =`

**0**

or Mathcad 14:

$$a(n) := \tan\left(\frac{1}{n \cdot \sqrt{n}}\right)$$

$$\lim_{n \to \infty} \ a(n) \to 0$$

or Mathematica 8:

In[47]:= `a[n_] := Tan[1 / (n * Sqrt[n])]`

In[48]:= `Limit[a[n], n → Infinity]`

Out[48]= 0|

or Maple 15:

$$a := n \to \tan\left(\frac{1}{n \cdot \sqrt{n}}\right)$$

$$n \to \tan\left(\frac{1}{n\sqrt{n}}\right)$$

$$\lim_{n \to \infty} (a(n))$$

$$0$$

and adding the fact that

$$\frac{1}{n\sqrt{n}} > \frac{1}{(n+1)\sqrt{n+1}}$$

and the tangent function is increasing on $\left(-\frac{\pi}{2}, \frac{\pi}{2}\right)$ it results that

$$\tan\frac{1}{n\sqrt{n}} > \tan\frac{1}{(n+1)\sqrt{n+1}}$$

namely $(a_n)_n$ is monotone decreasing; hence using the Leibnitz's test it follows that the alternating series is convergent.

7. Test the convergence of the positive series:

$$\sum_{n\geq 1}\left(\sqrt{n}-\sqrt{n-1}\right).$$

**Computer solution.**
Applying the Raabe's and Duhamel's test in Matlab 7.9:
**>>syms n**
**>> a=@(n)sqrt(n)-sqrt(n-1);**
**>> limit(n\*(a(n)/a(n+1)-1),inf)**
ans =
**1/2**
or in Mathcad 14:

$$a(n) := \sqrt{n} - \sqrt{n-1}$$

$$\lim_{n \to \infty}\left[n\cdot\left(\frac{a(n)}{a(n+1)} - 1\right)\right] \to \frac{1}{2}$$

or in Mathematica 8:

In[28]:= **a[n_] := Sqrt[n] - Sqrt[n - 1]**

In[29]:= **Limit[n \* (a[n] / a[n + 1] - 1), n → Infinity]**

Out[29]= $\dfrac{1}{2}$

or in Maple15:

$$a := n \to \sqrt{n} - \sqrt{n-1}$$

$$n \to \sqrt{n} - \sqrt{n-1}$$

$$\lim_{n \to \infty} n\cdot\left(\frac{a(n)}{a(n+1)} - 1\right)$$

$$\frac{1}{2}$$

we can deduce that the positive series is divergent.

8. Is the positive series

$$\sum_{n \geq 1} \left( \frac{6n^2 + 7n + 5}{2n^2 + 5n + 9} \right)^n$$

convergent?

**Computer solution.**

Applying the **root test** in Matlab 7.9:

```
>> syms n
>> a=@(n)((6*n^2+7*n+5)/(2*n^2+5*n+9))^n;
>> limit(a(n)^(1/n),inf)
ans =
3
```

and in Mathcad 14:

$$a(n) := \left( \frac{6 \cdot n^2 + 7 \cdot n + 5}{2 \cdot n^2 + 5 \cdot n + 9} \right)^n$$

$$\lim_{n \to \infty} a(n)^{\frac{1}{n}} \quad \text{simplify} \to 3$$

and in Mathematica 8:

```
In[26]:= a[n_] := ((6*n^2 + 7*n + 5) / (2*n^2 + 5*n + 9))^n

In[27]:= Limit[a[n]^(1/n), n → Infinity]

Out[27]= 3
```

and in Maple 15:

$$a := n \to \left( \frac{6 \cdot n^2 + 7 \cdot n + 5}{2 \cdot n^2 + 5 \cdot n + 9} \right)^n$$

$$n \to \left( \frac{6 n^2 + 7 n + 5}{2 n^2 + 5 n + 9} \right)^n$$

$$\lim_{n \to \infty} a(n)^{\frac{1}{n}}$$

3

we can decide that the positive series is divergent.

9. What can you tell about the nature of the series

$$\sum_{n\geq 1}\frac{1\cdot 3\cdots\cdots(2n-1)}{2\cdot 5\cdots\cdots(3n-1)}?$$

10.Use the tests of convergence of the positive series to establish the nature of the series:

a) $\displaystyle\sum_{n\geq 1}\frac{n^2}{2^n}$

b) $\displaystyle\sum_{n\geq 1}\frac{1}{n\sqrt[n]{n}}.$

**Computer solution.**
a) Applying the **ratio test** in Matlab 7.9:
>> **syms n**
>> **a=@(n)n^2/(2^n);**
>> **limit(a(n+1)/a(n),inf)**
**ans =**
**1/2**
or in Mathcad 14:

$$a(n) := \frac{n^2}{2^n}$$

$$\lim_{n\to\infty}\frac{a(n+1)}{a(n)} \to \frac{1}{2}$$

or in Mathematica 8:

In[30]:= **a[n_] := n^2 / 2^n**

In[31]:= **Limit[a[n + 1] / a[n], n → Infinity]**

Out[31]= $\dfrac{1}{2}$

or in Maple 15:

$$a := n \to \frac{n^2}{2^n}$$

$$n \to \frac{n^2}{2^n}$$

$$\lim_{n \to \infty} \frac{a(n+1)}{a(n)}$$

$$\frac{1}{2}$$

we can say that the positive series is convergent.

# 2
# Power Series

## 2.1 Region of Convergence

**Definition 2.1** (see [15], p. 281). A series of the form

$$\sum_{n=0}^{\infty} a_n x^n, a_n \in \mathbb{R}, \ n = 0, 1, 2 \ldots \tag{2.1}$$

is called **power series**.

**Definition 2.2** (see [15], p. 281). For any power series there exists a number $\rho \geq 0$ , called **radius of convergence** such that:

a) the power series is absolutely convergent on the interval $(-\rho, \rho)$ ;

b) the power series is divergent for $|x| > \rho$.

**Definition 2.3** (see [15], p. 282). The interval $(-\rho, \rho)$ constitutes the **interval of convergence** corresponding to the power series from (2.1).

**Proposition 2.4** (see [8] ). The interval of convergence can be ordinarily determined with the help of d'Alembert's and Cauchy's test, in the following two steps:

*Step 1.* One computes the number

$$\lambda = \lim_{n \to \infty} \sqrt[n]{|a_n|}, \tag{2.2}$$

G.A. Anastassiou and I.F. Iatan: Intelligent Routines, ISRL 39, pp. 41–70.
springerlink.com        © Springer-Verlag Berlin Heidelberg 2013

or

$$\lambda = \lim_{n\to\infty} \left| \frac{a_{n+1}}{a_n} \right|. \tag{2.3}$$

*Step 2.* If:

a) $\lambda = \infty$ then $\rho = 0$;
b) $\lambda = 0$ then $\rho = \infty$;
c) $0 < \lambda < \infty$ then $\rho = \frac{1}{\lambda}$.

**Definition 2.5** (see [8] ).The **set of convergence** (**region of convergence**) corresponding to the power series from (2.1) is the set of values of the argument $x$ for which the power series converges.

**Example 2.6.** Find the set of convergence of the power series:

a) $\displaystyle\sum_{n\geq 1} \frac{x^n}{n \cdot 2^n}$

b) $\displaystyle\sum_{n\geq 1} \frac{x^n}{n}$

c) $\displaystyle\sum_{n\geq 1} n! x^n$

d) $\displaystyle\sum_{n\geq 1} \frac{x^n}{2^n + 3^n}$

e) $\displaystyle\sum_{n\geq 1} (-1)^n \frac{n+1}{n^2+n+1} x^n$

f) $\displaystyle\sum_{n\geq 1} \frac{n x^n}{(n!)^2}$

g) $\displaystyle\sum_{n\geq 1} \left(1 + \frac{1}{n}\right)^{n^2+n} x^n.$

**Solutions.**
a) We shall have

$$a_n = \frac{1}{n \cdot 2^n}$$

therefore

$$\lambda = \lim_{n\to\infty} \left| \frac{a_{n+1}}{a_n} \right| = \lim_{n\to\infty} \frac{1}{(n+1) \cdot 2^{n+1}} \cdot n \cdot 2^n = \frac{1}{2} \lim_{n\to\infty} \frac{n}{n+1} = \frac{1}{2},$$

namely the radius of convergence is $\rho = 2$ and the interval of convergence is $(-2, 2)$.

We shall test the convergence at the end-points of the interval of convergence:

- when $x = -2$, the series

$$\sum_{n \geq 1} \frac{(-2)^n}{n \cdot 2^n} = \sum_{n \geq 1} \frac{(-1)^n \cdot 2^n}{n \cdot 2^n} = \sum_{n \geq 1} (-1)^n \frac{1}{n}$$

converges (by Leibnitz's test);

- when $x = 2$, the series

$$\sum_{n \geq 1} \frac{2^n}{n \cdot 2^n} = \sum_{n \geq 1} \frac{1}{n}$$

diverges.

Hence, the set of convergence of the power series

$$\sum_{n \geq 1} \frac{x^n}{n \cdot 2^n}$$

is $[-2, 2)$.

b) For

$$a_n = \frac{1}{n^n}$$

we shall have:

$$\lambda = \lim_{n \to \infty} \sqrt[n]{|a_n|} = \lambda = \lim_{n \to \infty} \sqrt[n]{\left|\frac{1}{n^n}\right|} = \lim_{n \to \infty} \frac{1}{n} = 0;$$

hence, the radius of convergence is $\rho = \infty$ and the interval of convergence, respectively the set of convergence is $(-\infty, \infty)$.

c) As $a_n = n!$ it results that

$$\left|\frac{a_{n+1}}{a_n}\right| = \frac{(n+1)!}{n!} = n + 1 \xrightarrow[n \to \infty]{} \infty$$

and the radius of convergence is $\rho = 0$, respectively the set of convergence is $\{0\}$.

d) Denoting

$$a_n = \frac{1}{2^n + 3^n}$$

one obtains

$$\left|\frac{a_{n+1}}{a_n}\right| = \frac{2^n + 3^n}{2^{n+1} + 3^{n+1}} \xrightarrow[n\to\infty]{} \frac{1}{3};$$

therefore the radius of convergence is $\rho = 3$ and the interval of convergence is $(-3, 3)$.

We shall test the convergence at the end-points of the interval of convergence:

- when $x = -3$, the series

$$\sum_{n\geq 1} (-1)^n \frac{3^n}{2^n + 3^n}$$

is divergent;

- when $x = 3$, the series

$$\sum_{n\geq 1} \frac{3^n}{2^n + 3^n}$$

diverges.

Hence, the set of convergence of the power series

$$\sum_{n\geq 1} \frac{x^n}{2^n + 3^n}$$

is $(-3, 3)$.

e) Taking

$$a_n = (-1)^n \frac{n+1}{n^2 + n + 1}$$

it results

$$\left|\frac{a_{n+1}}{a_n}\right| = \frac{n+2}{(n+1)^2 + n + 2} \cdot \frac{n^2 + n + 1}{n+1} \xrightarrow[n\to\infty]{} 1;$$

hence the radius of convergence is $\rho = 1$ and the interval of convergence is $(-1, 1)$.

We can also obtain this result in Matlab 7.9:

```
>>syms n
>> a=@(n) (-1)^n*(n+1)/(n^2+n+1);
>> 1/simplify(limit(abs(a(n+1)/a(n)),n,inf))
ans=
1
```

or in Mathcad 14:

$$a(n) := (-1)^n \cdot \frac{n+1}{n^2+n+1} \qquad \frac{1}{\lim\limits_{n \to \infty} \left| \dfrac{a(n+1)}{a(n)} \right|} \to 1$$

or  using Mathematica 8:

```
In[1]:= a[n_] := (-1)^n * (n + 1) / (n^2 + n + 1)

In[2]:= 1 / Limit[Abs[a[n + 1] / a[n]], n → Infinity]

Out[2]= 1
```

or with Maple 15:

$$a := n \to (-1)^n \cdot \frac{n+1}{n^2+n+1} :$$

$$\frac{1}{\lim\limits_{n \to \infty} \left| \dfrac{a(n+1)}{a(n)} \right|}$$

$$1$$

We want to test the convergence at the end-points of the interval of convergence:

- when $x = -1$ one obtains the series

$$\sum_{n \geq 1} (-1)^{2n} \frac{n+1}{n^2+n+1} = \sum_{n \geq 1} \frac{n+1}{n^2+n+1}$$

which is divergent, using the comparison test I since

$$\frac{n+1}{n^2+n+1} \geq \frac{n+1}{n^2+2n+1} = \frac{1}{n+1}$$

and the fact that the series $\sum_{n \geq 1} \frac{1}{n+1}$ is divergent.

- when $x = 1$ one achieves the alternating series

$$\sum_{n \geq 1} (-1)^n \frac{n+1}{n^2+n+1}$$

that is convergent using the Leibnitz's test.

The set of convergence of the power series

$$\sum_{n\geq 1}(-1)^n \frac{n+1}{n^2+n+1}$$

is $(-1,1]$.

f) We notice that

$$a_n = \frac{n}{(n!)^2} = \frac{n}{n^2\left[(n-1)!\right]^2} = \frac{1}{n\left[(n-1)!\right]^2};$$

it will result

$$\left|\frac{a_{n+1}}{a_n}\right| = \frac{1}{(n+1)\cdot(n!)^2}\cdot n\left[(n-1)!\right]^2$$

$$= \frac{1}{n^2(n+1)\cdot\left[(n-1)!\right]^2}\cdot n\left[(n-1)!\right]^2 = \frac{1}{n(n+1)}\xrightarrow[n\to\infty]{}0.$$

Hence the radius of convergence is $\rho = \infty$ and the interval of convergence, respectively the set of convergence is $(-\infty,\infty)$.

g) We shall have

$$a_n = \left(1+\frac{1}{n}\right)^{n^2+n},$$

therefore

$$\lambda = \lim_{n\to\infty}\sqrt[n]{|a_n|} = \lim_{n\to\infty}\left(1+\frac{1}{n}\right)^{\frac{n^2+n}{n}} = \lim_{n\to\infty}\left(1+\frac{1}{n}\right)^{n+1} = e,$$

i.e. the radius of convergence is $\rho = \frac{1}{e}$ and the interval of convergence is $\left(-\frac{1}{e},\frac{1}{e}\right)$.

The following Matlab 7.9 sequence helps us to obtain the radius of convergence:

```
>>syms n
>> 1/limit(((1+1/n)^(n^2+n))^(1/n),n,inf)
ans =
1/exp(1)
```

We can also get the radius of convergence using Mathcad 14:

$$a(n) := \left(1 + \frac{1}{n}\right)^{n^2+n}$$

$$\frac{1}{\lim\limits_{n \to \infty} a(n)^{\frac{1}{n}}} \to e^{-1}$$

or Mathematica 8:

In[73]:= a[n_] := (1 + 1 / n) ^ (n^2 + n)

In[75]:= 1 / Limit[a[n] ^ (1 / n), n → Infinity]

Out[75]= $\dfrac{1}{e}$

or Maple 15:

$$a := n \to \left(1 + \frac{1}{n}\right)^{n^2+n}$$

$$n \to \left(1 + \frac{1}{n}\right)^{n^2+n}$$

$$\frac{1}{\lim\limits_{n \to \infty} a(n)^{\frac{1}{n}}}$$

$$\frac{1}{e}$$

We shall test the convergence at the end-points of the interval of convergence:

- when $x = -\frac{1}{e}$ it results the series

$$\sum_{n \geq 1} \left(-\frac{1}{e}\right)^n \left(1 + \frac{1}{n}\right)^{n^2+n} = \sum_{n \geq 1} (-1)^n \frac{1}{e^n} \left(1 + \frac{1}{n}\right)^{n^2+n},$$

which diverges as

$$\lim_{n\to\infty} \frac{1}{e^n}\left(1+\frac{1}{n}\right)^{n^2+n} = e^{\frac{1}{2}} \neq 0,$$

see the Matlab7.9. commands:

```
>> syms n
>> limit(((1+1/n)^(n^2+n))/exp(n),n,inf)
ans=
exp(1/2)
```

The limit can be also computed using Mathcad 14:

$$\lim_{n\to\infty}\left[\frac{1}{e^n}\cdot\left(1+\frac{1}{n}\right)^{n^2+n}\right] \rightarrow \sqrt{e}$$

or Mathematica 8:

```
In[1]:= a[n_] := (1/E^n) * ((1+1/n)^(n^2 + n));

In[2]:= Limit[a[n], n → Infinity]

Out[2]= √e
```

or Maple 15:

$$\lim_{n\to\infty}\frac{1}{e^n}\cdot\left(1+\frac{1}{n}\right)^{n^2+n}$$

$$e^{\frac{1}{2}}$$

- when $x = \frac{1}{e}$, the series

$$\sum_{n\geq 1}\frac{1}{e^n}\left(1+\frac{1}{n}\right)^{n^2+n}$$

diverges, according to the root test: if $c_n = \frac{1}{e^n} \left(1 + \frac{1}{n}\right)^{n^2+n}$ then

$$\sqrt[n]{c_n} = \sqrt[n]{\frac{1}{e^n} \left(1 + \frac{1}{n}\right)^{n^2+n}} = \frac{1}{e} \left(1 + \frac{1}{n}\right)^{n+1} \geq 1.$$

Hence, the set of convergence of the power series

$$\sum_{n \geq 1} \left(1 + \frac{1}{n}\right)^{n^2+n} x^n$$

is $\left(-\frac{1}{e}, \frac{1}{e}\right)$.

## 2.2   Taylor and Mac Laurin Series

### 2.2.1   Expanding a Function in a Power Series

**Theorem 2.7** (see [8]). If a function $f(x)$ can be expanded in some neighbourhood $|x - a| < \rho$ of the point $a$ in a series of powers of $x - a$, namely $f(x)$ has a power series representation at the point $a$

$$f(x) = \sum_{n=0}^{\infty} c_n (x - a)^n \tag{2.4}$$

then its coefficients are given by the formula

$$c_n = \frac{f^{(n)}(a)}{n!}. \tag{2.5}$$

**Definition 2.8** (see [8]). The power series of the form

$$f(x) = \sum_{n=0}^{\infty} \frac{f^{(n)}(a)}{n!} (x - a)^n \tag{2.6}$$

$$= f(a) + (x - a) f'(a) + \frac{(x - a)^2}{2!} f''(a) + \ldots + \frac{(x - a)^n}{n!} f^{(n)}(a) + \cdots$$

is called the **Taylor series** and

$$R_n(x) = f(x) - T_n(x) \tag{2.7}$$

means the **remainder of the Taylor series**, where $T_n(x)$ is the $n$- th degree **Taylor polynomial** of $f$ at the point $a$:

$$T_n(x) = f(a) + (x - a) f'(a) + \frac{(x-a)^2}{2!} f''(a) + \ldots + \frac{(x-a)^n}{n!} f^{(n)}(a).$$
(2.8)

**Remark 2.9** (see [15], p. 282). In order to evaluate the remainder, one can employ the formula:

$$R_n(x) = \frac{(x-a)^{n+1}}{(n+1)!} f^{(n+1)}(a + \theta(x-a)), \ 0 < \theta < 1.$$
(2.9)

**Definition 2.10** (see [15], p. 282). The power series of the form

$$f(x) = \sum_{n=0}^{\infty} \frac{f^{(n)}(0)}{n!} x^n = f(0) + xf'(0) + \frac{x^2}{2!} f''(0) + \ldots + \frac{x^n}{n!} f^{(n)}(0) + \cdots$$
(2.10)

namely the particular case of the Taylor series for $a = 0$ is called the **Mac Laurin series**.

**Example 2.11.** Expand the function $f(x)$ in a series of powers of $x$:

$$\text{a)} \ \ f(x) = \frac{1}{2} \ln \frac{1+x}{1-x}, \ x \in (-1, 1)$$

$$\text{b)} \ \ f(x) = \cos^3 x, \ x \in \mathbb{R}$$

$$\text{c)} \ f(x) = \frac{3x - 5}{x^2 - 4x - 3}, \ x \in \mathbb{R} \setminus \{1, 3\}.$$

**Solutions.**
a) One can notice that

$$f(x) = \frac{1}{2} \left[ \underbrace{\ln(1+x)}_{f_1(x)} - \underbrace{\ln(1-x)}_{f_2(x)} \right] = \frac{1}{2} [f_1(x) - f_2(x)], \ x \in (-1, 1)$$

and

$$\begin{cases} f_1'(x) = \frac{1}{1+x} \\ f_1''(x) = -\frac{1}{(1+x)^2} \\ f_1'''(x) = \frac{2}{(1+x)^3} \\ f_1^{(4)}(x) = -\frac{6}{(1+x)^4} \end{cases}, \quad \begin{cases} f_2'(x) = -\frac{1}{1-x} \\ f_2''(x) = -\frac{1}{(1-x)^2} \\ f_2'''(x) = -\frac{2}{(1-x)^3} \\ f_2^{(4)}(x) = -\frac{6}{(1-x)^4}. \end{cases}$$

We shall expand both the function $f_1(x)$ and $f_2(x)$ in a Mac Laurin series of $x$:

$$f_1(x) = 0 + x \cdot 1 + \frac{x^2}{2!} \cdot (-1) + \frac{x^3}{3!} \cdot 2 + \frac{x^4}{4!} \cdot (-6) + \cdots$$

$$= x - \frac{x^2}{2!} + \frac{2x^3}{3!} - \frac{6x^4}{4!} + \cdots,$$

$$f_2(x) = -x - \frac{x^2}{2!} \cdot 1 - \frac{x^3}{3!} \cdot 2 - \frac{x^4}{4!} \cdot 6 + \cdots = -x - \frac{x^2}{2!} - \frac{2x^3}{3!} - \frac{6x^4}{4!} + \cdots;$$

hence

$$f(x) = \frac{1}{2}\left( x - \frac{x^2}{2!} + \frac{2x^3}{3!} - \frac{6x^4}{4!} + \cdots + x + \frac{x^2}{2!} + \frac{2x^3}{3!} + \frac{6x^4}{4!} + \cdots \right)$$

$$= \frac{1}{2}\left( 2x + 2 \cdot \frac{2x^3}{3!} + \cdots \right) = x + \frac{x^3}{3!} + \cdots = \sum_{n \geq 0} \frac{x^{2n+1}}{2n+1}.$$

We shall achieve the first eleven terms from the power series expansion of the function $f(x)$ of $x$, using the following Matlab 7.9 sequence:

```
>>syms x
>> taylor(1/2*log((1+x)/(1-x)),12)
ans =
x+1/3*x^3+1/5*x^5+1/7*x^7+1/9*x^9+1/11*x^11
```

or Mathcad 14:

$$\frac{1}{2} \cdot \ln\left( \frac{1+x}{1-x} \right) \quad \text{series}, 11 \quad \rightarrow x + \frac{x^3}{3} + \frac{x^5}{5} + \frac{x^7}{7} + \frac{x^9}{9} + \frac{x^{11}}{11}$$

or Mathematica 8:

In[89]:= **Series[1 / 2 \* Log[ (1 + x) / (1 - x)], {x, 0, 11}]**

Out[89]= $x + \dfrac{x^3}{3} + \dfrac{x^5}{5} + \dfrac{x^7}{7} + \dfrac{x^9}{9} + \dfrac{x^{11}}{11} + O[x]^{12}$

or Maple 15:

$$\text{series}\left( \frac{1}{2} \cdot \ln\left( \frac{1+x}{1-x} \right), x = 0, 12 \right)$$

$$x + \frac{1}{3}x^3 + \frac{1}{5}x^5 + \frac{1}{7}x^7 + \frac{1}{9}x^9 + \frac{1}{11}x^{11} + O(x^{12})$$

b) We shall deduce that:

$$f(x) = (\cos x \cdot \cos x)\cos x = \frac{\cos 2x + 1}{2}\cos x = \frac{1}{2}\cos 2x \cos x + \frac{1}{2}\cos x$$

$$= \frac{1}{2} \cdot \frac{1}{2}(\cos 3x + \cos x) + \frac{1}{2}\cos x = \frac{1}{4}\cos 3x + \frac{1}{4}\cos x + \frac{1}{2}\cos x = \frac{1}{4}\cos 3x + \frac{3}{4}\cos x;$$

therefore

$$\begin{cases} f'(x) = \frac{1}{4} \cdot 3 \cdot (-\sin 3x) - \frac{3}{4}\sin x = -\frac{3}{4}\sin 3x - \frac{3}{4}\sin x \\[2mm] f''(x) = -\frac{9}{4}\cos 3x - \frac{3}{4}\cos x \\[2mm] f'''(x) = \frac{27}{4}\sin 3x + \frac{3}{4}\sin x \end{cases}$$

and then we shall obtain the expanding of the function $f(x)$ in a series of powers of $x$:

$$f(x) = \sum_{n=0}^{\infty} \frac{f^{(n)}(0)}{n!}x^n = 1 + x \cdot 0 - \frac{x^2}{2!} \cdot \left(-\frac{9}{4} - \frac{3}{4}\right) + \frac{x^3}{3!} \cdot 0 + \frac{x^4}{4!} \cdot \left(\frac{27 \cdot 3}{4} + \frac{3}{4}\right) + \cdots$$

$$= -\frac{x^2}{2!} \cdot \frac{9^1 + 3}{4} + \frac{x^4}{4!} \cdot \frac{9^2 + 3}{4} + \cdots = \sum_{n \geq 0}(-1)^n \frac{x^{2n}}{(2n)!} \cdot (9^n + 3).$$

c) One can notice that

$$f(x) = \frac{3x - 5}{x^2 - 4x - 3} = \underbrace{\frac{1}{x-1}}_{f_1(x)} + \underbrace{\frac{2}{x-3}}_{f_2(x)} = f_1(x) + f_2(x),\ x \in \mathbb{R} \setminus \{1, 3\}$$

and

$$\begin{cases} f_1'(x) = -\frac{1}{(x-1)^2} \\[2mm] f_1''(x) = \frac{2}{(x-1)^3} \\[2mm] f_1'''(x) = -\frac{6}{(x-1)^4} \end{cases} , \qquad \begin{cases} f_2'(x) = -\frac{2}{(x-3)^2} \\[2mm] f_2''(x) = \frac{4}{(x-3)^3} \\[2mm] f_2'''(x) = -\frac{12}{(x-3)^4}. \end{cases}$$

Expand both the function $f_1(x)$ and $f_2(x)$ in a Mac Laurin series of $x$ one obtains:

$$f_1(x) = -1 - x - \frac{x^2}{2!} \cdot 2 - \frac{x^3}{3!} \cdot 6 + \cdots = -1 - x - x^2 - x^3 + \cdots = -\sum_{n \geq 0} x^n,$$

$$f_2(x) = -\frac{2}{3} - x \cdot \frac{2}{3^2} - \frac{x^2}{2!} \cdot \frac{4^2}{3^3} - \frac{x^3}{3!} \cdot \frac{12}{3^4} + \cdots$$

$$= -\frac{2}{3} - \frac{2x}{3^2} - \frac{2x^2}{3^3} - \frac{2x^3}{3^4} \cdots = -\sum_{n \geq 0} \frac{2}{3^{n+1}}x^n;$$

hence

$$f(x) = -\sum_{n\geq 0}\left(1 + \frac{2}{3^{n+1}}\right)x^n.$$

**Example 2.12.** Compute the approximate value for $\ln 1.2$ by expanding the function $f(x) = \ln(1+x)$ in a Taylor series of the 3rd order, in the point 0. Evaluate the committed error.

  **Solution.**

  As

$$\begin{cases} f'(x) = \frac{1}{1+x} \\[2mm] f''(x) = -\frac{1}{(1+x)^2} \\[2mm] f'''(x) = \frac{2}{(1+x)^3} \\[2mm] f^{(4)}(x) = -\frac{6}{(1+x)^4}. \end{cases}$$

Using the 3-th degree Taylor formula, we shall have

$$\ln(1+x) = \frac{x}{1!} + \frac{x^2}{2!}\cdot(-1) - \frac{x^3}{3!}\cdot 2 + R_3(x),$$

where

$$R_3(x) = \frac{x^4}{4!}\cdot f^{(4)}(\theta x) = \frac{x^4}{4!}\cdot\left(-\frac{6}{(1+\theta x)^4}\right) = -\frac{1}{4}\cdot\frac{x^4}{(1+\theta x)^4},\ \theta\in(0,1).$$

We shall achieve the approximate value for $\ln 1.2$ by expanding the function $f(x) = \ln(1+x)$ in a Taylor series of the 3-th order, in the point 0:

$$\ln 1.2 \approx \frac{0.2}{1!} - \frac{0.2^2}{2!} + \frac{0.2^3}{3!}\cdot 2 = 0.2 - \frac{0.04}{2} + \frac{0.008}{3} = 0.1826666667.$$

We shall check the result using Matlab 7.9:
>> **vpa(log(1.2),10)**
**ans =**
**0.1823215568**
or Mathcad 14:

$$\ln(1.2) = 0.1823215568$$

or Mathematica 8:

In[94]:= **SetAccuracy[Log[1.2], 11]**

Out[94]= **0.1823215568**

or Maple 15:

$\ln(1.2)$

0.1823215568

The committed error will be

$$R_3(0.2) = -\frac{1}{4} \cdot \frac{0.2^4}{(1 + 0.2 \cdot \theta)^4}, \ \theta \in (0,1);$$

hence

$$|R_3(0.2)| = \frac{1}{4} \cdot \frac{0.2^4}{(1 + 0.2 \cdot \theta)^4} \underset{\theta \in (0,1)}{<} \frac{0.2^4}{4} = 0.0004.$$

**Example 2.13.** Compute the number $\sqrt{e}$ to five decimals by expanding the function $f(x) = e^x$ in a series of powers of $x$.

**Solution.**

Using the $n$- th degree Mac Laurin formula, we shall have

$$e^x = 1 + \frac{x}{1!} + \frac{x^2}{2!} + \cdots + \frac{x^n}{n!} + \frac{x^{n+1}}{(n+1)!} \cdot e^{\theta x}, \ \theta \in (0,1);$$

substituting $x = \frac{1}{2}$ in the previous relation we shall have:

$$e^{1/2} = 1 + \frac{1}{2 \cdot 1!} + \frac{1}{2^2 \cdot 2!} + \cdots + \frac{1}{2^n \cdot n!} + \frac{1}{2^{n+1} \cdot (n+1)!} \cdot e^{\theta/2}, \ \theta \in (0,1).$$

Therefore,

$$R_n\left(\frac{1}{2}\right) = \frac{1}{2^{n+1} \cdot (n+1)!} \cdot e^{\theta/2} \underset{\theta \in (0,1)}{<} \frac{1}{2^{n+1} \cdot (n+1)!} \cdot e^{1/2}$$

$$< \frac{1}{2^{n+1} \cdot (n+1)!} \cdot 2 = \frac{1}{2^n \cdot (n+1)!}.$$

As we have to calculate the number $\sqrt{e}$ to five decimals it result that

$$\frac{1}{2^n \cdot (n+1)!} < \frac{1}{10^6} \iff 2^n \cdot (n+1)! > 10^6 \iff n \geq 7,$$

i.e. for $n \geq 7$ we can calculate $\sqrt{e}$ to five decimals; thus:

$$\sqrt{e} \approx 1 + \frac{1}{2 \cdot 1!} + \frac{1}{2^2 \cdot 2!} + \cdots + \frac{1}{2^7 \cdot 7!} = 1.6487345.$$

We shall check the result in Matlab 7.9:
>>vpa(exp(1/2),7))
ans=
**1.648721**
or in Mathcad 14:

$$\sqrt{e} = 1.6487213$$

or Mathematica 8:

In[100]:= **SetAccuracy[E^ (1 / 2) , 7]**

Out[100]= **1.648721**

or Maple 15:

*Digits* := 7

7

*evalf*$\left(\sqrt{e}\right)$

1.648721

**Example 2.14.** Use the Taylor formula to calculate the following limits:

$$\text{a) } \lim_{x \to 0} \frac{\tan x - \sin x}{x^3}$$

$$\text{b) } \lim_{x \to a} \frac{\sin x - \sin a}{x - a}$$

$$\text{c) } \lim_{x \to 0} \frac{\cos x - \cos 2x}{1 - \cos x}.$$

**Solutions.**
a) Supposing that

$$f(x) = \tan x - \sin x$$

as

$$\begin{cases} f'(x) = \frac{1}{\cos^2 x} - \cos x \\ f''(x) = \left(1 + \frac{2}{\cos^3 x}\right)\sin x \\ f'''(x) = \left(1 + \frac{2}{\cos^3 x}\right)\cos x + \frac{6\sin^2 x}{\cos^4 x} \\ f^{(4)}(x) = \left(\frac{4}{\cos^3 x} - 1\right)\sin x + 12 \cdot \frac{(\cos^2 x + 2\sin^2 x)\sin x}{\cos^5 x} \end{cases}$$

and expanding the function $f(x)$ in a Taylor series of the 3rd order, in the point 0 it results:

$$f(x) \approx 0 + \frac{x}{1!}\cdot 0 + \frac{x^2}{2!}\cdot 0 + \frac{x^3}{3!}\cdot 3 + R_3(x),$$

where

$$R_3(x) = \frac{x^4}{4!}\left[\left(\frac{4}{\cos^3 \theta x} - 1\right)\sin \theta x + 12 \cdot \frac{(\cos^2 \theta x + 2\sin^2 \theta x)\sin \theta x}{\cos^5 \theta x}\right], \ \theta \in (0,1);$$

hence

$$f(x) \approx \frac{x^3}{2} + \frac{x^4}{4!}\left[\left(\frac{4}{\cos^3 \theta x} - 1\right)\sin \theta x + 12 \cdot \frac{(\cos^2 \theta x + 2\sin^2 \theta x)\sin \theta x}{\cos^5 \theta x}\right], \ \theta \in (0,1);$$

therefore

$$\lim_{x\to 0}\frac{\tan x - \sin x}{x^3} = \lim_{x\to 0}\frac{f(x)}{x^3}$$

$$= \lim_{x\to 0}\frac{1}{x^3}\left\{\frac{x^3}{2} + \frac{x^4}{4!}\left[\left(\frac{4}{\cos^3 \theta x} - 1\right)\sin \theta x + 12 \cdot \frac{(\cos^2 \theta x + 2\sin^2 \theta x)\sin \theta x}{\cos^5 \theta x}\right]\right\} = \frac{1}{2}.$$

We can check this result in Matlab 7.9:
```
>> syms x
>> limit((tan(x)-sin(x))/x^3,x,0)
ans =
1/2
```
or in Mathcad 14:

$$\lim_{x\to 0}\frac{\tan(x) - \sin(x)}{x^3} \to \frac{1}{2}$$

or Mathematica 8:

In[109]:= **Limit[ (Tan[x] - Sin[x]) / x^3, x → 0]**

Out[109]= $\dfrac{1}{2}$

or Maple 15:

$$\lim_{x \to 0} \frac{\tan(x) - \sin(x)}{x^3}$$

$$\frac{1}{2}$$

b) We shall expand the function $f(x) = \sin x$ in a Taylor series of the first order, in the point $a$; taking into account that

$$\begin{cases} f_1'(x) = \cos x \\ f_1''(x) = -\sin x \end{cases}, \quad \begin{cases} f_1'(a) = \cos a \\ f_1''(a) = -\sin a \end{cases}$$

it results:

$$f(x) \approx \sin a + \frac{x - a}{1!} \cdot \cos a + R_1(x),$$

where

$$R_1(x) = \frac{(x - a)^2}{2!} \cdot f_1''(a + \theta(x - a))$$

$$= \frac{(x - a)^2}{2!} \cdot (-\sin(a + \theta(x - a))), \ \theta \in (0, 1).$$

Hence,

$$\lim_{x \to a} \frac{\sin x - \sin a}{x - a} = \lim_{x \to a} \frac{f(x) - \sin a}{x - a}$$

$$= \lim_{x \to a} \frac{1}{x - a} \left[ (x - a) \cos a + \frac{(x - a)^2}{2!} \cdot (-\sin(a + \theta(x - a))) \right] = \cos a.$$

We can prove the correctness of this result in Matlab 7.9:
>> **syms x a**
>> **limit((sin(x)-sin(a))/(x-a),x,a)**

ans =
cos(a)

or in Mathcad 14:

$$\lim_{x \to a} \frac{\sin(x) - \sin(a)}{x - a} \to \cos(a)$$

or in Mathematica 8:

In[110]:= **Limit[ (Sin[x] - Sin[a]) / (x - a), x → a]**

Out[110]= **Cos[a]**

or in Maple 15:

$$\lim_{x \to a} \frac{\sin(x) - \sin(a)}{x - a}$$

$$\cos(a)$$

c) We shall expand the functions $f(x) = \cos x$ and $g(x) = \cos 2x$ in a Taylor series of the 3rd order, in the point 0; taking into account that

$$\begin{cases} f_1'(x) = -\sin x \\ f_1''(x) = -\cos x \\ f_1'''(x) = \sin x \\ f_1^{(4)}(x) = \cos x \end{cases}, \quad \begin{cases} f_2'(x) = -2\sin 2x \\ f_2''(x) = -4\cos 2x \\ f_2'''(x) = 8\sin 2x \\ f_2^{(4)}(x) = 16\cos 2x \end{cases}$$

it results:

$$f(x) \approx 1 + \frac{x}{1!} \cdot 0 + \frac{x^2}{2!} \cdot (-1) + \frac{x^3}{3!} \cdot 0 + R_3 f(x),$$

where

$$R_3 f(x) = \frac{x^4}{4!} \cdot f^{(4)}(\theta x) = \frac{x^4}{4!} \cdot \cos \theta x, \ \theta \in (0, 1)$$

and

$$g(x) \approx 1 + \frac{x}{1!} \cdot 0 + \frac{x^2}{2!} \cdot (-4) + \frac{x^3}{3!} \cdot 0 + R_3 g(x),$$

where

$$R_3g\left(x\right) = \frac{x^4}{4!} \cdot g^{(4)}\left(\theta x\right) = \frac{x^4}{4!} \cdot 16\cos 2\theta x, \ \theta \in (0,1).$$

We shall deduce that:

$$\lim_{x\to 0} \frac{\cos x - \cos 2x}{1 - \cos x} = \lim_{x\to 0} \frac{1 - \frac{x^2}{2!} + \frac{x^4}{4!} \cdot \cos\theta x - 1 + \frac{4x^2}{2!} - \frac{x^4}{4!} \cdot 16\cos 2\theta x}{1 - \left(1 - \frac{x^2}{2!} + \frac{x^4}{4!} \cdot \cos\theta x\right)}$$

$$= \lim_{x\to 0} \frac{\frac{3x^2}{2!} + \frac{x^4}{4!}\left(\cos\theta x - 16\cos 2\theta x\right)}{\frac{x^2}{2!} - \frac{x^4}{4!} \cdot \cos\theta x}$$

$$= \lim_{x\to 0} \frac{\frac{3}{2!} + \frac{x^2}{4!}\left(\cos\theta x - 16\cos 2\theta x\right)}{\frac{1}{2!} - \frac{x^2}{4!} \cdot \cos\theta x} = \frac{3}{2} \cdot 2 = 3.$$

This result can be also achieved in Matlab 7.9:
>>**syms x**
>>**limit((cos(x)-cos(2*x))/(1-cos(x)),x,0)**
ans=
3
or in Mathcad 14:

$$\lim_{x \to 0} \frac{\cos(x) - \cos(2 \cdot x)}{1 - \cos(x)} \to 3$$

or in Mathematica 8:

In[111]:= **Limit[(Cos[x] - Cos[2 * x]) / (1 - Cos[x]), x → 0]**

Out[111]= 3

or in Maple 15:

$$\lim_{x \to 0} \frac{\cos(x) - \cos(2 \cdot x)}{1 - \cos(x)}$$

3

## 2.3  Sum of a Power Series

**Definition 2.15** (see [8]). The function

$$f(x) = \lim_{n\to\infty} f_n(x), \qquad (2.11)$$

where

$$f_n(x) = \sum_{k=0}^{n} a_k x^k, \ a_k \in \mathbb{R}, \ k = \overline{0,n} \qquad (2.12)$$

and $x$ belongs to the region of convergence is called the **sum** of the power series and

$$R_n(x) = f(x) - f_n(x) \qquad (2.13)$$

its **remainder**.

**Theorem 2.16** (see [8]). If the power series $\sum_{n=0}^{\infty} c_n(x-a)^n$ has the radius of convergence $\rho > 0$, then the function $f$ defined by

$$f(x) = \lim_{n\to\infty} \sum_{k=0}^{n} c_k(x-a)^k \qquad (2.14)$$

is differentiable on the interval $(a - \rho, \ a + \rho)$ and

$$f'(x) = \sum_{n=0}^{\infty} nc_n(x-a)^{n-1}, \qquad (2.15)$$

$$\int f(x) = C + \sum_{n=0}^{\infty} c_n \frac{(x-a)^{n+1}}{n+1}; \qquad (2.16)$$

the interval of convergence for the power series from (2.15) and (2.16) is $(a - \rho, \ a + \rho)$.

**Example 2.17.** Find the set of convergence and the sum of the power series:

$$\sum_{n\geq 0} (-1)^n \frac{x^{3n+1}}{3n+1}.$$

**Solution.**
We denote

$$a_n = \frac{1}{3n+1};$$

therefore

$$\lambda = \lim_{n \to \infty} \left| \frac{a_{n+1}}{a_n} \right| = \lim_{n \to \infty} \frac{3n+1}{3n+4} = 1,$$

namely the radius of convergence is and the interval of convergence is $(-1, 1)$.

At the end-points of the interval of convergence we shall have:

- when $x = -1$, the series

$$\sum_{n \geq 1} (-1)^{4n+1} \frac{1}{3n+1} = - \sum_{n \geq 1} \frac{1}{3n+1}$$

diverges;

- when $x = 1$, the series

$$\sum_{n \geq 1} (-1)^n \frac{1}{3n+1}$$

converges ( by Leibnitz's test).

Hence, the set of convergence of the power series

$$\sum_{n \geq 0} (-1)^n \frac{x^{3n+1}}{3n+1}$$

is $(-1, 1]$.

Let be $f(x)$ the sum of our power series, namely:

$$f(x) = \lim_{n \to \infty} \sum_{k=0}^{n} (-1)^k \frac{x^{3k+1}}{3k+1}. \tag{2.17}$$

Using the Theorem 2.16 one deduces

$$f'(x) = \sum_{n \geq 0} (-1)^n \frac{3n+1}{3n+1} \cdot x^{3n} = \sum_{n \geq 0} (-1)^n x^{3n}$$

$$= 1 - x^3 + x^6 + \cdots + (-1)^n x^{3n} + \cdots$$

namely

$$f'(x) = \frac{1}{1+x^3} \text{ for } |x| < 1.$$

Therefore

$$\int f'(x) \, dx = \int \frac{1}{1+x^3} dx$$

and

$$f(x) = \int \frac{1}{3} \left[ \frac{1}{x+1} - \frac{x-2}{x^2-x+1} \right] dx$$

$$= \frac{1}{3} \int \frac{dx}{x+1} - \frac{1}{6} \int \frac{2x-1}{x^2-x+1} dx + \frac{1}{2} \int \frac{dx}{x^2-x+1}$$

$$= \frac{1}{3} \ln(x+1) - \frac{1}{6} \ln(x^2-x+1) + \frac{1}{2} \int \frac{dx}{\left(x-\frac{1}{2}\right)^2 + \left(\frac{\sqrt{3}}{2}\right)^2}$$

$$= \frac{1}{3} \ln(x+1) - \frac{1}{6} \ln(x^2-x+1) + \frac{1}{2} \cdot \frac{2}{\sqrt{3}} \arctan \frac{x-\frac{1}{2}}{\frac{\sqrt{3}}{2}} + C;$$

hence

$$f(x) = \frac{1}{3} \ln(x+1) - \frac{1}{6} \ln(x^2-x+1) + \frac{1}{\sqrt{3}} \arctan \frac{2x-1}{\sqrt{3}} + C.$$

We shall have

$$f(0) = \frac{1}{\sqrt{3}} \arctan\left(\frac{-1}{\sqrt{3}}\right) + C = \frac{1}{\sqrt{3}} \cdot \left(-\frac{\pi}{6}\right) + C = -\frac{\pi}{6\sqrt{3}} + C.$$

Taking into account that from (2.17) one deduces $f(0) = 0$ we shall have $C = \frac{\pi}{6\sqrt{3}}$ and the sum of our power series will be

$$f(x) = \frac{1}{3} \ln(x+1) - \frac{1}{6} \ln(x^2-x+1) + \frac{1}{\sqrt{3}} \arctan \frac{2x-1}{\sqrt{3}} + \frac{\pi}{6\sqrt{3}}.$$

**Example 2.18.** Find the sum of the power series:

$$\text{a) } x + \frac{x^2}{2} + \frac{x^3}{3} + \cdots + \frac{x^n}{n} + \cdots$$

$$\text{b) } \sum_{n \geq 1} n^2 x^n, \ x \in (-1,1).$$

**Solutions.**
a) As for this power series

$$a_n = \frac{1}{n}$$

using (2.3) it results that the radius of convergence will be $\rho = 1$ and the interval of convergence $(-1,1)$.

When:

- $x = -1$ one obtains the series

$$\sum_{n \geq 1} (-1)^n \frac{1}{n},$$

which converges (see the Leibnitz's test);

- $x = 1$ one obtains the series

$$\sum_{n \geq 1} \frac{1}{n},$$

which diverges.

The set of convergence for the power series $\sum_{n \geq 1} \frac{x^n}{n}$ will be $[-1, 1)$.

Let be the sum of our power series. Using the Theorem 2.16 one deduces:

$$f'(x) = \sum_{n \geq 1} \frac{n \cdot x^{n-1}}{n} = \sum_{n \geq 1} x^{n-1} = 1 + x + x^2 + \cdots + x^{n-1} + \cdots ; \quad (2.18)$$

we can notice that this series represents the expansion of the function $\frac{1}{1-x}$ in a series of powers of $x$, namely

$$f'(x) = \frac{1}{1 - x}, \quad (\forall) \ x \in [-1, 1); \qquad (2.19)$$

therefore

$$f(x) = \int \frac{1}{1 - x} dx = -\int (\ln(1 - x))' \, dx = \ln(1 - x).$$

We can also compute the sum of the power series $\sum_{n \geq 1} \frac{x^n}{n}$ using the following Matlab 7.9 sequence:

```
>>syms x n
>> symsum(x^n/n,n,1,inf)
ans =
-log(1-x)
```

or in Mathcad 14:

$$\sum_{n=1}^{\infty} \frac{x^n}{n} \rightarrow \begin{vmatrix} -\ln(1 - x) & \text{if } x \neq 1 \wedge |x| \leq 1 \\ \infty & \text{if } 1 \leq x \end{vmatrix}$$

or in Mathematica 8:

```
In[112]:= Sum[x^n/n, {n, 1, Infinity}]

Out[112]= -Log[1 - x]
```

or in Maple 15:

$$\sum_{n=1}^{\infty} \frac{x^n}{n}$$

$$-\ln(1 - x)$$

b) Let be $f(x)$ the sum of the power series

$$\sum_{n\geq 1} x^n = x + x^2 + x^3 + \cdots + x^n + \cdots ;$$

one can notice that this series represents the expansion of the function $\frac{1}{1-x}$ in a series of powers of $x$, namely

$$\sum_{n\geq 1} x^n = \frac{1}{1 - x}. \tag{2.20}$$

Using the Theorem 2.16 one deduces

$$f'(x) = \sum_{n\geq 1} nx^{n-1}. \tag{2.21}$$

Deriving the relation (2.20) in its both members we have:

$$\sum_{n\geq 1} nx^{n-1} = \frac{1}{(1 - x)^2}. \tag{2.22}$$

Taking into account the relations (2.21) and (2.20) one obtains:

$$f'(x) = \sum_{n\geq 1} nx^{n-1} = \frac{1}{(1 - x)^2}. \tag{2.23}$$

We shall multiply by $x$ the both members of the relation (2.23); hence

$$\sum_{n\geq 1} nx^n = \frac{x}{(1 - x)^2}. \tag{2.24}$$

Deriving the relation (2.24) in its both members one deduces:

$$\sum_{n\geq 1} n^2 x^{n-1} = \left( \frac{x}{(1 - x)^2} \right)' = \frac{1 + x}{(1 - x)^3}. \tag{2.25}$$

Multiplying by $x$ the both members of the relation (2.25) it results

$$\sum_{n\geq 1} n^2 x^n = \frac{x(1+x)}{(1-x)^3}. \tag{2.26}$$

Deriving the relation (2.26) in its both members we shall have:

$$\sum_{n\geq 1} n^3 x^{n-1} = \left(\frac{x(1+x)}{(1-x)^3}\right)' = \frac{x^2 + 4x + 1}{(1-x)^4}. \tag{2.27}$$

Let be $g(x)$ the sum of the power series $\sum_{n\geq 1} n^2 x^n$; using the Theorem 2.16 one deduces

$$g'(x) = \sum_{n\geq 1} n^3 x^{n-1} \overset{(2.27)}{=} \frac{x^2 + 4x + 1}{(1-x)^4}; \tag{2.28}$$

therefore

$$g(x) = \int \frac{x^2 + 4x + 1}{(1-x)^4} dx = \frac{x(1+x)}{(1-x)^3}.$$

## 2.4 Problems

1. Which is the set of convergence of the power series

$$\sum_{n\geq 1} \frac{2^n}{2n+1} \cdot x^n ?$$

**Computer solution.**
For

$$a_n = \frac{2^n}{2n+1}$$

we achieve the radius of convergence in Matlab 7.9:

```
>>syms n
>> a=@(n) 2^n/(2*n+1);
>> 1/limit(abs(a(n+1)/a(n)),n,inf))
ans=
1/2
```

or in Mathcad 14:

$$a(n) := \frac{2^n}{2 \cdot n + 1}$$

$$\lim_{n \to \infty} \frac{1}{\left| \dfrac{a(n+1)}{a(n)} \right|} \to \frac{1}{2}$$

or using Mathematica 8:

```
In[6]:= a[n_] := 2^n / (2 * n + 1)

In[7]:= 1 / Limit[Abs[a[n + 1] / a[n]], n -> Infinity]

Out[7]= 1/2
```

or with Maple 15:

$$a := n \to \frac{2^n}{(2 \cdot n + 1)}$$

$$n \to \frac{2^n}{2n + 1}$$

$$\lim_{n \to \infty} \frac{1}{\left| \dfrac{a(n+1)}{a(n)} \right|}$$

$$\frac{1}{2}$$

2. Find the set of convergence of the power series

$$\sum_{n \geq 0} a_n \cdot x^n,$$

in the following two cases:

- $a_0, a_1, \cdots, a_n, \cdots$ is an arithmetic progression, with the common ratio $r$, $r \neq 0$;

- $a_0, a_1, \cdots, a_n, \cdots$ is a geometric progression, with the common ratio $r$, $r \neq 0$, $r \neq 1$.

3. Expand the function

$$f(x) = \sqrt{1 + x}, \; x \geq -1$$

in a series of powers of $x$.

4. Find the power series expansion of the following function of $x$:

$$f(x) = \frac{3}{-2x^2 + x + 1}, \; x \in \mathbb{R} \setminus \left\{ 1, \; -\frac{1}{2} \right\}.$$

**Computer solution.**

The power series expansion of the 7-th order, of the function $f(x)$ is achieved in Matlab 7.9:

```
>>syms x
>> taylor(3/(-2*x^2+x+1),8)

ans =
- 255*x^7 + 129*x^6 - 63*x^5 + 33*x^4 - 15*x^3 + 9*x^2 -
3*x + 3
```

or in Mathcad 14:

$$\frac{3}{-2 \cdot x^2 + x + 1} \; series, 8 \; \rightarrow 3 - 3 \cdot x + 9 \cdot x^2 - 15 \cdot x^3 + 33 \cdot x^4 - 63 \cdot x^5 + 129 \cdot x^6 - 255 \cdot x^7$$

or using Mathematica 8:

In[2]:= **Series[3 / (-2 \* x^2 + x + 1) , {x, 0, 7}]**

Out[2]= $3 - 3x + 9x^2 - 15x^3 + 33x^4 - 63x^5 + 129x^6 - 255x^7 + O[x]^8$

or with Maple 15:

$$series\left( \frac{3}{-2 \cdot x^2 + x + 1}, x = 0, 8 \right)$$

$$3 - 3x + 9x^2 - 15x^3 + 33x^4 - 63x^5 + 129x^6 - 255x^7 + O(x^8)$$

5. Compute the approximate value for $\sqrt{1.01}$ by expanding the function $f(x) = \sqrt{1+x^2}$ in a Taylor series of the 3-th order, in the point 0. Evaluate the committed error.

6. Calculate the number $\sqrt[4]{19}$ to three decimals.

7. Find the interval of convergence of the power series and test the convergence at the end-points of the interval of convergence:

$$a) \sum_{n\geq 0} \frac{n}{n+1} x^n$$

$$b) \sum_{n\geq 0} (-1)^n \frac{x^{2n+1}}{2n+1}$$

$$c) \sum_{n\geq 1} \frac{n!}{n^n} x^n.$$

8. Find the sum of the power series:

$$\sum_{n\geq 1} \frac{n^2}{6^n} (3^n - 2^n).$$

**Computer solution.**
We find the sum of the power series in Matlab 7.9:
>>**syms n**
>>**symsum(n^2*(3^n-2^n)/6^n,1,Inf)**
**ans=**
**9/2**
or in Mathcad 14:

$$\sum_{n=1}^{\infty} \frac{n^2 \cdot \left(3^n - 2^n\right)}{6^n} \to \frac{9}{2}$$

or using Mathematica 8:

In[2]:= **Sum [n^2 * (3^n – 2^n) / 6^n, {n, 1, Infinity}]**

Out[2]= $\frac{9}{2}$

and in Maple 15:

$$\sum_{n=1}^{\infty} \frac{n^2 \cdot (3^n - 2^n)}{6^n}$$

$$\frac{9}{2}$$

9. Use the Taylor formula to calculate the limit

$$\lim_{x \to 1} \frac{\ln x}{1 - x}.$$

**Computer solution.**
We shall obtain in Matlab 7.9:
```
>>syms x
>>limit(log(x)/(1-x),1)
ans=
-1
```
and in Mathcad 14:

$$\lim_{x \to 1} \frac{\ln (x)}{1 - x} \to -1$$

or using Mathematica 8:

```
In[22]:= Limit[Log[x] / (1 - x), x → 1]

Out[22]= -1
```

or in Maple 15:

$$\lim_{x \to 1} \frac{\log(x)}{1 - x}$$

$$-1$$

10.Find the limit:

$$\lim_{x \to 0} \frac{x^2 \cos^2 x - \sin^2 x}{(\arcsin x)^4}.$$

**Computer solution.**

We shall have in Matlab 7.9:
>>**syms x**
>>**limit((x^2\*cos(x)^2-sin(x)^2)/(asin(x)^4),0)**
**ans=**
**-2/3**
in Mathcad 14:

$$\lim_{x \to 0} \frac{x^2 \cdot \cos(x)^2 - \sin(x)^2}{a\sin(x)^4} \to -\frac{2}{3}$$

or using Mathematica 8:

```
In[1]:= Limit[ (x^2 * Cos[x]^2 - Sin[x]^2) / (ArcSin[x]^4) , x -> 0]
```

$$Out[1]= -\frac{2}{3}$$

or in Maple 15:

$$\lim_{x \to 0} \frac{x^2 \cdot \cos(x)^2 - \sin(x)^2}{\arcsin(x)^4}$$

$$-\frac{2}{3}$$

# 3
# Differentiation Theory of the Functions

## 3.1 Partial Derivatives and Differentiable Functions of Several Variables

### 3.1.1 Partial Derivatives

**Definition 3.1** (see [8]). A variable quantity $z$ is called a **single-valued function** of the two variables $x$, $y$, if there corresponds to each set of their values $(x, y)$ in a given range a unique value of $z$. The variables $x$ and $y$ are called **arguments** or **independent variables**. The functional relation is denoted by

$$z = f(x, y), \; f : \mathbb{R}^2 \to \mathbb{R}. \tag{3.1}$$

We can define functions of three or more arguments in the same way.
**Definition 3.2** (see [41], p. 166). If $z = f(x, y)$, then assuming, for example, that $y$ is constant, we get the derivative

$$\frac{\partial z}{\partial x} = \frac{\partial f}{\partial x} = \lim_{\substack{x \to a \\ x \neq a}} \frac{f(x, b) - f(a, b)}{x - a} = f'_x(x, y), \tag{3.2}$$

which is called the **partial derivative** of the function $z$ with respect to the variable $x$.

In a similar way, we define and denote the partial derivative of the function $z$ with respect to the variable $y$.

G.A. Anastassiou and I.F. Iatan: Intelligent Routines, ISRL 39, pp. 71–156.
springerlink.com                     © Springer-Verlag Berlin Heidelberg 2013

**Definition 3.3** (see [15], p. 144). If they exist, the partial derivatives of the functions $f'_x$ and $f'_y$ they are known as **second order partial derivatives** and one denotes by:

$$\begin{cases} f''_{x^2} = (f'_x)'_x = \frac{\partial}{\partial x}\left(\frac{\partial f}{\partial x}\right) = \frac{\partial^2 f}{\partial x^2} \\ f''_{xy} = (f'_x)'_y = \frac{\partial}{\partial y}\left(\frac{\partial f}{\partial x}\right) = \frac{\partial^2 f}{\partial y \partial x} \\ f''_{yx} = (f'_y)'_x = \frac{\partial}{\partial x}\left(\frac{\partial f}{\partial y}\right) = \frac{\partial^2 f}{\partial x \partial y} \\ f''_{y^2} = (f'_y)'_y = \frac{\partial}{\partial y}\left(\frac{\partial f}{\partial y}\right) = \frac{\partial^2 f}{\partial y^2}. \end{cases} \tag{3.3}$$

**Proposition 3.4 (Schwarz's criterion,** see [15], p. 144). If the function $f$ has the mixed second order partial derivatives $f''_{xy}$ and $f''_{yx}$ in a neighborhood of the point $(a,b) \in A \subset \mathbb{R}^2$ and if $f''_{xy}$ and $f''_{yx}$ are continuous in $(a,b)$ then

$$f''_{xy}(a,b) = f''_{yx}(a,b).$$

**Example 3.5.** Let $f : \mathbb{R}^2 \to \mathbb{R}$, be specified as

$$f(x,y) = \sqrt{\sin^2 x + \sin^2 y}.$$

Find the first partial derivatives of the function $f$ at the points $\left(\frac{\pi}{4},0\right)$ and $\left(\frac{\pi}{4},\frac{\pi}{4}\right)$.

**Solution.**

We shall have

$$f'_x(x,y) = \frac{\partial f}{\partial x}(x,y) = \frac{2\sin x \cos x}{2\sqrt{\sin^2 x + \sin^2 y}} = \frac{\sin x \cos x}{\sqrt{\sin^2 x + \sin^2 y}}$$

$$f'_x\left(\frac{\pi}{4},0\right) = \frac{\sqrt{2}}{2} = 0.7071; \quad f'_x\left(\frac{\pi}{4},\frac{\pi}{4}\right) = \frac{1}{2} = 0.5$$

$$f'_y(x,y) = \frac{\partial f}{\partial y}(x,y) = \frac{2\sin y \cos y}{2\sqrt{\sin^2 x + \sin^2 y}} = \frac{\sin y \cos y}{\sqrt{\sin^2 x + \sin^2 y}}$$

$$f'_y\left(\frac{\pi}{4},0\right) = 0; \quad f'_y\left(\frac{\pi}{4},\frac{\pi}{4}\right) = \frac{1}{2} = 0.5.$$

We can also compute these first partial derivatives of the function $f$ in Matlab 7.9:

```
>> syms x y
>> f=@(x,y)sqrt(sin(x)^2+sin(y)^2);
>> s=diff(f(x,y),x);
>> ss=subs(s,{x,y},{pi/4,0})
ss =
 0.7071
>> s1=subs(s,{x,y},{pi/4,pi/4})
```

s1 =
  0.5000
>> t=diff(f(x,y),y);
>> tt=subs(t,{x,y},{pi/4,0})
tt =
  0
>> t1=subs(t,{x,y},{pi/4,pi/4})
t1 =
  0.5000
or in Mathcad 14:

$$f(x,y) := \sqrt{\sin(x)^2 + \sin(y)^2}$$

$$x0 := \frac{\pi}{4} \quad y0 := 0 \quad x1 := \frac{\pi}{4} \quad y1 := \frac{\pi}{4}$$

$$\frac{\partial}{\partial x0} f(x0,y0) \text{ simplify} \rightarrow \frac{\sqrt{2}}{2} \qquad \frac{\partial}{\partial x1} f(x1,y1) \rightarrow \frac{1}{2}$$

$$\frac{\partial}{\partial y0} f(x0,y0) \rightarrow 0 \qquad \frac{\partial}{\partial y1} f(x1,y1) \rightarrow \frac{1}{2}$$

or with Mathematica 8:

In[4]:= f[x_, y_] := Sqrt[Sin[x]^2 + Sin[y]^2];

In[5]:= SetAccuracy[D[f[x, y], x] /. {x → Pi / 4, y → 0}, 4]

Out[5]= 0.7071

In[6]:= SetAccuracy[D[f[x, y], x] /. {x → Pi / 4, y → Pi / 4}, 4]

Out[6]= 0.500

In[7]:= SetAccuracy[D[f[x, y], y] /. {x → Pi / 4, y → 0}, 4]

Out[7]= 0. × 10^{-4}

In[8]:= SetAccuracy[D[f[x, y], y] /. {x → Pi / 4, y → Pi / 4}, 4]

Out[8]= 0.500

or using Maple 15:

$$f := (x, y) \rightarrow \sqrt{\sin(x)^2 + \sin(y)^2}$$

$$(x, y) \rightarrow \sqrt{\sin(x)^2 + \sin(y)^2}$$

$$evalf\left( \left. \frac{\partial}{\partial x} f(x, y) \right|_{x = \frac{\pi}{4}, y = 0} \right)$$

$$0.7071067810$$

$$evalf\left( \left. \frac{\partial}{\partial x} f(x, y) \right|_{x = \frac{\pi}{4}, y = \frac{\pi}{4}} \right)$$

$$0.5000000000$$

$$evalf\left( \left. \frac{\partial}{\partial y} f(x, y) \right|_{x = \frac{\pi}{4}, y = 0} \right)$$

$$0.$$

$$evalf\left( \left. \frac{\partial}{\partial y} f(x, y) \right|_{x = \frac{\pi}{4}, y = \frac{\pi}{4}} \right)$$

$$0.5000000000$$

**Example 3.6.** Let $g : \mathbb{R}^3 \rightarrow \mathbb{R}$, be specified as

$$g(x, y, z) = xe^{yz}.$$

Find the first partial derivatives of the function $g$ at the point $(1, 1, 1)$.

**Solution.**

Thinking of $y$ and $z$ as constants, the partial derivative of $g$ with respect to $x$ is:

$$g'_x = \frac{\partial g}{\partial x} = e^{yz}.$$

Thinking of $x$ and $z$ as constants, the partial derivative of $g$ with respect to $y$ is:

$$g'_y = \frac{\partial g}{\partial x} = xze^{yz}.$$

Thinking of $x$ and $y$ as constants, the partial derivative of $g$ with respect to $z$ is:

$$g'_z = \frac{\partial g}{\partial z} = xye^{yz}.$$

Hence

$$\begin{cases} g'_x\,(1,1,1) = e = 2.7183 \\ g'_y\,(1,1,1) = e = 2.7183 \\ g'_z\,(1,1,1) = e = 2.7183. \end{cases}$$

In order to find the first partial derivatives of the function at the point $(1,1,1)$ we can use the following Matlab 7.9 sequence:

```
>> syms x y z
>> g=@(x,y,z) x*exp(y*z);
>> s=diff(g(x,y,z),x);
>> s1=subs(s,{x,y,z},{1,1,1})
s1 =
 2.7183
>> t=diff(g(x,y,z),y);
>> t1=subs(t,{x,y,z},{1,1,1})
t1 =
 2.7183
>> u=diff(g(x,y,z),z);
>> u1=subs(u,{x,y,z},{1,1,1})
u1 =
 2.7183
```

or in Mathcad 14:

$$g(x,y,z) := x \cdot e^{y \cdot z} \quad x0 := 1 \quad y0 := 1 \quad z0 := 1$$

$$\frac{\partial}{\partial x0}\, g(x0,y0,z0) = 2.7183 \qquad \frac{\partial}{\partial y0}\, g(x0,y0,z0) = 2.7183$$

$$\frac{\partial}{\partial z0}\, g(x0,y0,z0) = 2.7183$$

or with Mathematica 8:

In[64]:= `g[x_, y_, z_] := x * E^(y * z)`

In[65]:= `SetAccuracy[D[g[x, y, z], x] /. {x → 1, y → 1, z → 1}, 5]`

Out[65]= `2.7183`

In[66]:= `SetAccuracy[D[g[x, y, z], y] /. {x → 1, y → 1, z → 1}, 5]`

Out[66]= `2.7183`

In[67]:= `SetAccuracy[D[g[x, y, z], z] /. {x → 1, y → 1, z → 1}, 5]`

Out[67]= `2.7183`

or using Maple 15:

$$g := (x, y, z) \rightarrow x \cdot e^{y \cdot z}$$

$$(x, y, z) \rightarrow x e^{y z}$$

$$Digits := 5$$

$$5$$

$$evalf\left( \left. \frac{\partial}{\partial x} g(x, y, z) \right|_{x=1, y=1, z=1} \right)$$

$$2.7183$$

$$evalf\left( \left. \frac{\partial}{\partial y} g(x, y, z) \right|_{x=1, y=1, z=1} \right)$$

$$2.7183$$

$$evalf\left( \left. \frac{\partial}{\partial z} g(x, y, z) \right|_{x=1, y=1, z=1} \right)$$

$$2.7183$$

**Example 3.7.** Get the first and the second partial derivatives for the following functions:

$$\text{a) } f(x, y) = \arctan \frac{x+y}{1-xy}.$$

$$b) \ g\left(x, y, z\right) = y^{x^z}.$$

**Solutions.**

a) We calculate the first partial derivatives of the function $f$. One finds that:

$$f'_x = \frac{\partial f}{\partial x} = \frac{1}{1 + \left(\frac{x+y}{1-xy}\right)^2} \cdot \frac{\partial}{\partial x}\left(\frac{x+y}{1-xy}\right)$$

$$= \frac{(1-xy)^2}{(1-xy)^2 + (x+y)^2} \cdot \frac{1 - xy + y\,(x+y)}{(1-xy)^2}$$

$$= \frac{1 - xy + xy + y^2}{1 - 2xy + x^2y^2 + x^2 + 2xy + y^2} = \frac{1 + y^2}{1 + x^2 + y^2\,(1+x^2)}$$

$$= \frac{1 + y^2}{(1+y^2)\,(1+x^2)} = \frac{1}{1+x^2}$$

and

$$f'_y = \frac{\partial f}{\partial y} = \frac{1}{1 + \left(\frac{x+y}{1-xy}\right)^2} \cdot \frac{\partial}{\partial y}\left(\frac{x+y}{1-xy}\right)$$

$$= \frac{(1-xy)^2}{(1+x^2)\,(1+y^2)} \cdot \frac{1 - xy + x\,(x+y)}{(1-xy)^2}$$

$$= \frac{1 - xy + x^2 + xy}{(1+x^2)\,(1+y^2)} = \frac{1 + x^2}{(1+x^2)\,(1+y^2)} = \frac{1}{1+y^2}.$$

Differentiating these expressions we shall obtain the second partial derivatives of $f$:

$$f''_{x^2} = \left(f'_x\right)'_x = \frac{\partial}{\partial x}\left(\frac{\partial f}{\partial x}\right) = \frac{\partial^2 f}{\partial x^2} = -\frac{2x}{(1+x^2)^2},$$

$$f''_{y^2} = \left(f'_y\right)'_y = \frac{\partial}{\partial y}\left(\frac{\partial f}{\partial y}\right) = \frac{\partial^2 f}{\partial y^2} = -\frac{2y}{(1+y^2)^2},$$

$$f''_{xy} = \left(f'_x\right)'_y = \frac{\partial}{\partial y}\left(\frac{\partial f}{\partial x}\right) = \frac{\partial^2 f}{\partial y \partial x} = 0,$$

$$f''_{yx} = \left(f'_y\right)'_x = \frac{\partial}{\partial x}\left(\frac{\partial f}{\partial y}\right) = \frac{\partial^2 f}{\partial x \partial y} = 0.$$

We can notice that the mixed partials derivatives in both orders are equal:

$$\frac{\partial^2 f}{\partial x \partial y} = \frac{\partial^2 f}{\partial y \partial x}.$$

We could also say that using Schwarz's criterion (see the Proposition 3.4).

b) We shall have:

$$g'_x = \frac{\partial g}{\partial x} = y^{x^z} \ln y \cdot \frac{\partial}{\partial x} (x^z) = y^{x^z} z x^{z-1} \ln y,$$

$$g'_y = \frac{\partial g}{\partial y} = x^z y^{x^z-1},$$

$$g'_z = \frac{\partial g}{\partial z} = y^{x^z} \ln y \cdot \frac{\partial}{\partial z} (x^z) = y^{x^z} \ln y \cdot x^z \ln x \cdot \frac{\partial}{\partial z} (z) = y^{x^z} x^z \ln x \ln y,$$

$$g''_{x^2} = \frac{\partial}{\partial x} \left( \frac{\partial g}{\partial x} \right) = \frac{\partial^2 g}{\partial x^2} = \frac{\partial}{\partial x} \left( y^{x^z} \right) z x^{z-1} \ln y + y^{x^z} z (z-1) x^{z-2} \ln y$$

$$= y^{x^z} z x^{z-1} \ln y \cdot z x^{z-1} \ln y + y^{x^z} z (z-1) x^{z-2} \ln y$$

$$= y^{x^z} z x^{z-2} \ln y (z x^z \ln y + z - 1),$$

$$g''_{y^2} = \frac{\partial}{\partial y} \left( \frac{\partial g}{\partial y} \right) = \frac{\partial^2 g}{\partial y^2} = x^z \frac{\partial}{\partial y} \left( y^{x^z-1} \right) = x^z (x^z - 1) y^{x^z-2},$$

$$g''_{z^2} = \frac{\partial}{\partial z} \left( \frac{\partial g}{\partial z} \right) = \frac{\partial^2 g}{\partial z^2} = \frac{\partial}{\partial z} \left( y^{x^z} \right) x^z \ln x \ln y + y^{x^z} \frac{\partial}{\partial z} (x^z) \ln x \ln y$$

$$= y^{x^z} x^z \ln x \ln y \cdot x^z \ln x \ln y + y^{x^z} x^z \ln x \ln x \ln y$$

$$= y^{x^z} x^z \ln^2 x \ln y (x^z \ln y + 1),$$

$$g''_{xy} = (g'_x)'_y = \frac{\partial}{\partial y} \left( \frac{\partial g}{\partial x} \right) = \frac{\partial^2 g}{\partial y \partial x} = \frac{\partial}{\partial y} \left( y^{x^z} \right) z x^{z-1} \ln y + y^{x^z} z x^{z-1} \cdot \frac{1}{y}$$

$$= x^z y^{x^z-1} z x^{z-1} \ln y + z x^{z-1} y^{x^z-1}$$

$$= z x^{z-1} y^{x^z-1} (x^z \ln y + 1),$$

$$g''_{xz} = (g'_x)'_z = \frac{\partial}{\partial z} \left( \frac{\partial g}{\partial x} \right) = \frac{\partial^2 g}{\partial z \partial x}$$

$$= \frac{\partial}{\partial z} \left( y^{x^z} \right) z x^{z-1} \ln y + y^{x^z} z x^{z-1} \ln y + y^{x^z} z \frac{\partial}{\partial z} (x^{z-1}) \ln y$$

$$= y^{x^z} x^z \ln x \ln y \cdot z x^{z-1} \ln y + y^{x^z} x^{z-1} \ln y + y^{x^z} z x^{z-1} \ln x \ln y$$

$$= y^{x^z} z x^{z-1} \ln y (z x^z \ln x \ln y + 1 + z \ln x),$$

$$g''_{yz} = (g'_y)'_z = \frac{\partial}{\partial z}\left(\frac{\partial g}{\partial y}\right) = \frac{\partial^2 g}{\partial z \partial y} = \frac{\partial}{\partial z}(x^z)\, y^{x^z - 1} + x^z \frac{\partial}{\partial z}\left(y^{x^z - 1}\right)$$

$$= x^z \ln x \cdot y^{x^z - 1} + x^z y^{x^z - 1} \ln y \cdot \frac{\partial}{\partial z}(x^z - 1)$$

$$= x^z \ln x \cdot y^{x^z - 1} + x^z y^{x^z - 1} \ln y \cdot x^z \ln x$$

$$= x^z y^{x^z - 1} \ln x \cdot (1 + x^z \ln y).$$

We can compute these partial derivatives in Matlab 7.9:

```
>> syms x y z
>> g=@(x,y,z) y^(x^z);
>> s=diff(g,x)
s =
x^(z - 1)*y^(x^z)*z*log(y)
>> t=diff(g,y)
t =
x^z*y^(x^z - 1)
>> u=diff(g,z)
u =
x^z*y^(x^z)*log(x)*log(y)
>> factor(diff(s,x))
ans =
(y^(x^z)*log(y)*x^z*z*(z + x^z*z*log(y) - 1))/x^2
>> factor(diff(t,y))
ans =
x^z*y^(x^z - 2)*(x^z - 1)
>> factor(diff(u,z))
ans =
y^(x^z)*log(x)^2*log(y)*x^z*(x^z*log(y) + 1)
>> simplify(diff(s,y))
ans =
(x^z*y^(x^z)*z*(x^z*log(y) + 1))/(x*y)
>> simplify(diff(s,z))
ans =
(x^z*y^(x^z)*log(y)*(z*log(x) + x^z*z*log(x)*log(y) + 1))/x
>> simplify(diff(t,z))
ans =
x^z*y^(x^z - 1)*log(x)*(x^z*log(y) + 1)
```

or in Mathcad 14:

$$g(x,y,z) := y^{x^z}$$

$$\frac{\partial}{\partial x}g(x,y,z) \text{ simplify } \rightarrow x^{z-1} \cdot y^{x^z} \cdot z \cdot \ln(y)$$

$$\frac{\partial}{\partial y}g(x,y,z) \text{ simplify } \rightarrow x^z \cdot y^{x^z-1}$$

$$\frac{\partial}{\partial z}g(x,y,z) \text{ simplify } \rightarrow x^z \cdot y^{x^z} \cdot \ln(x) \cdot \ln(y)$$

$$\frac{\partial^2}{\partial x^2}g(x,y,z) \text{ simplify } \rightarrow \frac{x^z \cdot y^{x^z} \cdot z \cdot \ln(y) \cdot \left(z + x^z \cdot z \cdot \ln(y) - 1\right)}{x^2}$$

$$\frac{\partial^2}{\partial y^2}g(x,y,z) \text{ simplify } \rightarrow x^z \cdot y^{x^z-2} \cdot \left(x^z - 1\right)$$

$$\frac{\partial^2}{\partial z^2}g(x,y,z) \text{ simplify } \rightarrow x^z \cdot y^{x^z} \cdot \ln(x)^2 \cdot \ln(y) \cdot \left(x^z \cdot \ln(y) + 1\right)$$

$$\frac{\partial}{\partial y}\left(\frac{\partial}{\partial x}g(x,y,z)\right) \text{ simplify } \rightarrow \frac{x^z \cdot y^{x^z} \cdot z \cdot \left(x^z \cdot \ln(y) + 1\right)}{x \cdot y}$$

$$\frac{\partial}{\partial z}\left(\frac{\partial}{\partial x}g(x,y,z)\right) \text{ simplify } \rightarrow \frac{x^z \cdot y^{x^z} \cdot \ln(y) \cdot \left(z \cdot \ln(x) + x^z \cdot z \cdot \ln(x) \cdot \ln(y) + 1\right)}{x}$$

$$\frac{\partial}{\partial z}\left(\frac{\partial}{\partial y}g(x,y,z)\right) \text{ simplify } \rightarrow x^z \cdot y^{x^z-1} \cdot \ln(x) \cdot \left(x^z \cdot \ln(y) + 1\right)$$

or with Mathematica 8:

In[68]:= g[x_, y_, z_] := y^x^z

In[69]:= D[g[x, y, z], x]

Out[69]= $x^{-1+z} \, y^{x^z} \, z \, Log[y]$

In[70]:= D[g[x, y, z], y]

Out[70]= $x^z \, y^{-1+x^z}$

In[71]:= D[g[x, y, z], z]

Out[71]= $x^z \, y^{x^z} \, Log[x] \, Log[y]$

In[73]:= Simplify[D[g[x, y, z], {x, 2}]]

Out[73]= $x^{-2+z} \, y^{x^z} \, z \, Log[y] \left(-1 + z + x^z \, z \, Log[y]\right)$

In[74]:= Simplify[D[g[x, y, z], {y, 2}]]

Out[74]= $x^z \left(-1 + x^z\right) y^{-2+x^z}$

In[75]:= Simplify[D[g[x, y, z], {z, 2}]]

Out[75]= $x^z \, y^{x^z} \, Log[x]^2 \, Log[y] \left(1 + x^z \, Log[y]\right)$

In[76]:= Simplify[D[g[x, y, z], x, y]]

Out[76]= $x^{-1+z} \, y^{-1+x^z} \, z \left(1 + x^z \, Log[y]\right)$

In[77]:= Simplify[D[g[x, y, z], x, z]]

Out[77]= $x^{-1+z} \, y^{x^z} \, Log[y] \left(1 + z \, Log[x] \left(1 + x^z \, Log[y]\right)\right)$

In[78]:= Simplify[D[g[x, y, z], y, z]]

Out[78]= $x^z \, y^{-1+x^z} \, Log[x] \left(1 + x^z \, Log[y]\right)$

or using Maple 15:

$g := (x, y, z) \rightarrow y^{x^z}$

$$(x, y, z) \rightarrow y^{x^z}$$

$simplify\left( \dfrac{\partial}{\partial x} g(x, y, z) \right)$

$$y^{x^z} x^{z-1} z \ln(y)$$

$simplify\left( \dfrac{\partial}{\partial y} g(x, y, z) \right)$

$$y^{x^z - 1} x^z$$

$simplify\left( \dfrac{\partial}{\partial z} g(x, y, z) \right)$

$$y^{x^z} x^z \ln(x) \ln(y)$$

$simplify\left( \dfrac{\partial^2}{\partial x^2} g(x, y, z) \right)$

$$\ln(y) \, y^{x^z} z \left( x^{2z-2} z \ln(y) + x^{z-2} z - x^{z-2} \right)$$

$simplify\left( \dfrac{\partial^2}{\partial y^2} g(x, y, z) \right)$

$$-y^{x^z - 2} \left( -x^{2z} + x^z \right)$$

$simplify\left( \dfrac{\partial^2}{\partial z^2} g(x, y, z) \right)$

$$\ln(x)^2 \ln(y) \, y^{x^z} \left( x^{2z} \ln(y) + x^z \right)$$

$simplify\left( \dfrac{\partial^2}{\partial y \partial x} g(x, y, z) \right)$

$$y^{x^z - 1} z \left( x^{2z-1} \ln(y) + x^{z-1} \right)$$

$simplify\left( \dfrac{\partial^2}{\partial z \partial x} g(x, y, z) \right)$

$$\ln(y) \, y^{x^z} \left( x^{2z-1} z \ln(y) \ln(x) + x^{z-1} z \ln(x) + x^{z-1} \right)$$

$simplify\left( \dfrac{\partial^2}{\partial z \partial y} g(x, y, z) \right)$

$$\ln(x) \, y^{x^z - 1} \left( x^{2z} \ln(y) + x^z \right)$$

## 3.1.2   The Total Differential of a Function

**Defintion 3.8** (see [8]). The **total increment** of a function $z = f(x, y)$ is the difference

$$\Delta z = \Delta f(x, y) = f(x + \Delta x, y + \Delta y) - f(x, y). \qquad (3.4)$$

**Defintion 3.9** (see [8]). The **total** (or **exact**) **differential** of a function $z = f(x, y)$ is the principal part of the total increment $\Delta z$, which is linear with respect to the increments in the arguments $\Delta x$ and $\Delta y$.

**Defintion 3.10** (see [8]). A function definitely has a total differential if its partial derivatives are continuous. If a function has a total differential, then it is called **differentiable**.

**Remark 3.11.** The differentials of independent variables coincide with their increments, that is $\Delta x = \mathrm{d}x$, $\Delta y = \mathrm{d}y$.

**Proposition 3.12** (see [37], p. 133 and [41], p. 175). The differentials of the the first and respectively of the second order for a function of $n$ variables $f(x_1, x_2, \ldots, x_n)$ are computed from the formula:

$$\mathrm{d}f(a) = \frac{\partial f}{\partial x_1}(a)\,\mathrm{d}x_1 + \ldots + \frac{\partial f}{\partial x_n}(a)\,\mathrm{d}x_n, \qquad (3.5)$$

$$\mathrm{d}^2 f(a) = \sum_{i=1}^{n}\sum_{j=1}^{n} \frac{\partial^2 f}{\partial x_i \partial x_j}(a)\,\mathrm{d}x_i \mathrm{d}x_j, \qquad (3.6)$$

where $a = (a_1, a_2, \ldots, a_n) \in \mathbb{R}^n$.

**Example 3.13.** Given the function

$$f(x, y) = x^2 - xy + 2y^2 + 3x - 5y,$$

calculate its differentials of the the first and the second order $\mathrm{d}f(1, 1)$ and $\mathrm{d}^2 f(1, 1)$.

   **Solution.**

   Using (3.5) and (3.6), for our function one gets:

$$\mathrm{d}f(1, 1) = \frac{\partial f}{\partial x}(1, 1)\,\mathrm{d}x + \frac{\partial f}{\partial y}(1, 1)\,\mathrm{d}y$$

and

$$\mathrm{d}^2 f(1, 1) = \frac{\partial^2 f}{\partial x^2}(1, 1)\,\mathrm{d}x^2 + 2\frac{\partial^2 f}{\partial x \partial y}(1, 1)\,\mathrm{d}x\mathrm{d}y + \frac{\partial^2 f}{\partial y^2}(1, 1)\,\mathrm{d}y^2.$$

   As we have

$$\begin{cases} \frac{\partial f}{\partial x} = 2x - y + 3 \implies \frac{\partial f}{\partial x}(1,1) = 4 \\ \frac{\partial f}{\partial y} = -x + 4y - 5 \implies \frac{\partial f}{\partial y}(1,1) = -2 \\ \frac{\partial^2 f}{\partial x^2} = 2 \implies \frac{\partial^2 f}{\partial x^2}(1,1) = 2 \\ \frac{\partial^2 f}{\partial y^2} = 4 \implies \frac{\partial^2 f}{\partial y^2}(1,1) = 4 \\ \frac{\partial^2 f}{\partial x \partial y} = -1 \implies \frac{\partial^2 f}{\partial x \partial y}(1,1) = -1 \end{cases}$$

it results that

$$df(1,1) = 4dx - 2dy$$

and

$$d^2 f(1,1) = 2dx^2 - 2dxdy + 4dy^2.$$

We shall calculate $df(1,1)$ and $d^2 f(1,1)$ using Matlab 7.9:

```
>> f=@(x,y) x^2-x*y+2*y^2+3*x-5*y;
>> syms x y dx dy dx2 dy2 dxdy
>> u=diff(f,x)*dx+diff(f,y)*dy ;
>> subs(u,{x,y},{1,1})
ans =
4*dx-2*dy
>> w=dif(f,x,2)*dx2+2*diff(f,x,y)*dxdy+diff(f,y,2)*dy2 ;
>> subs(w,{x,y},{1,1})
ans =
2*dx2-4*dxdy+4*dy2
```

or in Mathcad 14:

$$f(x,y) := x^2 - x \cdot y + 2 \cdot y^2 + 3 \cdot x - 5y$$

$$\frac{\partial}{\partial x} f(x,y) \cdot dx + \frac{\partial}{\partial y} f(x,y) \cdot dy \text{ substitute}, x = 1, y = 1 \rightarrow 4 \cdot dx - 2 \cdot dy$$

$$\frac{d^2}{dx^2} f(x,y) \cdot dx^2 + 2 \cdot \frac{\partial}{\partial x} \frac{\partial}{\partial y} f(x,y) \cdot dxdy + \frac{d^2}{dy^2} f(x,y) \cdot dy^2 \text{ substitute}, x = 1, y = 1 \rightarrow 2 \cdot dx^2 + 4 \cdot dy^2 - 2 \cdot dxdy$$

and with Mathematica 8:

```
In[25]:= u := Dt[x^2 - x*y + 2*y^2 + 3*x - 5*y];

In[26]:= u /. Dt[x] -> dx;

In[27]:= % /. Dt[y] -> dy;

In[28]:= % /. x -> 1;

In[29]:= % /. y -> 1
Out[29]= 4 dx - 2 dy

In[30]:= Simplify[Dt[u]];

In[31]:= % /. Dt[x] -> dx;

In[32]:= % /. Dt[y] -> dy;

In[33]:= % /. Dt[dx] -> 0;

In[34]:= % /. Dt[dy] -> 0;

In[35]:= % /. x -> 1;

In[36]:= % /. y -> 1
Out[36]= 2 dx² - 2 dx dy + 4 dy²
```

and with Maple 15:

$$f := (x, y) \rightarrow x^2 - x \cdot y + 2 \cdot y^2 + 3 \cdot x - 5 \cdot y :$$
$$u := \frac{\partial}{\partial x} f(x, y) \cdot dx + \frac{\partial}{\partial y} f(x, y) \cdot dy$$

$$(2x - y + 3) \, dx + (-x + 4y - 5) \, dy$$

$$u \Big|_{x=1, y=1}$$

$$4 \, dx - 2 \, dy$$

$$w := \frac{\partial^2}{\partial x^2} f(x, y) \cdot dx^2 + 2 \cdot \frac{\partial^2}{\partial x \partial y} f(x, y) \cdot dx dy + \frac{\partial^2}{\partial y^2} f(x, y) \cdot dy^2 :$$

$$w \Big|_{x=1, y=1}$$

$$2 \, dx^2 - 2 \, dx dy + 4 \, dy^2$$

**Example 3.14.** Evaluate $df(3, 4, 5)$ for the function

$$f(x, y, z) = \frac{z}{\sqrt{x^2 + y^2}}.$$

**Solution.**
Using the relation (3.5) one obtains

$$df(3, 4, 5) = \frac{\partial f}{\partial x}(3, 4, 5)\,dx + \frac{\partial f}{\partial y}(3, 4, 5)\,dy + \frac{\partial f}{\partial z}(3, 4, 5)\,dz.$$

We shall compute

$$\begin{cases} \frac{\partial f}{\partial x} = \frac{-\frac{2x}{2\sqrt{x^2+y^2}}}{x^2+y^2} = -\frac{xz}{(x^2+y^2)^{3/2}} \implies \frac{\partial f}{\partial x}(3, 4, 5) = -\frac{3}{25} \\[2mm] \frac{\partial f}{\partial y} = -\frac{yz}{(x^2+y^2)^{3/2}} \implies \frac{\partial f}{\partial y}(3, 4, 5) = -\frac{4}{25} \\[2mm] \frac{\partial f}{\partial z} = \frac{\sqrt{x^2+y^2}}{x^2+y^2} = \frac{1}{\sqrt{x^2+y^2}} \implies \frac{\partial f}{\partial z}(3, 4, 5) = \frac{1}{5}. \end{cases}$$

Finally,

$$df(3, 4, 5) = -\frac{3}{25}dx - \frac{4}{25}dy + \frac{1}{5}dz = -\frac{1}{5}\left(\frac{3}{5}dx + \frac{4}{5}dy - 5dz\right).$$

We can also obtain this result in Maple 7.9:

```
>> f=@(x,y,z) z/sqrt(x^2+y^2);
>> syms x y z dx dy dz
>> u=diff(f,x)*dx+diff(f,y)*dy+diff(f,z)*dz ;
>> subs(u,{x,y,z},{3,4,5})
ans =
dz/5-(4*dy)/25-(3*dx)/25
```

or using Mathcad 14:

$$f(x,y,z) := \frac{z}{\sqrt{x^2 + y^2}}$$

$$\frac{\partial}{\partial x}f(x,y,z)\cdot dx + \frac{\partial}{\partial y}f(x,y,z)\cdot dy + \frac{\partial}{\partial z}f(x,y,z)\cdot dz \text{ substitute}, x = 3, y = 4, z = 5 \rightarrow \frac{dz}{5} - \frac{4 \cdot dy}{25} - \frac{3 \cdot dx}{25}$$

or with Mathematica 8:

In[26]:= **u := Dt[z / Sqrt[x^2 + y^2]];**

In[27]:= **u /. Dt[x] -> dx;**

In[28]:= **% /. Dt[y] -> dy;**

In[29]:= **% /. Dt[z] -> dz;**

In[30]:= **% /. x → 3;**

In[31]:= **% /. y → 4;**

In[32]:= **% /. z → 5**

Out[32]= $\dfrac{1}{50}$ (-6 dx - 8 dy) + $\dfrac{dz}{5}$

In[33]:= **Simplify[%]**

Out[33]= $\dfrac{1}{25}$ (-3 dx - 4 dy + 5 dz)

or in Maple 15:

$$f := (x, y, z) \rightarrow \frac{z}{\sqrt{x^2 + y^2}} :$$
$$u := \frac{\partial}{\partial x} f(x, y, z) dx + \frac{\partial}{\partial y} f(x, y, z) \cdot dy + \frac{\partial}{\partial z} f(x, y, z) \cdot dz :$$
$$simplify\left(u\Big|_{x=3, y=4, z=5}\right)$$

$$-\frac{3}{25} dx - \frac{4}{25} dy + \frac{1}{5} dz$$

**Example 3.15.** Evaluate $d^2 f$ if the function $f$ is:

$$f(x, y, z) = \sqrt{x^2 + y^2 + z^2}.$$

**Solution.**
Taking into account the relation (3.6) we shall have:

$$d^2 f(x, y, z) = \frac{\partial^2 f}{\partial x^2}(x, y, z) dx^2 + \frac{\partial^2 f}{\partial y^2}(x, y, z) dy^2 + \frac{\partial^2 f}{\partial z^2}(x, y, z) dz^2$$

$$+2\frac{\partial^2 f}{\partial x \partial y}(x, y, z) dx dy + 2\frac{\partial^2 f}{\partial x \partial z}(x, y, z) dx dz + 2\frac{\partial^2 f}{\partial y \partial z}(x, y, z) dy dz,$$

where

$$
\begin{cases}
\frac{\partial f}{\partial x}(x,y,z) = \frac{x}{\sqrt{x^2+y^2+z^2}} \\[2mm]
\frac{\partial f}{\partial y}(x,y,z) = \frac{y}{\sqrt{x^2+y^2+z^2}} \\[2mm]
\frac{\partial f}{\partial z}(x,y,z) = \frac{z}{\sqrt{x^2+y^2+z^2}}
\end{cases}
$$

$$
\begin{aligned}
\frac{\partial^2 f}{\partial x^2}(x,y,z) &= \frac{\sqrt{x^2+y^2+z^2} - \frac{2x^2}{2\sqrt{x^2+y^2+z^2}}}{x^2+y^2+z^2} \\[2mm]
&= \frac{x^2+y^2+z^2-x^2}{\left(x^2+y^2+z^2\right)^{3/2}} \\[2mm]
&= \frac{y^2+z^2}{\left(x^2+y^2+z^2\right)^{3/2}},
\end{aligned}
$$

$$
\frac{\partial^2 f}{\partial y^2}(x,y,z) = \frac{x^2+z^2}{\left(x^2+y^2+z^2\right)^{3/2}},
$$

$$
\frac{\partial^2 f}{\partial z^2}(x,y,z) = \frac{x^2+y^2}{\left(x^2+y^2+z^2\right)^{3/2}},
$$

$$
\frac{\partial^2 f}{\partial x \partial y}(x,y,z) = -\frac{xy}{\left(x^2+y^2+z^2\right)^{3/2}},
$$

$$
\frac{\partial^2 f}{\partial x \partial z}(x,y,z) = -\frac{xz}{\left(x^2+y^2+z^2\right)^{3/2}},
$$

$$
\frac{\partial^2 f}{\partial y \partial z}(x,y,z) = -\frac{yz}{\left(x^2+y^2+z^2\right)^{3/2}}.
$$

Therefore

$$
\begin{aligned}
\mathrm{d}^2 f(x,y,z) &= \frac{y^2+z^2}{\left(x^2+y^2+z^2\right)^{3/2}}\mathrm{d}x^2 + \frac{x^2+z^2}{\left(x^2+y^2+z^2\right)^{3/2}}\mathrm{d}y^2 \\[2mm]
&\quad + \frac{x^2+y^2}{\left(x^2+y^2+z^2\right)^{3/2}}\mathrm{d}z^2 - \frac{2xy}{\left(x^2+y^2+z^2\right)^{3/2}}\mathrm{d}x\mathrm{d}y \\[2mm]
&\quad - \frac{2xz}{\left(x^2+y^2+z^2\right)^{3/2}}\mathrm{d}x\mathrm{d}z - \frac{2yz}{\left(x^2+y^2+z^2\right)^{3/2}}\mathrm{d}y\mathrm{d}z \\[2mm]
&= \frac{1}{\left(x^2+y^2+z^2\right)^{3/2}}\left[\left(y^2+z^2\right)\mathrm{d}x^2 + \left(x^2+z^2\right)\mathrm{d}y^2\right. \\[2mm]
&\quad \left. + \left(x^2+y^2\right)\mathrm{d}z^2 - 2xy\,\mathrm{d}x\mathrm{d}y - 2xz\,\mathrm{d}x\mathrm{d}z - 2yz\,\mathrm{d}y\mathrm{d}z\right].
\end{aligned}
$$

We shall obtain this expression of $d^2 f(x, y, z)$ using Matlab 7.9:
>> f=@(x,y,z) sqrt(x^2+y^2+z^2);
>> syms x y z dx2 dy2 dz2 dxdy dxdz dydz
>> u=dif(f,x,2)*dx2+diff(f,y,2)*dy2+diff(f,z,2)*dz2 ;
>>v=2*(diff(f,x,y)*dxdy+diff(f,x,z)*dxdz+diff(f,y,z)*dydz);
>> simplify(u+v)
ans =
(dx2*(x^2 + y^2 + z^2) - dy2*y^2 - dz2*z^2 - dx2*x^2 +
dy2*(x^2 + y^2 + z^2) + dz2*(x^2 + y^2 + z^2) +
2*dxdy*y*(x^2 + y^2 + z^2) + 2*dxdz*z*(x^2 + y^2 + z^2)
+ 2*dydz*z*(x^2 + y^2 + z^2))/(x^2 + y^2 + z^2)^(3/2)
and Mathcad 14:

$$f(x,y,z) := \sqrt{x^2 + y^2 + z^2}$$

$$\frac{d^2}{dx^2}f(x,y,z) \cdot dx^2 + \frac{d^2}{dy^2}f(x,y,z) \cdot dy^2 + \frac{d^2}{dz^2}f(x,y,z) \cdot dz^2 + 2\frac{\partial}{\partial x}\frac{\partial}{\partial y}f(x,y,z) \cdot dxdy +$$

$$2\frac{\partial}{\partial x}\frac{\partial}{\partial z}f(x,y,z) \cdot dxdz + 2\frac{\partial}{\partial y}\frac{\partial}{\partial z}f(x,y,z) \cdot dydz \text{ simplify } \rightarrow$$

$$\frac{dx^2 \cdot y^2 + dx^2 \cdot z^2 + dy^2 \cdot x^2 + dy^2 \cdot z^2 + dz^2 \cdot x^2 + dz^2 \cdot y^2 - 2 \cdot dxdy \cdot x \cdot y - 2 \cdot dxdz \cdot x \cdot z - 2 \cdot dydz \cdot y \cdot z}{\left(x^2 + y^2 + z^2\right)^{\frac{3}{2}}}$$

and Mathematica 8:

In[58]:= u := Dt[Sqrt[x^2 + y^2 + z^2]];

In[59]:= Dt[u];

In[60]:= % /. Dt[x] -> dx;

In[61]:= % /. Dt[y] -> dy;

In[62]:= % /. Dt[z] -> dz;

In[63]:= % /. Dt[dx] → 0;

In[64]:= % /. Dt[dy] → 0;

In[65]:= % /. Dt[dz] → 0;

In[66]:= Simplify[%]

$$\text{Out[66]=} \quad \frac{dz^2 \left(x^2 + y^2\right) - 2\,dx\,dz\,x\,z - 2\,dy\,y\,(dx\,x + dz\,z) + dy^2 \left(x^2 + z^2\right) + dx^2 \left(y^2 + z^2\right)}{\left(x^2 + y^2 + z^2\right)^{3/2}}$$

and Maple 15:

$$f := (x, y, z) \to \sqrt{x^2 + y^2 + z^2} :$$

$$u := \frac{\partial^2}{\partial x^2} f(x, y, z) dx^2 + \frac{\partial^2}{\partial y^2} f(x, y, z) \cdot dy^2 + \frac{\partial^2}{\partial z^2} f(x, y, z) \cdot dz^2 :$$

$$v := 2 \cdot \frac{\partial^2}{\partial x \partial y} f(x, y, z) dx dy + 2 \cdot \frac{\partial^2}{\partial x \partial z} f(x, y, z) dx dz + 2 \cdot \frac{\partial^2}{\partial y \partial z} f(x, y, z) dy dz :$$

$$simplify(u + v)$$

$$\frac{dx^2 y^2 + dx^2 z^2 + dy^2 x^2 + dy^2 z^2 + dz^2 x^2 + dz^2 y^2 - 2xy\,dx dy - 2xz\,dx dz - 2yz\,dy dz}{\left(x^2 + y^2 + z^2\right)^{3/2}}$$

## 3.1.3 Applying the Total Differential of a Function to Approximate Calculations

**Proposition 3.16** (see [8]). The total differential of a function can be approximated by:

$$f(x_0 + h, y_0 + h, z_0 + r) - f(x_0, y_0, z_0) \qquad (3.7)$$
$$\cong df(x_0, y_0, z_0)(h, k, r) = \frac{\partial f}{\partial x}(x_0, y_0, z_0) \cdot h + \frac{\partial f}{\partial y}(x_0, y_0, z_0) \cdot k + \frac{\partial f}{\partial z}(x_0, y_0, z_0) \cdot r.$$

For a function of two variables, the formula (3.7) will become

$$f(x_0 + h, y_0 + h) - f(x_0, y_0) \cong df(x_0, y_0)(h, k) = \frac{\partial f}{\partial x}(x_0, y_0) \cdot h + \frac{\partial f}{\partial y}(x_0, y_0) \cdot k. \quad (3.8)$$

**Example 3.17.** Calculate approximately the following expressions by replacing the variation of a function with its differential:

a) $1.002 \cdot 2.003^2 \cdot 3.004^3$;

b) $\sin 29° \tan 46°$;

c) $\sqrt{1.02^3 + 1.97^3}$.

**Solutions.**

a) We shall consider the function $f(x, y, z) = xy^2z^3$ and we shall use the relation (3.7), taking

$$\begin{cases} x_0 = 1, \quad y_0 = 2, \quad z_0 = 1 \\ h = 0.002, \, k = 0.003, \, r = 0.004. \end{cases}$$

We shall get

$$f(1.002, 2.003, 3.004) \cong f(1, 2, 3)$$
$$= \frac{\partial f}{\partial x}(1, 2, 3) \cdot 0.002 + \frac{\partial f}{\partial y}(1, 2, 3) \cdot 0.003 + \frac{\partial f}{\partial z}(1, 2, 3) \cdot 0.004,$$

where

$$\begin{cases} \frac{\partial f}{\partial x} = y^2z^3 \\ \frac{\partial f}{\partial y} = 2xyz^3 \\ \frac{\partial f}{\partial z} = 3xy^2z^2. \end{cases}$$

Therefore

$$f(1.002, 2.003, 3.004) \cong 1 \cdot 2^2 \cdot 3^3 + 2^2 \cdot 3^3 \cdot 0.002 + 2 \cdot 2 \cdot 3^3 \cdot 0.003 + 3 \cdot 2^2 \cdot 3^2 \cdot 0.004$$
$$= 2^2 \cdot 3^3 (1 + 0.002 + 0.003 + 0.004) = 108 \cdot 1.009 = 108.972.$$

b) Supposing that

$$f(x, y) = \sin x \tan y$$

and

$$\begin{cases} x_0 = 30°, \ y_0 = 45° \\ h = -1°, \ \ k = 1° \end{cases}$$

using (3.8) we shall have:

$$f(29°, 46°) \approx f(30°, 45°) + \frac{\partial f}{\partial x}(30°, 45°) \cdot (-1°) + \frac{\partial f}{\partial y}(30°, 45°) \cdot 1°,$$

where

$$\begin{cases} \frac{\partial f}{\partial x} = \cos x \tan y \\ \frac{\partial f}{\partial y} = \sin x \cdot \frac{1}{\cos^2 y}. \end{cases}$$

It will result that

$$f(29°, 46°) \approx \frac{1}{2} \cdot 1 + \frac{\sqrt{3}}{2} \cdot 1 \cdot \left(-\frac{\pi}{180}\right) + \frac{1}{2} \cdot 2 \cdot \frac{\pi}{180} = 0.5023.$$

We shall check this result using Matlab 7.9:
>> **sin(degtorad(29))\*tan(degtorad(46))**
**ans =**
 **0.5020**
or Mathcad 14:

$$\sin(29 \cdot \deg) \cdot \tan(46 \cdot \deg) = 0.502$$

or Mathematica 8:

In[8]:= **SetPrecision[Sin[29 Degree] \* Tan[46 Degree], 5]**

Out[8]= **0.5020**

or Maple 15:

*Digits* := 4

$$\text{evalf}(\sin(\text{convert}(29 \ \text{degrees}, \text{radians})) \cdot \tan(\text{convert}(46 \ \text{degrees}, \text{radians})))$$

$$4$$

$$0.5024$$

c) One assumes that

$$f(x,y) = \sqrt{x^3 + y^3} = \left(x^3 + y^3\right)^{1/2}$$

and

$$\begin{cases} x_0 = 1, \quad y_0 = 2 \\ h = 0.02, \; k = -0.03. \end{cases}$$

Applying the formula (3.8) we shall obtain

$$f(1.02, 1.97) \approx f(1,2) + \frac{\partial f}{\partial x}(1,2) \cdot 0.02 + \frac{\partial f}{\partial y}(1,2) \cdot (-0.03),$$

where

$$\begin{cases} \frac{\partial f}{\partial x} = \frac{1}{2} \cdot 3x^2 \left(x^3 + y^3\right)^{-1/2} \\ \frac{\partial f}{\partial y} = \frac{1}{2} \cdot 3y^2 \left(x^3 + y^3\right)^{-1/2}. \end{cases}$$

Hence,

$$f(1.02, 1.97) \approx 3 + \frac{1}{2} \cdot 3 \cdot \left(1^3 + 2^3\right)^{-1/2} \cdot 0.02 + \frac{1}{2} \cdot 3 \cdot 2^2 \cdot \left(1^3 + 2^3\right)^{-1/2} \cdot (-0.03) = 2.95.$$

### 3.1.4  The Functional Determinant

**Definition 3.18** (see [37], p. 138). For a vector function $f : \mathbb{R}^3 \to \mathbb{R}^3$

$$f(x,y,z) = \left(f_1(x,y,z), \; f_2(x,y,z), \; f_3(x,y,z)\right),$$

where $f_i : \mathbb{R}^3 \to \mathbb{R}$, $(\forall)\, i = \overline{1,3}$, the **Jacobian matrix** attached to the vector function $f$ in the point $a \in \mathbb{R}^3$ is the matrix

$$J_f(a) = \begin{pmatrix} \frac{\partial f_1}{\partial x}(a) & \frac{\partial f_1}{\partial y}(a) & \frac{\partial f_1}{\partial z}(a) \\ \frac{\partial f_2}{\partial x}(a) & \frac{\partial f_2}{\partial y}(a) & \frac{\partial f_2}{\partial z}(a) \\ \frac{\partial f_3}{\partial x}(a) & \frac{\partial f_3}{\partial y}(a) & \frac{\partial f_3}{\partial z}(a) \end{pmatrix}. \tag{3.9}$$

**Proposition 3.19** (see [37], p. 138). The approximate value of the vector function $f : \mathbb{R}^3 \to \mathbb{R}^3$ in the point $a \in \mathbb{R}^3$ one finds using the formula:

$$f(x_0 + h, y_0 + h, z_0 + r) - f(x_0, y_0, z_0) \tag{3.10}$$

$$\approx df(x_0, y_0, z_0)(h,k,r) = \left(J_f(x_0, y_0, z_0) \cdot \begin{pmatrix} h \\ k \\ r \end{pmatrix}\right)^t.$$

**Example 3.20.** Determine the Jacobian matrix $Jf(-1,0,-1)$ attached to the vector function $f : \mathbb{R}^3 \to \mathbb{R}^3$, defined by:

$$f(x,y,z) = \left((x+y+z)^2,\ 2x+y-2z, 3x^2+2xy-12xz-18zy\right).$$

**Solution.**

In the case of our function we have:

$$\begin{cases} f_1(x,y,z) = (x+y+z)^2 \\ f_2(x,y,z) = 2x+y-2z \\ f_3(x,y,z) = 3x^2+2xy-12xz-18zy; \end{cases}$$

therefore

$$Jf(-1,0,-1) = \begin{pmatrix} 2(x+y+z) & 2(x+y+z) & 2(x+y+z) \\ 2 & 1 & -2 \\ 6x+2y-12z & 2x-18z & -12x-18y \end{pmatrix}_{|(-1,0,-1)}$$

$$= \begin{pmatrix} -4 & -4 & -4 \\ 2 & 1 & -2 \\ 6 & 16 & 12 \end{pmatrix}.$$

We can also achieve this matrix using the Matlab 7.9 sequence :

```
>> syms x y z
>> u=[x y z];
>> f=[(x+y+z)^2 2*x+y-2*z 3*x^2+2*x*y-12*x*z-18*z*y];
>> J=jacobian(f,u)
J =
[ 2*x + 2*y + 2*z, 2*x + 2*y + 2*z, 2*x + 2*y + 2*z]
[ 2, 1, -2]
[ 6*x + 2*y - 12*z, 2*x - 18*z, - 12*x - 18*y]

>> v=[-1 0 -1];
>> subs(J,u,v)
ans =
-4 -4 -4
2 1 -2
6 16 12
```

or using Mathcad 14:

$$f(x) := \begin{bmatrix} \left(x_0 + x_1 + x_2\right)^2 \\ 2 \cdot x_0 + x_1 - 2 \cdot x_2 \\ 3 \cdot \left(x_0\right)^2 + 2 \cdot x_0 \cdot x_1 - 12 \cdot x_0 \cdot x_2 - 18 \cdot x_2 \cdot x_1 \end{bmatrix} \qquad v := \begin{pmatrix} -1 \\ 0 \\ -1 \end{pmatrix}$$

$$\mathrm{Jacob}(f(v), v) = \begin{pmatrix} -4 & -4 & -4 \\ 2 & 1 & -2 \\ 6 & 16 & 12 \end{pmatrix}$$

or with Mathematica 8:

In[20]:= D[{(x + y + z)^2, 2*x + y - 2*z, 3*x^2 + 2*x*y - 12*x*z - 18*z*y}, {{x, y, z}}]

Out[20]= {{2 (x + y + z), 2 (x + y + z), 2 (x + y + z)}, {2, 1, -2}, {6 x + 2 y - 12 z, 2 x - 18 z, -12 x - 18 y}}

In[21]:= % /. x → -1

Out[21]= {{2 (-1 + y + z), 2 (-1 + y + z), 2 (-1 + y + z)}, {2, 1, -2}, {-6 + 2 y - 12 z, -2 - 18 z, 12 - 18 y}}

In[22]:= % /. y → 0

Out[22]= {{2 (-1 + z), 2 (-1 + z), 2 (-1 + z)}, {2, 1, -2}, {-6 - 12 z, -2 - 18 z, 12}}

In[23]:= % /. z → -1

Out[23]= {{-4, -4, -4}, {2, 1, -2}, {6, 16, 12}}

or in Maple 15:

*with( VectorCalculus)* :
*evalf(Jacobian([ (x + y + z)², 2·x + y − 2·z, 3·x² + 2·x·y − 12·x·z − 18·z·y], [x, y, z] = [−1, 0, −1]))*

$$\begin{bmatrix} -4. & -4. & -4. \\ 2. & 1. & -2. \\ 6. & 16. & 12. \end{bmatrix}$$

**Example 3.21.** Find the approximate value of the vector function $f$ : $\mathbb{R}^3 \to \mathbb{R}^3$, given as

$$f\left(x,y,z\right) = \left((x+y+z)^2,\ 2x+y-2z,\ 3x^2+2xy-12xz-18zy\right),$$

in the point $(-1.05, 0.01, -1.03)$.

**Solution.**

For our function we have

$$\begin{cases} x_0 = -1, & y_0 = 0, & z_0 = -1 \\ h = -0.05, & k = 0.01, & r = -0.03 \end{cases}$$

and using (3.10):

$$f\left(-1.05, 0.01, -1.03\right) - f\left(-1, 0, -1\right)$$

$$\approx \mathrm{d}f\left(-1,0,-1\right)\left(-0.05, 0.01, -0.03\right) = \left(\begin{pmatrix} -4 & -4 & -4 \\ 2 & 1 & -2 \\ 6 & 16 & 12 \end{pmatrix} \cdot \begin{pmatrix} -0.05 \\ 0.01 \\ -0.03 \end{pmatrix}\right)^t;$$

therefore

$$f\left(-1.05, 0.01, -1.03\right) \approx (4, 0, -9)$$
$$+ \left(-4\left(-0.05 + 0.01 - 0.03\right),\ 2\cdot(-0.05) + 0.01 + 2\cdot 0.03,\ 6\cdot(-0.05) + 16\cdot 0.01 - 12\cdot 0.03\right),$$

i.e.

$$f\left(-1.05, 0.01, -1.03\right) \approx (4.28, -0.03, -9.5).$$

We shall check this result in Matlab 7.9:

```
>>f=@(x)[(x(1)+x(2)+x(3))^2; 2*x(1)+x(2)-2*x(3);
3*x(1)^2+2*x(1)*x(2)-12*x(1)*x(3)-18*x(3)*x(2)];
>>v=[-1.05; 0.01; -1.03];
>> digits(3)
>> vpa(f(v))
  ans =
  4.28
  -0.03
  -9.51
```

or in Mathcad 14:

$$f(x) := \begin{bmatrix} \left(x_0 + x_1 + x_2\right)^2 \\ 2 \cdot x_0 + x_1 - 2 \cdot x_2 \\ 3 \cdot \left(x_0\right)^2 + 2 \cdot x_0 \cdot x_1 - 12 \cdot x_0 \cdot x_2 - 18 x_2 \cdot x_1 \end{bmatrix}$$

$$v := \begin{pmatrix} -1.05 \\ 0.01 \\ -1.03 \end{pmatrix} \qquad f(v) = \begin{pmatrix} 4.28 \\ -0.03 \\ -9.51 \end{pmatrix}$$

or with Mathematica 8:

In[1]:=Function[{x, y, z}, {(x + y + z)^2, 2*x + y - 2*z, 3*x^2 + 2*x*y - 12*x*z - 18*z*y} ][-1.05, 0.01, -1.03]
Out[1]:={4.2849, -0.03, -9.5061}
In[2]:=SetPrecision[%, 3]
Out[2]:={4.28, -0.0300, -9.51}

and Maple 15:

$f := (x, y, z) \rightarrow \left[(x+y+z)^2, 2 \cdot x + y - 2 \cdot z, 3 \cdot x^2 + 2 \cdot x \cdot y - 12 \cdot x \cdot z - 18 \cdot z \cdot y\right]:$
$Digits := 3$

$f(-1.05, 0.01, -1.03)$

$[4.28, -0.03, -9.54]$

**Definition 3.22** (see [15], p. 161). Let $F : \mathbb{R}^3 \rightarrow \mathbb{R}^3$ be a vector function

$$F(x, y, z) = (f_1(x, y, z), \; f_2(x, y, z), \; f_3(x, y, z)),$$

where $f_i : \mathbb{R}^3 \rightarrow \mathbb{R}$, $(\forall) i = \overline{1,3}$. The **functional determinant** of the functions $f_1$, $f_2$, $f_3$ is:

$$\frac{D(f_1, f_2, f_3)}{D(x, y, z)} = |J_F(x, y, z)| = \begin{vmatrix} \frac{\partial f_1}{\partial x} & \frac{\partial f_1}{\partial y} & \frac{\partial f_1}{\partial z} \\ \frac{\partial f_2}{\partial x} & \frac{\partial f_2}{\partial y} & \frac{\partial f_2}{\partial z} \\ \frac{\partial f_3}{\partial x} & \frac{\partial f_3}{\partial y} & \frac{\partial f_3}{\partial z} \end{vmatrix}. \tag{3.11}$$

**Example 3.23.** Let $F : D \rightarrow \mathbb{R}^3$, $D \subset [0, \infty) \times \mathbb{R}^2$ defined as

$$F(\rho, \theta, \varphi) = (f_1(\rho, \theta, \varphi), \; f_2(\rho, \theta, \varphi), \; f_3(\rho, \theta, \varphi))$$
$$= (\rho \sin \theta \cos \varphi, \; \rho \sin \theta \sin \varphi, \; \rho \cos \theta).$$

Calculate the functional determinant (the Jacobian) of the functions $f_1$, $f_2$, $f_3$, namely

$$\frac{D\left(f_1,\ f_2,\ f_3\right)}{D\left(\rho,\theta,\varphi\right)}.$$

**Solution.**
Using the formula (3.11) corresponding to the functional determinant of the functions $f_1$, $f_2$, $f_3$ one gets:

$$\frac{D\left(f_1,\ f_2,\ f_3\right)}{D\left(\rho,\theta,\varphi\right)} = \begin{vmatrix} \frac{\partial f_1}{\partial \rho} & \frac{\partial f_1}{\partial \theta} & \frac{\partial f_1}{\partial \varphi} \\ \frac{\partial f_2}{\partial \rho} & \frac{\partial f_2}{\partial \theta} & \frac{\partial f_2}{\partial \varphi} \\ \frac{\partial f_3}{\partial \rho} & \frac{\partial f_3}{\partial \theta} & \frac{\partial f_3}{\partial \varphi} \end{vmatrix} = \begin{vmatrix} \sin\theta\cos\varphi & \rho\cos\theta\cos\varphi & -\rho\sin\theta\sin\varphi \\ \sin\theta\sin\varphi & \rho\cos\theta\sin\varphi & \rho\sin\theta\cos\varphi \\ \cos\theta & -\rho\sin\theta & 0 \end{vmatrix}$$

$$= \rho^2\sin\theta \cdot \begin{vmatrix} \sin\theta\cos\varphi & \cos\theta\cos\varphi & -\sin\varphi \\ \sin\theta\sin\varphi & \cos\theta\sin\varphi & \cos\varphi \\ \cos\theta & -\sin\theta & 0 \end{vmatrix}$$

$$= \rho^2\sin\theta \left[\cos\theta \begin{vmatrix} \cos\theta\cos\varphi & -\sin\varphi \\ \cos\theta\sin\varphi & \cos\varphi \end{vmatrix} + \sin\theta \begin{vmatrix} \sin\theta\cos\varphi & -\sin\varphi \\ \sin\theta\sin\varphi & \cos\varphi \end{vmatrix}\right];$$

hence

$$\frac{D\left(f_1,\ f_2,\ f_3\right)}{D\left(\rho,\theta,\varphi\right)} = \rho^2\sin\theta$$

$$\cdot \left[\cos\theta\left(\cos\theta\cos^2\varphi + \cos\theta\sin^2\varphi\right) + \sin\theta\left(\sin\theta\cos^2\varphi + \sin\theta\sin^2\varphi\right)\right]$$
$$= \rho^2\sin\theta \cdot \left(\cos^2\theta + \sin^2\theta\right) = \rho^2\sin\theta.$$

In Matlab 7.9 we shall have:
```
>> syms rho th phi
>> F=[rho*sin(th)*cos(phi) rho*sin(th)*sin(phi) rho*cos(th)]
F =
[rho *sin(th)*cos(phi), rho *sin(th)*sin(phi), rho *cos(th)]
>> v=[rho th phi];
>> J=jacobian(F,v)
J =
[ sin(th)*cos(phi), rho *cos(th)*cos(phi), - rho *sin(th)*sin(phi)]
[ sin(th)*sin(phi), rho *cos(th)*sin(phi), rho *sin(th)*cos(phi)]
[ cos(th), -rho *sin(th), 0]
>> simplify(det(J))
ans =
sin(th)*rho^2
```
and in Mathcad 14:

$$f(x) := \begin{pmatrix} x_0 \cdot \sin(x_1) \cdot \cos(x_2) \\ x_0 \cdot \sin(x_1) \cdot \sin(x_2) \\ x_0 \cdot \cos(x_1) \end{pmatrix}$$

$$|\text{Jacob}(f(x), x)| \text{ simplify } \rightarrow \sin(x_1) \cdot (x_0)^2$$

and with Mathematica 8:

In[58]:= **Det[D[{ρ * Sin[θ] * Cos[φ], ρ * Sin[θ] * Sin[φ], ρ * Cos[θ]}, {{ρ, θ, φ}}]];**

In[59]:= **Simplify[%]**

Out[59]= $ρ^2$ Sin[θ]

and using Maple 15:

*with( VectorCalculus ) :*
*M, d := Jacobian( [ ρ · sin(θ) · cos(φ), ρ · sin(θ) · sin(φ), ρ · cos(θ) ], [ ρ, θ, φ], 'determinant') :*
*simplify(d, trig)*

$$\sin(θ)\, ρ^2$$

## 3.1.5  Homogeneous Functions

**Definition 3.24** (see [41], p. 180). The function $f : F \rightarrow \mathbb{R}$, $E \subseteq \mathbb{R}^3$ is called an **homogeneous function of degree** $m$, if

$$f(tx, ty, tz) = t^m f(x, y, z), \quad (\forall)\ t \in \mathbb{R} \setminus \{0\}. \tag{3.12}$$

**Proposition 3.25** (see [41], p. 180). If $f(x, y, z)$ is an homogeneous function of degree $m$ and it is differentiable on the set E, then the Euler's relation is verified:

$$x\frac{\partial f}{\partial x} + y\frac{\partial f}{\partial y} + z\frac{\partial f}{\partial z} = mf(x, y, z). \tag{3.13}$$

**Example 3.26.** Prove that the function

$$f(x, y, z) = \frac{x^2 + 2xz - y^2}{x^2 + y^2 + z^2}$$

is an homogeneous function and verify for it the Euler's relation.

**Solution.**

We shall compute

$$f(tx, ty, tz) = \frac{t^2 x^2 + 2t^2 xz - t^2 y^2}{t^2 (x^2 + y^2 + z^2)}$$

$$= \frac{x^2 + 2xz - y^2}{x^2 + y^2 + z^2} = t^0 \cdot \frac{x^2 + 2xz - y^2}{x^2 + y^2 + z^2},$$

therefore, $f$ is an homogeneous function of degree $m = 0$.

As

$$\begin{cases} \frac{\partial f}{\partial x} = (-2) \cdot \frac{-2xy^2 - xz^2 + zx^2 - zy^2 - z^3}{(x^2 + y^2 + z^2)^2} \\[4mm] \frac{\partial f}{\partial y} = (-2y) \cdot \frac{2x^2 + z^2 + 2xz}{(x^2 + y^2 + z^2)^2} \\[4mm] \frac{\partial f}{\partial z} = 2 \cdot \frac{x^3 + xy^2 - xz^2 - zx^2 + zy^2}{(x^2 + y^2 + z^2)^2} \end{cases}$$

it results that:

$$x \frac{\partial f}{\partial x} + y \frac{\partial f}{\partial y} + z \frac{\partial f}{\partial z} = 0 = mf(x, y, z).$$

In Matlab 7.9 we can solve this problem in the following way:
```
function r=f(x,y,z)
r=(x^2+2*x*z-y^2)/(x^2+y^2+z^2);
end
```
One saves the file with f.m then in the command line one writes:
```
>> syms x y z t m
>> l=f(x,y,z)
l =
(x^2+2*x*z-y^2)/(x^2+y^2+z^2)
>> u=simplify(f(x*t,y*t,z*t))
u =
(x^2+2*x*z-y^2)/(x^2+y^2+z^2)
>> r=u/l
r=
1
>> m=eval(solve('t^m=r', m))
m =
0
```
We verify the Euler's relation from (3.13).
```
>>w=x*diff(f(x,y,z),x)+ y*diff(f(x,y,z),y)+ z*diff(f(x,y,z),z);
>>simplify(w)
ans=
0
```

>>w==m*f(x,y,z)

ans=

1

If we shall use Mathcad 14:

$$f(x,y,z) := \frac{x^2 + 2 \cdot x \cdot z - y^2}{x^2 + y^2 + z^2} \qquad r(x,y,z,t) := \frac{f(x \cdot t, y \cdot t, z \cdot t)}{f(x,y,z)}$$

$$m := t^m = r(x,y,z,t) \; \text{solve}, m \; \rightarrow 0$$

$$w(x,y,z) := x \cdot \frac{d}{dx} f(x,y,z) + y \cdot \frac{d}{dy} f(x,y,z) + z \cdot \frac{d}{dz} f(x,y,z)$$

$$w(x,y,z) = m \cdot f(x,y,z) \rightarrow 1$$

Solving with Mathematica 8 we have:

In[14]:= f[x_, y_, z_] := (x^2 + 2*x*z - y^2) / (x^2 + y^2 + z^2);

In[15]:= u := Simplify[f[x*t, y*t, z*t]];

In[16]:= l := f[x, y, z]

In[17]:= r := u / l;

In[18]:= m := Log[r] / Log[t]

In[19]:= w := Simplify[x*D[f[x, y, z], x] + y*D[f[x, y, z], y] + z*D[f[x, y, z], z]]

In[20]:= w == m*f[x, y, z]

Out[20]= True

In Maple 15 we shall have:

$f := (x, y, z) \rightarrow \dfrac{x^2 + 2 \cdot x \cdot z - y^2}{x^2 + y^2 + z^2}$ :

$l := f(x, y, z)$ :

$u := simplify(f(x \cdot t, y \cdot t, z \cdot t))$ :

$r := \dfrac{u}{l}$ :

$m := solve(t^m = r, m)$

$$0$$

$w := simplify(x \cdot diff(f(x, y, z), x) + y \cdot diff(f(x, y, z), y) + z \cdot diff(f(x, y, z), z))$

$$0$$

$evalb(w = m \cdot f(x, y, z))$

$$true$$

## 3.2   Derivation and Differentiation of Composite Functions of Several Variables

*The case of one independent variable* (see [8]): if $z = f(x, y)$ is a differentiable function of the arguments $x$ and $y$, which, in turn, are differentiable functions of an independent variable $t$,

$$\begin{cases} x = \varphi(t) \\ y = \psi(t) \end{cases}$$

then the derivative of the composite function $z = f(\varphi(t), \psi(t))$ may be computed from the formula:

$$\frac{\partial z}{\partial t} = \frac{\partial z}{\partial x} \cdot \frac{\partial x}{\partial t} + \frac{\partial z}{\partial y} \cdot \frac{\partial y}{\partial t}. \tag{3.14}$$

*The case of several independent variables* (see [8]): if $z$ is is a composite function of several independent variables, for instance, $z = f(x, y)$, where

$$\begin{cases} x = \varphi(u, v) \\ y = \psi(u, v) \end{cases}$$

($u$ and $v$ being independent variables), then the partial derivatives with respect to $u$ and $v$ are given by

$$\begin{cases} \frac{\partial z}{\partial u} = \frac{\partial z}{\partial x} \cdot \frac{\partial x}{\partial u} + \frac{\partial z}{\partial y} \cdot \frac{\partial y}{\partial u} \\ \frac{\partial z}{\partial v} = \frac{\partial z}{\partial x} \cdot \frac{\partial x}{\partial v} + \frac{\partial z}{\partial y} \cdot \frac{\partial y}{\partial v}. \end{cases} \tag{3.15}$$

**Example 3.27.** Prove that the function

$$f\left(x,y\right) = xy + x\varphi\left(\frac{y}{x}\right)$$

satisfies the equation :

$$xf'_x + yf'_y = xy + f.$$

**Solution.**

Denoting

$$u\left(x,y\right) = \frac{y}{x}$$

using the formula (3.14) we shall have

$$f'_x = \frac{\partial f}{\partial x} = y + \varphi\left(\frac{y}{x}\right) + x\frac{\partial\varphi}{\partial u}\cdot\frac{\partial u}{\partial x} = y + \varphi\left(\frac{y}{x}\right) - x\cdot\frac{y}{x^2}\cdot\frac{\partial\varphi}{\partial u}$$

$$f'_y = \frac{\partial f}{\partial y} = x + x\frac{\partial\varphi}{\partial u}\cdot\frac{\partial u}{\partial y} = x + x\cdot\frac{1}{x}\cdot\frac{\partial\varphi}{\partial u}.$$

Finally, it results

$$xf'_x + yf'_y = \underbrace{xy + x\varphi\left(\frac{y}{x}\right) - x^2\cdot\frac{y}{x^2}\cdot\frac{\partial\varphi}{\partial u}}_{f} + xy + y\cdot\frac{\partial\varphi}{\partial u}$$

$$= f - y\cdot\frac{\partial\varphi}{\partial u} + xy + y\cdot\frac{\partial\varphi}{\partial u} = xy + f.$$

We shall check this equality using Mathcad 14:

$$f(x,y,\varphi) := x{\cdot}y + x{\cdot}\varphi\left(\frac{y}{x}\right)$$

$$w1(x,y,\varphi) := x{\cdot}\left(\frac{d}{dx}f(x,y,\varphi)\right) + y{\cdot}\left(\frac{d}{dy}f(x,y,\varphi)\right) \text{ simplify } \rightarrow x{\cdot}\left(2{\cdot}y + \varphi\left(\frac{y}{x}\right)\right)$$

$$w(x,y,\varphi) := w1(x,y,\varphi) \text{ expand } \rightarrow x{\cdot}\varphi\left(\frac{y}{x}\right) + 2{\cdot}x{\cdot}y$$

$$w(x,y,\varphi) = x{\cdot}y + f(x,y,\varphi) \rightarrow 1$$

or Mathematica 8:

In[28]:= f[x_, y_] := x * y + x * φ[y / x]

In[29]:= w := Expand[Simplify[x * D[f[x, y], x] + y * D[f[x, y], y]]]

In[30]:= w == x * y + f[x, y]

Out[30]= True

or Maple 15:

$$f := (x, y) \rightarrow x \cdot y + x \cdot \varphi\left(\frac{y}{x}\right):$$

$$w := evala\left(Expand\left(simplify\left(x \cdot \frac{\partial}{\partial x} f(x, y) + y \cdot \frac{\partial}{\partial y} f(x, y)\right)\right)\right)$$

$$2xy + x\varphi\left(\frac{y}{x}\right)$$

$$evalb(w = x \cdot y + f(x, y))$$

$$true$$

We can not do that in Matlab 7.9.

**Example 3.28.** Let $f : \mathbb{R}^2 \rightarrow \mathbb{R}$ be a function of class $C^2$ on $\mathbb{R}^2$, be specified as

$$f(x, y) = g(xy, x + y).$$

Compute:

$$\frac{\partial f}{\partial x}, \frac{\partial f}{\partial y}, \frac{\partial^2 f}{\partial x \partial y}.$$

**Solution.**

Denoting

$$\begin{cases} u = xy \\ v = x + y \end{cases}$$

one gets

$$\frac{\partial f}{\partial x} = \frac{\partial g}{\partial u} \cdot \frac{\partial u}{\partial x} + \frac{\partial g}{\partial v} \cdot \frac{\partial v}{\partial x} = y\frac{\partial g}{\partial u} + \frac{\partial g}{\partial v},$$

$$\frac{\partial f}{\partial y} = \frac{\partial g}{\partial u} \cdot \frac{\partial u}{\partial y} + \frac{\partial g}{\partial v} \cdot \frac{\partial v}{\partial y} = x\frac{\partial g}{\partial u} + \frac{\partial g}{\partial v},$$

$$\frac{\partial^2 f}{\partial x \partial y} = \frac{\partial}{\partial x}\left(\frac{\partial f}{\partial y}\right) = \frac{\partial}{\partial x}\left(x\underbrace{\frac{\partial g}{\partial u}}_{F} + \underbrace{\frac{\partial g}{\partial v}}_{G}\right) = \frac{\partial g}{\partial u} + x\frac{\partial F}{\partial x} + \frac{\partial G}{\partial x}$$

$$= \frac{\partial g}{\partial u} + x\left(\frac{\partial F}{\partial u} \cdot \frac{\partial u}{\partial x} + \frac{\partial F}{\partial v} \cdot \frac{\partial v}{\partial x}\right) + \frac{\partial G}{\partial u} \cdot \frac{\partial u}{\partial x} + \frac{\partial G}{\partial v} \cdot \frac{\partial v}{\partial x}$$

$$= \frac{\partial g}{\partial u} + x\left(y\frac{\partial^2 g}{\partial u^2} + \frac{\partial^2 g}{\partial u \partial v}\right) + y\frac{\partial^2 g}{\partial u \partial v} + \frac{\partial^2 g}{\partial v^2}$$

$$= xy\frac{\partial^2 g}{\partial u^2} + (x+y)\frac{\partial^2 g}{\partial u \partial v} + \frac{\partial^2 g}{\partial v^2} + \frac{\partial g}{\partial u}.$$

We shall also obtain a computer solution using Mathcad 14:

$$f(x,y,g) := g(x \cdot y, x + y)$$

$$\frac{\partial}{\partial x} f(x,y,g) \to y \cdot \left|\begin{matrix} x0 \leftarrow x \cdot y \\ \frac{d}{dx0} g(x0, x+y) \end{matrix}\right. \quad + \left|\begin{matrix} x0 \leftarrow x+y \\ \frac{d}{dx0} g(x \cdot y, x0) \end{matrix}\right.$$

$$\frac{\partial}{\partial y} f(x,y,g) \to x \cdot \left|\begin{matrix} x0 \leftarrow x \cdot y \\ \frac{d}{dx0} g(x0, x+y) \end{matrix}\right. \quad + \left|\begin{matrix} x0 \leftarrow x+y \\ \frac{d}{dx0} g(x \cdot y, x0) \end{matrix}\right.$$

$$\frac{\partial}{\partial x}\left(\frac{\partial}{\partial y} f(x,y,g)\right) \to \left|\begin{matrix} x0 \leftarrow x+y \\ \frac{d^2}{dx0^2} g(x \cdot y, x0) \end{matrix}\right. \quad + \left|\begin{matrix} x0 \leftarrow x \cdot y \\ \frac{d}{dx0} g(x0, x+y) \end{matrix}\right. \quad + x \cdot y \cdot \left|\begin{matrix} x0 \leftarrow x \cdot y \\ \frac{d^2}{dx0^2} g(x0, x+y) \end{matrix}\right.$$

or Mathematica 8:

In[81]:= f[x_, y_] := g[x*y, x+y]

In[82]:= D[f[x, y], x]

Out[82]= g$^{(0,1)}$ [x y, x + y] + y g$^{(1,0)}$ [x y, x + y]

In[83]:= D[f[x, y], y]

Out[83]= g$^{(0,1)}$ [x y, x + y] + x g$^{(1,0)}$ [x y, x + y]

In[84]:= Collect[D[f[x, y], x, y], g$^{(1,1)}$ [x y, x + y]]

Out[84]= g$^{(0,2)}$ [x y, x + y] + g$^{(1,0)}$ [x y, x + y] + (x + y) g$^{(1,1)}$ [x y, x + y] + x y g$^{(2,0)}$ [x y, x + y]

or Maple 15:

$f := (x, y) \rightarrow g(x \cdot y, x + y):$

$\dfrac{\partial}{\partial x} f(x, y)$

$$D_1(g)(xy, x+y) y + D_2(g)(xy, x+y)$$

$\dfrac{\partial}{\partial y} f(x, y)$

$$D_1(g)(xy, x+y) x + D_2(g)(xy, x+y)$$

$collect\left( \dfrac{\partial}{\partial x} \dfrac{\partial}{\partial y} f(x, y), D_{1,2}(g)(xy, x+y) \right)$

$$(x+y) D_{1,2}(g)(xy, x+y) + D_{1,1}(g)(xy, x+y) yx + D_1(g)(xy, x+y) + D_{2,2}(g)(xy, x+y)$$

**Example 3.29.** Let $f$ be a function of class $C^2$ on $\mathbb{R}^2$ and g: $\mathbb{R}^2 \to \mathbb{R}$, be specified as

$$g(x, y) = f(x^2 + y^2, x^2 - y^2).$$

Find the second partial derivatives of the function $f$.

**Solution.**

Denoting

$$\begin{cases} u = x^2 + y^2 \\ v = x^2 - y^2 \end{cases}$$

we shall have

$$\frac{\partial g}{\partial x} = \frac{\partial f}{\partial u} \cdot \frac{\partial u}{\partial x} + \frac{\partial f}{\partial v} \cdot \frac{\partial v}{\partial x} = 2x \frac{\partial f}{\partial u} + 2x \frac{\partial f}{\partial v} = 2x \left( \frac{\partial f}{\partial u} + \frac{\partial f}{\partial v} \right),$$

$$\frac{\partial g}{\partial y} = \frac{\partial f}{\partial u} \cdot \frac{\partial u}{\partial y} + \frac{\partial f}{\partial v} \cdot \frac{\partial v}{\partial y} = 2y \frac{\partial f}{\partial u} - 2y \frac{\partial f}{\partial v} = 2y \left( \frac{\partial f}{\partial u} - \frac{\partial f}{\partial v} \right),$$

$$\frac{\partial^2 g}{\partial x^2} = \frac{\partial}{\partial x} \left( \frac{\partial g}{\partial x} \right) = \frac{\partial}{\partial x} \left( 2x \left( \underbrace{\frac{\partial f}{\partial u}}_{F} + \underbrace{\frac{\partial f}{\partial v}}_{G} \right) \right) = 2 \left( \frac{\partial f}{\partial u} + \frac{\partial f}{\partial v} \right) + 2x \left( \frac{\partial F}{\partial x} + \frac{\partial G}{\partial x} \right)$$

$$= 2 \left( \frac{\partial f}{\partial u} + \frac{\partial f}{\partial v} \right) + 2x \left( \frac{\partial F}{\partial u} \cdot \frac{\partial u}{\partial x} + \frac{\partial F}{\partial v} \cdot \frac{\partial v}{\partial x} + \frac{\partial G}{\partial u} \cdot \frac{\partial u}{\partial x} + \frac{\partial G}{\partial v} \cdot \frac{\partial v}{\partial x} \right)$$

$$= 2 \left( \frac{\partial f}{\partial u} + \frac{\partial f}{\partial v} \right) + 2x \left( 2x \frac{\partial^2 f}{\partial u^2} + 2x \frac{\partial^2 f}{\partial u \partial v} + 2x \frac{\partial^2 f}{\partial u \partial v} + 2x \frac{\partial^2 f}{\partial v^2} \right),$$

i.e.

$$\frac{\partial^2 g}{\partial x^2} = 2 \left( \frac{\partial f}{\partial u} + \frac{\partial f}{\partial v} \right) + 4x^2 \left( \frac{\partial^2 f}{\partial u^2} + 2 \frac{\partial^2 f}{\partial u \partial v} + \frac{\partial^2 f}{\partial v^2} \right).$$

Similarly,

$$\frac{\partial^2 g}{\partial y^2} = \frac{\partial}{\partial y} \left( \frac{\partial g}{\partial y} \right) = \frac{\partial}{\partial y} \left( 2y \left( \underbrace{\frac{\partial f}{\partial u}}_{F} - \underbrace{\frac{\partial f}{\partial v}}_{G} \right) \right) = 2 \left( \frac{\partial f}{\partial u} - \frac{\partial f}{\partial v} \right) + 2y \left( \frac{\partial F}{\partial y} - \frac{\partial G}{\partial y} \right)$$

$$= 2 \left( \frac{\partial f}{\partial u} - \frac{\partial f}{\partial v} \right) + 2y \left( \frac{\partial F}{\partial u} \cdot \frac{\partial u}{\partial y} + \frac{\partial F}{\partial v} \cdot \frac{\partial v}{\partial y} - \frac{\partial G}{\partial u} \cdot \frac{\partial u}{\partial y} - \frac{\partial G}{\partial v} \cdot \frac{\partial v}{\partial y} \right)$$

$$= 2 \left( \frac{\partial f}{\partial u} - \frac{\partial f}{\partial v} \right) + 2y \left( 2y \frac{\partial^2 f}{\partial u^2} - 2y \frac{\partial^2 f}{\partial u \partial v} - 2y \frac{\partial^2 f}{\partial u \partial v} + 2y \frac{\partial^2 f}{\partial v^2} \right),$$

i.e.

$$\frac{\partial^2 g}{\partial y^2} = 2 \left( \frac{\partial f}{\partial u} - \frac{\partial f}{\partial v} \right) + 4y^2 \left( \frac{\partial^2 f}{\partial u^2} - 2 \frac{\partial^2 f}{\partial u \partial v} + \frac{\partial^2 f}{\partial v^2} \right)$$

and

$$\frac{\partial^2 g}{\partial x \partial y} = \frac{\partial}{\partial x}\left(\frac{\partial g}{\partial y}\right) = \frac{\partial}{\partial x}\left(2y\left(\underbrace{\frac{\partial f}{\partial u}}_{F} - \underbrace{\frac{\partial f}{\partial v}}_{G}\right)\right) = 2y\left(\frac{\partial F}{\partial x} - \frac{\partial G}{\partial x}\right)$$

$$= 2y\left(\frac{\partial F}{\partial u}\cdot\frac{\partial u}{\partial x} + \frac{\partial F}{\partial v}\cdot\frac{\partial v}{\partial x} - \frac{\partial G}{\partial u}\cdot\frac{\partial u}{\partial x} - \frac{\partial G}{\partial v}\cdot\frac{\partial v}{\partial x}\right)$$

$$= 2y\left(2x\frac{\partial^2 f}{\partial u^2} + 2x\frac{\partial^2 f}{\partial u \partial v} - 2x\frac{\partial^2 f}{\partial u \partial v} - 2x\frac{\partial^2 f}{\partial v^2}\right) = 4xy\left(\frac{\partial^2 f}{\partial u^2} - \frac{\partial^2 f}{\partial v^2}\right).$$

We shall also give a computer solution using Mathcad 14:

$$g(x,y,f) := f\left(x^2 + y^2, x^2 - y^2\right)$$

$$\frac{\partial^2}{\partial x^2}g(x,y,f) \text{ collect, } x \rightarrow \left(4\cdot\left|\begin{matrix}x0 \leftarrow x^2 - y^2\\ \frac{d^2}{dx0^2}f\left(x^2 + y^2, x0\right)\end{matrix}\right| + 4\cdot\left|\begin{matrix}x0 \leftarrow x^2 + y^2\\ \frac{d^2}{dx0^2}f\left(x0, x^2 - y^2\right)\end{matrix}\right|\right)\cdot x^2 + 2\cdot\left|\begin{matrix}x0 \leftarrow x^2 - y^2\\ \frac{d}{dx0}f\left(x^2 + y^2, x0\right)\end{matrix}\right| + 2\cdot\left|\begin{matrix}x0 \leftarrow x^2 + y^2\\ \frac{d}{dx0}f\left(x0, x^2 - y^2\right)\end{matrix}\right|$$

$$\frac{\partial^2}{\partial y^2}g(x,y,f) \text{ collect, } y \rightarrow \left(4\cdot\left|\begin{matrix}x0 \leftarrow x^2 - y^2\\ \frac{d^2}{dx0^2}f\left(x^2 + y^2, x0\right)\end{matrix}\right| + 4\cdot\left|\begin{matrix}x0 \leftarrow x^2 + y^2\\ \frac{d^2}{dx0^2}f\left(x0, x^2 - y^2\right)\end{matrix}\right|\right)\cdot y^2 + 2\cdot\left|\begin{matrix}x0 \leftarrow x^2 + y^2\\ \frac{d}{dx0}f\left(x0, x^2 - y^2\right)\end{matrix}\right| - 2\cdot\left|\begin{matrix}x0 \leftarrow x^2 - y^2\\ \frac{d}{dx0}f\left(x^2 + y^2, x0\right)\end{matrix}\right|$$

$$\frac{\partial}{\partial x}\frac{\partial}{\partial y}g(x,y,f) \rightarrow 4\cdot x\cdot y\cdot\left|\begin{matrix}x0 \leftarrow x^2 + y^2\\ \frac{d^2}{dx0^2}f\left(x0, x^2 - y^2\right)\end{matrix}\right| - 4\cdot x\cdot y\cdot\left|\begin{matrix}x0 \leftarrow x^2 - y^2\\ \frac{d^2}{dx0^2}f\left(x^2 + y^2, x0\right)\end{matrix}\right|$$

and Mathematica 8:

In[9]:= $g[x\_, y\_] := f[x^2 + y^2, x^2 - y^2]$

In[10]:= $Simplify[D[g[x, y], \{x, 2\}]]$

Out[10]= $2 \left( f^{(0,1)} \left[ x^2 + y^2, x^2 - y^2 \right] + 2 x^2 f^{(0,2)} \left[ x^2 + y^2, x^2 - y^2 \right] + \right.$
$\left. f^{(1,0)} \left[ x^2 + y^2, x^2 - y^2 \right] + 4 x^2 f^{(1,1)} \left[ x^2 + y^2, x^2 - y^2 \right] + 2 x^2 f^{(2,0)} \left[ x^2 + y^2, x^2 - y^2 \right] \right)$

In[11]:= $Simplify[D[g[x, y], \{y, 2\}]]$

Out[11]= $2 \left( -f^{(0,1)} \left[ x^2 + y^2, x^2 - y^2 \right] + 2 y^2 f^{(0,2)} \left[ x^2 + y^2, x^2 - y^2 \right] + \right.$
$\left. f^{(1,0)} \left[ x^2 + y^2, x^2 - y^2 \right] - 4 y^2 f^{(1,1)} \left[ x^2 + y^2, x^2 - y^2 \right] + 2 y^2 f^{(2,0)} \left[ x^2 + y^2, x^2 - y^2 \right] \right)$

In[12]:= $Simplify[D[g[x, y], x, y]]$

Out[12]= $4 x y \left( -f^{(0,2)} \left[ x^2 + y^2, x^2 - y^2 \right] + f^{(2,0)} \left[ x^2 + y^2, x^2 - y^2 \right] \right)$

and Maple 15:

$g := (x, y) \rightarrow f(x^2 + y^2, x^2 - y^2) :$

$collect\left( \frac{\partial^2}{\partial x^2} g(x, y), x \right) :$

$collect\left( \frac{\partial^2}{\partial y^2} g(x, y), y \right) :$

$factor\left( \frac{\partial^2}{\partial x \partial y} g(x, y) \right)$

$4 x y \left( D_{1,1}(f) \left( x^2 + y^2, (x - y)(x + y) \right) - D_{2,2}(f) \left( x^2 + y^2, (x - y)(x + y) \right) \right)$

We can not do that in Matlab 7.9.

**Example 3.30.** Compute $df$ for

$$f(x, y) = g \left( 1 + xy, x^2 + y^2 \right), \quad (\forall) \ (x, y) \in \mathbb{R}^2.$$

**Solution.**

Using (3.5), the total differential of the function $z = f(x, y)$ is computed from the formula

$$dz = \frac{\partial z}{\partial x} dx + \frac{\partial z}{\partial y} dy. \tag{3.16}$$

Denoting

$$\begin{cases} u = 1 + xy \\ v = x^2 + y^2 \end{cases}$$

it results

$$\frac{\partial f}{\partial x} = \frac{\partial g}{\partial u} \cdot \frac{\partial u}{\partial x} + \frac{\partial g}{\partial v} \cdot \frac{\partial v}{\partial x} = y\frac{\partial g}{\partial u} + 2x\frac{\partial g}{\partial v},$$

$$\frac{\partial f}{\partial y} = \frac{\partial g}{\partial u} \cdot \frac{\partial u}{\partial y} + \frac{\partial g}{\partial v} \cdot \frac{\partial v}{\partial y} = x\frac{\partial g}{\partial u} + 2y\frac{\partial g}{\partial v};$$

the total differential of the function $z = f(x, y)$ will be computed using the formula (3.16):

$$df = \frac{\partial f}{\partial x}dx + \frac{\partial f}{\partial y}dy = \left( y\frac{\partial g}{\partial u} + 2x\frac{\partial g}{\partial v} \right) dx + \left( x\frac{\partial g}{\partial u} + 2y\frac{\partial g}{\partial v} \right) dy.$$

We can also obtain this result using Mathcad 14:

$$f(x,y,g) := g\left(1 + x \cdot y, x^2 + y^2\right)$$

$$df(x,y,dx,dy,g) := \frac{d}{dx}f(x,y,g) \cdot dx + \frac{d}{dy}f(x,y,g) \cdot dy$$

$$df(x,y,dx,dy,g) \text{ collect, } dx \rightarrow \left( 2 \cdot x \cdot \left| \begin{matrix} x0 \leftarrow x^2 + y^2 \\ \frac{d}{dx0}g(x \cdot y + 1, x0) \end{matrix} \right. + y \cdot \left| \begin{matrix} x0 \leftarrow x \cdot y + 1 \\ \frac{d}{dx0}g(x0, x^2 + y^2) \end{matrix} \right. \right) \cdot dx + dy \cdot \left( x \cdot \left| \begin{matrix} x0 \leftarrow x \cdot y + 1 \\ \frac{d}{dx0}g(x0, x^2 + y^2) \end{matrix} \right. + 2 \cdot y \cdot \left| \begin{matrix} x0 \leftarrow x^2 + y^2 \\ \frac{d}{dx0}g(x \cdot y + 1, x0) \end{matrix} \right. \right)$$

and Mathematica 8:

```
In[47]:= f[x_, y_] := g[1 + x*y, x^2 + y^2];

In[48]:= Simplify[Dt[f[x, y]]];

In[49]:= % /. Dt[x] -> dx;

In[50]:= % /. Dt[y] -> dy;

In[51]:= Collect[Collect[%, dx], dy]

Out[51]= dy (2 y g^(0,1) [1 + xy, x^2 + y^2] + x g^(1,0) [1 + xy, x^2 + y^2]) + dx (2 x g^(0,1) [1 + xy, x^2 + y^2] + y g^(1,0) [1 + xy, x^2 + y^2])
```

and Maple 15:

$$f := (x, y) \rightarrow g\left(1 + x \cdot y, x^2 + y^2\right):$$

$$df := \frac{\partial}{\partial x} f(x, y) \cdot dx + \frac{\partial}{\partial y} f(x, y) \cdot dy$$

$$\left(D_1(g)\left(1 + xy, x^2 + y^2\right) y + 2 D_2(g)\left(1 + xy, x^2 + y^2\right) x\right) dx$$
$$+ \left(D_1(g)\left(1 + xy, x^2 + y^2\right) x + 2 D_2(g)\left(1 + xy, x^2 + y^2\right) y\right) dy$$

**Example 3.31.** Compute $df$ and $d^2 f$ for the function

$$f(x, y) = g\left(x + y, x^2 + y^2\right), \quad (\forall) \ (x, y) \in \mathbb{R}^2.$$

**Solution.**
Denoting

$$\begin{cases} u = x + y \\ v = x^2 + y^2 \end{cases}$$

it results

$$\frac{\partial f}{\partial x} = \frac{\partial g}{\partial u} \cdot \frac{\partial u}{\partial x} + \frac{\partial g}{\partial v} \cdot \frac{\partial v}{\partial x} = \frac{\partial g}{\partial u} + 2x \frac{\partial g}{\partial v},$$

$$\frac{\partial f}{\partial y} = \frac{\partial g}{\partial u} \cdot \frac{\partial u}{\partial y} + \frac{\partial g}{\partial v} \cdot \frac{\partial v}{\partial y} = \frac{\partial g}{\partial u} + 2y \frac{\partial g}{\partial v};$$

the total differential of the function $z = f(x, y)$ will be computed using the formula (3.16):

$$df = \frac{\partial f}{\partial x} dx + \frac{\partial f}{\partial y} dy = \left(\frac{\partial g}{\partial u} + 2x \frac{\partial g}{\partial v}\right) dx + \left(\frac{\partial g}{\partial u} + 2y \frac{\partial g}{\partial v}\right) dy.$$

If $z = f(x, y)$, where $x$ and $y$ are independent variables then, using (3.6), the second differential of the function $z$ is computed from the formula

$$d^2 z = \frac{\partial^2 z}{\partial x^2} dx^2 + 2 \frac{\partial^2 z}{\partial x \partial y} dx dy + \frac{\partial^2 z}{\partial y^2} dy^2; \tag{3.17}$$

therefore, in our case

$$d^2 f = \frac{\partial^2 f}{\partial x^2} dx^2 + 2 \frac{\partial^2 f}{\partial x \partial y} dx dy + \frac{\partial^2 f}{\partial y^2} dy^2,$$

where

$$\frac{\partial^2 f}{\partial x^2} = \frac{\partial}{\partial x}\left(\frac{\partial f}{\partial x}\right) = \frac{\partial}{\partial x}\left(\underbrace{\frac{\partial g}{\partial u}}_{F} + 2x\underbrace{\frac{\partial g}{\partial v}}_{G}\right) = \frac{\partial F}{\partial x} + 2\frac{\partial g}{\partial v} + 2x\frac{\partial G}{\partial x}$$

$$= \frac{\partial F}{\partial u}\cdot\frac{\partial u}{\partial x} + \frac{\partial F}{\partial v}\cdot\frac{\partial v}{\partial x} + 2\frac{\partial g}{\partial v} + 2x\left(\frac{\partial G}{\partial u}\cdot\frac{\partial u}{\partial x} + \frac{\partial G}{\partial v}\cdot\frac{\partial v}{\partial x}\right)$$

$$= \frac{\partial^2 g}{\partial u^2} + 2x\frac{\partial^2 g}{\partial u\partial v} + 2\frac{\partial g}{\partial v} + 2x\left(\frac{\partial^2 g}{\partial u\partial v} + 2x\frac{\partial^2 g}{\partial v^2}\right),$$

namely

$$\frac{\partial^2 f}{\partial x^2} = \frac{\partial^2 g}{\partial u^2} + 4x\frac{\partial^2 g}{\partial u\partial v} + 4x^2\frac{\partial^2 g}{\partial v^2} + 2\frac{\partial g}{\partial v};$$

$$\frac{\partial^2 f}{\partial y^2} = \frac{\partial}{\partial y}\left(\frac{\partial f}{\partial y}\right) = \frac{\partial}{\partial y}\left(\underbrace{\frac{\partial g}{\partial u}}_{F} + 2y\underbrace{\frac{\partial g}{\partial v}}_{G}\right) = \frac{\partial F}{\partial y} + 2\frac{\partial g}{\partial v} + 2y\frac{\partial G}{\partial y}$$

$$= \frac{\partial F}{\partial u}\cdot\frac{\partial u}{\partial y} + \frac{\partial F}{\partial v}\cdot\frac{\partial v}{\partial y} + 2\frac{\partial g}{\partial v} + 2y\left(\frac{\partial G}{\partial u}\cdot\frac{\partial u}{\partial y} + \frac{\partial G}{\partial v}\cdot\frac{\partial v}{\partial y}\right)$$

$$= \frac{\partial^2 g}{\partial u^2} + 2y\frac{\partial^2 g}{\partial u\partial v} + 2\frac{\partial g}{\partial v} + 2y\left(\frac{\partial^2 g}{\partial u\partial v} + 2y\frac{\partial^2 g}{\partial v^2}\right),$$

i.e.

$$\frac{\partial^2 f}{\partial y^2} = \frac{\partial^2 g}{\partial u^2} + 4y\frac{\partial^2 g}{\partial u\partial v} + 4y^2\frac{\partial^2 g}{\partial v^2} + 2\frac{\partial g}{\partial v};$$

$$\frac{\partial^2 f}{\partial x\partial y} = \frac{\partial}{\partial x}\left(\frac{\partial f}{\partial y}\right) = \frac{\partial}{\partial x}\left(\underbrace{\frac{\partial g}{\partial u}}_{F} + 2y\underbrace{\frac{\partial g}{\partial v}}_{G}\right) = \frac{\partial F}{\partial x} + 2y\frac{\partial G}{\partial x}$$

$$= \frac{\partial F}{\partial u}\cdot\frac{\partial u}{\partial x} + \frac{\partial F}{\partial v}\cdot\frac{\partial v}{\partial x} + 2y\left(\frac{\partial G}{\partial u}\cdot\frac{\partial u}{\partial x} + \frac{\partial G}{\partial v}\cdot\frac{\partial v}{\partial x}\right)$$

$$= \frac{\partial^2 g}{\partial u^2} + 2x\frac{\partial^2 g}{\partial u\partial v} + 2y\left(\frac{\partial^2 g}{\partial u\partial v} + 2x\frac{\partial^2 g}{\partial v^2}\right),$$

namely

$$\frac{\partial^2 f}{\partial x\partial y} = \frac{\partial^2 g}{\partial u^2} + 2\left(x + y\right)\frac{\partial^2 g}{\partial u\partial v} + 4xy\frac{\partial^2 g}{\partial v^2}.$$

Hence,

$$d^2 f = \left( \frac{\partial^2 g}{\partial u^2} + 4x \frac{\partial^2 g}{\partial u \partial v} + 4x^2 \frac{\partial^2 g}{\partial v^2} + 2 \frac{\partial g}{\partial v} \right) dx^2 + 2 \left[ \frac{\partial^2 g}{\partial u^2} + 2(x+y) \frac{\partial^2 g}{\partial u \partial v} + 4xy \frac{\partial^2 g}{\partial v^2} \right] dx\,dy$$
$$+ \left( \frac{\partial^2 g}{\partial u^2} + 4y \frac{\partial^2 g}{\partial u \partial v} + 4y^2 \frac{\partial^2 g}{\partial v^2} + 2 \frac{\partial g}{\partial v} \right) dy^2.$$

We shall give a computer solution using Mathcad 14:

$$f(x,y,g) := g\left(x + y, x^2 + y^2\right)$$

$$df(x,y,dx,dy,g) := \frac{d}{dx} f(x,y,g) \cdot dx + \frac{d}{dy} f(x,y,g) \cdot dy$$

$$df(x,y,dx,dy,g) \text{ collect}, dx \rightarrow \left( 2 \cdot x \cdot \left| x0 \leftarrow x^2 + y^2 \right| \cdot \left| x0 \leftarrow x + y \right| \right) \cdot dx + dy \cdot \left( 2 \cdot y \cdot \left| x0 \leftarrow x^2 + y^2 \right| \cdot \left| x0 \leftarrow x + y \right| \right)$$
$$\left| \frac{d}{dx0} g(x+y,x0) \right| \left| \frac{d}{dx0} g\left(x0, x^2 + y^2\right) \right| \qquad \left| \frac{d}{dx0} g(x+y,x0) \right| \left| \frac{d}{dx0} g\left(x0, x^2 + y^2\right) \right|$$

$$ddf(x,y,ddx,ddy,dxdy,g) := \frac{d^2}{dx^2} f(x,y,g) \cdot ddx + 2 \cdot \frac{d}{dx} \frac{d}{dy} f(x,y,g) \cdot dxdy + \frac{d^2}{dy^2} f(x,y,g) \cdot ddy$$

$$ddf(x,y,ddx,ddy,dxdy,g) \text{ collect}, ddx \rightarrow \left( 4 \cdot x^2 \cdot \left| x0 \leftarrow x^2 + y^2 \right| \cdot \left| x0 \leftarrow x + y \right| + 2 \cdot \left| x0 \leftarrow x^2 + y^2 \right| \right) \cdot ddx +$$
$$\left| \frac{d^2}{dx0^2} g(x+y,x0) \right| \left| \frac{d^2}{dx0^2} g\left(x0, x^2 + y^2\right) \right| \left| \frac{d}{dx0} g(x+y,x0) \right|$$

$$ddy \cdot \left( 4 \cdot y^2 \cdot \left| x0 \leftarrow x^2 + y^2 \right| \cdot \left| x0 \leftarrow x + y \right| + 2 \cdot \left| x0 \leftarrow x^2 + y^2 \right| \right) + dxdy \cdot \left( 8 \cdot x \cdot y \cdot \left| x0 \leftarrow x^2 + y^2 \right| + 2 \cdot \left| x0 \leftarrow x + y \right| \right)$$
$$\left| \frac{d^2}{dx0^2} g(x+y,x0) \right| \left| \frac{d^2}{dx0^2} g\left(x0, x^2 + y^2\right) \right| \left| \frac{d}{dx0} g(x+y,x0) \right| \qquad \left| \frac{d^2}{dx0^2} g(x+y,x0) \right| \left| \frac{d^2}{dx0^2} g\left(x0, x^2 + y^2\right) \right|$$

and Mathematica 8:

In[817]:= f[x_, y_] := g[x + y, x^2 + y^2];

In[818]:= Dt[f[x, y]];

In[819]:= % /. Dt[x] -> dx;

In[820]:= % /. Dt[y] -> dy;

In[821]:= Collect[Collect[%, dx], dy]

Out[821]= $dx \left(2 x g^{(0,1)}\left[x+y, x^2+y^2\right]+g^{(1,0)}\left[x+y, x^2+y^2\right]\right)+dy \left(2 y g^{(0,1)}\left[x+y, x^2+y^2\right]+g^{(1,0)}\left[x+y, x^2+y^2\right]\right)$

In[822]:= Dt[Dt[f[x, y]]];

In[823]:= % /. Dt[x] -> dx;

In[824]:= % /. Dt[y] -> dy;

In[825]:= % /. Dt[x]^2 -> dx^2;

In[826]:= % /. Dt[y]^2 -> dy^2;

In[827]:= % /. Dt[Dt[x]] -> dx^2;

In[828]:= % /. Dt[Dt[y]] -> dy^2;

In[829]:= % /. Dt[dx] -> 0;

In[830]:= % /. Dt[dy] -> 0;

In[831]:= Collect[Collect[Collect[Expand[%], dx*dy], dx^2], dy^2]

$$dx^2 \left(2 g^{(0,1)}\left[x+y, x^2+y^2\right]+4 x^2 g^{(0,2)}\left[x+y, x^2+y^2\right]+4 x g^{(1,1)}\left[x+y, x^2+y^2\right]+g^{(2,0)}\left[x+y, x^2+y^2\right]\right)$$
$$+dy^2 \left(2 g^{(0,1)}\left[x+y, x^2+y^2\right]+4 y^2 g^{(0,2)}\left[x+y, x^2+y^2\right]+4 y g^{(1,1)}\left[x+y, x^2+y^2\right]+g^{(2,0)}\left[x+y, x^2+y^2\right]\right)$$
$$+dx dy \left(8 x y g^{(0,2)}\left[x+y, x^2+y^2\right]+4 x g^{(1,1)}\left[x+y, x^2+y^2\right]+4 y g^{(1,1)}\left[x+y, x^2+y^2\right]+2 g^{(2,0)}\left[x+y, x^2+y^2\right]\right)$$

and in Maple 15:

$f := (x, y) \rightarrow g(x + y, x^2 + y^2):$

$df := \dfrac{\partial}{\partial x} f(x, y) \cdot dx + \dfrac{\partial}{\partial y} f(x, y) \cdot dy$

$\left(D_1(g)(x + y, x^2 + y^2) + 2D_2(g)(x + y, x^2 + y^2) x\right) dx + \left(D_1(g)(x + y, x^2 + y^2) + 2D_2(g)(x\right.$

$\left. + y, x^2 + y^2) y\right) dy$

$ddf := \dfrac{\partial^2}{\partial x^2} f(x, y) \cdot dx^2 + 2 \cdot \dfrac{\partial^2}{\partial x \partial y} f(x, y) \cdot dx dy + \dfrac{\partial^2}{\partial y^2} f(x, y) \cdot dy^2$

$\left(D_{1,1}(g)(x + y, x^2 + y^2) + 2D_{1,2}(g)(x + y, x^2 + y^2) x + 2\left(D_{1,2}(g)(x + y, x^2 + y^2)\right.\right.$

$\left. + 2D_{2,2}(g)(x + y, x^2 + y^2) x\right) x + 2D_2(g)(x + y, x^2 + y^2))\, dx^2 + 2\left(D_{1,1}(g)(x + y, x^2 + y^2)\right.$

$+ 2D_{1,2}(g)(x + y, x^2 + y^2) y + 2\left(D_{1,2}(g)(x + y, x^2 + y^2) + 2D_{2,2}(g)(x + y, x^2 + y^2) y\right) x\right)$

$dx dy + \left(D_{1,1}(g)(x + y, x^2 + y^2) + 2D_{1,2}(g)(x + y, x^2 + y^2) y + 2\left(D_{1,2}(g)(x + y, x^2 + y^2)\right.\right.$

$\left. + 2D_{2,2}(g)(x + y, x^2 + y^2) y\right) y + 2D_2(g)(x + y, x^2 + y^2))\, dy^2$

**Example 3.32.** Calculate the expression:

$$E = f''_{x^2} + f''_{xy} + f''_{y^2}$$

if

$$f(u, v) = \ln\left(u^2 + v^2\right),$$

where

$$\begin{cases} u(x, y) = xy \\ v(x, y) = x^2 - y^2. \end{cases}$$

**Solution.**
We shall compute:

$$f'_x = \frac{\partial f}{\partial x} = \frac{\partial f}{\partial u} \cdot \frac{\partial u}{\partial x} + \frac{\partial f}{\partial v} \cdot \frac{\partial v}{\partial x} = y\frac{\partial f}{\partial u} + 2x\frac{\partial f}{\partial v},$$

$$f''_{x^2} = \frac{\partial^2 f}{\partial x^2} = \frac{\partial}{\partial x}\left(\frac{\partial f}{\partial x}\right) = \frac{\partial}{\partial x}\left(\underbrace{y\frac{\partial f}{\partial u}}_{F} + \underbrace{2x\frac{\partial f}{\partial v}}_{G}\right) = y\frac{\partial F}{\partial x} + 2\frac{\partial f}{\partial v} + 2x\frac{\partial G}{\partial x}$$

$$= y\left(\frac{\partial F}{\partial u}\cdot\frac{\partial u}{\partial x} + \frac{\partial F}{\partial v}\cdot\frac{\partial v}{\partial x}\right) + 2\frac{\partial f}{\partial v} + 2x\left(\frac{\partial G}{\partial u}\cdot\frac{\partial u}{\partial x} + \frac{\partial G}{\partial v}\cdot\frac{\partial v}{\partial x}\right)$$

$$= y\left(y\frac{\partial^2 f}{\partial u^2} + 2x\frac{\partial^2 f}{\partial u\partial v}\right) + 2\frac{\partial f}{\partial v} + 2x\left(y\frac{\partial^2 f}{\partial u\partial v} + 2x\frac{\partial^2 f}{\partial v^2}\right),$$

namely

$$f''_{x^2} = y^2\frac{\partial^2 f}{\partial u^2} + 4xy\frac{\partial^2 f}{\partial u\partial v} + 4x^2\frac{\partial^2 f}{\partial v^2} + 2\frac{\partial f}{\partial v};$$

$$f'_y = \frac{\partial f}{\partial y} = \frac{\partial f}{\partial u}\cdot\frac{\partial u}{\partial y} + \frac{\partial f}{\partial v}\cdot\frac{\partial v}{\partial y} = x\frac{\partial f}{\partial u} - 2y\frac{\partial f}{\partial v},$$

$$f''_{y^2} = \frac{\partial^2 f}{\partial y^2} = \frac{\partial}{\partial y}\left(\frac{\partial f}{\partial y}\right) = \frac{\partial}{\partial y}\left(\underbrace{x\frac{\partial f}{\partial u}}_{F} - \underbrace{2y\frac{\partial f}{\partial v}}_{G}\right) = x\frac{\partial F}{\partial y} - 2\frac{\partial f}{\partial v} - 2y\frac{\partial G}{\partial y}$$

$$= x\left(\frac{\partial F}{\partial u}\cdot\frac{\partial u}{\partial y} + \frac{\partial F}{\partial v}\cdot\frac{\partial v}{\partial y}\right) - 2\frac{\partial f}{\partial v} - 2y\left(\frac{\partial G}{\partial u}\cdot\frac{\partial u}{\partial y} + \frac{\partial G}{\partial v}\cdot\frac{\partial v}{\partial y}\right)$$

$$= x\left(x\frac{\partial^2 f}{\partial u^2} - 2y\frac{\partial^2 f}{\partial u\partial v}\right) - 2\frac{\partial f}{\partial v} - 2y\left(x\frac{\partial^2 f}{\partial u\partial v} - 2y\frac{\partial^2 f}{\partial v^2}\right),$$

namely

$$f''_{y^2} = x^2\frac{\partial^2 f}{\partial u^2} - 4xy\frac{\partial^2 f}{\partial u\partial v} + 4y^2\frac{\partial^2 f}{\partial v^2} - 2\frac{\partial f}{\partial v};$$

$$f''_{xy} = (f'_x)'_y = \frac{\partial}{\partial y}\left(\frac{\partial f}{\partial x}\right) = \frac{\partial}{\partial y}\left(\underbrace{y\frac{\partial f}{\partial u}}_{F} + \underbrace{2x\frac{\partial g}{\partial v}}_{G}\right) = \frac{\partial f}{\partial u} + y\frac{\partial F}{\partial y} + 2x\frac{\partial G}{\partial y}$$

$$= \frac{\partial f}{\partial u} + y\left(\frac{\partial F}{\partial u}\cdot\frac{\partial u}{\partial y} + \frac{\partial F}{\partial v}\cdot\frac{\partial v}{\partial y}\right) + 2x\left(\frac{\partial G}{\partial u}\cdot\frac{\partial u}{\partial y} + \frac{\partial G}{\partial v}\cdot\frac{\partial v}{\partial y}\right)$$

$$= \frac{\partial f}{\partial u} + y\left(x\frac{\partial^2 f}{\partial u^2} - 2y\frac{\partial^2 f}{\partial u\partial v}\right) + 2x\left(x\frac{\partial^2 f}{\partial u\partial v} - 2y\frac{\partial^2 f}{\partial v^2}\right),$$

namely,

$$f''_{xy} = xy \frac{\partial^2 f}{\partial u^2} + 2\left(x^2 - y^2\right) \frac{\partial^2 f}{\partial u \partial v} - 4xy \frac{\partial^2 f}{\partial v^2} + \frac{\partial f}{\partial u}.$$

We shall achieve:

$$E = y^2 \frac{\partial^2 f}{\partial u^2} + 4xy \frac{\partial^2 f}{\partial u \partial v} + 4x^2 \frac{\partial^2 f}{\partial v^2} + 2 \frac{\partial f}{\partial v} + xy \frac{\partial^2 f}{\partial u^2} + 2\left(x^2 - y^2\right) \frac{\partial^2 f}{\partial u \partial v}$$

$$- 4xy \frac{\partial^2 f}{\partial v^2} + \frac{\partial f}{\partial u} + x^2 \frac{\partial^2 f}{\partial u^2} - 4xy \frac{\partial^2 f}{\partial u \partial v} + 4y^2 \frac{\partial^2 f}{\partial v^2} - 2 \frac{\partial f}{\partial v};$$

therefore

$$E = \left(x^2 + xy + y^2\right) \frac{\partial^2 f}{\partial u^2} + 2\left(x^2 - y^2\right) \frac{\partial^2 f}{\partial u \partial v} + 4\left(x^2 - xy + y^2\right) \frac{\partial^2 f}{\partial v^2} + \frac{\partial f}{\partial u},$$

where

$$\frac{\partial f}{\partial u} = \frac{2u}{u^2 + v^2} = \frac{2xy}{x^2 y^2 + \left(x^2 - y^2\right)^2} = \frac{2xy}{x^4 - x^2 y^2 + y^4},$$

$$\frac{\partial^2 f}{\partial u^2} = \frac{\partial}{\partial u}\left(\frac{\partial f}{\partial u}\right) = \frac{2\left(u^2 + v^2\right) - 2u \cdot 2u}{\left(u^2 + v^2\right)^2} = 2 \cdot \frac{v^2 - u^2}{\left(u^2 + v^2\right)^2} = 2 \cdot \frac{\left(x^2 - y^2\right)^2 - x^2 y^2}{\left(x^4 - x^2 y^2 + y^4\right)^2},$$

$$\frac{\partial f}{\partial v} = \frac{2v}{u^2 + v^2} = \frac{2\left(x^2 - y^2\right)}{x^4 - x^2 y^2 + y^4},$$

$$\frac{\partial^2 f}{\partial v^2} = \frac{\partial}{\partial v}\left(\frac{\partial f}{\partial v}\right) = \frac{2\left(u^2 + v^2\right) - 2v \cdot 2v}{\left(u^2 + v^2\right)^2} = 2 \cdot \frac{u^2 - v^2}{\left(u^2 + v^2\right)^2} = 2 \cdot \frac{x^2 y^2 - \left(x^2 - y^2\right)^2}{\left(x^4 - x^2 y^2 + y^4\right)^2},$$

$$\frac{\partial^2 f}{\partial u \partial v} = \frac{\partial}{\partial u}\left(\frac{\partial f}{\partial v}\right) = 2v \cdot \frac{-2u}{\left(u^2 + v^2\right)^2} = -\frac{4uv}{\left(u^2 + v^2\right)^2} = -\frac{4xy\left(x^2 - y^2\right)}{\left(x^4 - x^2 y^2 + y^4\right)^2}.$$

Finally, it results

$$E = 2\left(-x^2 - xy - y^2 + 4x^2 + 4y^2 - 4xy\right) \cdot \frac{x^2 y^2 - x^4 + 2x^2 y^2 - y^4}{\left(x^4 - x^2 y^2 + y^4\right)^2}$$

$$- \frac{8xy\left(x^2 - y^2\right)}{\left(x^4 - x^2 y^2 + y^4\right)^2} + \frac{2xy}{x^4 - x^2 y^2 + y^4}$$

$$= \frac{2}{\left(x^4 - x^2 y^2 + y^4\right)^2} \cdot \left[\left(3x^2 - 5xy + 3y^2\right)\left(-x^4 + 3x^2 y^2 - y^4\right) - 4xy\left(x^2 - y^2\right)^2 \right.$$

$$\left. + xy\left(x^4 - x^2 y^2 + y^4\right)\right]$$

and after some elementary calculations one gets:

$$E = \frac{2}{(x^4 - x^2 y^2 + y^4)^2} \cdot \left[ 3 \left( -x^6 + 2x^4 y^2 + 2x^2 y^4 - y^6 \right) + 2xy \left( x^4 - 4x^2 y^2 + y^4 \right) \right].$$

We can also achieve this result using Matlab 7.9:

```
>> f=@(x,y)log((x*y)^2+(x^2-y^2)^2);
>> syms x y
>> simplify(diff(f,x,2)+diff(diff(f,y),x)+diff(f,y,2))
ans =
(2*(- 3*x^6 + 2*x^5*y + 6*x^4*y^2 - 8*x^3*y^3 +
    6*x^2*y^4 + 2*x*y^5 - 3*y^6))/(x^4 - x^2*y^2 + y^4)^2
```

or in Mathcad 14:

$$u(x,y) := x \cdot y$$

$$v(x,y) := x^2 - y^2$$

$$f(u,v,x,y) := \ln\!\left( u(x,y)^2 + v(x,y)^2 \right)$$

$$E(u,v,x,y) := \frac{d^2}{dx^2} f(u,v,x,y) + \frac{d}{dx}\!\left( \frac{d}{dy} f(u,v,x,y) \right) + \frac{d^2}{dy^2} f(u,v,x,y)$$

$$E(u,v,x,y) \text{ simplify} \rightarrow \frac{2 \cdot \left( 2 \cdot x^5 \cdot y - 3 \cdot x^6 + 6 \cdot x^4 \cdot y^2 - 8 \cdot x^3 \cdot y^3 + 6 \cdot x^2 \cdot y^4 + 2 \cdot x \cdot y^5 - 3 \cdot y^6 \right)}{\left( x^4 - x^2 \cdot y^2 + y^4 \right)^2}$$

or with Mathematica 8:

```
In[129]:= f[x_, y_] := Log[(x*y)^2 + (x^2 - y^2)^2]

In[130]:= D[f[x, y], {x, 2}] + D[f[x, y], x, y] + D[f[x, y], {y, 2}];

In[131]:= Simplify[%]
```

$$Out[131]= \frac{2 \left( -3 x^6 + 2 x^5 y + 6 x^4 y^2 - 8 x^3 y^3 + 6 x^2 y^4 + 2 x y^5 - 3 y^6 \right)}{\left( x^4 - x^2 y^2 + y^4 \right)^2}$$

or Maple 15:

$u := (x, y) \to x \cdot y:$
$v := (x, y) \to x^2 - y^2:$

$f := (u, v) \to \ln\left(u(x, y)^2 + v(x, y)^2\right)$

$$(u, v) \to \ln\left(u(x, y)^2 + v(x, y)^2\right)$$

$evala\left( \dfrac{\partial^2}{\partial x^2} f(u, v) + \dfrac{\partial^2}{\partial x \partial y} f(u, v) + \dfrac{\partial^2}{\partial y^2} f(u, v) \right)$

$$-\frac{2\left(-6y^4 x^2 - 6x^4 y^2 + 3y^6 + 3x^6 + 8x^3 y^3 - 2x^5 y - 2xy^5\right)}{\left(-x^2 y^2 + x^4 + y^4\right)^2}$$

## 3.3   Change of Variables

**Example 3.33.** What happens to the equation

$$x\frac{\partial z}{\partial x} - y\frac{\partial z}{\partial y} = 0 \tag{3.18}$$

if one makes the change of independent variables

$$\begin{cases} u = \frac{x}{y} \\ v = \ln\frac{y}{x} \end{cases}, \quad xy > 0?$$

**Solution.**
We shall calculate:

$$\frac{\partial z}{\partial x} = \frac{\partial z}{\partial u} \cdot \frac{\partial u}{\partial x} + \frac{\partial z}{\partial v} \cdot \frac{\partial v}{\partial x} = \frac{1}{y} \cdot \frac{\partial z}{\partial u} - \frac{\frac{y}{x^2}}{\frac{y}{x}} \cdot \frac{\partial z}{\partial v},$$

namely

$$\frac{\partial z}{\partial x} = \frac{1}{y} \cdot \frac{\partial z}{\partial u} - \frac{1}{x} \cdot \frac{\partial z}{\partial v}; \tag{3.19}$$

$$\frac{\partial z}{\partial y} = \frac{\partial z}{\partial u} \cdot \frac{\partial u}{\partial y} + \frac{\partial z}{\partial v} \cdot \frac{\partial v}{\partial y} = -\frac{x}{y^2} \cdot \frac{\partial z}{\partial u} + \frac{\frac{1}{x}}{\frac{y}{x}} \cdot \frac{\partial z}{\partial v},$$

namely

$$\frac{\partial z}{\partial y} = -\frac{x}{y^2} \cdot \frac{\partial z}{\partial u} + \frac{1}{y} \cdot \frac{\partial z}{\partial v}. \tag{3.20}$$

Substituting (3.19) and (3.20) into (3.18) it results

$$x\left(\frac{1}{y}\cdot\frac{\partial z}{\partial u}-\frac{1}{x}\cdot\frac{\partial z}{\partial v}\right)-y\left(-\frac{x}{y^2}\cdot\frac{\partial z}{\partial u}+\frac{1}{y}\cdot\frac{\partial z}{\partial v}\right)=0\Longleftrightarrow$$

$$\Longleftrightarrow 2\left(\frac{x}{y}\cdot\frac{\partial z}{\partial u}-\frac{\partial z}{\partial v}\right)=0\Longleftrightarrow\frac{x}{y}\cdot\frac{\partial z}{\partial u}-\frac{\partial z}{\partial v}=0$$

such that the initial equation becomes:

$$u\frac{\partial z}{\partial u}-\frac{\partial z}{\partial v}=0.$$

Using Mathematica 8, the left hand side of the initial equation becomes:

```
In[340]:= z[x_, y_] := r[x, y]

In[341]:= u[x_, y_] := x / y

In[342]:= v[x_, y_] := Log[y / x]

In[343]:= b = Expand[Simplify[x*D[z[u[x, y], v[x, y]], x] - y*D[z[u[x, y], v[x, y]], y]]];

In[344]:= b /. Log[y / x] → v;

In[345]:= % /. x → u*y

Out[345]= -2 r^(0,1) [u, v] + 2 u r^(1,0) [u, v]
```

or Maple 15:

$$z := (x, y) \rightarrow w(x, y) :$$
$$u := (x, y) \rightarrow \frac{x}{y} :$$
$$v := (x, y) \rightarrow \ln\left(\frac{y}{x}\right) :$$
$$b := expand\left(x\cdot\frac{\partial}{\partial x}z(u(x, y), v(x, y)) - y\cdot\frac{\partial}{\partial y}z(u(x, y), v(x, y))\right) :$$
$$b\bigg|_{x = u\cdot y,\ \ln\left(\frac{y}{x}\right) = v}$$

$$2\,u\,D_1(w)(u, v) - 2\,D_2(w)(u, v)$$

**Example 3.34.** What becomes the equation

$$\frac{\partial^2 z}{\partial x^2} + y^2 \frac{\partial^2 z}{\partial y^2} - y \frac{\partial^2 z}{\partial x \partial y} + \frac{\partial z}{\partial x} + 2y \frac{\partial z}{\partial y} = 0 \qquad (3.21)$$

after the change of independent variables:

$$\begin{cases} x = u - v \\ y = e^{u+v} \end{cases}, \quad (\forall) \ (u, v) \in \mathbb{R}^2 ?$$

**Solution.**
We shall calculate:

$$\frac{\partial z}{\partial u} = \frac{\partial z}{\partial x} \cdot \frac{\partial x}{\partial u} + \frac{\partial z}{\partial y} \cdot \frac{\partial y}{\partial u} = \frac{\partial z}{\partial x} + e^{u+v} \cdot \frac{\partial z}{\partial y},$$

$$\frac{\partial z}{\partial v} = \frac{\partial z}{\partial x} \cdot \frac{\partial x}{\partial v} + \frac{\partial z}{\partial y} \cdot \frac{\partial y}{\partial v} = -\frac{\partial z}{\partial x} + e^{u+v} \cdot \frac{\partial z}{\partial y}.$$

We have to solve the system

$$\begin{cases} \frac{\partial z}{\partial x} + e^{u+v} \cdot \frac{\partial z}{\partial y} = \frac{\partial z}{\partial u} \\ -\frac{\partial z}{\partial x} + e^{u+v} \cdot \frac{\partial z}{\partial y} = \frac{\partial z}{\partial v} \end{cases} \qquad (3.22)$$

in order to find $\frac{\partial z}{\partial x}$ and $\frac{\partial z}{\partial y}$.
As the determinant

$$\Delta = \begin{vmatrix} 1 & e^{u+v} \\ -1 & e^{u+v} \end{vmatrix} = 2e^{u+v} \neq 0, \quad (\forall) \ (u, v) \in \mathbb{R}^2$$

we can solve the system (3.22) using the Cramer's method:

$$\frac{\partial z}{\partial x} = \frac{\begin{vmatrix} \frac{\partial z}{\partial u} & e^{u+v} \\ \frac{\partial z}{\partial v} & e^{u+v} \end{vmatrix}}{\Delta} = \frac{e^{u+v} \left( \frac{\partial z}{\partial u} - \frac{\partial z}{\partial v} \right)}{2e^{u+v}} = \frac{1}{2} \left( \frac{\partial z}{\partial u} - \frac{\partial z}{\partial v} \right),$$

$$\frac{\partial z}{\partial y} = \frac{\begin{vmatrix} 1 & \frac{\partial z}{\partial u} \\ -1 & \frac{\partial z}{\partial v} \end{vmatrix}}{\Delta} = \frac{1}{2e^{u+v}} \left( \frac{\partial z}{\partial u} + \frac{\partial z}{\partial v} \right).$$

Therefore:

$$\frac{\partial^2 z}{\partial x^2} = \frac{\partial}{\partial x}\left(\frac{\partial z}{\partial x}\right) = \frac{\partial}{\partial x}\left(\underbrace{\frac{1}{2}\left(\frac{\partial z}{\partial u} - \frac{\partial z}{\partial v}\right)}_{F}\right) = \frac{1}{2}\frac{\partial F}{\partial x}$$

$$= \frac{1}{2}\left[\frac{1}{2}\left(\frac{\partial F}{\partial u} - \frac{\partial F}{\partial v}\right)\right] = \frac{1}{4}\left[\frac{\partial}{\partial u}\left(\frac{\partial z}{\partial u} - \frac{\partial z}{\partial v}\right) - \frac{\partial}{\partial v}\left(\frac{\partial z}{\partial u} - \frac{\partial z}{\partial v}\right)\right]$$

$$= \frac{1}{4}\left(\frac{\partial^2 z}{\partial u^2} - 2\frac{\partial^2 z}{\partial u \partial v} + \frac{\partial^2 z}{\partial v^2}\right),$$

$$\frac{\partial^2 z}{\partial y^2} = \frac{\partial}{\partial y}\left(\frac{\partial z}{\partial y}\right) = \frac{\partial}{\partial y}\left(\underbrace{\frac{1}{2e^{u+v}}\left(\frac{\partial z}{\partial u} + \frac{\partial z}{\partial v}\right)}_{F}\right) = \frac{\partial F}{\partial y} = \frac{1}{2e^{u+v}}\left(\frac{\partial F}{\partial v} + \frac{\partial F}{\partial u}\right)$$

$$= \frac{1}{2e^{u+v}}\left\{\frac{\partial}{\partial v}\left[\frac{1}{2e^{u+v}}\left(\frac{\partial z}{\partial v} + \frac{\partial z}{\partial u}\right)\right] + \frac{\partial}{\partial u}\left[\frac{1}{2e^{u+v}}\left(\frac{\partial z}{\partial v} + \frac{\partial z}{\partial u}\right)\right]\right\}$$

$$= \frac{1}{2e^{u+v}}\left[-\frac{1}{2e^{u+v}}\left(\frac{\partial z}{\partial v} + \frac{\partial z}{\partial u}\right) + \frac{1}{2e^{u+v}}\cdot\frac{\partial}{\partial v}\left(\frac{\partial z}{\partial v} + \frac{\partial z}{\partial u}\right)\right.$$

$$\left. -\frac{1}{2e^{u+v}}\left(\frac{\partial z}{\partial v} + \frac{\partial z}{\partial u}\right) + \frac{1}{2e^{u+v}}\cdot\frac{\partial}{\partial u}\left(\frac{\partial z}{\partial v} + \frac{\partial z}{\partial u}\right)\right]$$

$$= \frac{1}{2e^{u+v}}\left[-\frac{1}{e^{u+v}}\left(\frac{\partial z}{\partial v} + \frac{\partial z}{\partial u}\right) + \frac{1}{2e^{u+v}}\left(\frac{\partial^2 z}{\partial v^2} + \frac{\partial^2 z}{\partial u \partial v} + \frac{\partial^2 z}{\partial u \partial v} + \frac{\partial^2 z}{\partial u^2}\right)\right]$$

$$= \frac{1}{2e^{2(u+v)}}\left[-\frac{\partial z}{\partial v} - \frac{\partial z}{\partial u} + \frac{1}{2}\left(\frac{\partial^2 z}{\partial v^2} + 2\frac{\partial^2 z}{\partial u \partial v} + \frac{\partial^2 z}{\partial u^2}\right)\right]$$

$$= \frac{1}{4e^{2(u+v)}}\left(-2\frac{\partial z}{\partial v} - 2\frac{\partial z}{\partial u} + \frac{\partial^2 z}{\partial v^2} + 2\frac{\partial^2 z}{\partial u \partial v} + \frac{\partial^2 z}{\partial u^2}\right),$$

$$\frac{\partial^2 z}{\partial x \partial y} = \frac{\partial}{\partial x}\left(\frac{\partial z}{\partial y}\right) = \frac{\partial}{\partial x}\left(\underbrace{\frac{1}{2e^{u+v}}\left(\frac{\partial z}{\partial u} + \frac{\partial z}{\partial v}\right)}_{F}\right) = \frac{\partial F}{\partial x} = \frac{1}{2}\left(\frac{\partial F}{\partial u} - \frac{\partial F}{\partial v}\right)$$

$$= \frac{1}{2}\left\{\frac{\partial}{\partial u}\left[\frac{1}{2e^{u+v}}\left(\frac{\partial z}{\partial v} + \frac{\partial z}{\partial u}\right)\right] - \frac{\partial}{\partial v}\left[\frac{1}{2e^{u+v}}\left(\frac{\partial z}{\partial v} + \frac{\partial z}{\partial u}\right)\right]\right\}$$

$$= \frac{1}{2}\left[-\frac{1}{2e^{u+v}}\left(\frac{\partial z}{\partial v} + \frac{\partial z}{\partial u}\right) + \frac{1}{2e^{u+v}}\cdot\frac{\partial}{\partial u}\left(\frac{\partial z}{\partial v} + \frac{\partial z}{\partial u}\right)\right.$$

$$\left. +\frac{1}{2e^{u+v}}\left(\frac{\partial z}{\partial v} + \frac{\partial z}{\partial u}\right) - \frac{1}{2e^{u+v}}\cdot\frac{\partial}{\partial v}\left(\frac{\partial z}{\partial v} + \frac{\partial z}{\partial u}\right)\right]$$

$$= \frac{1}{4e^{u+v}}\left(\frac{\partial^2 z}{\partial u \partial v} + \frac{\partial^2 z}{\partial u^2} - \frac{\partial^2 z}{\partial v^2} - \frac{\partial^2 z}{\partial u \partial v}\right) = \frac{1}{4e^{u+v}}\left(\frac{\partial^2 z}{\partial u^2} - \frac{\partial^2 z}{\partial v^2}\right).$$

Substituting the expression of the second partial derivatives in the equation (3.21) we shall have:

$$\frac{1}{4}\left(\frac{\partial^2 z}{\partial u^2} - 2\frac{\partial^2 z}{\partial u \partial v} + \frac{\partial^2 z}{\partial v^2}\right) + e^{2(u+v)} \cdot \frac{1}{4e^{2(u+v)}}\left(-2\frac{\partial z}{\partial v} - 2\frac{\partial z}{\partial u} + \frac{\partial^2 z}{\partial v^2} + 2\frac{\partial^2 z}{\partial u \partial v} + \frac{\partial^2 z}{\partial u^2}\right)$$

$$-e^{u+v} \cdot \frac{1}{4e^{u+v}}\left(\frac{\partial^2 z}{\partial u^2} - \frac{\partial^2 z}{\partial v^2}\right) + \frac{1}{2}\left(\frac{\partial z}{\partial u} - \frac{\partial z}{\partial v}\right) + 2e^{u+v} \cdot \frac{1}{2e^{u+v}}\left(\frac{\partial z}{\partial u} + \frac{\partial z}{\partial v}\right) = 0 \iff$$

$$\frac{1}{4}\left(\frac{\partial^2 z}{\partial u^2} - 2\frac{\partial^2 z}{\partial u \partial v} + \frac{\partial^2 z}{\partial v^2} - 2\frac{\partial z}{\partial v} - \frac{\partial z}{\partial u} + \frac{\partial^2 z}{\partial v^2} + 2\frac{\partial^2 z}{\partial u \partial v} + \frac{\partial^2 z}{\partial u^2} - \frac{\partial^2 z}{\partial u^2} + \frac{\partial^2 z}{\partial v^2}\right)$$

$$+\frac{1}{2}\frac{\partial z}{\partial u} - \frac{1}{2}\frac{\partial z}{\partial v} + \frac{\partial z}{\partial v} + \frac{\partial z}{\partial u} = 0 \iff$$

$$\frac{3}{4}\frac{\partial^2 z}{\partial v^2} - \frac{1}{2} \cdot \frac{\partial z}{\partial v} - \frac{1}{2} \cdot \frac{\partial z}{\partial u} + \frac{1}{4}\frac{\partial^2 z}{\partial u^2} + \frac{3}{2}\frac{\partial z}{\partial u} + \frac{1}{2}\frac{\partial z}{\partial v} = 0,$$

namely

$$\frac{3}{4}\frac{\partial^2 z}{\partial v^2} + \frac{1}{4}\frac{\partial^2 z}{\partial u^2} + \frac{\partial z}{\partial u} = 0 \iff 4\frac{\partial z}{\partial u} + \frac{\partial^2 z}{\partial u^2} + 3\frac{\partial^2 z}{\partial v^2} = 0.$$

**Example 3.35.** Transform the expression

$$E = x^2\frac{\partial^2 z}{\partial x^2} + 2xy\frac{\partial^2 z}{\partial x \partial y} + y^2\frac{\partial^2 z}{\partial y^2}$$

by switching to the polar coordinates:

$$\begin{cases} x = \rho\cos\theta \\ y = \rho\sin\theta \end{cases}, \ \rho > 0, \ \theta \in [0, 2\pi].$$

**Solution.**
We shall have:

$$\frac{\partial z}{\partial \rho} = \frac{\partial z}{\partial x} \cdot \frac{\partial x}{\partial \rho} + \frac{\partial z}{\partial y} \cdot \frac{\partial y}{\partial \rho} = \cos\theta\frac{\partial z}{\partial x} + \sin\theta\frac{\partial z}{\partial y},$$

$$\frac{\partial z}{\partial \theta} = \frac{\partial z}{\partial x} \cdot \frac{\partial x}{\partial \theta} + \frac{\partial z}{\partial y} \cdot \frac{\partial y}{\partial \theta} = -\rho\sin\theta\frac{\partial z}{\partial x} + \rho\cos\theta\frac{\partial z}{\partial y}.$$

One obtains the system:

$$\begin{cases} \cos\theta \dfrac{\partial z}{\partial x} + \sin\theta \dfrac{\partial z}{\partial y} = \dfrac{\partial z}{\partial \rho} \\[2mm] -\rho\sin\theta \dfrac{\partial z}{\partial x} + \rho\cos\theta \dfrac{\partial z}{\partial y} = \dfrac{\partial z}{\partial \theta}. \end{cases} \qquad (3.23)$$

Since

$$\Delta = \begin{vmatrix} \cos\theta & \sin\theta \\ -\rho\sin\theta & \rho\cos\theta \end{vmatrix} = \rho \neq 0$$

we can solve the system (3.23) using the Cramer's method:

$$\frac{\partial z}{\partial x} = \frac{\begin{vmatrix} \frac{\partial z}{\partial \rho} & \sin\theta \\ \frac{\partial z}{\partial \theta} & \rho\cos\theta \end{vmatrix}}{\Delta} = \frac{1}{\rho}\left(\rho\cos\theta \frac{\partial z}{\partial \rho} - \sin\theta \frac{\partial z}{\partial \theta}\right) = \cos\theta \frac{\partial z}{\partial \rho} - \frac{\sin\theta}{\rho}\frac{\partial z}{\partial \theta},$$

$$\frac{\partial z}{\partial y} = \frac{\begin{vmatrix} \cos\theta & \frac{\partial z}{\partial \rho} \\ -\rho\sin\theta & \frac{\partial z}{\partial \theta} \end{vmatrix}}{\Delta} = \frac{1}{\rho}\left(\rho\sin\theta \frac{\partial z}{\partial \rho} + \cos\theta \frac{\partial z}{\partial \theta}\right) = \sin\theta \frac{\partial z}{\partial \rho} + \frac{\cos\theta}{\rho}\frac{\partial z}{\partial \theta}.$$

Therefore:

$$\frac{\partial^2 z}{\partial x^2} = \frac{\partial}{\partial x}\underbrace{\left(\cos\theta \frac{\partial z}{\partial \rho} - \frac{\sin\theta}{\rho}\frac{\partial z}{\partial \theta}\right)}_{F} = \frac{\partial F}{\partial x} = \cos\theta \frac{\partial F}{\partial \rho} - \frac{\sin\theta}{\rho}\frac{\partial F}{\partial \theta}$$

$$= \cos\theta \frac{\partial}{\partial \rho}\left(\cos\theta \frac{\partial z}{\partial \rho} - \frac{\sin\theta}{\rho}\frac{\partial z}{\partial \theta}\right) - \frac{\sin\theta}{\rho}\frac{\partial}{\partial \theta}\left(\cos\theta \frac{\partial z}{\partial \rho} - \frac{\sin\theta}{\rho}\frac{\partial z}{\partial \theta}\right)$$

$$= \cos\theta \left(\cos\theta \frac{\partial^2 z}{\partial \rho^2} + \frac{\sin\theta}{\rho^2}\frac{\partial z}{\partial \theta} - \frac{\sin\theta}{\rho}\frac{\partial^2 z}{\partial \rho\partial \theta}\right)$$

$$\quad - \frac{\sin\theta}{\rho}\left(\cos\theta \frac{\partial^2 z}{\partial \rho\partial \theta} - \sin\theta \frac{\partial z}{\partial \rho} - \frac{\sin\theta}{\rho}\frac{\partial^2 z}{\partial \theta^2} - \frac{\cos\theta}{\rho}\frac{\partial z}{\partial \theta}\right)$$

$$= \cos^2\theta \frac{\partial^2 z}{\partial \rho^2} + \frac{\sin\theta\cos\theta}{\rho^2}\frac{\partial z}{\partial \theta} - \frac{\sin\theta\cos\theta}{\rho}\frac{\partial^2 z}{\partial \rho\partial \theta} - \frac{\sin\theta\cos\theta}{\rho}\frac{\partial^2 z}{\partial \rho\partial \theta}$$

$$\quad + \frac{\sin^2\theta}{\rho}\frac{\partial z}{\partial \rho} + \frac{\sin^2\theta}{\rho^2}\frac{\partial^2 z}{\partial \theta^2} + \frac{\sin\theta\cos\theta}{\rho^2}\frac{\partial z}{\partial \theta}$$

$$= \cos^2\theta \frac{\partial^2 z}{\partial \rho^2} - 2\frac{\sin\theta\cos\theta}{\rho}\frac{\partial^2 z}{\partial \rho\partial \theta} + \frac{\sin^2\theta}{\rho^2}\frac{\partial^2 z}{\partial \theta^2} + 2\frac{\sin\theta\cos\theta}{\rho^2}\frac{\partial z}{\partial \theta} + \frac{\sin^2\theta}{\rho}\frac{\partial z}{\partial \rho},$$

$$\frac{\partial^2 z}{\partial y^2} = \frac{\partial}{\partial y} \underbrace{\left( \sin\theta \frac{\partial z}{\partial \rho} + \frac{\cos\theta}{\rho} \frac{\partial z}{\partial \theta} \right)}_{F} = \frac{\partial F}{\partial y} = \sin\theta \frac{\partial F}{\partial \rho} + \frac{\cos\theta}{\rho} \frac{\partial F}{\partial \theta}$$

$$= \sin\theta \frac{\partial}{\partial \rho} \left( \sin\theta \frac{\partial z}{\partial \rho} + \frac{\cos\theta}{\rho} \frac{\partial z}{\partial \theta} \right) + \frac{\cos\theta}{\rho} \frac{\partial}{\partial \theta} \left( \sin\theta \frac{\partial z}{\partial \rho} + \frac{\cos\theta}{\rho} \frac{\partial z}{\partial \theta} \right)$$

$$= \sin\theta \left( \sin\theta \frac{\partial^2 z}{\partial \rho^2} + \frac{\cos\theta}{\rho} \frac{\partial^2 z}{\partial \rho \partial \theta} - \frac{\cos\theta}{\rho^2} \frac{\partial z}{\partial \theta} \right)$$

$$+ \frac{\cos\theta}{\rho} \left( \sin\theta \frac{\partial^2 z}{\partial \rho \partial \theta} + \cos\theta \frac{\partial z}{\partial \rho} + \frac{\cos\theta}{\rho} \frac{\partial^2 z}{\partial \theta^2} - \frac{\sin\theta}{\rho} \frac{\partial z}{\partial \theta} \right)$$

$$= \sin^2\theta \frac{\partial^2 z}{\partial \rho^2} + \frac{\sin\theta \cos\theta}{\rho} \frac{\partial^2 z}{\partial \rho \partial \theta} - \frac{\sin\theta \cos\theta}{\rho^2} \frac{\partial z}{\partial \theta} + \frac{\sin\theta \cos\theta}{\rho} \frac{\partial^2 z}{\partial \rho \partial \theta}$$

$$+ \frac{\cos^2\theta}{\rho} \frac{\partial z}{\partial \rho} + \frac{\cos^2\theta}{\rho^2} \frac{\partial^2 z}{\partial \theta^2} - \frac{\sin\theta \cos\theta}{\rho^2} \frac{\partial z}{\partial \theta}$$

$$= \sin^2\theta \frac{\partial^2 z}{\partial \rho^2} + 2 \frac{\sin\theta \cos\theta}{\rho} \frac{\partial^2 z}{\partial \rho \partial \theta} + \frac{\cos^2\theta}{\rho^2} \frac{\partial^2 z}{\partial \theta^2} + \frac{\cos^2\theta}{\rho} \frac{\partial z}{\partial \rho} - 2 \frac{\sin\theta \cos\theta}{\rho^2} \frac{\partial z}{\partial \theta},$$

$$\frac{\partial^2 z}{\partial x \partial y} = \frac{\partial}{\partial x} \underbrace{\left( \sin\theta \frac{\partial z}{\partial \rho} + \frac{\cos\theta}{\rho} \frac{\partial z}{\partial \theta} \right)}_{F} = \frac{\partial F}{\partial x} = \cos\theta \frac{\partial F}{\partial \rho} - \frac{\sin\theta}{\rho} \frac{\partial F}{\partial \theta}$$

$$= \cos\theta \frac{\partial}{\partial \rho} \left( \sin\theta \frac{\partial z}{\partial \rho} + \frac{\cos\theta}{\rho} \frac{\partial z}{\partial \theta} \right) - \frac{\sin\theta}{\rho} \frac{\partial}{\partial \theta} \left( \sin\theta \frac{\partial z}{\partial \rho} + \frac{\cos\theta}{\rho} \frac{\partial z}{\partial \theta} \right)$$

$$= \cos\theta \left( \sin\theta \frac{\partial^2 z}{\partial \rho^2} + \frac{\cos\theta}{\rho} \frac{\partial^2 z}{\partial \rho \partial \theta} - \frac{\cos\theta}{\rho^2} \frac{\partial z}{\partial \theta} \right)$$

$$- \frac{\sin\theta}{\rho} \left( \sin\theta \frac{\partial^2 z}{\partial \rho \partial \theta} + \cos\theta \frac{\partial z}{\partial \rho} + \frac{\cos\theta}{\rho} \frac{\partial^2 z}{\partial \theta^2} - \frac{\sin\theta}{\rho} \frac{\partial z}{\partial \theta} \right)$$

$$= \sin\theta \cos\theta \frac{\partial^2 z}{\partial \rho^2} + \frac{\cos^2\theta}{\rho} \frac{\partial^2 z}{\partial \rho \partial \theta} - \frac{\cos^2\theta}{\rho^2} \frac{\partial z}{\partial \theta} - \frac{\sin^2\theta}{\rho} \frac{\partial^2 z}{\partial \rho \partial \theta}$$

$$- \frac{\sin\theta \cos\theta}{\rho} \frac{\partial z}{\partial \rho} - \frac{\sin\theta \cos\theta}{\rho^2} \frac{\partial^2 z}{\partial \theta^2} + \frac{\sin^2\theta}{\rho^2} \frac{\partial z}{\partial \theta}$$

$$= \sin\theta \cos\theta \frac{\partial^2 z}{\partial \rho^2} + \frac{\cos^2\theta - \sin^2\theta}{\rho} \frac{\partial^2 z}{\partial \rho \partial \theta} - \frac{\sin\theta \cos\theta}{\rho^2} \frac{\partial^2 z}{\partial \theta^2}$$

$$- \frac{\sin\theta \cos\theta}{\rho} \frac{\partial z}{\partial \rho} - \frac{\cos^2\theta - \sin^2\theta}{\rho^2} \frac{\partial z}{\partial \theta}.$$

We shall have:

$$E = \rho^2 \cos^2\theta \left( \cos^2\theta \frac{\partial^2 z}{\partial\rho^2} - 2\frac{\sin\theta\cos\theta}{\rho}\frac{\partial^2 z}{\partial\rho\partial\theta} + \frac{\sin^2\theta}{\rho^2}\frac{\partial^2 z}{\partial\theta^2} + 2\frac{\sin\theta\cos\theta}{\rho^2}\frac{\partial z}{\partial\theta} + \frac{\sin^2\theta}{\rho}\frac{\partial z}{\partial\rho} \right)$$

$$+ 2\rho^2 \sin\theta\cos\theta \left( \sin\theta\cos\theta \frac{\partial^2 z}{\partial\rho^2} + \frac{\cos^2\theta - \sin^2\theta}{\rho}\frac{\partial^2 z}{\partial\rho\partial\theta} - \frac{\sin\theta\cos\theta}{\rho^2}\frac{\partial^2 z}{\partial\theta^2} \right.$$

$$\left. - \frac{\sin\theta\cos\theta}{\rho}\frac{\partial z}{\partial\rho} - \frac{\cos^2\theta - \sin^2\theta}{\rho^2}\frac{\partial z}{\partial\theta} \right) + \rho^2 \sin^2\theta \left( \sin^2\theta\frac{\partial^2 z}{\partial\rho^2} + 2\frac{\sin\theta\cos\theta}{\rho}\frac{\partial^2 z}{\partial\rho\partial\theta} \right.$$

$$\left. + \frac{\cos^2\theta}{\rho^2}\frac{\partial^2 z}{\partial\theta^2} + \frac{\cos^2\theta}{\rho}\frac{\partial z}{\partial\rho} - 2\frac{\sin\theta\cos\theta}{\rho^2}\frac{\partial z}{\partial\theta} \right)$$

$$= \rho^2 \cos^4\theta \frac{\partial^2 z}{\partial\rho^2} - 2\rho\sin\theta\cos^3\theta \frac{\partial^2 z}{\partial\rho\partial\theta} + \sin^2\theta\cos^2\theta\frac{\partial^2 z}{\partial\theta^2} + 2\sin\theta\cos^3\theta\frac{\partial z}{\partial\theta}$$

$$+ \rho\sin^2\theta\cos^2\theta\frac{\partial z}{\partial\rho} + 2\rho^2\sin^2\theta\cos^2\theta\frac{\partial^2 z}{\partial\rho^2} + 2\rho\sin\theta\cos^3\theta\frac{\partial^2 z}{\partial\rho\partial\theta}$$

$$- 2\rho\sin^3\theta\cos\theta\frac{\partial^2 z}{\partial\rho\partial\theta} - 2\sin^2\theta\cos^2\theta\frac{\partial^2 z}{\partial\theta^2} - 2\rho\sin^2\theta\cos^2\theta\frac{\partial z}{\partial\rho}$$

$$+ 2\sin^3\theta\cos\theta\frac{\partial z}{\partial\theta} - 2\sin\theta\cos^3\theta\frac{\partial z}{\partial\theta} + \rho^2\sin^4\theta\frac{\partial^2 z}{\partial\rho^2} + 2\rho\sin^3\theta\cos\theta\frac{\partial^2 z}{\partial\rho\partial\theta}$$

$$+ \sin^2\theta\cos^2\theta\frac{\partial^2 z}{\partial\theta^2} + \rho\sin^2\theta\cos^2\theta\frac{\partial z}{\partial\rho} - 2\sin^3\theta\cos\theta\frac{\partial z}{\partial\theta}$$

$$= \rho^2 \left( \cos^2\theta + \sin^2\theta \right)^2 \frac{\partial^2 z}{\partial\rho^2};$$

hence

$$E = \rho^2 \frac{\partial^2 z}{\partial\rho^2}.$$

## 3.4   Taylor's Formula for Functions of Two Variables

**Definition 3.36** (see [8]). Let $f : \mathbb{R}^2 \to \mathbb{R}$ be a function of two variables which have continuous partial derivatives of all orders up to and including the $(n+1)$-th in the neighborhood of a point $(a, b) \in \mathbb{R}^2$. The **Taylor's formula around the point** $(a, b)$, for $f(x, y)$ is:

$$f(x,y) = f(a,b) + \frac{1}{1!}\left[(x-a)\frac{\partial f}{\partial x}(a,b) + (y-b)\frac{\partial f}{\partial y}(a,b)\right] + \frac{1}{2!} \qquad (3.24)$$

$$\cdot\left[(x-a)^2\frac{\partial^2 f}{\partial x^2}(a,b) + 2(x-a)(y-b)\frac{\partial^2 f}{\partial x\partial y}(a,b) + (y-b)^2\frac{\partial^2 f}{\partial y^2}(a,b)\right]$$

$$+\frac{1}{3!}\left[(x-a)^3\frac{\partial^3 f}{\partial x^3}(a,b) + 3(x-a)^2(y-b)\frac{\partial^3 f}{\partial^2 x\partial y}(a,b)\right.$$

$$\left. + 3(x-a)(y-b)^2\frac{\partial^3 f}{\partial x\partial^2 y}(a,b) + (y-b)^3\frac{\partial^3 f}{\partial y^3}(a,b)\right]$$

$$+\cdots+\frac{1}{n!}\left[(x-a)\frac{\partial}{\partial x} + (y-b)\frac{\partial}{\partial y}(a,b)\right]^n f(a,b) + R_n(x,y),$$

where

$$R_n(x,y) = \frac{1}{(n+1)!}\left[(x-a)\frac{\partial}{\partial x} + (y-b)\frac{\partial}{\partial y}(a,b)\right]^{n+1} f(\xi,\eta),$$

$$\begin{cases} \xi = a + \theta(x-a) \\ \eta = b + \theta(x-b) \end{cases}, \quad \theta \in (0,1).$$

**Remark 3.37** (see [8]). The particular case of (3.24), when $a = b = 0$ is called the **Mac Laurin's formula**.

**Proposition 3.38** (see [15], p. 144). Let $f : \mathbb{R}^2 \to \mathbb{R}$ be a function of two variables which have continuous partial derivatives of all orders up to and including the $(n+1)$-th in the neighborhood of a point $(x_0, y_0) \in \mathbb{R}^2$.

In a different notation, the formula (3.24) may be written in the form:

$$f(x_0 + h, y_0 + k) = f(x_0, y_0) + \frac{1}{1!}\left[h\frac{\partial f}{\partial x}(x_0, y_0) + k\frac{\partial f}{\partial y}(x_0, y_0)\right] + \qquad (3.25)$$

$$+\frac{1}{2!}\left[h^2\frac{\partial^2 f}{\partial x^2}(x_0, y_0) + 2hk\frac{\partial^2 f}{\partial x\partial y}(x_0, y_0) + k^2\frac{\partial^2 f}{\partial y^2}(x_0, y_0)\right]$$

$$+\cdots+\frac{1}{n!}\left[h\frac{\partial}{\partial x} + k\frac{\partial}{\partial y}\right]^n f(x_0, y_0)$$

$$+\frac{1}{(n+1)!}\left[h\frac{\partial}{\partial x} + k\frac{\partial}{\partial y}\right]^{n+1} f(x_0 + \theta h, y_0 + \theta k), \theta \in (0,1).$$

**Example 3.39.** Write the Taylor's formula of the third order, for the function

$$f(x,y) = e^x \sin y$$

around the point $(0,0) \in \mathbb{R}^2$.

**Solution.**

We shall compute:

- the first partial derivatives:

$$\begin{cases} \frac{\partial f}{\partial x} = e^x \sin y \implies \frac{\partial f}{\partial x}(0,0) = 0 \\ \frac{\partial f}{\partial y} = e^x \cos y \implies \frac{\partial f}{\partial y}(0,0) = 1, \end{cases}$$

- the second partial derivatives:

$$\begin{cases} \frac{\partial^2 f}{\partial x^2} = e^x \sin y \implies \frac{\partial^2 f}{\partial x^2}(0,0) = 0 \\ \frac{\partial^2 f}{\partial x \partial y} = e^x \cos y \implies \frac{\partial^2 f}{\partial x \partial y}(0,0) = 1 \\ \frac{\partial^2 f}{\partial y^2} = -e^x \sin y \implies \frac{\partial^2 f}{\partial y^2}(0,0) = 0, \end{cases}$$

- the third partial derivatives:

$$\begin{cases} \frac{\partial^3 f}{\partial x^3} = e^x \sin y \implies \frac{\partial^3 f}{\partial x^3}(0,0) = 0 \\ \frac{\partial^3 f}{\partial x^2 \partial y} = e^x \cos y \implies \frac{\partial^3 f}{\partial x^2 \partial y}(0,0) = 1 \\ \frac{\partial^3 f}{\partial x \partial y^2} = -e^x \sin y \implies \frac{\partial^3 f}{\partial x \partial y^2}(0,0) = 0 \\ \frac{\partial^3 f}{\partial y^3} = -e^x \cos y \implies \frac{\partial^3 f}{\partial y^3}(0,0) = -1, \end{cases}$$

- the fourth partial derivatives:

$$\begin{cases} \frac{\partial^4 f}{\partial x^4} = e^x \sin y \\ \frac{\partial^4 f}{\partial x^3 \partial y} = e^x \cos y \\ \frac{\partial^4 f}{\partial x^2 \partial y^2} = -e^x \sin y \\ \frac{\partial^4 f}{\partial x \partial y^3} = -e^x \cos y \\ \frac{\partial^4 f}{\partial y^4} = e^x \sin y. \end{cases}$$

The Taylor's formula of the third order, around the point $(0,0) \in \mathbb{R}^2$, for the function $f(x,y)$ will be:

$$f(x,y) = 0 + \frac{1}{1!}(x \cdot 0 + y \cdot 1) + \frac{1}{2!}(x^2 \cdot 0 + 2xy \cdot 1 + y^2 \cdot 0)$$

$$+ \frac{1}{3!}\left[x^3 \cdot 0 + 3x^2 y \cdot 1 + 3xy^2 \cdot 0 + y^3 \cdot (-1)\right]$$

$$+ \frac{1}{4!}\left[x^4 e^{\theta x} \sin \theta y + 4x^3 y e^{\theta x} \cos \theta y + 6x^2 y^2 \left(-e^{\theta x} \sin \theta y\right)\right.$$

$$+ 4xy^3 \left(-e^{\theta x} \cos \theta y\right) + y^4 \left(-e^{\theta x} \sin \theta y\right)] , \quad \theta \in (0,1)$$

i.e.

$$f(x,y) = y + xy + \frac{1}{2}x^2 y - \frac{1}{6}y^3$$

$$+ \frac{1}{4!} \left[x^4 \sin \theta y + 4x^3 y \cos \theta y - 6x^2 y^2 \sin \theta y - 4xy^3 \cos \theta y - y^4 \sin \theta y\right] , \quad \theta \in (0,1).$$

We shall give a computer solution in Mathcad 14:

$$e^x \cdot \sin(y) \text{ series}, x, y, 3 \; \rightarrow \; y - \frac{y^3}{6} + x \cdot y + \frac{x^2 \cdot y}{2}$$

and using Maple 15:

$mtaylor(\exp(x) \cdot \sin(y), [x, y], 4)$

$$y + xy - \frac{1}{6}y^3 + \frac{1}{2}x^2 y$$

**Example 3.40.** Write the Taylor's formula of the third order, for the function

$$f(x,y) = -x^2 + 2xy + 3y^2 - 6x - 2y - 4$$

around the point $(-2,1) \in \mathbb{R}^2$.

**Solution.**

We shall compute:

- the first partial derivatives:

$$\begin{cases} \frac{\partial f}{\partial x} = -2x + 2y - 6 \implies \frac{\partial f}{\partial x}(-2,1) = 0 \\ \frac{\partial f}{\partial y} = 2x + 6y - 2 \implies \frac{\partial f}{\partial y}(-2,1) = 0, \end{cases}$$

- the second partial derivatives:

$$\begin{cases} \frac{\partial^2 f}{\partial x^2} = -2 \implies \frac{\partial^2 f}{\partial x^2}(-2,1) = -2 \\ \frac{\partial^2 f}{\partial x \partial y} = 2 \implies \frac{\partial^2 f}{\partial x \partial y}(-2,1) = 2 \\ \frac{\partial^2 f}{\partial y^2} = 6 \implies \frac{\partial^2 f}{\partial y^2}(-2,1) = 6, \end{cases}$$

- the third partial derivatives:

$$\frac{\partial^3 f}{\partial x^3} = 0 \implies \frac{\partial^3 f}{\partial x^3}(0,0) = 0$$
$$\frac{\partial^3 f}{\partial x^2 \partial y} = 0 \implies \frac{\partial^3 f}{\partial x^2 \partial y}(0,0) = 0$$
$$\frac{\partial^3 f}{\partial x \partial y^2} = 0 \implies \frac{\partial^3 f}{\partial x \partial y^2}(0,0) = 0$$
$$\frac{\partial^3 f}{\partial y^3} = 0 \implies \frac{\partial^3 f}{\partial y^3}(0,0) = 0.$$

The Taylor's formula of the $n$-th order, around the point $(-2, 1) \in \mathbb{R}^2$, for the function $f(x, y)$ will be:

$$f(x,y) = 1 + \frac{1}{1!}[(x+2) \cdot 0 + (y-1) \cdot 0]$$
$$+ \frac{1}{2!}\left[\left((x+2)^2 \cdot (-2) + 2(x+2)(y-1) \cdot 2 + (y-1)^2 \cdot 6\right)\right]$$
$$= 1 - (x+2)^2 + 2(x+2)(y-1) + 3(y-1)^2.$$

We can also obtain this result with Mathcad 14:

$-x^2 + 2 \cdot x \cdot y + 3 \cdot y^2 - 6 \cdot x - 2 \cdot y - 4 \text{ series}, x \equiv -2, y \equiv 1, 8 \rightarrow 1 + 3 \cdot (y-1)^2 + 2 \cdot (x+2) \cdot (y-1) - (x+2)^2$

and Maple 15:

$mtaylor(-x^2 + 2 \cdot x \cdot y + 3 \cdot y^2 - 6 \cdot x - 2 \cdot y - 4, [x=-2, y=1], 8)$
$$1 - (x+2)^2 + 2(x+2)(y-1) + 3(y-1)^2$$

**Example 3.41.** Use the Taylor's formula of the second order to find the approximate value for $(0.95)^{2.01}$.

**Solution.**
We have to take the function $f(x, y) = x^y$.
Applying the formula (3.25) for

$$\begin{cases} x_0 = 1, & y_0 = 2 \\ h = -0.05, & k = 0.01 \end{cases}$$

one obtains:

$$f(1 - 0.05, 2 + 0.01) \approx f(1,2) + \frac{1}{1!}\left[(-0.05)\frac{\partial f}{\partial x}(1,2) + 0.01\frac{\partial f}{\partial y}(1,2)\right]$$

$$+ \frac{1}{2!}\left[\left((-0.05)^2\frac{\partial^2 f}{\partial x^2}(1,2) + 2(-0.05) \cdot 0.01\frac{\partial^2 f}{\partial x \partial y}(1,2) + 0.01^2\frac{\partial f}{\partial y}(1,2)\right)\right].$$

As the successive partial derivatives are

$$\begin{cases} \frac{\partial f}{\partial x} = yx^{y-1} \\ \frac{\partial f}{\partial y} = x^y \ln x \end{cases}$$

and

$$\begin{cases} \frac{\partial^2 f}{\partial x^2} = y(y-1)x^{y-2} \\ \frac{\partial^2 f}{\partial x \partial y} = yx^{y-1} \ln x + x^y \cdot \frac{1}{x} = x^{y-1}(y \ln x + 1) \\ \frac{\partial^2 f}{\partial y^2} = x^y \ln x \cdot \ln x = x^y \ln^2 x \end{cases}$$

we deduce that:

$$f(0.95, 2.01) \approx 1 - 0.1 + \frac{1}{2!}(2 \cdot 0.0025 - 0.001) = 0.9 + \frac{1}{2}(0.005 - 0.001) = 0.902.$$

We shall check this result in Matlab 7.9:
```
>> f=@(x,y)x^y;
>> vpa(f(0.95,2.01),3)
ans =
0.902
```
or in Mathcad 14:

$$f(x,y) := x^y$$

$$f(0.95, 2.01) = 0.902$$

or using Mathematica 8:

```
In[51]:= f[x_, y_] := x^y

In[52]:= SetPrecision[f[0.95, 2.01], 3]

Out[52]= 0.902
```

or with Maple 15:

$$f := (x, y) \to x^y$$

$$(x, y) \to x^y$$

$$Digits := 3$$

$$3$$

$$f(0.95, 2.01)$$

$$0.902$$

We also want to find the approximation error using the relation

$$R_2 = \frac{1}{3!} \left[ h^3 \frac{\partial^3 f}{\partial x^3} (x_0 + \theta h, \ y_0 + \theta h) + 3h^2 k \frac{\partial^3 f}{\partial^2 x \partial y} (x_0 + \theta h, \ y_0 + \theta h) \right.$$

$$\left. + 3hk^2 \frac{\partial^3 f}{\partial x \partial^2 y} (x_0 + \theta h, \ y_0 + \theta h) + k^3 \frac{\partial^3 f}{\partial y^3} (x_0 + \theta h, \ y_0 + \theta h) \right], \quad \theta \in (0,1).$$

In order to do that we have to calculate the third partial derivatives of the function $f$:

$$\begin{cases} \frac{\partial^3 f}{\partial x^3} = y\,(y-1)\,(y-2)\,x^{y-2} \\ \frac{\partial^3 f}{\partial x^2 \partial y} = (y-1)\,x^{y-2}\,(y \ln x + 1) + x^{y-1} \cdot \frac{y}{x} = x^{y-2}\,[(y-1)\,y \ln x + y - 1 + y] \\ \qquad = x^{y-2}\,[(y-1)\,y \ln x + 2y - 1] \\ \frac{\partial^3 f}{\partial x \partial y^2} = yx^{y-1} \ln^2 x + x^y \cdot \frac{2}{x} \ln x = x^{y-1} \ln x\,(y \ln x + 2) \\ \frac{\partial^3 f}{\partial y^3} = x^y \ln x \cdot \ln^2 x = x^y \ln^3 x. \end{cases}$$

Hence,

$$R_2 < \frac{1}{3!} \left[ (-0.05)^3 \cdot 2\,(2-1)\,(2-2) \cdot 1^{2-3} + 3\,(-0.05)^2 \cdot 0.01 \cdot 1^{2-2}\,(1 \cdot 2 \ln 1 + 4 - 1) \right.$$

$$\left. + 3\,(-0.05) \cdot 0.01^2 \cdot 1^{2-1} \ln 1\,(2 \ln 1 + 2) + 1^2 \cdot \ln^3 1 \cdot (0.01)^3 \right]$$

$$= \frac{1}{3!} \cdot (9 \cdot 0.000025) = 0.0000375.$$

We can prove that in Matlab 7.9:

```
>> x0=1;y0=2;h=-0.05;k=0.01;
>> t=0:0.1:1;
>> f=@(x,y)x^y;
>> syms x y
>> f1=diff(f(x,y),x,3)
f1 =
x^(y - 3)*y*(y - 1)*(y - 2)
>> f4=diff(f(x,y),y,3)
f4 =
x^y*log(x)^3
>> f2=simplify(diff(diff(f(x,y),y),x,2))
f2 =
x^(y - 2)*(2*y + y^2*log(x) - y*log(x) - 1)
>> f3=simplify(diff(diff(f(x,y),x),y,2))
f3 =
```

x^(y - 1)*log(x)*(y*log(x) + 2)
>> R2=(1/6)*(h^3*subs(f1,{x,y},{x0+t*h,y0+t*k})+
3*h^2*k*subs(f2,{x,y},{x0+t*h,y0+t*k})+
3*h*k^2*subs(f3,{x,y},{x0+t*h,y0+t*k})+k^3*subs(f4,{x,y},
{x0+t*h,y0+t*k}))
R2 =
 1.0e-004 *
 Columns 1 through 6
 0.3750 0.3738 0.3726 0.3714 0.3701 0.3689
 Columns 7 through 11
 0.3676 0.3663 0.3649 0.3635 0.3622
>> plot(t,R2)

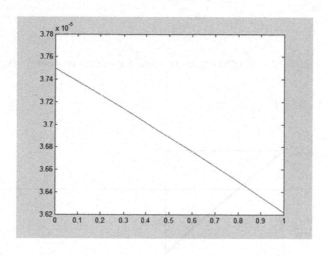

>>r2=R2(1)
r2 =
 3.7500e-005
or in Mathcad 14:

$f(x,y) := x^y$

$x0 := 1$    $y0 := 2$    $h := -0.05$    $k := 0.01$

$f1(x,y) := \dfrac{d^3}{dx^3} f(x,y)$            $f4(x,y) := \dfrac{d^3}{dy^3} f(x,y)$

$f2(x,y) := \dfrac{d^2}{dx^2}\dfrac{d}{dy} f(x,y)$         $f3(x,y) := \dfrac{d^2}{dy^2}\dfrac{d}{dx} f(x,y)$

$R2(t) := \dfrac{1}{6}\left( h^3 \cdot f1(x0 + t\cdot h, y0 + t\cdot k) + 3\cdot h^2 \cdot k\cdot f2(x0 + t\cdot h, y0 + t\cdot k) + 3\cdot h\cdot k^2\cdot f3(x0 + t\cdot h, y0 + t\cdot k) + k^3 \cdot f4(x0 + t\cdot h, y0 + t\cdot k) \right)$

$t := 0, 0.1 .. 1$

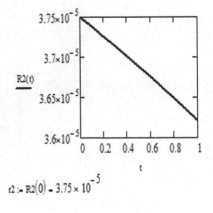

$r2 := R2(0) - 3.75 \times 10^{-5}$

or in Mathematica 8:

```
In[1]:= x0 = 1; y0 = 2; h = -0.05; k = 0.01;
```

```
In[2]:= t = Range[0, 1, 0.1];
```

```
In[3]:= f[x_, y_] := x^y;
```

```
In[4]:= f1 = D[f[x, y], {x, 3}]
```

Out[4]= $x^{-3+y}$ (-2 + y) (-1 + y) y

```
In[5]:= f4 = D[f[x, y], {y, 3}]
```

Out[5]= $x^y$ Log[x]$^3$

```
In[6]:= f2 = Simplify[D[D[f[x, y], y], {x, 2}]]
```

Out[6]= $x^{-2+y}$ (-1 + 2 y + (-1 + y) y Log[x])

```
In[7]:= f3 = Simplify[D[D[f[x, y], y], x], {y, 2}]]
```

Out[7]= $x^{-1+y}$ Log[x] (2 + y Log[x])

```
In[8]:= ff1 = f1 /. {x → x0 + t*h, y → y0 + t*k}; ff2 = f2 /. {x → x0 + t*h, y → y0 + t*k};
```

```
In[9]:= ff3 = f3 /. {x → x0 + t*h, y → y0 + t*k}; ff4 = f4 /. {x → x0 + t*h, y → y0 + t*k};
```

```
In[10]:= R2 = 1/6 * (h^3 * ff1 + 3 * h^2 * k * ff2 + 3 * h * k^2 * ff3 + k^3 * ff4);
```

```
In[11]:= ListLinePlot[{R2}]
```

Out[11]= 

```
In[12]:= r2 = R2[[1]]
```

Out[12]= 0.0000375

or with Maple 15:

$x0 := 1 : y0 := 2 : h := -0.05 : k := 0.01 :$

$f := x^y :$

$f1 := factor\left( simplify\left( \dfrac{\partial^3}{\partial x^3} f \right) \right) :$

$f4 := factor\left( simplify\left( \dfrac{\partial^3}{\partial y^3} f \right) \right) :$

$f2 := factor\left( simplify\left( \dfrac{\partial}{\partial y}\left( \dfrac{\partial^2}{\partial x^2} \right) f \right) \right) :$

$f3 := factor\left( simplify\left( \dfrac{\partial}{\partial x}\left( \dfrac{\partial^2}{\partial y^2} \right) f \right) \right) :$

$ff1 := f1\big|_{x = x0 + t \cdot h,\, y = y0 + t \cdot k} :$

$ff2 := f2\big|_{x = x0 + t \cdot h,\, y = y0 + t \cdot k} :$

$ff3 := f3\big|_{x = x0 + t \cdot h,\, y = y0 + t \cdot k} :$

$ff4 := f4\big|_{x = x0 + t \cdot h,\, y = y0 + t \cdot k} :$

$R2 := t \rightarrow \dfrac{1}{6} \cdot \left( h^3 \cdot ff1 + 3 \cdot h^2 \cdot k \cdot ff2 + 3 \cdot h \cdot k^2 \cdot ff3 + k^3 \cdot ff4 \right) :$

$plot(R2(t), t = 0 .. 1)$

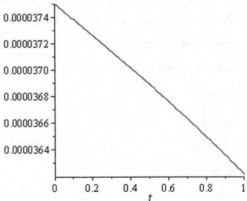

$Digits := 3$

$$3$$

$R2(t)\Big|_{t=0}$

$$0.0000375$$

**Example 3.42.** Use the Taylor's formula of the second order to find the approximate value for $\sqrt{1.03} \cdot \sqrt[3]{0.98}$.

**Solution.**
One takes the function:

$$f(x, y) = \sqrt{x} \cdot \sqrt[3]{y} = x^{\frac{1}{2}} \cdot y^{\frac{1}{3}}.$$

Applying the formula (3.25) for

$$\begin{cases} x_0 = 1, & y_0 = 1 \\ h = 0.03, & k = -0.02 \end{cases}$$

one obtains:

$$f(1 + 0.03, 1 - 0.02) \approx f(1,1) + \frac{1}{1!}\left[0.03\frac{\partial f}{\partial x}(1,1) - 0.02\frac{\partial f}{\partial y}(1,1)\right]$$

$$+\frac{1}{2!}\left[\left((0.03)^2 \frac{\partial^2 f}{\partial x^2}(1,1) - 2 \cdot 0.03 \cdot 0.02\frac{\partial^2 f}{\partial x \partial y}(1,1) + (-0.02)^2 \frac{\partial f}{\partial y}(1,1)\right)\right].$$

We shall calculate the successive partial derivatives:

$$\begin{cases} \frac{\partial f}{\partial x} = \frac{1}{2}x^{-\frac{1}{2}} \cdot y^{\frac{1}{3}} \\ \frac{\partial f}{\partial y} = \frac{1}{3}x^{\frac{1}{2}} \cdot y^{-\frac{2}{3}} \end{cases}$$

$$\begin{cases} \frac{\partial^2 f}{\partial x^2} = -\frac{1}{4}x^{-\frac{3}{2}} \cdot y^{\frac{1}{3}} \\ \frac{\partial^2 f}{\partial x \partial y} = \frac{1}{6}x^{-\frac{1}{2}} \cdot y^{-\frac{2}{3}} \\ \frac{\partial^2 f}{\partial y^2} = -\frac{2}{9}x^{\frac{1}{2}} \cdot y^{-\frac{5}{3}} \end{cases}$$

$$\begin{cases} \frac{\partial^3 f}{\partial x^3} = \frac{3}{8}x^{-\frac{5}{2}} \cdot y^{\frac{1}{3}} \\ \frac{\partial^3 f}{\partial x^2 \partial y} = -\frac{1}{12}x^{-\frac{3}{2}} \cdot y-\frac{2}{3} \\ \frac{\partial^3 f}{\partial x \partial y^2} = -\frac{1}{9}x^{-\frac{1}{2}} \cdot y^{-\frac{5}{3}} \\ \frac{\partial^3 f}{\partial y^3} = \frac{10}{27}x^{\frac{1}{2}} \cdot y^{-\frac{8}{3}}. \end{cases}$$

We shall deduce that:

$$f(1.03, 0.98) \approx 1 + \frac{1}{1!}\left(0.03 \cdot \frac{1}{2} - 0.02 \cdot \frac{1}{3}\right)$$

$$+\frac{1}{2!}\left(-0.03^2 \cdot \frac{1}{4} - 2 \cdot 0.03 \cdot 0.02 \cdot \frac{1}{6} + (-0.02)^2 \cdot \left(-\frac{2}{9}\right)\right) = 1.0083 - 0.00026 = 1.00804.$$

We shall check this result in Maple 7.9:
```
>> f=@(x,y) x^(1/2)*y^(1/3);
>> vpa(f(1.03,0.98),6)
ans =
1.00808
```
or in Mathcad 14:

$$f(x,y) := x^{\frac{1}{2}} \cdot y^{\frac{1}{3}}$$

$$f(1.03, 0.98) = 1.00808$$

or with Mathematica 8:

```
In[63]:= f[x_, y_] := x^(1/2) * y^(1/3)

In[64]:= SetPrecision[f[1.03, 0.98], 6]

Out[64]= 1.00808
```

or using Maple 15:

$$f := (x, y) \rightarrow x^{\frac{1}{2}} \cdot y^{\frac{1}{3}} :$$
Digits := 6 :
$f(1.03, 0.98)$

$$1.00808$$

The approximation error will be:

$$R_2 < \frac{1}{3!}\left[0.03^3 \cdot \frac{3}{8} - 3 \cdot 0.03^2 \cdot 0.02 \cdot \left(-\frac{1}{12}\right)\right.$$

$$\left. + 3 \cdot 0.03 \cdot (-0.02)^2 \cdot \left(-\frac{1}{9}\right) + (-0.02)^3 \cdot \frac{10}{27}\right] = 0.000001277006.$$

We can verify this result in Matlab 7.9:

```
>> x0=1;y0=1;h=0.03;k=-0.02;
>> t=0:0.1:1;
>> f=@(x,y)x^(1/2)*y^(1/3);
>> syms x y
>> f1=diff(f(x,y),x,3)
 f1 =
(3*y^(1/3))/(8*x^(5/2))
>> f4=diff(f(x,y),y,3)
f4 =
(10*x^(1/2))/(27*y^(8/3))
>> f2=simplify(diff(diff(f(x,y),y),x,2))
f2 =
-1/(12*x^(3/2)*y^(2/3))
>> f3=simplify(diff(diff(f(x,y),x),y,2))
f3 =
-1/(9*x^(1/2)*y^(5/3))
>> R2=(1/6)*(h^3*subs(f1,{x,y},{x0+t*h,y0+t*k})+
3*h^2*k*subs(f2,{x,y},{x0+t*h,y0+t*k})+
3*h*k^2*subs(f3,{x,y},{x0+t*h,y0+t*k})+
k^3*subs(f4,{x,y},{x0+t*h,y0+t*k})) ;
>> plot(t,R2)
```

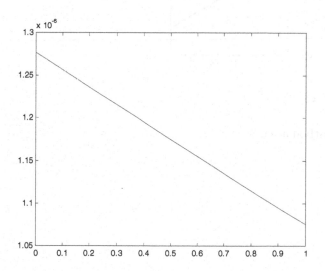

```
>> r2=R2(1)
r2 =
 1.2770e-006
```
or in Mathcad 14:

$$f(x,y) := x^{\frac{1}{2}} \cdot y^{\frac{1}{3}}$$

$x0 := 1$    $y0 := 1$    $h := 0.03$    $k := -0.02$

$$f1(x,y) := \frac{d^3}{dx^3}f(x,y) \qquad\qquad f4(x,y) := \frac{d^3}{dy^3}f(x,y)$$

$$f2(x,y) := \frac{d^2}{dx^2}\frac{d}{dy}f(x,y) \qquad\qquad f3(x,y) := \frac{d^2}{dy^2}\frac{d}{dx}f(x,y)$$

$$R2(t) := \frac{1}{6}\left( h^3 \cdot f1(x0+t \cdot h, y0+t \cdot k) + 3 \cdot h^2 \cdot k \cdot f2(x0+t \cdot h, y0+t \cdot k) + 3 \cdot h \cdot k^2 \cdot f3(x0+t \cdot h, y0+t \cdot k) + k^3 \cdot f4(x0+t \cdot h, y0+t \cdot k) \right)$$

$t := 0, 0.1 .. 1$

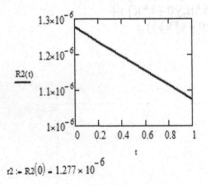

$r2 := R2(0) - 1.277 \times 10^{-6}$

and using Mathematica 8:

In[1]:= x0 = 1; y0 = 1; h = 0.03; k = -0.02; t = Range[0, 1, 0.1];

In[2]:= f[x_, y_] := x^(1/2) * y^(1/3);

In[3]:= f1 = D[f[x, y], {x, 3}]

Out[3]= $\dfrac{3 \, y^{1/3}}{8 \, x^{5/2}}$

In[4]:= f4 = D[f[x, y], {y, 3}]

Out[4]= $\dfrac{10 \sqrt{x}}{27 \, y^{8/3}}$

In[5]:= f2 = Simplify[D[D[f[x, y], y], {x, 2}]]

Out[5]= $-\dfrac{1}{12 \, x^{3/2} \, y^{2/3}}$

In[6]:= f3 = Simplify[D[D[f[x, y], x], {y, 2}]]

Out[6]= $-\dfrac{1}{9 \sqrt{x} \, y^{5/3}}$

In[7]:= ff1 = f1 /. {x → x0 + t*h, y → y0 + t*k}; ff2 = f2 /. {x → x0 + t*h, y → y0 + t*k};

In[8]:= ff3 = f3 /. {x → x0 + t*h, y → y0 + t*k}; ff4 = f4 /. {x → x0 + t*h, y → y0 + t*k};

In[9]:= R2 = 1/6 * (h^3 * ff1 + 3 * h^2 * k * ff2 + 3 * h * k^2 * ff3 + k^3 * ff4);

In[10]:= ListLinePlot[{R2}]

Out[10]=

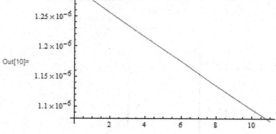

In[11]:= r2 = R2[[1]]

Out[11]= $1.27701 \times 10^{-6}$

and in Maple 15:

$x0 := 1 : y0 := 1 : h := 0.03 : k := -0.02 :$

$f := x^{\frac{1}{2}} \cdot y^{\frac{1}{3}} :$

$f1 := factor\left(simplify\left(\dfrac{\partial^3}{\partial x^3} f\right)\right) :$

$f4 := factor\left(simplify\left(\dfrac{\partial^3}{\partial y^3} f\right)\right) :$

$f2 := factor\left(simplify\left(\dfrac{\partial}{\partial y}\dfrac{\partial^2}{\partial x^2} f\right)\right) :$

$f3 := factor\left(simplify\left(\dfrac{\partial}{\partial x}\dfrac{\partial^2}{\partial y^2} f\right)\right) :$

$ff1 := f1\Big|_{x=x0+t\cdot h,\, y=y0+t\cdot k} :$

$ff2 := f2\Big|_{x=x0+t\cdot h,\, y=y0+t\cdot k} :$

$ff3 := f3\Big|_{x=x0+t\cdot h,\, y=y0+t\cdot k} :$

$ff4 := f4\Big|_{x=x0+t\cdot h,\, y=y0+t\cdot k} :$

$R2 := t \to \dfrac{1}{6}\cdot\left(h^3\cdot ff1 + 3\cdot h^2\cdot k\cdot ff2 + 3\cdot h\cdot k^2\cdot ff3 + k^3\cdot ff4\right) :$

$plot(R2(t), t = 0..1)$

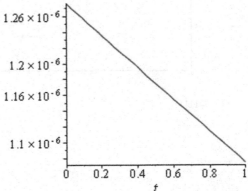

$Digits := 6 :$

$R2(t)\Big|_{t=0}$

0.00000127701

## 3.5  Problems

1. Find the second partial derivatives of the function

$$z = \arctan \frac{x}{y}.$$

2. Find

$$\begin{cases} f_x' \left(1, 2, 0\right) \\ f_y' \left(1, 2, 0\right) \\ f_z' \left(1, 2, 0\right) \end{cases}$$

if

$$f \left(x, y, z\right) = \ln \left(xy + z\right).$$

3. Show that

$$\frac{\partial u}{\partial x} + \frac{\partial u}{\partial y} + \frac{\partial u}{\partial z} = 1$$

if

$$u = x + \frac{x - y}{y - z}.$$

4. Calculate

$$\begin{vmatrix} \frac{\partial x}{\partial r} & \frac{\partial x}{\partial \varphi} \\ \frac{\partial y}{\partial r} & \frac{\partial y}{\partial \varphi} \end{vmatrix}$$

if

$$\begin{cases} x = ar \cos \varphi \\ y = br \sin \varphi. \end{cases}$$

**Computer solution.**
   We shall compute the determinant in Matlab 7.9:

```
>> syms r phi a b
>> x=a*r*cos(phi);
>> y=b*r*sin(phi);
>> F=[x y];
>> v=[r phi];
>> J=jacobian(F,v)
J =
```

```
[ a*cos(phi), -a*r*sin(phi)]
[ b*sin(phi), b*r*cos(phi)]
>> simplify(det(J))
ans =
a*b*r
```

and in Mathcad 14:

$$f(x,a,b) := \begin{pmatrix} a \cdot x_0 \cdot \cos(x_1) \\ b \cdot x_0 \cdot \sin(x_1) \end{pmatrix}$$

$$|\text{Jacob}(f(x,a,b),x)| \text{ simplify} \rightarrow a \cdot b \cdot x_0$$

and using Mathematica 8:

```
In[4]:= Det[D[{a * r * Cos[φ], b * r * Sin[φ]}, {{r, φ}}]];

In[5]:= Simplify[%]

Out[5]= a b r
```

and with Maple 15:

*with*(*VectorCalculus*) :
$M, d := Jacobian([a \cdot r \cdot \cos(\varphi), b \cdot r \cdot \sin(\varphi)], [r, \varphi], 'determinant')$ :
*simplify*(*d, trig*)

$$a b r$$

5. Find

$$\begin{cases} df(1,2) \\ d^2 f(1,2) \end{cases}$$

if

$$f(x,y) = x^2 + xy + y^2 - 4\ln x - 10\ln y.$$

**Computer solution.**
We have found that

$$df(1,2) = 0$$

$$d^2 f(1,2) = 6dx^2 + 4dxdy + \frac{9}{2}dy^2$$

with Matlab 7.9:

```
>> f=@(x,y) x^2+x*y+y^2-4*log(x)-10*log(y)
f =
 @(x,y)x^2+x*y+y^2-4*log(x)-10*log(y)
>> syms x y dx dy dx2 dy2 dxdy
>> u=diff(f,x)*dx+diff(f,y)*dy
u =
dx*(2*x + y - 4/x) + dy*(x + 2*y - 10/y)
>> subs(u,{x,y},{1,2})
ans =
 0
>> w=diff(f,x,2)*dx2+2*diff(f,x,y)*dxdy+diff(f,y,2)*dy2
w =
dxdy*(2*x + 4*y - 20/y) + dx2*(4/x^2 + 2) + dy2*(10/y^2
+ 2)
>> subs(w,{x,y},{1,2})
ans =
6*dx2 + (9*dy2)/2
```

or using Mathcad 14:

$$f(x,y) := x^2 + x \cdot y + y^2 - 4 \cdot \ln(x) - 10 \cdot \ln(y)$$

$$w(dx, dy) := \frac{\partial}{\partial x} f(x,y) \cdot dx + \frac{\partial}{\partial y} f(x,y) \cdot dy \text{ substitute}, x = 1, y = 2 \rightarrow 0$$

$$\frac{\partial^2}{\partial x^2} f(x,y) \cdot dx^2 + 2 \cdot \frac{\partial}{\partial x} \frac{\partial}{\partial y} f(x,y) \cdot dxdy + \frac{\partial^2}{\partial y^2} f(x,y) \cdot dy^2 \text{ substitute}, x = 1, y = 2 \rightarrow 6 \cdot dx^2 + \frac{9 \cdot dy^2}{2} + 2 \cdot dxdy$$

or with Mathematica 8:

```
In[296]:= u := Dt[x^2 + x * y + y^2 - 4 * Log[x] - 10 * Log[y]];

In[297]:= u /. Dt[x] -> dx;

In[298]:= % /. Dt[y] -> dy;

In[299]:= % /. x → 1;

In[300]:= % /. y → 2

Out[300]= 0

In[301]:= Simplify[Dt[u]];

In[302]:= % /. Dt[x] -> dx;

In[303]:= % /. Dt[y] -> dy;

In[304]:= % /. Dt[dx] → 0;

In[305]:= % /. Dt[dy] → 0;

In[306]:= % /. x → 1;

In[307]:= % /. y → 2
```

$$\text{Out[307]= } 6 \, dx^2 + 2 \, dx \, dy + \frac{9 \, dy^2}{2}$$

or in Maple 15:

$f := (x, y) \rightarrow x^2 + x \cdot y + y^2 - 4 \cdot \ln(x) - 10 \cdot \ln(y) :$

$u := \dfrac{\partial}{\partial x} f(x, y) dx + \dfrac{\partial}{\partial y} f(x, y) dy :$

$u \Big|_{x=1, y=2}$

$$0$$

$w := \dfrac{\partial^2}{\partial x^2} f(x, y) dx^2 + 2 \cdot \dfrac{\partial^2}{\partial x \partial y} f(x, y) dx dy + \dfrac{\partial^2}{\partial y^2} f(x, y) dy^2 :$

$w \Big|_{x=1, y=2}$

$$6 \, dx^2 + 2 \, dx dy + \frac{9}{2} \, dy^2$$

6. Compute the function $df$ and $d^2 f$ for the function:

$$f(x, y, z) = \cos(x + 2y + 3z), \quad (\forall) \ (x, y, z) \in \mathbb{R}^3.$$

7. Prove that the function

$$f(x, y, z) = \frac{1}{\sqrt{x^2 + y^2 + z^2}}, \quad (x, y, z) \neq (0, 0, 0)$$

satisfies the Lapace equation:

$$\Delta \equiv f''_{x^2} + f''_{y^2} + f''_{z^2} = 0.$$

8. Find the expression of

$$E = 1 + y'^2$$

in the polar coordinates:

$$\begin{cases} x = r \cos \theta \\ y = r \sin \theta. \end{cases}$$

**Computer solution.**
We shall compute the expression $E$ in Matlab 7.9:

```
>> x=@(r,th)r*cos(th);
>> y=@(r,th)r*sin(th);
>> syms r th dr dth
>>yp=(diff(y(r,th),r)*dr+diff(y(r,th),th)*dth)/(diff(x(r,th),r)
   *dr+diff(x(r,th),th)*dth);
```

>> syms rp
>> simplify(1+subs(yp,dr,rp*dth)^2)
ans =
(r^2 + rp^2)/(rp*cos(th) - r*sin(th))^2
and in Mathcad 14:

$x(r, \theta) := r \cdot \cos(\theta)$

$y(r, \theta) := r \cdot \sin(\theta)$

$$yp(r, \theta, dr, d\theta) := \frac{\dfrac{d}{dr} y(r, \theta) \cdot dr + \dfrac{d}{d\theta} y(r, \theta) \cdot d\theta}{\dfrac{d}{dr} x(r, \theta) \cdot dr + \dfrac{d}{d\theta} x(r, \theta) \cdot d\theta}$$

$$m(r, rp, \theta) := yp(r, \theta, dr, d\theta) \text{ substitute, } dr = rp \cdot d\theta \;\rightarrow\; \frac{r \cdot \cos(\theta) + rp \cdot \sin(\theta)}{rp \cdot \cos(\theta) - r \cdot \sin(\theta)}$$

$$1 + m(r, rp, \theta)^2 \text{ factor} \;\rightarrow\; \frac{\left(r^2 + rp^2\right) \cdot \left(\cos(\theta)^2 + \sin(\theta)^2\right)}{(rp \cdot \cos(\theta) - r \cdot \sin(\theta))^2} \text{ simplify} \;\rightarrow\; \frac{r^2 + rp^2}{(rp \cdot \cos(\theta) - r \cdot \sin(\theta))^2}$$

and with Mathematica 8:

In[56]:= $x[r\_, \theta\_] := r * \text{Cos}[\theta]$

In[57]:= $y[r\_, \theta\_] := r * \text{Sin}[\theta]$

In[58]:= $u := (D[y[r, \theta], r] * dr + D[y[r, \theta], \theta] * d\theta) / (D[x[r, \theta], r] * dr + D[x[r, \theta], \theta] * d\theta)$

In[59]:= $u /. dr \rightarrow R * d\theta;$

In[60]:= $v = \text{Simplify}[\%];$

In[61]:= $\text{Simplify}[1 + v^2]$

Out[61]= $\dfrac{r^2 + R^2}{(R \cos[\theta] - r \sin[\theta])^2}$

and in Maple 15:

$x := (r, \theta) \rightarrow r \cdot \cos(\theta) :$
$y := (r, \theta) \rightarrow r \cdot \sin(\theta) :$

$$v := \left. \frac{\dfrac{\partial}{\partial r} y(r, \theta) dr + \dfrac{\partial}{\partial \theta} y(r, \theta) d\theta}{\dfrac{\partial}{\partial r} x(r, \theta) dr + \dfrac{\partial}{\partial \theta} x(r, \theta) d\theta} \right|_{dr = R \cdot d\theta} :$$

$u := factor(1 + v^2)$

$$\frac{(r^2 + R^2)\left(\cos(\theta)^2 + \sin(\theta)^2\right)}{\left(\cos(\theta) R - r \sin(\theta)\right)^2}$$

9. Prove that the function

$$f(x, y) = \frac{x^3 + y^3}{x^2 + y^2}, \quad (x, y) \neq (0, 0)$$

is an homogeneous function and check the Euler's relation for it.
**Computer solution.**
We shall solve this problem in Matlab 7.9:
function r = f(x,y)
r=(x^3+y^3)/(x^2+y^2);
end
One saves the file with f.m then in the command line one writes:
>> syms x y t m
>> l=f(x,y);
>> u=simplify(f(x*t,y*t));
>> r=u/l;
>> m=eval(log(r)/log(t))
m =
 1
We verify the Euler's relation from (3.13).
>> w=simplify(x*diff(f(x,y),x)+y*diff(f(x,y),y));
>> w==m*f(x,y)
ans =
 1
in Mathcad 14:

$$f(x,y) := \frac{x^3 + y^3}{x^2 + y^2} \qquad\qquad r(x,y,t) := \frac{f(x \cdot t, y \cdot t)}{f(x,y)}$$

$$\underset{\mathrm{\sim}}{m} := t^m = r(x,y,t) \; \text{solve}, m \;\rightarrow\; 1$$

$$w(x,y) := x \cdot \frac{d}{dx} f(x,y) + y \cdot \frac{d}{dy} f(x,y) \; \text{simplify} \;\rightarrow\; \frac{x^3 + y^3}{x^2 + y^2}$$

$$w(x,y) = m \cdot f(x,y) \rightarrow 1$$

with Mathematica 8:

```
In[117]:= f[x_, y_] := (x^3 + y^3) / (x^2 + y^2)

In[118]:= u = Simplify[f[x * t, y * t]];

In[119]:= l := f[x, y]

In[120]:= r := u / l

In[121]:= m = Log[r] / Log[t]

Out[121]= 1

In[122]:= w := Simplify[x * D[f[x, y], x] + y * D[f[x, y], y]]

In[123]:= w == m * f[x, y]

Out[123]= True
```

using Maple 15:

$$f := (x, y) \rightarrow \frac{x^3 + y^3}{x^2 + y^2}:$$

$$l := f(x, y):$$

$$u := simplify(f(x \cdot t, y \cdot t)):$$

$$r := simplify\left(\frac{u}{l}\right):$$

$$m := \frac{\ln(r)}{\ln(t)}$$

1

$$w := simplify(x \cdot diff(f(x, y), x) + y \cdot diff(f(x, y), y)):$$

$$evalb(w = m \cdot f(x, y))$$

true

10. Prove that the function

$$f(x, y, z) = \frac{ax + by + cz}{\sqrt{x^2 + y^2 + z^2}}, \quad (\forall) \ (x, y, z) \in \mathbb{R}^3 \backslash \{(0, 0, 0)\}$$

checks the relation

$$x\frac{\partial f}{\partial x} + y\frac{\partial f}{\partial y} + z\frac{\partial f}{\partial z} = 0.$$

11. Write the Taylor's formula for the function

$$f(x, y) = e^{x+y}$$

around the point $(-1, 1) \in \mathbb{R}^2$.
   **Computer solution.**
   We shall use Mathcad 14:

$$e^{x+y} \ series, x = -1, y = 1, 3 \ \rightarrow 1 + y + \frac{(y-1)^2}{2} + x + (x + 1) \cdot (y - 1) + \frac{(x+1)^2}{2}$$

and Maple 15:

$$mtaylor(\exp(x + y), [x = -1, y = 1], 3)$$

$$1 + x + y + \frac{1}{2}(1 + x)^2 + (y - 1)(1 + x) + \frac{1}{2}(y - 1)^2$$

12.  Find $\frac{\partial z}{\partial t}$ of

$$z = e^{3x+2y},$$

where

$$\begin{cases} x = \cos t \\ y = t^2. \end{cases}$$

13.  Show that the function

$$z = \varphi\left(x^2 + y^2\right)$$

satisfies the equation

$$y\frac{\partial z}{\partial x} - x\frac{\partial z}{\partial y} = 0.$$

14.  What do we obtain from

$$x^2\frac{\partial^2 z}{\partial x^2} + 2xy\frac{\partial^2 z}{\partial x \partial y} + y^2\frac{\partial^2 z}{\partial y^2} = 0$$

if one makes the change of independent variables

$$\begin{cases} x = u \\ y = uv \ ? \end{cases}$$

15.  Show that the function

$$z = x\varphi\left(\frac{y}{x}\right) + \psi\left(\frac{y}{x}\right)$$

satisfies the equation

$$x^2\frac{\partial^2 z}{\partial x^2} + 2xy\frac{\partial^2 z}{\partial x \partial y} + y^2\frac{\partial^2 z}{\partial y^2} = 0.$$

**Computer solution.**
We shall check the relation using Mathcad 14:

$$z(x,y,\varphi,\psi) := x \cdot \varphi\left(\frac{y}{x}\right) + \psi\left(\frac{y}{x}\right)$$

$$w1(x,y,\varphi,\psi) := x^2 \cdot \frac{d^2}{dx^2}z(x,y,\varphi,\psi) + 2 \cdot xy \cdot \frac{d}{dx}\frac{d}{dy}z(x,y,\varphi,\psi) + y^2 \cdot \frac{d^2}{dy^2}z(x,y,\varphi,\psi)$$

$$w1(x,y,\varphi,\psi) \text{ expand } \rightarrow 0$$

and in Mathematica 8:

In[27]:= z[x_, y_] := x*φ (y/x) + ψ (y/x)

In[28]:= Expand[x^2*D[z[x, y], {x, 2}] + 2*x*y*D[z[x, y], x, y] + y^2*D[z[x, y], {y, 2}]]

Out[28]= 0

and with Maple 15:

$$z := (x, y) \rightarrow x \cdot \varphi\left(\frac{y}{x}\right) + \psi\left(\frac{y}{x}\right)$$

$$(x, y) \rightarrow x \varphi\left(\frac{y}{x}\right) + \psi\left(\frac{y}{x}\right)$$

$$evala\left( Expand\left( x^2 \cdot \frac{\partial^2}{\partial x^2} z(x, y) + 2 \cdot x \cdot y \cdot \frac{\partial^2}{\partial x \partial y} z(x, y) + y^2 \cdot \frac{\partial^2}{\partial y^2} z(x, y) \right) \right)$$

$$0$$

16. Write the first four terms of a power series expansion in $x$ and $y$ of the function

$$f(x, y) = (1 + x)^{1+y}.$$

**Computer solution.**
We shall find the first four terms of a power series expansion using Mathcad 14:

$$(1 + x)^{1+y} \ series, x, y, 4 \rightarrow 1 + x + x \cdot y + \frac{x^2 \cdot y}{2}$$

and with Maple 15:

$$mtaylor\left((1 + x)^{1+y}, [x, y], 4\right)$$

$$1 + x + yx + \frac{1}{2}yx^2$$

17. Show that the function

$$u\left(x,t\right) = A\sin\left(a\lambda t + \varphi\right)\sin\lambda x$$

satisfies the equation of oscillations of a string

$$\frac{\partial^2 u}{\partial t^2} = a^2 \frac{\partial^2 u}{\partial x^2}.$$

**Computer solution.**
We shall prove that equality in Mathcad 14:

$$u(x,t,A,a,\varphi,\lambda) := A\cdot\sin(a\cdot\lambda\cdot t + \varphi)\cdot\sin(\lambda\cdot x)$$

$$\frac{d^2}{dt^2}u(x,t,A,a,\varphi,\lambda) - a^2\cdot\frac{d^2}{dx^2}u(x,t,A,a,\varphi,\lambda) \text{ expand } \rightarrow 0$$

and using Mathematica 8:

In[35]:= $\mathbf{u[x\_,\ t\_]\ :=\ A*Sin[a*\lambda*t+\varphi]*Sin[\lambda*x]}$

In[36]:= $\mathbf{Expand[D[u[x,\ t],\ \{t,\ 2\}]\ -\ a\char`\^2*D[u[x,\ t],\ \{x,\ 2\}]]}$

Out[36]= 0

and with Maple 15:

$$u := (x,t) \rightarrow A\cdot\sin(a\cdot\lambda\cdot t + \varphi)\cdot\sin(\lambda\cdot x)$$

$$(x,t)\rightarrow A\sin(a\lambda t + \varphi)\sin(\lambda x)$$

$$evala\left(Expand\left(\frac{\partial^2}{\partial t^2}u(x,t)-a^2\cdot\frac{\partial^2}{\partial x^2}u(x,t)\right)\right)$$

$$0$$

18.  Find the partial derivative $\frac{\partial z}{\partial x}$ if

$$z = \varphi\left(x^2 + y^2\right).$$

**Computer solution.**
We shall find that partial derivative in Mathcad 14:

$$z(x,y,\varphi) := \varphi\!\left(x^2 + y^2\right)$$

$$\frac{d}{dx}z(x,y,\varphi) \to 2 \cdot x \cdot \left| x0 \leftarrow x^2 + y^2 \right.$$

$$\left.\frac{d}{dx0}\varphi(x0)\right.$$

and with Mathematica 8:

In[65]:= $\mathbf{z[x\_, y\_] := \varphi[x\hat{}2 + y\hat{}2]}$

In[66]:= $\mathbf{D[z[x, y], x]}$

Out[66]= $2 \times \varphi'\left[x^2 + y^2\right]$

and using Maple 15:

$z := (x,y) \to \varphi\!\left(x^2 + y^2\right):$
$\mathit{diff}(z(x,y),x)$

$$2\,\mathrm{D}(\varphi)\left(x^2 + y^2\right)x$$

19. Take the equation of the oscillations of a string

$$\frac{\partial^2 u}{\partial t^2} = a^2\frac{\partial^2 u}{\partial x^2}, \ a \neq 0$$

and change it to the new independent variables $a$ and $b$, where

$$\begin{cases} \alpha = x - at \\ \beta = x + at. \end{cases}$$

20. Transform the following equation to new independent variables $u$ and $v$ :

$$x\frac{\partial z}{\partial x} + y\frac{\partial z}{\partial y} - z = 0,$$

if

$$\begin{cases} u = x \\ v = \frac{y}{x}. \end{cases}$$

## Computer solution.

We shall see what becomes the left member of the initial equation using Mathematica 8:

In[27]:= z[x_, y_] := r[x, y]

In[28]:= u[x_, y_] := x;

In[29]:= v[x_, y_] := y/x;

In[30]:= a = Expand[x*D[z[u[x, y], v[x, y]], x] + y*D[z[u[x, y], v[x, y]], y] -
        z[u[x, y], v[x, y]]];

In[31]:= a /. x → u;

In[32]:= % /. y/x → v

Out[32]= $-r\left[u, \dfrac{y}{u}\right] + u\, r^{(1,0)}\left[u, \dfrac{y}{u}\right]$

and Maple 15:

$z := (x, y) \rightarrow r(x, y) :$
$u := (x, y) \rightarrow x :$
$v := (x, y) \rightarrow \dfrac{y}{x} :$
$b := simplify\left(x \cdot \dfrac{\partial}{\partial x} z(u(x, y), v(x, y)) + y \cdot \dfrac{\partial}{\partial y} z(u(x, y), v(x, y)) - z(u(x, y), v(x, y))\right) :$
$b\Big|_{x = u,\, \frac{y}{x} = v}$

$$D_1(r)(u, v)\, u - r(u, v)$$

# 4
# Fundamentals of Field Theory

## 4.1 Derivative in a Given Direction of a Function

**Definition 4.1** (see [8]). The **derivative** of a differentiable function $z = f(x, y)$ **in the direction** $\vec{s} = \overrightarrow{PP_1}$ is

$$\frac{\partial z}{\partial \vec{s}} = \frac{\partial z}{\partial x} \cos \alpha + \frac{\partial z}{\partial y} \cos \beta, \tag{4.1}$$

where $\alpha$ is the angle formed by the vector $\vec{s}$ with the $Ox$- axis and $\beta$ is the angle formed by the vector $\vec{s}$ with the $Oy$ -axis (see Fig. 4.1).

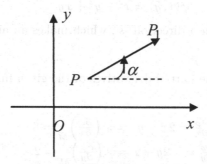

**Fig. 4.1**

In a similar fashion, we define the derivative in a given direction $\vec{s}$ for a function of three variables $u = f(x, y, z)$. In this case,

G.A. Anastassiou and I.F. Iatan: Intelligent Routines, ISRL 39, pp. 157–186.
springerlink.com                  © Springer-Verlag Berlin Heidelberg 2013

$$\frac{\partial u}{\partial \vec{s}} = \frac{\partial u}{\partial x}\cos\alpha + \frac{\partial u}{\partial y}\cos\beta + \frac{\partial u}{\partial z}\cos\gamma, \qquad (4.2)$$

where $\cos\alpha, \cos\beta, \cos\gamma$ are the direction cosines of the direction $\vec{s}$. The directional derivative characterizes the rate of change of the function in the given direction.

**Definition 4.2** (see [44], p. 591). A scalar function defined for any point of the space is called a **scalar field** if it makes a correspondence in which to each point $P(x, y, z)$ respectively to each position vector $\vec{r}$ from a given domain is assigned a scalar $\varphi(x, y, z) = \varphi(\vec{r})$; for example, the temperature and the density of a body are some scalar fields.

**Remark 4.3** (see [44], p. 591). In the vector analysis, the scalar fields associate to each point in space a scalar. Scalar fields can be represented by the level surfaces $\varphi(x, y, z) = const$ or the level straight lines $\varphi(x, y) = const$. For example, on maps the straight lines of equal temperature (the isotherms) are some level straight lines.

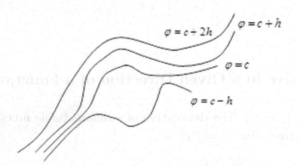

**Fig. 4.2** A scalar field

**Example 4.4.** Find the derivative of the scalar field

$$\varphi(x, y) = x^2 - y^2 + xy$$

at the point $M(2, 2)$ in a direction $\vec{s}$, which makes an angle of $30°$ with the $Ox$- axis.

**Solution.**

We shall compute the partial derivatives of the given function and their values at the point $M$:

$$\frac{\partial\varphi}{\partial x} = 2x + y \implies \left(\frac{\partial\varphi}{\partial x}\right)_M = 6$$
$$\frac{\partial\varphi}{\partial y} = -2y + x \implies \left(\frac{\partial\varphi}{\partial y}\right)_M = 2.$$

Here

$$\begin{cases} \cos\alpha = \cos 30° = \frac{\sqrt{3}}{2} \\ \cos\beta = \cos\left(\frac{\pi}{2} - \alpha\right) = \sin\alpha = \frac{1}{2}, \end{cases}$$

therefore

$$\frac{\partial \varphi}{\partial \vec{s}}(M) = 3\sqrt{3} - 1 = 4.196.$$

We shall achieve this result using Matlab 7.9:

```
>> f=@(x,y)x^2-y^2+x*y;
>> c1=cos(degtorad(30));
>> c2=cos(degtorad(60));
>> syms x y
>> subs(diff(f(x,y),x),{x,y},{2,2})*c1+subs(diff(f(x,y),y),
   {x,y},{2,2})*c2
ans =
 4.1962
```

and in Mathcad 14:

$$\varphi(x,y) := x^2 - y^2 + x \cdot y$$

$$x0 := 2 \quad y0 := 2$$

$$\frac{\partial}{\partial x0} \varphi(x0,y0) \cdot \cos(30 \cdot deg) + \frac{\partial}{\partial y0} \varphi(x0,y0) \cdot \cos(60 \cdot deg) = 4.196$$

and in Mathematica 8:

```
In[10]:= φ[x_, y_] := x^2 - y^2 + x*y

In[11]:= (D[φ[x, y], x] /. {x → 2, y → 2}) *Cos[30 Degree] +
         (D[φ[x, y], y] /. {x → 2, y → 2}) *Cos[60 Degree]

Out[11]= -1 + 3 √3
```

and with Maple 15:

```
with(Student[MultivariateCalculus]):
c1 := cos(convert(30 degrees, radians)):
c2 := cos(convert(60 degrees, radians)):
DirectionalDerivative(x²-y² + x·y, [x, y] = [2, 2], [c1, c2])
```

$$3\sqrt{3} - 1$$

**Example 4.5.** Let

$$\varphi(x, y, z) = \arcsin \frac{z}{\sqrt{x^2 + y^2}}$$

be a scalar field.

Find the derivative of the function $\varphi$ at the point $M(1,1,1)$ in the direction $\overrightarrow{MN}$ knowing that $N(2,3,-2)$.

**Solution.**

As

$$\overrightarrow{MN} = \overrightarrow{i} + 2\overrightarrow{j} - 3\overrightarrow{k}$$

one finds that

$$\frac{\overrightarrow{MN}}{\left\|\overrightarrow{MN}\right\|} = \frac{\overrightarrow{i} + 2\overrightarrow{j} - 3\overrightarrow{k}}{\sqrt{1^2 + 2^2 + 3^2}} = \frac{1}{\sqrt{14}}\overrightarrow{i} + \frac{2}{\sqrt{14}}\overrightarrow{j} - \frac{3}{\sqrt{14}}\overrightarrow{k};$$

therefore

$$\frac{\partial\varphi}{\partial\overrightarrow{s}}(M) = \frac{\partial\varphi}{\partial x}(1,1,1)\cdot\frac{1}{\sqrt{14}} + \frac{\partial\varphi}{\partial y}(1,1,1)\cdot\frac{2}{\sqrt{14}} + \frac{\partial\varphi}{\partial z}(1,1,1)\cdot\left(-\frac{3}{\sqrt{14}}\right).$$

Computing the partial derivatives and their values at the point $M$

$$\frac{\partial\varphi}{\partial x} = \frac{-\frac{1}{2}\cdot 2xz\cdot\left(x^2+y^2\right)^{-\frac{3}{2}}}{\sqrt{1-\frac{z^2}{x^2+y^2}}} = \frac{xz}{(x^2+y^2)\sqrt{x^2+y^2-z^2}} \implies \frac{\partial\varphi}{\partial x}(M) = -\frac{1}{2}$$

$$\frac{\partial\varphi}{\partial y} = \frac{-\frac{1}{2}\cdot 2yz\cdot\left(x^2+y^2\right)^{-\frac{3}{2}}}{\sqrt{1-\frac{z^2}{x^2+y^2}}} = \frac{yz}{(x^2+y^2)\sqrt{x^2+y^2-z^2}} \implies \frac{\partial\varphi}{\partial y}(M) = -\frac{1}{2}$$

$$\frac{\partial\varphi}{\partial z} = \frac{\sqrt{x^2+y^2}}{\sqrt{1-\frac{z^2}{x^2+y^2}}} = \frac{1}{\sqrt{x^2+y^2-z^2}} \implies \frac{\partial\varphi}{\partial z}(M) = 1;$$

one obtains:

$$\frac{\partial\varphi}{\partial\overrightarrow{s}}(M) = -\frac{1}{2}\cdot\frac{1}{\sqrt{14}} - \frac{1}{2}\cdot\frac{2}{\sqrt{14}} - \frac{3}{\sqrt{14}} = -\frac{9}{2\sqrt{14}} = -1.2027.$$

We need to pass the following steps in order to achieve this result in Matlab 7.9:

*Step 1.* Determine the analytical expression of the direction $\overrightarrow{s} = \overrightarrow{MN}$:

$$\overrightarrow{MN} = (x_N - x_M)\overrightarrow{i} + (y_N - y_M)\overrightarrow{j} + (z_N - z_M)\overrightarrow{k}.$$

>> **M=[1 1 1];**
>> **N=[2 3 -2];**
>> **s=N-M**
s =
  1 2 -3

*Step 2.* Determine the direction cosines of the direction $\overrightarrow{s}$.

```
>> w=norm(s);
>>u=s/w
u=
 0.2673 0.5345 -0.8018
```

*Step 3.* Determine the derivative of $\varphi$ at the point M in the direction $\vec{s}$.

```
>> syms x y z
>> phi=asin(z/sqrt(x^2+y^2));
>> d1=diff(phi,x);
>> g=subs(subs(subs(d1,x,1),y,1),z,1);
>> d2=diff(phi,y);
>> h=subs(subs(subs(d2,x,1),y,1),z,1);
>> d3=diff(phi,z);
>> k=subs(subs(subs(d3,x,1),y,1),z,1)
>> d=dot(u,[g h k])
d =
 -1.2027
```

We shall obtain this result using Mathcad 14:

$$\varphi(x,y,z) := asin\left(\frac{z}{\sqrt{x^2+y^2}}\right) \qquad M := \begin{pmatrix} 1 \\ 1 \\ 1 \end{pmatrix} \qquad N := \begin{pmatrix} 2 \\ 3 \\ -2 \end{pmatrix}$$

$$x0 := 1 \qquad y0 := 1 \qquad z0 := 1 \qquad u := \frac{N - M}{|N - M|}$$

$$v := \begin{pmatrix} \dfrac{\partial}{\partial x0}\,\varphi(x0,y0,z0) \\[2ex] \dfrac{\partial}{\partial y0}\,\varphi(x0,y0,z0) \\[2ex] \dfrac{\partial}{\partial z0}\,\varphi(x0,y0,z0) \end{pmatrix} \qquad\qquad u \cdot v = -1.2027$$

and Mathematica 8:

```
In[63]:= m := {1, 1, 1};
```

```
In[64]:= n := {2, 3, -2};
```

```
In[65]:= s := n - m
```

```
In[66]:= u := s / Norm[s]
```

```
In[67]:= φ[x_, y_, z_] := ArcSin[z / Sqrt[x^2 + y^2]];
```

```
In[68]:= w := {D[φ[x, y, z], x] /. {x → 1, y → 1, z → 1},
          D[φ[x, y, z], y] /. {x → 1, y → 1, z → 1}, D[φ[x, y, z], z] /. {x → 1, y → 1, z → 1}}
```

```
In[69]:= SetPrecision[u.w, 5]
```

```
Out[69]= -1.2027
```

and in Maple 15:

$M := \langle 1, 1, 1 \rangle :$
$N := \langle 2, 3, -2 \rangle :$
$with(\, VectorCalculus) :$
$SetCoordinates(\, cartesian_{x,y,z}) :$
$$u := \frac{N - M}{Norm(N - M, 2)} :$$
$with(\, Student[\, MultivariateCalculus]) :$
$Digits := 5 :$

$$evalf\left( DirectionalDerivative\left( \arcsin\left( \frac{z}{\sqrt{x^2 + y^2}} \right), [x, y, z] = [1, 1, 1], [u_1, u_2, u_3] \right) \right)$$

$$-1.2027$$

## 4.2   Differential Operators

**Definition 4.6** (see [44], p. 591). If through the function $v = v(\vec{r})$, respectively $v = v(x, y, z)$ is assigned to each point of the space a vector $v = v(x, y, z) = v(\vec{r})$, then the function $v = v(\vec{r})$ defines a **vector field**.

For example, the force fields $F = F(\vec{r})$ or the electricity fields $E(\vec{r})$ are some vector fields.

They are plotted by arrows taken at different points $\vec{r}$ of the space whose vector directions and lengths represents the vector $v(\vec{r})$(see Fig. 4.3).

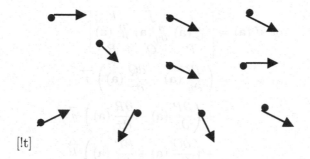

**Fig. 4.3** A vector field

**Definition 4.7** (see [44], p. 592). The vector

$$\operatorname{grad} \varphi\left(P\right) = \frac{\partial \varphi}{\partial x} \cdot \overrightarrow{i} + \frac{\partial \varphi}{\partial y} \cdot \overrightarrow{j} + \frac{\partial \varphi}{\partial z} \cdot \overrightarrow{k} \equiv \nabla u, \qquad (4.3)$$

where

$$\nabla = \overrightarrow{i} \frac{\partial}{\partial x} + \overrightarrow{j} \frac{\partial}{\partial y} + \overrightarrow{k} \frac{\partial}{\partial z} \qquad (4.4)$$

is the Hamiltonian operator (del or nabla), is called the **gradient** of the field

$$u = \varphi\left(P\right) = \varphi\left(x, y, z\right),$$

at the given point $P$.

**Remark 4.8** (see [44], p. 593). By introducing the gradient, a scalar field $\varphi = \varphi\left(\overrightarrow{r}\right)$ was obtained from a vector field $\operatorname{grad} \varphi$. However, the converse is not true, i.e. a vector field can not always be considered the gradient of a scalar field. The vector fields $v\left(\overrightarrow{r}\right)$ for which this is possible are called conservative or potential and $\varphi$ is the potential of the field $v$.

**Definition 4.9** (see [44], p. 594). The **divergence** (div) of a vector field

$$\overrightarrow{v} = P\left(x, y, z\right) \overrightarrow{i} + Q\left(x, y, z\right) \overrightarrow{j} + R\left(x, y, z\right) \overrightarrow{k}$$

in a point $\mathbf{a}\left(a_x, a_y, a_z\right)$ from $\mathbb{R}^3$ is the scalar

$$\operatorname{div} \overrightarrow{v}\left(\mathbf{a}\right) = \frac{\partial P}{\partial x}\left(\mathbf{a}\right) + \frac{\partial Q}{\partial y}\left(\mathbf{a}\right) + \frac{\partial R}{\partial z}\left(\mathbf{a}\right) = \nabla \mathbf{a}, \qquad (4.5)$$

i.e. the divergence is a scalar field, which is derived from a vector field.

**Definition 4.10** (see [2], p. 8). The **rotation** (curl) of a vector field

$$\overrightarrow{v} = P\left(x, y, z\right) \overrightarrow{i} + Q\left(x, y, z\right) \overrightarrow{j} + R\left(x, y, z\right) \overrightarrow{k}$$

in a point $\mathbf{a}\left(a_x, a_y, a_z\right)$ from $\mathbb{R}^3$ is the scalar

$$\text{rot } \vec{v}\,(\mathbf{a}) = \begin{vmatrix} \vec{i} & \vec{j} & \vec{k} \\ \frac{\partial}{\partial x}\,(\mathbf{a}) & \frac{\partial}{\partial y}\,(\mathbf{a}) & \frac{\partial}{\partial z}\,(\mathbf{a}) \\ P & Q & R \end{vmatrix} \tag{4.6}$$

$$= \left( \frac{\partial R}{\partial y}\,(\mathbf{a}) - \frac{\partial Q}{\partial z}\,(\mathbf{a}) \right) \vec{i}$$

$$+ \left( \frac{\partial P}{\partial z}\,(\mathbf{a}) - \frac{\partial R}{\partial x}\,(\mathbf{a}) \right) \vec{j}$$

$$+ \left( \frac{\partial Q}{\partial x}\,(\mathbf{a}) - \frac{\partial P}{\partial y}\,(\mathbf{a}) \right) \vec{k}.$$

**Proposition 4.11** (see [44], p. 596). The gradient, the divergence and the curl of a field are some notions which are independent of the chosen coordinate system.

From this reason, it is said that these quantities are invariant under a change of coordinates.

**Definition 4.12** (see [8]). The vector field $\vec{v}$ is called a **potential** if

$$\vec{v} = \text{grad}\,\varphi, \tag{4.7}$$

where $\varphi$ is a scalar function (the **potential** of the field).

**Proposition 4.13** (see [8] and [2], p. 11). For the potential property of a field

$$\vec{v} = P\,(x, y, z)\,\vec{i} + Q\,(x, y, z)\,\vec{j} + R\,(x, y, z)\,\vec{k},$$

given in a simply-connected domain, it is necessary and sufficient that to be *non-rotational*, i.e., $\text{rot } \vec{v} = \vec{0}$. Then, there exists a potential $\varphi$ defined by

$$d\varphi = P\,(x, y, z)\,dx + Q\,(x, y, z)\,dy + R\,(x, y, z)\,dz. \tag{4.8}$$

**Example 4.14.** Build a vector field of the gradient of the following scalar fields:

$$\text{a) } \varphi\,(x, y, z) = x^4 + xy^2 + y^3 + x^2 z + 1$$
$$\text{b) } \varphi\,(x, y, z) = xyz \cdot e^{x^2 + y^2 + z^2}.$$

**Solutions.**
a) One obtains:

$$\text{grad}\,\varphi = \left( 4x^3 + y^2 + 2xz \right) \vec{i} + \left( 2xy + 3y^2 \right) \vec{j} + x^2 \vec{k}.$$

In Matlab 7.9 we shall have:
```
>> syms x y z
>> f=x^4+x*y^2+y^3+x^2*z+1 ;
>>v=[x y z] ;
>> jacobian(f,v)
 ans =
 [ 4*x^3+y^2+2*x*z, 2*x*y+3*y^2, x^2]
```
We shall compute this gradient using Mathcad 14:

$$\varphi(x,y,z) := x^4 + x \cdot y^2 + y^3 + x^2 \cdot z + 1$$

$$\begin{pmatrix} \dfrac{\partial}{\partial x} \varphi(x,y,z) \\[2mm] \dfrac{\partial}{\partial y} \varphi(x,y,z) \\[2mm] \dfrac{\partial}{\partial z} \varphi(x,y,z) \end{pmatrix} \rightarrow \begin{pmatrix} 4 \cdot x^3 + 2 \cdot z \cdot x + y^2 \\[2mm] 3 \cdot y^2 + 2 \cdot x \cdot y \\[2mm] x^2 \end{pmatrix}$$

and with Mathematica 8:

In[30]:= **Needs["VectorAnalysis`"]**

In[31]:= φ = **Xx^4 + Xx * Yy^2 + Yy^3 + Xx^2 * Zz + 1;**

In[32]:= **Grad[φ]**

Out[32]= $\left\{ 4\,Xx^3 + Yy^2 + 2\,Xx\,Zz,\ 2\,Xx\,Yy + 3\,Yy^2,\ Xx^2 \right\}$

and in Maple 15:

*with(VectorCalculus) :*
$g := Gradient(x^4 + x \cdot y^2 + y^3 + x^2 \cdot z + 1, [x, y, z]) :$
$v := g_1, g_2, g_3$

$$4x^3 + y^2 + 2xz,\ 2xy + 3y^2,\ x^2$$

b) It will result that:

$$\text{grad}\,\varphi = \left(yze^{x^2+y^2+z^2} + 2x \cdot xyz \cdot e^{x^2+y^2+z^2}\right)\vec{i}$$

$$+ \left(xze^{x^2+y^2+z^2} + 2y \cdot xyze^{x^2+y^2+z^2}\right)\vec{j}$$

$$+ \left(xye^{x^2+y^2+z^2} + 2z \cdot xyze^{x^2+y^2+z^2}\right)\vec{k}$$

$$= \left[(yz + 2x^2yz)\,\vec{i} + (xz + 2xy^2z)\,\vec{j} + (xy + 2xyz^2)\,\vec{k}\right]e^{x^2+y^2+z^2}.$$

**Example 4.15.** If one denotes:

$$\begin{cases} \vec{r} = x\,\vec{i} + y\,\vec{j} + z\,\vec{k} \\ r = \|\vec{r}\| = \sqrt{x^2 + y^2 + z^2} \end{cases}$$

find the gradient of the following scalar fields:

$$\text{a) } \varphi(r) = r^n$$
$$\text{b) } \varphi(r) = \ln r.$$

**Solutions.**

a) We shall compute the partial derivative of the field $\varphi$:

$$\frac{\partial\varphi}{\partial x} = n \cdot \left(\sqrt{x^2 + y^2 + z^2}\right)^{n-1} \cdot \frac{2x}{2\sqrt{x^2 + y^2 + z^2}}$$

$$= nx\left(\sqrt{x^2 + y^2 + z^2}\right)^{n-2} = nxr^{n-2},$$

$$\frac{\partial\varphi}{\partial y} = n \cdot \left(\sqrt{x^2 + y^2 + z^2}\right)^{n-1} \cdot \frac{2y}{2\sqrt{x^2 + y^2 + z^2}}$$

$$= ny\left(\sqrt{x^2 + y^2 + z^2}\right)^{n-2} = nyr^{n-2},$$

$$\frac{\partial\varphi}{\partial z} = n \cdot \left(\sqrt{x^2 + y^2 + z^2}\right)^{n-1} \cdot \frac{2z}{2\sqrt{x^2 + y^2 + z^2}}$$

$$= nz\left(\sqrt{x^2 + y^2 + z^2}\right)^{n-2} = nzr^{n-2};$$

hence

$$\text{grad}\,\varphi = nxr^{n-2}\vec{i} + nyr^{n-2}\vec{j} + nzr^{n-2}\vec{k} = nr^{n-2}\left(x\,\vec{i} + y\,\vec{j} + z\,\vec{k}\right) = nr^{n-2}\vec{r}.$$

In Matlab 7.9 we have to use the following commands:

```
>> syms x y z n r
>> phi=sqrt(x^2+y^2+z^2)^n ;
```

```
>> v=[x y z];
>> c=jacobian(phi,v) ;
>>simplify(subs(subs(c,sqrt(x^2+y^2+z^2),r),
   x^2+y^2+z^2,r^2))
 ans =
 [ r^(n-2)*n*x, r^(n-2)*n*y, r^(n-2)*n*z]
```

while in Mathcad 14:

$$\varphi(x,y,z,n) := \left(\sqrt{x^2 + y^2 + z^2}\right)^n$$

$$w(x,y,z,n) := \begin{pmatrix} \dfrac{\partial}{\partial x}\,\varphi(x,y,z,n) \\[2mm] \dfrac{\partial}{\partial y}\,\varphi(x,y,z,n) \\[2mm] \dfrac{\partial}{\partial z}\,\varphi(x,y,z,n) \end{pmatrix} \text{ simplify } \rightarrow \begin{bmatrix} n\cdot x\cdot\left(x^2 + y^2 + z^2\right)^{\frac{n}{2}-1} \\[2mm] n\cdot y\cdot\left(x^2 + y^2 + z^2\right)^{\frac{n}{2}-1} \\[2mm] n\cdot z\cdot\left(x^2 + y^2 + z^2\right)^{\frac{n}{2}-1} \end{bmatrix}$$

$$w(x,y,z,n) \text{ substitute,}\sqrt{x^2 + y^2 + z^2} = r, x^2 + y^2 + z^2 = r^2 \rightarrow \begin{bmatrix} \dfrac{n\cdot x\cdot\left(r^2\right)^{\frac{n}{2}}}{r^2} \\[3mm] \dfrac{n\cdot y\cdot\left(r^2\right)^{\frac{n}{2}}}{r^2} \\[3mm] \dfrac{n\cdot z\cdot\left(r^2\right)^{\frac{n}{2}}}{r^2} \end{bmatrix}$$

and with Mathematica 8:

In[111]:= **Needs["VectorAnalysis`"]**

In[112]:= **φ = (Xx^2 + Yy^2 + Zz^2)^(Nn/2);**

In[113]:= **b := Simplify[Grad[φ]];**

In[114]:= **b /. $\left(Xx^2 + Yy^2 + Zz^2\right)$^(1/2) -> r;**

In[115]:= **% /. $Xx^2 + Yy^2 + Zz^2$ -> r^2**

Out[115]= $\left\{Nn\left(r^2\right)^{-1+\frac{Nn}{2}} Xx,\ Nn\left(r^2\right)^{-1+\frac{Nn}{2}} Yy,\ Nn\left(r^2\right)^{-1+\frac{Nn}{2}} Zz\right\}$

and using Maple 15:

*with(VectorCalculus) :*

$g := simplify\left(Gradient\left(\left(\sqrt{x^2+y^2+z^2}\right)^n, [x,y,z]\right)\Big|_{\sqrt{x^2+y^2+z^2}=r,\ x^2+y^2+z^2=r^2}\right) :$

$v := g_1, g_2, g_3$

$$r^{n-2}nx,\ r^{n-2}ny,\ r^{n-2}nz$$

b) It shall result:

$$\operatorname{grad}\varphi = \frac{x}{x^2+y^2+z^2}\,\vec{i} + \frac{y}{x^2+y^2+z^2}\,\vec{j} + \frac{z}{x^2+y^2+z^2}\,\vec{k}$$

$$= \frac{1}{x^2+y^2+z^2}\left(x\,\vec{i} + y\,\vec{j} + z\,\vec{k}\right) = \frac{\vec{r}}{\|r^2\|}.$$

In this case, we need the following Matlab 7.9 sequence :
```
>> syms x y z r
>> phi=log(sqrt(x^2+y^2+z^2)) ;
>>v=[x y z];
>> c=jacobian(phi,v)
 c =
[ 1/(x^2+y^2+z^2)*x, 1/(x^2+y^2+z^2)*y,
1/(x^2+y^2+z^2)*z]
>> simplify(subs(c,x^2+y^2+z^2,r^2))
 ans =
[ 1/r^2*x, 1/r^2*y, 1/r^2*z]
```
We shall also compute the gradient in Mathcad 14:

$$\varphi(x,y,z) := \ln\left(\sqrt{x^2 + y^2 + z^2}\right)$$

$$w(x,y,z) := \begin{pmatrix} \dfrac{\partial}{\partial x}\,\varphi(x,y,z) \\[2mm] \dfrac{\partial}{\partial y}\,\varphi(x,y,z) \\[2mm] \dfrac{\partial}{\partial z}\,\varphi(x,y,z) \end{pmatrix} \text{simplify} \rightarrow \begin{pmatrix} \dfrac{x}{x^2 + y^2 + z^2} \\[2mm] \dfrac{y}{x^2 + y^2 + z^2} \\[2mm] \dfrac{z}{x^2 + y^2 + z^2} \end{pmatrix}$$

$$w(x,y,z) \ \text{substitute}, x^2 + y^2 + z^2 = r^2 \rightarrow \begin{pmatrix} \dfrac{x}{r^2} \\[2mm] \dfrac{y}{r^2} \\[2mm] \dfrac{z}{r^2} \end{pmatrix}$$

and with Mathematica 8:

```
In[116]:= Needs["VectorAnalysis`"]

In[117]:= φ = Log[Sqrt[Xx^2 + Yy^2 + Zz^2]];

In[118]:= b := Simplify[Grad[φ]];

In[119]:= b /. Xx² + Yy² + Zz² -> r^2

Out[119]= { Xx/r², Yy/r², Zz/r² }
```

and using Maple 15:

with(*VectorCalculus*) :

$$g := simplify\left( Gradient\left( \ln\left(\sqrt{x^2+y^2+z^2}\right), [x, y, z]\right)\right|_{\sqrt{x^2+y^2+z^2} = r, x^2+y^2+z^2 = r^2}\right):$$

$$v := g_1, g_2, g_3$$

$$\frac{x}{r^2}, \frac{y}{r^2}, \frac{z}{r^2}$$

**Example 4.16.** Find the gradient of the field

$$\varphi(x, y, z) = f\left(x + y + z, \; x^2 + y^2 + z^2\right).$$

**Solution.**
Denoting

$$\begin{cases} x + y + z = u \\ x^2 + y^2 + z^2 = v \end{cases}$$

we shall obtain:

$$\frac{\partial\varphi}{\partial x} = \frac{\partial\varphi}{\partial u} \cdot \frac{\partial u}{\partial x} + \frac{\partial\varphi}{\partial v} \cdot \frac{\partial v}{\partial x} = \frac{\partial\varphi}{\partial u} + 2x\frac{\partial\varphi}{\partial v},$$

$$\frac{\partial\varphi}{\partial y} = \frac{\partial\varphi}{\partial u} \cdot \frac{\partial u}{\partial y} + \frac{\partial\varphi}{\partial v} \cdot \frac{\partial v}{\partial y} = \frac{\partial\varphi}{\partial u} + 2y\frac{\partial\varphi}{\partial v},$$

$$\frac{\partial\varphi}{\partial z} = \frac{\partial\varphi}{\partial u} \cdot \frac{\partial u}{\partial z} + \frac{\partial\varphi}{\partial v} \cdot \frac{\partial v}{\partial z} = \frac{\partial\varphi}{\partial u} + 2z\frac{\partial\varphi}{\partial v};$$

hence

$$\mathrm{grad}\,\varphi = \left(\frac{\partial\varphi}{\partial u} + 2x\frac{\partial\varphi}{\partial v}\right)\vec{i} + \left(\frac{\partial\varphi}{\partial u} + 2y\frac{\partial\varphi}{\partial v}\right)\vec{j} + \left(\frac{\partial\varphi}{\partial u} + 2z\frac{\partial\varphi}{\partial v}\right)\vec{k}$$

$$= \left(\vec{i} + \vec{j} + \vec{k}\right)\frac{\partial\varphi}{\partial u} + 2\vec{r}\frac{\partial\varphi}{\partial v}.$$

We can also solve this exercise in Matlab 7.9:

```
>>syms x y z
>> f=[x+y+z; x^2+y^2+z^2] ;
>>t=[x y z] ;
>> jacobian(f,t)
 ans =
 [ 1, 1, 1]
 [ 2*x, 2*y, 2*z]
```

and in Mathcad 14:

$$f(x) := \begin{bmatrix} x_0 + x_1 + x_2 \\ \left(x_0\right)^2 + \left(x_1\right)^2 + \left(x_2\right)^2 \end{bmatrix}$$

$$\text{Jacob}(f(x),x) \rightarrow \begin{pmatrix} 1 & 1 & 1 \\ 2 \cdot x_0 & 2 \cdot x_1 & 2 \cdot x_2 \end{pmatrix}$$

and using Mathematica 8:

In[122]:= `D[{x + y + z, x^2 + y^2 + z^2}, {{x, y, z}}]`

Out[122]= `{{1, 1, 1}, {2 x, 2 y, 2 z}}`

and with Maple 15:

*with( VectorCalculus)* :
*Jacobian*$\left([x + y + z, x^2 + y^2 + z^2], [x, y, z]\right)$

$$\begin{bmatrix} 1 & 1 & 1 \\ 2x & 2y & 2z \end{bmatrix}$$

**Example 4.17.** Compute the divergence and the curl of the vector field

$$\vec{v} = xz^3 \vec{i} - 2x^2 yz \vec{j} + 2yz^4 \vec{k}$$

in the point $M(1, -1, 1)$.
    **Solution.**
    As

$$\begin{cases} P(x, y, z) = xz^3 \\ Q(x, y, z) = -2x^2 yz \iff \\ R(x, y, z) = 2yz^4 \end{cases}$$

$$\begin{cases} \frac{\partial P}{\partial x} = z^3 \\ \frac{\partial Q}{\partial y} = -2x^2 z \\ \frac{\partial R}{\partial z} = 8yz^3 \end{cases}$$

we shall obtain

$$\text{div } \vec{v}(1, -1, 1) = \frac{\partial P}{\partial x}(1, -1, 1) + \frac{\partial Q}{\partial y}(1, -1, 1) + \frac{\partial R}{\partial z}(1, -1, 1)$$
$$= 1 - 2 - 8 = -9.$$

Computing

$$\begin{cases} \frac{\partial R}{\partial y} = 2z^4 \\ \frac{\partial Q}{\partial z} = -2x^2y \\ \frac{\partial P}{\partial z} = 3xz^2 \\ \frac{\partial R}{\partial x} = 0 \\ \frac{\partial Q}{\partial x} = -4xyz \\ \frac{\partial P}{\partial y} = 0 \end{cases} \iff$$

$$\begin{cases} \frac{\partial R}{\partial y}(1,-1,1) = 2 \\ \frac{\partial Q}{\partial z}(1,-1,1) = 2 \\ \frac{\partial P}{\partial z}(1,-1,1) = 3 \\ \frac{\partial R}{\partial x}(1,-1,1) = 0 \\ \frac{\partial Q}{\partial x}(1,-1,1) = 4 \\ \frac{\partial P}{\partial y}(1,-1,1) = 0 \end{cases}$$

it results

$$\text{rot } \vec{v}(1,-1,1) = \left( \frac{\partial R}{\partial y}(1,-1,1) - \frac{\partial Q}{\partial z}(1,-1,1) \right) \vec{i}$$
$$+ \left( \frac{\partial P}{\partial z}(1,-1,1) - \frac{\partial R}{\partial x}(1,-1,1) \right) \vec{j}$$
$$+ \left( \frac{\partial Q}{\partial x}(1,-1,1) - \frac{\partial P}{\partial y}(1,-1,1) \right) \vec{k}$$
$$= (2-2)\vec{i} + (3-0)\vec{j} + (4-0)\vec{k} = 3\vec{j} + 4\vec{k}.$$

In Matlab 7.9, we shall have:

```
>> syms x y z
>> P=@(x,y,z)x*z^3;
>> Q=@(x,y,z)-2*x^2*y*z;
>> R=@(x,y,z) 2*y*z^4 ;
>> a=subs(diff(P(x,y,z),x),{x,y,z},{1,1,1}) ;
>> b=subs(diff(Q(x,y,z),y),{x,y,z},{1,1,1}) ;
>> c=subs(diff(R(x,y,z),z),{x,y,z},{1,1,1}) ;
>> div=a+b+c
div =
 -9
>> u1=subs(diff(R(x,y,z),y),{x,y,z},{1,1,1}) ;
>>u2=subs(diff(Q(x,y,z),z),{x,y,z},{1,1,1}) ;
>>u=u1-u2 ;
>> v1=subs(diff(P(x,y,z),z),{x,y,z},{1,1,1}) ;
>>v2=subs(diff(R(x,y,z),x),{x,y,z},{1,1,1}) ;
>>v=v1-v2 ;
```

```
>> w1=subs(diff(Q(x,y,z),x),{x,y,z},{1,1,1}) ;
>>w2=subs(diff(P(x,y,z),y),{x,y,z},{1,1,1}) ;
>>w =w1-w2;
>> rot=[u v w]
rot =
  0 3 4
```

We shall also solve this problem in Mathcad 14:

$$f(x) := \begin{bmatrix} x_0 \cdot (x_2)^3 \\ -2 \cdot (x_0)^2 \cdot x_1 \cdot x_2 \\ 2 \cdot x_1 \cdot (x_2)^4 \end{bmatrix} \qquad u := \begin{pmatrix} 1 \\ -1 \\ 1 \end{pmatrix}$$

$$v(x) := \text{Jacob}(f(x),x)$$

$$w(x) := \begin{pmatrix} v(x)_{0,0} \\ v(x)_{1,1} \\ v(x)_{2,2} \end{pmatrix}$$

$$\sum_{i=0}^{2} w(x)_i \ \ \text{substitute}, x_0 = u_0, x_1 = u_1, x_2 = u_2 \ \rightarrow \ -9$$

$$y(a,b,c) := \begin{pmatrix} f(x)_0 \\ f(x)_1 \\ f(x)_2 \end{pmatrix} \ \text{substitute}, x_0 = a, x_1 = b, x_2 = c \ \rightarrow \begin{pmatrix} a \cdot c^3 \\ -2 \cdot a^2 \cdot b \cdot c \\ 2 \cdot b \cdot c^4 \end{pmatrix}$$

$$\begin{pmatrix} \frac{\partial}{\partial b} y(a,b,c)_2 - \frac{\partial}{\partial c} y(a,b,c)_1 \\ \frac{\partial}{\partial c} y(a,b,c)_0 - \frac{\partial}{\partial a} y(a,b,c)_2 \\ \frac{\partial}{\partial a} y(a,b,c)_1 - \frac{\partial}{\partial b} y(a,b,c)_0 \end{pmatrix} \ \text{substitute}, a = u_0, b = u_1, c = u_2 \ \rightarrow \begin{pmatrix} 0 \\ 3 \\ 4 \end{pmatrix}$$

and with Mathematica 8:

In[141]:= **Needs["VectorAnalysis`"]**

In[142]:= **v := {Xx * Zz^3, -2 * Xx^2 * Yy * Zz, 2 * Yy * Zz^4}**

In[143]:= **Div[v];**

In[144]:= **% /. {Xx → 1, Yy → -1, Zz → 1}**

Out[144]= **-9**

In[145]:= **Curl[v]**

Out[145]= $\left\{2\,Xx^2\,Yy + 2\,Zz^4,\ 3\,Xx\,Zz^2,\ -4\,Xx\,Yy\,Zz\right\}$

In[146]:= **% /. {Xx → 1, Yy → -1, Zz → 1}**

Out[146]= **{0, 3, 4}**

and using Maple 15:

*with(VectorCalculus)* :
*SetCoordinates*('*cartesian*'$_{x,y,z}$) :
$F := VectorField(\langle x \cdot z^3, -2 \cdot x^2 \cdot y \cdot z, 2 \cdot y \cdot z^4 \rangle)$ :
$Divergence(F)\big|_{x=1,\,y=-1,\,z=1}$

$$-9$$

$r := Curl(F)\big|_{x=1,\,y=-1,\,z=1}$

$$3\,\underset{y}{e} + 4\,\underset{z}{e}$$

*rot* := $r_1, r_2, r_3$

$$0, 3, 4$$

**Example 4.18.** Let be the vector field

$$\vec{v} = \sin x\,\vec{i} - y e^{y^2}\,\vec{j} + \cos 2z\,\vec{k}.$$

Find:
a) div $\vec{v}$;
b) rot $\vec{v}$;
c) the scalar potential $\varphi(x, y, z)$ of this field.
**Solutions.**

a) It results that

$$\operatorname{div} \vec{v} = \frac{\partial P}{\partial x} + \frac{\partial Q}{\partial y} + \frac{\partial R}{\partial z},$$

where

$$\begin{cases} P\left(x,y,z\right) = \sin x \\ Q\left(x,y,z\right) = ye^{y^2} \\ R\left(x,y,z\right) = \cos 2z; \end{cases}$$

therefore

$$\operatorname{div} \vec{v} = \cos x + e^{y^2} + 2ye^{y^2} - 2\sin 2z.$$

b) The rotation of the vector field $\vec{v}$ will be

$$\operatorname{rot} \vec{v} = \left(\frac{\partial R}{\partial y} - \frac{\partial Q}{\partial z}\right) \vec{i} + \left(\frac{\partial P}{\partial z} - \frac{\partial R}{\partial x}\right) \vec{j} + \left(\frac{\partial Q}{\partial x} - \frac{\partial P}{\partial y}\right) \vec{k}$$

$$= (0-0)\,\vec{i} + (0-0)\,\vec{j} + (0-0)\,\vec{k} = \vec{0}.$$

c) As $\operatorname{rot} \vec{v} = \vec{0}$ it results that there is a scalar function $\varphi\left(x,y,z\right)$ which constitutes the potential of the field such that:

$$d\varphi\left(x,y,z\right) = \sin x dx + ye^{y^2}\,dy + \cos 2z dz$$

one deduces that

$$\begin{cases} \frac{\partial \varphi}{\partial x} = \sin x \\ \frac{\partial \varphi}{\partial y} = ye^{y^2} \\ \frac{\partial \varphi}{\partial z} = \cos 2z. \end{cases} \tag{4.9}$$

From the first equation of the relation (4.9) one obtains:

$$\varphi\left(x,y,z\right) = -\cos x + C\left(y,z\right). \tag{4.10}$$

Using the relation (4.10) and the second equation of the relation (4.9) we shall have:

$$ye^{y^2} = \frac{\partial C}{\partial y}\left(y,z\right) = C'\left(y,z\right),$$

namely

$$C\left(y,z\right) = \int ye^{y^2}\,dy + C\left(z\right) = \frac{1}{2}e^{y^2} + C\left(z\right). \tag{4.11}$$

Substituting (4.11) into (4.10) one obtains:

$$\varphi(x, y, z) = -\cos x + \frac{1}{2}e^{y^2} + C(z).\qquad(4.12)$$

From (4.9) and (4.12) we shall deduce:

$$\cos 2z = \frac{\partial C}{\partial z}(z) = C'(z);$$

hence

$$C(z) = \int \cos 2z dz + K = \frac{1}{2}\sin 2z + K.\qquad(4.13)$$

Substituting (4.13) into (4.12) it results that

$$\varphi(x, y, z) = -\cos x + \frac{1}{2}e^{y^2} + \frac{1}{2}\sin 2z + K.$$

We can also solve this problem in Matlab 7.9:

```
>> syms x y z
>> P=sin(x);
>> Q=y*exp(y^2);
>> R=cos(2*z);
>> div=diff(P,x)+diff(Q,y)+diff(R,z)
div =
 cos(x)+exp(y^2)+2*y^2*exp(y^2)-2*sin(2*z)
>> r1=diff(R,y)-diff(Q,z);
>> r2=diff(P,z)-diff(R,x);
>> r3=diff(Q,x)-diff(P,y);
>> rot=[r1 r2 r3]
rot =
 [ 0, 0, 0]
>> syms K
>> phi=int(P,x)+int(Q,y)+int(R,z)+K
phi =
 -cos(x)+1/2*exp(y^2)+1/2*sin(2*z)+K
```

Solving this problem with other softwares, we shall obtain in Mathcad 14:

$$f(x) := \begin{bmatrix} \sin(x_0) \\ x_1 \cdot e^{(x_1)^2} \\ \cos(2 \cdot x_2) \end{bmatrix} \qquad v(x) := \text{Jacob}(f(x),x)$$

$$w(x) := \begin{pmatrix} v(x)_{0,0} \\ v(x)_{1,1} \\ v(x)_{2,2} \end{pmatrix}$$

Divergence

$$\sum_{i=0}^{2} w(x)_i \rightarrow 2 \cdot e^{(x_1)^2} \cdot (x_1)^2 + \cos(x_0) - 2 \cdot \sin(2 \cdot x_2) + e^{(x_1)^2}$$

$$y(a,b,c) := \begin{pmatrix} f(x)_0 \\ f(x)_1 \\ f(x)_2 \end{pmatrix} \text{substitute}, x_0 = a, x_1 = b, x_2 = c \rightarrow \begin{pmatrix} \sin(a) \\ b \cdot e^{b^2} \\ \cos(2 \cdot c) \end{pmatrix}$$

Curl

$$\begin{pmatrix} \dfrac{\partial}{\partial b} y(a,b,c)_2 - \dfrac{\partial}{\partial c} y(a,b,c)_1 \\[2ex] \dfrac{\partial}{\partial c} y(a,b,c)_0 - \dfrac{\partial}{\partial a} y(a,b,c)_2 \\[2ex] \dfrac{\partial}{\partial a} y(a,b,c)_1 - \dfrac{\partial}{\partial b} y(a,b,c)_0 \end{pmatrix} \rightarrow \begin{pmatrix} 0 \\ 0 \\ 0 \end{pmatrix}$$

Scalar potential

$$\varphi(a,b,c) := \int y(a,b,c)_0 \, da + \int y(a,b,c)_1 \, db + \int y(a,b,c)_2 \, dc \rightarrow \frac{\sin(2 \cdot c)}{2} + \frac{e^{b^2}}{2} - \cos(a)$$

and with Mathematica 8:

In[183]:= **Needs["VectorAnalysis`"]**

In[184]:= **v := {Sin[Xx], Yy*Exp[Yy^2], Cos[2*Zz]}**

In[185]:= **Div[v]**

Out[185]= $e^{Yy^2} + 2 e^{Yy^2} Yy^2 + Cos[Xx] - 2 Sin[2 Zz]$

In[186]:= **Curl[v]**

Out[186]= {0, 0, 0}

In[187]:= **Integrate[v[[1]], Xx] + Integrate[v[[2]], Yy] + Integrate[v[[3]], Zz]**

Out[187]= $\dfrac{e^{Yy^2}}{2} - Cos[Xx] + \dfrac{1}{2} Sin[2 Zz]$

and in Maple 15:

*with(VectorCalculus) :*
*SetCoordinates('cartesian'$_{x,y,z}$) :*

$F := VectorField\left(\left\langle \sin(x), y \cdot e^{y^2}, \cos(2 \cdot z)\right\rangle\right) :$
*Divergence(F)*

$$\cos(x) + e^{y^2} + 2y^2 e^{y^2} - 2\sin(2z)$$

*Curl(F)*

$$0\underset{x}{e}$$

*with(linalg) :*

$v := \left[F_1, F_2, F_3\right]$

$$\left[\sin(x), y e^{y^2}, \cos(2z)\right]$$

*potential(v, [x, y, z],'φ')*

$$true$$

φ

$$-\cos(x) + \dfrac{1}{2} e^{y^2} + \dfrac{1}{2}\sin(2z)$$

## 4.3    Problems

1. Find the derivative of the function

$$z = 2x^2 - 3y^2$$

at the point $M(1,0)$ in a direction, which makes an angle of $120°$ with the $Ox$ -axis.

2. Find the gradient of the field

$$\varphi(x, y, z) = \arctan \frac{x + y + z - xyz}{1 - xy - yz - xz}.$$

3. Which is the gradient of the field

$$\varphi(x, y, z) = f\left(x^2 + y^2 + z^2\right)?$$

4. Compute the divergence of the vector field

$$\overrightarrow{v} = xyz\,\overrightarrow{i} + y^2\,(z - x)\,\overrightarrow{j} + \frac{z}{x - y}\,\overrightarrow{k}, \ x \neq y$$

in the point $M(1,0,-1)$.
   **Computer solution.**
   We shall achieve the divergence vector using Matlab 7.9:
   ```
   >> syms x y z
   >> P=@(x,y,z)x*y*z;
   >> Q=@(x,y,z)y^2*(z-x);
   >> R=@(x,y,z)z/(x-y);
   >> a=subs(diff(P(x,y,z),x),{x,y,z},{1,0,-1});
   >> b=subs(diff(Q(x,y,z),y),{x,y,z},{1,0,-1});
   >> c=subs(diff(R(x,y,z),z),{x,y,z},{1,0,-1});
   >> div=a+b+c
   div =
   1
   ```
   and in Mathcad 14:

$$f(x) := \begin{bmatrix} x_0 \cdot x_1 \cdot x_2 \\ (x_1)^2 \cdot (x_2 - x_0) \\ \dfrac{x_2}{x_0 - x_1} \end{bmatrix} \qquad u := \begin{pmatrix} 1 \\ 0 \\ -1 \end{pmatrix}$$

$$v(x) := \text{Jacob}(f(x), x)$$

$$w(x) := \begin{pmatrix} v(x)_{0,0} \\ v(x)_{1,1} \\ v(x)_{2,2} \end{pmatrix}$$

$$\sum_{i=0}^{2} w(x)_i, \text{ substitute}, x_0 = u_0, x_1 = u_1, x_2 = u_2 \to 1$$

and with Mathematica 8:

```
In[1]:= Needs ["VectorAnalysis`"]

In[2]:= v := {Xx * Yy * Zz, Yy^2 * (Zz - Xx) , Zz / (Xx - Yy)}

In[3]:= Div [v];

In[4]:= % /. {Xx → 1, Yy → 0, Zz → -1}

Out[4]= 1
```

and Maple 15:

*with*(*VectorCalculus*) :
*SetCoordinates*('*cartesian*'$_{x, y, z}$) :

$$F := \textit{VectorField}\left(\left\langle x \cdot y \cdot z, y^2 \cdot (z - x), \frac{z}{x - y}\right\rangle\right) :$$

$$\textit{Divergence}(F)\Big|_{x = 1, y = 0, z = -1}$$

$$1$$

5. Derive the formulas:

a) $\operatorname{grad}(\varphi_1\varphi_2) = \varphi_1 \operatorname{grad}(\varphi_2) + \varphi_1 \operatorname{grad}(\varphi_2)$

b) $\operatorname{rot}(\varphi\vec{v}) = \varphi \operatorname{rot}\vec{v} - \vec{v} \times \operatorname{grad}\varphi$

c) $\operatorname{div}(\vec{v}_1 \times \vec{v}_2) = \vec{v}_2\operatorname{rot}\vec{v}_1 - \vec{v}_1\operatorname{rot}\vec{v}_2$

d) $\operatorname{rot}(\varphi\vec{v}) = \varphi \operatorname{div}\vec{v} + \vec{v} \operatorname{grad}\varphi$.

6. Check if the vector field $\varphi$ satisfy the equations:

$$a) \operatorname{rot}\operatorname{grad}\varphi = 0$$

$$b) \operatorname{div}\operatorname{rot}\varphi = 0.$$

**Computer solution.**
b) We shall prove this relation (i.e. div rot $\varphi = 0$) with Matematica 8:

In[22]:= **Needs["VectorAnalysis`"]**

In[23]:= **f[Xx_, Yy_, Zz_] := f1[Xx, Yy, Zz];**

In[24]:= **g[Xx_, Yy_, Zz_] := f2[Xx, Yy, Zz];**

In[25]:= **h[Xx_, Yy_, Zz_] := f3[Xx, Yy, Zz];**

In[26]:= **v := {f[Xx, Yy, Zz], g[Xx, Yy, Zz], h[Xx, Yy, Zz]}**

In[31]:= **Div[Curl[v]]**

Out[31]= 0

and using Maple 15:

*with(VectorCalculus)* :
*SetCoordinates('cartesian'$_{x,y,z}$)* :
$f := (x, y, z) \to f1(x, y, z)$ :
$g := (x, y, z) \to f2(x, y, z)$ :
$h := (x, y, z) \to f3(x, y, z)$ :
$G := VectorField(\langle f(x, y, z), g(x, y, z), h(x, y, z)\rangle)$ :
*Divergence(Curl(G))*

0

7. Let be the vector field

$$\vec{v} = \left(3x^2 + 2x\right)\vec{i} - \left(z - 3y^2\right)\vec{j} + \left(y + 2z\right)\vec{k}.$$

Find:

a) div $\vec{v}$;
b) rot $\vec{v}$;
c) the scalar potential $\varphi(x, y, z)$ of this field.
8. Find the derivative of the function

$$z = x^3 - 2x^2 y + xy^2 + 1$$

at the point $M(1, 2)$ in the direction from this point to the point $N(4, 6)$.
**Computer solution.**
We shall find this derivative in Matlab 7.9:

```
>> M=[1 2]; N=[4 6];s=N-M;
>> w=norm(s);u=s/w;
>> syms x y
>> z=x^3-2*x^2*y+x*y^2+1;
>> d1=subs(diff(z,x),{x,y},{1,2});
>> d2=subs(diff(z,y),{x,y},{1,2});
>> d=dot(u,[d1 d2])
d =
  1
```

and with Mathcad 14:

$$\varphi(x,y) := x^3 - 2 \cdot x^2 \cdot y + x \cdot y^2 + 1$$

$$M := \begin{pmatrix} 1 \\ 2 \end{pmatrix} \qquad N := \begin{pmatrix} 4 \\ 6 \end{pmatrix}$$

$$x0 := 1 \qquad y0 := 2 \qquad\qquad u := \frac{N - M}{|N - M|}$$

$$v := \begin{pmatrix} \dfrac{\partial}{\partial x0}\,\varphi(x0, y0) \\[2mm] \dfrac{\partial}{\partial y0}\,\varphi(x0, y0) \end{pmatrix} \qquad\qquad u \cdot v = 1$$

and using Mathematica 8:

```
In[39]:= m := {1, 2};

In[40]:= n := {4, 6};

In[41]:= s := n - m

In[42]:= u := s / Norm[s]

In[43]:= z[x_, y_] := x^3 - 2*x^2*y + x*y^2 + 1;

In[44]:= w := {D[z[x, y], x] /. {x → 1, y → 2},
              D[z[x, y], y] /. {x → 1, y → 2}}

In[45]:= SetPrecision[u.w, 3]

Out[45]= 1.00
```

and in Maple 15:

$M := \langle 1, 2 \rangle :$
$N := \langle 4, 6 \rangle :$
$with(VectorCalculus) :$
$SetCoordinates(cartesian_{x,y}) :$
$$u := \frac{N - M}{Norm(N - M, 2)} :$$
$with(Student[MultivariateCalculus]) :$
$Digits := 5 :$
$evalf\left(DirectionalDerivative(x^3 - 2 \cdot x^2 \cdot y + x \cdot y^2 + 1, [x, y] = [1, 2], [u_1, u_2])\right)$

$$1.$$

9. Evaluate the divergence and the rotation of the vector $\vec{v}$ if

$$\vec{v} = \vec{c} f(r),$$

where:

- $\vec{c}$ is a constant vector,
- $\vec{r} = x \vec{i} + y \vec{j} + z \vec{k}$,
- $r = \|\vec{r}\| = \sqrt{x^2 + y^2 + z^2}$.

## Computer solution.
We shall solve our problem using Mathematica 8:

In[213]:= **Needs** ["VectorAnalysis`"]

In[214]:= **w** := **f** [Sqrt[Xx^2 + Yy^2 + Zz^2]] * {c1, c2, c3}

In[215]:= **Simplify** [Div[w]]

$$
\text{Out[215]=}\quad \frac{(c1\,Xx + c2\,Yy + c3\,Zz)\; f'\left[\sqrt{Xx^2 + Yy^2 + Zz^2}\right]}{\sqrt{Xx^2 + Yy^2 + Zz^2}}
$$

In[216]:= **Simplify** [Curl[w]]

$$
\text{Out[216]=}\quad \left\{ \frac{(c3\,Yy - c2\,Zz)\; f'\left[\sqrt{Xx^2 + Yy^2 + Zz^2}\right]}{\sqrt{Xx^2 + Yy^2 + Zz^2}}, \right.
$$

$$
\left. \frac{(-c3\,Xx + c1\,Zz)\; f'\left[\sqrt{Xx^2 + Yy^2 + Zz^2}\right]}{\sqrt{Xx^2 + Yy^2 + Zz^2}}, \; \frac{(c2\,Xx - c1\,Yy)\; f'\left[\sqrt{Xx^2 + Yy^2 + Zz^2}\right]}{\sqrt{Xx^2 + Yy^2 + Zz^2}} \right\}
$$

and with Maple 15:

with( *VectorCalculus* ) :
SetCoordinates( *'cartesian'*$_{x,y,z}$ ) :
$F := f\!\left(\sqrt{x^2 + y^2 + z^2}\right) \cdot VectorField(\langle c1, c2, c3\rangle)$ :
simplify( *Divergence*(F))

$$
\frac{D(f)\left(\sqrt{x^2 + y^2 + z^2}\right)(xc1 + yc2 + zc3)}{\sqrt{x^2 + y^2 + z^2}}
$$

v := simplify( *Curl*(F))

$$
-\frac{D(f)\left(\sqrt{x^2+y^2+z^2}\right)(-yc3+zc2)}{\sqrt{x^2+y^2+z^2}}\bar{e}_x + \left(\frac{D(f)\left(\sqrt{x^2+y^2+z^2}\right)(zc1-xc3)}{\sqrt{x^2+y^2+z^2}}\right)\bar{e}_y
$$

$$
-\frac{D(f)\left(\sqrt{x^2+y^2+z^2}\right)(-xc2+yc1)}{\sqrt{x^2+y^2+z^2}}\bar{e}_z
$$

10.  Find div $\vec{v}$ and curl $\vec{v}$ for the central vector field

$$\vec{v}\,(P) = f\,(r)\,\frac{\vec{r}}{r},$$

$\vec{r}, r$ being defined at the previous problem, namely:

$$\begin{cases} \vec{r} = x\,\vec{i} + y\,\vec{j} + z\,\vec{k} \\ r = \|\vec{r}\| = \sqrt{x^2 + y^2 + z^2}. \end{cases}$$

**Computer solution.**
We shall compute div $\vec{v}$ and curl $\vec{v}$ in Mathematica 8:

In[204]:= **Needs["VectorAnalysis`"]**

In[205]:= **w := f[Sqrt[Xx^2 + Yy^2 + Zz^2]] \* {Xx, Yy, Zz} / Sqrt[Xx^2 + Yy^2 + Zz^2]**

In[206]:= **Curl[w]**

Out[206]= {0, 0, 0}

In[207]:= **Simplify[Div[w]]**

Out[207]= $\dfrac{2\,f\left[\sqrt{Xx^2 + Yy^2 + Zz^2}\,\right]}{\sqrt{Xx^2 + Yy^2 + Zz^2}} + f'\left[\sqrt{Xx^2 + Yy^2 + Zz^2}\,\right]$

and with Maple 15:

*with*(*VectorCalculus*) :

*SetCoordinates*('*cartesian*'$_{x,y,z}$) :

$$F := \frac{f\left(\sqrt{x^2+y^2+z^2}\right)}{\sqrt{x^2+y^2+z^2}} \cdot VectorField(\langle x,y,z\rangle) :$$

*simplify*(*Divergence*(*F*))

$$\frac{2f\left(\sqrt{x^2+y^2+z^2}\right) + D(f)\left(\sqrt{x^2+y^2+z^2}\right)\sqrt{x^2+y^2+z^2}}{\sqrt{x^2+y^2+z^2}}$$

*Curl*(*F*)

$$\begin{matrix} 0 \\ x \end{matrix}$$

# 5
# Implicit Functions

## 5.1 Derivative of Implicit Functions

If the relationship between $x$ and $y$ is given implicitly

$$F(x, y) = 0, \tag{5.1}$$

then, in order to find the derivative $y'_x = y'$ in the simplest cases, it is sufficient:

1) to calculate the derivative, with respect to $x$, of the left hand side of (5.1), taking $y$ as a function of $x$;

2) to equate this derivative to zero, i.e., to set

$$\frac{\partial F}{\partial x}(x, y) = 0; \tag{5.2}$$

3) to solve the resulting equation for $y$.

**Definition 5.1** (see [8]). A **derivative of the second order** or the **second derivative** of the function $y = f(x)$ is the derivative of its derivative, i.e.,

$$y'' = (y')'. \tag{5.3}$$

G.A. Anastassiou and I.F. Iatan: Intelligent Routines, ISRL 39, pp. 187–244.
springerlink.com        © Springer-Verlag Berlin Heidelberg 2013

**Remark 5.2** (see [8]). In general, the $n$-th derivative of a function $y = f(x)$, denoted $y^{(n)}$ is the derivative of the derivative of order $(n-1)$.

**Example 5.3.** Compute $y'$ and $y''$ for the implicit function $y = f(x)$, defined by the following equation:

$$xy + \ln y = 0, \ y > 0.$$

**Solution.**

We shall derivate the equation with respect to $x$; therefore

$$y + xy' + \frac{y'}{y} = 0, \ y > 0,$$

namely

$$y^2 + xyy' + y' = 0, \ y > 0.$$

It results

$$y' = -\frac{y^2}{xy + 1}, \ xy + 1 \neq 0.$$

Hence

$$y'' = -\frac{2yy'(xy+1) - y^2(y + xy')}{(xy+1)^2} = \frac{-2xy^2y' - 2yy' + y^3 + xy^2y'}{(xy+1)^2}$$

$$= \frac{-(xy^2 + 2y)y' + y^3}{(xy+1)^2} = \frac{y^3(2xy+3)}{(xy+1)^3}, \ xy + 1 \neq 0.$$

We can check these results only using Maple 15:

$f := x \cdot y + \ln(y) = 0$

$$xy + \ln(y) = 0$$

$yp := implicitdiff(f, y, x)$

$$-\frac{y^2}{xy+1}$$

$ys := factor(implicitdiff(f, y, x\$2))$

$$\frac{y^3(2xy+3)}{(xy+1)^3}$$

**Example 5.4.** Compute $y'$, $y''$, $y'''$ if the function $y(x)$ is defined implicit by the equation:

$$x^2 + xy + y^2 = 3.$$

**Solution.**

We deduce

$$2x + y + xy' + 2yy' = 0,$$

i.e.

$$(x + 2y)\, y' = -2x - y;$$

hence

$$y' = -\frac{2x + y}{x + 2y}.$$

Further

$$
\begin{aligned}
y'' &= \frac{(-2 + y')\,(x + 2y) + (1 + 2y')\,(2x + y)}{(x + 2y)^2} \\
&= \frac{-2x - 4y - xy' - 2yy' + 2x + y + 4xy' + 2yy'}{(x + 2y)^2} = \frac{-3y + 3xy'}{(x + 2y)^2} \\
&= (-3) \cdot \frac{y + x \cdot \frac{2x+y}{x+2y}}{(x + 2y)^2} = (-6) \cdot \frac{x^2 + xy + y^2}{(x + 2y)^3} = -\frac{18}{(x + 2y)^3}
\end{aligned}
$$

and, finally:

$$
\begin{aligned}
y''' &= 18 \cdot \frac{3\,(x + 2y)^2\,(1 + 2yy')}{(x + 2y)^6} = 54 \cdot \frac{1 - 2 \cdot \frac{2x+y}{x+2y}}{(x + 2y)^4} \\
&= 54 \cdot \frac{x + 2y - 4x - 2y}{(x + 2y)^5} = -\frac{162x}{(x + 2y)^5}.
\end{aligned}
$$

We shall see the same thing using Maple 15:

$$f := x^2 + x \cdot y + y^2 = 3$$

$$x^2 + xy + y^2 = 3$$

$$y1 := implicitdiff(f, y, x)$$

$$-\frac{2x+y}{x+2y}$$

$$y2 := factor(implicitdiff(f, y, x\$2))\Big|$$

$$x^2 + xy + y^2 = 3$$

$$-\frac{18}{(x+2y)^3}$$

$$y3 := factor(implicitdiff(f, y, x\$3))\Big|$$

$$x^2 + xy + y^2 = 3$$

$$-\frac{162x}{(x+2y)^5}$$

**Example 5.5.** Compute $z'_x$ and $z'_y$ for the implicit function $z = f(x, y)$ defined by:

$$x^3 + 2y^3 + z^3 - 3xyz - 2y + 3 = 0.$$

**Solution.**
One finds the derivate with respect to $x$:

$$3x^2 + 3z^2 \cdot z'_x - 3yz - 3xy \cdot z'_x = 0 \qquad | : 3$$

$$\Longleftrightarrow x^2 - yz + (z^2 - xy) \cdot z'_x = 0 \Longrightarrow z'_x = \frac{-x^2 + yz}{z^2 - xy}, \quad z^2 - xy \neq 0.$$

One finds the derivative with respect to $y$:

$$6y^2 + 3z^2 \cdot z'_y - 3xz - 3xy \cdot z'_y - 2 = 0$$

$$\Longleftrightarrow 3(z^2 - xy) \cdot z'_y + 6y^2 - 3xz - 2 = 0 \Longrightarrow z'_y = \frac{1}{3} \cdot \frac{2 - 6y^2 + 3xz}{z^2 - xy}, \quad z^2 - xy \neq 0.$$

We shall check these results in Maple 15:

$f := x^3 + 2 \cdot y^3 + z^3 - 3 \cdot x \cdot y \cdot z - 2 \cdot y + 3 = 0 :$
$implicitdiff(f, z, x)$

$$\frac{-x^2 + yz}{z^2 - xy}$$

$simplify(implicitdiff(f, z, y))$

$$-\frac{1}{3} \frac{-6y^2 + 3xz + 2}{-z^2 + xy}$$

**Example 5.6.** Find the first and the second partial derivatives of the implicit function $z = f(x, y)$ defined by:

$$x^2 + y^2 + z^2 - 1 = 0.$$

**Solution.**
One obtains:

$$2x + 2zz'_x = 0 \Longrightarrow z'_x = -\frac{x}{z}, \ z \neq 0$$

and

$$2y + 2zz'_y = 0 \Longrightarrow z'_y = -\frac{y}{z}, \ z \neq 0.$$

Therefore

$$z''_{x^2} = \left(z'_x\right)'_x = -\frac{z + xz'_x}{z^2} = \frac{-z - x \cdot \frac{x}{z}}{z^2} = \frac{-z^2 - x^2}{z^3}, \ z \neq 0.$$

Taking into account that

$$z^2 = 1 - x^2 - y^2$$

we have

$$z''_{x^2} = \frac{-1 + x^2 + y^2 - x^2}{z^3} = \frac{-1 + y^2}{z^3}, \ z \neq 0.$$

Similarly,

$$z''_{y^2} = \left(z'_y\right)'_y = -\frac{z + yz'_y}{z^2} = \frac{-z - y \cdot \frac{y}{z}}{z^2} = \frac{-z^2 - y^2}{z^3}$$

$$= \frac{-1 + x^2 + y^2 - y^2}{z^3} = \frac{-1 + x^2}{z^3}, \ z \neq 0$$

and

$$z''_{xy} = (z'_x)'_y = \frac{xz'_y}{z^2} = -\frac{xy}{z^3}, \ z \neq 0.$$

We shall also give a computer solution using Maple 15:

$f := x^2 + y^2 + z^2 - 1 = 0$ :
*implicitdiff*$(f, z, x)$

$$-\frac{x}{z}$$

*implicitdiff*$(f, z, y)$

$$-\frac{y}{z}$$

*implicitdiff*$(f, z, x\$2\ )\Big|_{x^2 + z^2 = 1 - y^2}$

$$-\frac{1-y^2}{z^3}$$

*implicitdiff*$(f, z, y\$2\ )\Big|_{y^2 + z^2 = 1 - x^2}$

$$-\frac{1-x^2}{z^3}$$

*implicitdiff*$(f, z, x, y\ )$

$$-\frac{xy}{z^3}$$

**Example 5.7.** Compute $z''_{xy}(1, -2)$ if $z(x, y)$ is a function defined implicitly by the equation:

$$x^2 + 2y^2 + 3z^3 + xy - z = 0$$

and $z(1, -2) = 1$.
   **Solution.**
   We shall have:

$$2x + 9z^2 z'_x + y - z'_x = 0,$$

i.e.

$$2x + y = z'_x \left(1 - 9z^2\right);$$

hence

$$z'_x = \frac{2x + y}{1 - 9z^2}.$$

Similarly,

$$4y + 9z^2 z'_y + x - z'_y = 0,$$

i.e.

$$z'_y = \frac{x + 4y}{1 - 9z^2}.$$

We deduce that:

$$z''_{xy} = (z'_x)'_y = \frac{1 - 9z^2 + 18zz'_y (2x + y)}{(1 - 9z^2)^2} = \frac{1 - 9z^2 + 18z (2x + y) \cdot \frac{x+4y}{1-9z^2}}{(1 - 9z^2)^2};$$

therefore

$$z''_{xy} = \frac{81z^4 - 18z^2 + 18z (2x + y) (x + 4y) + 1}{(1 - 9z^2)^3}$$

and

$$z''_{xy} (-1, 2) = -\frac{1}{8}.$$

We have the following computer solution with Maple 15:

$f := x^2 + 2 \cdot y^2 + 3 \cdot z^2 + x \cdot y - z = 0$

$$x^2 + 2y^2 + 3z^2 + xy - z = 0$$

*implicitdiff*$(f, z, x, y)\big|_{x = 1, y = -2, z = 1}$

$$-\frac{1}{8}$$

## 5.2   Differentiation of Implicit Functions

**Example 5.8.** Find the expression for d$z$ and d$^2z$ for the implicit function $z = f(x, y)$ defined by:

$$\ln z = x + y + z - 1, z > 0.$$

**Solution.**

We shall use the following formula (from the third chapter):

$$dz = \frac{\partial z}{\partial x}dx + \frac{\partial z}{\partial y}dy = z'_x dx + z'_y dy.$$

After we shall derivate the given equation with respect to $x$ we shall have:

$$\frac{z'_x}{z} = 1 + z'_x \implies z'_x = z + z z'_x \implies z'_x = -\frac{z}{z-1}, \ z \neq 1.$$

Similarly, we shall obtain

$$z'_y = -\frac{z}{z-1}, \ z \neq 1.$$

Hence

$$dz = -\frac{z}{z-1}(dx + dy).$$

It will results

$$z''_{x^2} = \frac{-z'_x(z-1) + z z'_x}{(z-1)^2} = \frac{z'_x(-z+1+z)}{(z-1)^2} = \frac{z'_x}{(z-1)^2} = -\frac{z}{(z-1)^3}, \ z \neq 1,$$

$$z''_{y^2} = -\frac{z}{(z-1)^3}, \ z \neq 1,$$

$$z''_{xy} = \frac{-z'_y(z-1) + z z'_y}{(z-1)^2} = \frac{-z'_y(z-z+1)}{(z-1)^2} = \frac{z'_y}{(z-1)^2} = -\frac{z}{(z-1)^3}, \ z \neq 1;$$

therefore, using a relation (from chapter 3) we shall have:

$$d^2 z = \frac{\partial^2 f}{\partial x^2}dx^2 + 2\frac{\partial^2 f}{\partial x \partial y}dxdy + \frac{\partial^2 f}{\partial y^2}dy^2 = z''_{x^2}dx^2 + 2z''_{xy}dxdy + z''_{y^2}dy^2$$

$$= -\frac{z}{(z-1)^3}(dx^2 + 2dxdy + dy^2).$$

A computer solution of this example can be given in Maple 15:

$f := x + y + z = \ln(z)$

$$x + y + z = \ln(z)$$

$dz := simplify(implicitdiff(f, z, x) \cdot dx + implicitdiff(f, z, y) \cdot dy)$

$$-\frac{z(dx + dy)}{z - 1}$$

$d := factor(implicitdiff(f, z, x\$2)dx^2 + 2 \cdot implicitdiff(f, z, x, y) \cdot dxdy + implicitdiff(f, z, y\$2)dy^2)$

$$-\frac{z(dx^2 + 2\,dxdy + dy^2)}{(z - 1)^3}$$

**Example 5.9.** Compute the first and the second differentials for the function $y = f(x)$ and $z = g(x)$, defined by system:

$$\begin{cases} x^2 + y^2 + 3z^2 = 1 \\ x^2 + y^2 - z^2 = 0. \end{cases}$$

**Solution.**

By differentiating of our system of equations we obtain:

$$\begin{cases} 2x\,dx + 2y\,dy + 6z\,dz = 0 \\ 2x\,dx + 2y\,dy - 2z\,dz = 0 \end{cases} \Longleftrightarrow \begin{cases} 2y\,dy + 6z\,dz = -2x\,dx \\ 2y\,dy - 2z\,dz = -2x\,dx \end{cases}$$

$$\Longleftrightarrow \begin{cases} dz = 0 \Longrightarrow d^2z = 0 \\ 2y\,dy = -2x\,dx, \end{cases}$$

such that

$$dy = -\frac{x}{y}dx, \quad y \neq 0.$$

We shall differentiate the previous relation:

$$d^2y = \frac{(-y\,dx + x\,dy)\,dx}{y^2} = \frac{1}{y^2}\left(-y\,dx - \frac{x^2}{y}dx\right)dx = -\frac{x^2 + y^2}{y^3}d^2x, \quad y \neq 0.$$

We shall compute $\frac{dy}{dx}$ and $\frac{d^2y}{d^2x}$ using Maple 15:

$$f := x^2 + y^2 + 3 \cdot z^2 = 1 :$$
$$g := x^2 + y^2 - z^2 = 0 :$$
$$implicitdiff(\{f, g\}, \{y, z\}, y, x)$$

$$-\frac{x}{y}$$

$$implicitdiff(\{f, g\}, \{z, y\}, y, x\$2)$$

$$-\frac{x^2 + y^2}{y^3}$$

**Example 5.10.** The functions $u(x,y)$ and $v(x,y)$ are defined implicit functions. Find the first and the second differentials if

$$\begin{cases} x + y + u + v - 1 = 0 \\ x^2 + y^2 + u^2 + v^2 - 1 = 0. \end{cases}$$

**Solution.**
We have:

$$\begin{cases} u + v = -x - y + 1 \\ u^2 + v^2 = -x^2 - y^2 + 1. \end{cases}$$

By differentiating one deduces:

$$\begin{cases} du + dv = -dx - dy \\ 2u\,du + 2v\,dv = -2x\,dx - 2y\,dy; \end{cases}$$

hence

$$dv = \frac{(u - x)\,dx + (u - y)\,dy}{v - u},$$

$$du = -dx - dy - dv = -dx - dy - \frac{(u - x)\,dx + (u - y)\,dy}{v - u}$$
$$= \frac{(v - x)\,dx + (v - y)\,dy}{u - v}.$$

We shall compute the second differentials of the functions and $u(x,y)$ and $v(x,y)$; therefore

$$d^2u = \frac{(u - v)\,[(dv - dx)\,dx + (dv - dy)\,dy] - (du - dv)\,[(v - x)\,dx + (v - y)\,dy]}{(u - v)^2}.$$

We can notice

$$(v - x)\,dx + (v - y)\,dy = (u - v)\,du.$$

It results

$$d^2u = \frac{(u - v)\left[dv\,(dx + dy) - dx^2 - dy^2\right] - (du - dv)(u - v)\,du}{(u - v)^2}$$

$$= \frac{(u - v)\left[dv\,(dx + dy) - dx^2 - dy^2 - (du - dv)\,du\right]}{(u - v)^2}.$$

As $dx + dy = -du - dv$ one obtains

$$d^2u = \frac{dv\,(-du - dv) - dx^2 - dy^2 - (du - dv)\,du}{u - v}$$

$$= \frac{-du\,dv - dv^2 - dx^2 - dy^2 - du^2 + du\,dv}{u - v},$$

namely

$$d^2u = -\frac{dx^2 + dy^2 + du^2 + dv^2}{u - v}$$

or

$$d^2u = \left(-\frac{2}{(u - v)^3}\right)\left[\left(u^2 - uv + v^2 - vx + x^2 - xu\right)dx^2\right.$$
$$\left. + \left(v^2 - vx - yv + 2xy - xu + u^2 - yu\right)dx\,dy + \left(u^2 - uv + v^2 - yv + y^2 - yu\right)dy^2\right].$$

Similarly,

$$d^2v = \frac{(v - u)\left[(du - dx)\,dx + (du - dy)\,dy\right] - (dv - dv)\left[(u - x)\,dx + (u - y)\,dy\right]}{(v - u)^2}.$$

But

$$(u - x)\,dx + (u - y)\,dy = (v - u)\,dv,$$

such that

$$d^2v = \frac{(v - u)\left[(du - dx)\,dx + (du - dy)\,dy\right] - (v - u)\,dv\,(dv - du)}{(u - v)^2}$$

$$= \frac{du\,(dx + dy) - dx^2 - dy^2 - (dv - du)\,dv}{v - u}$$

$$= \frac{du\,(-du - dv) - dx^2 - dy^2 - (dv - du)\,dv}{v - u}$$

$$= \frac{du^2 - du\,dv - dx^2 - dy^2 - dv^2 + du\,dv}{v - u} = \frac{dx^2 + dy^2 + du^2 + dv^2}{u - v}$$

or

$$d^2v = \frac{2}{(u-v)^3}\left[\left(u^2 - uv + v^2 - vx + x^2 - xu\right)dx^2\right.$$
$$\left. + \left(v^2 - vx - yv + 2xy - xu + u^2 - yu\right)dxdy + \left(u^2 - uv + v^2 - yv + y^2 - yu\right)dy^2\right].$$

We shall show a computer solution in Maple 15:

$f := x + y + u + v - 1 = 0 :$
$g := x^2 + y^2 + u^2 + v^2 - 1 = 0 :$
$ux := implicitdiff(\{f, g\}, \{u(x, y), v(x, y)\}, u, x) :$
$uy := implicitdiff(\{f, g\}, \{u(x, y), v(x, y)\}, u, y) :$
$du := ux \cdot dx + uy \cdot dy$

$$\frac{(v-x)\,dx}{u-v} + \frac{(v-y)\,dy}{u-v}$$

$vx := implicitdiff(\{f, g\}, \{u(x, y), v(x, y)\}, v, x) :$
$vy := implicitdiff(\{f, g\}, \{u(x, y), v(x, y)\}, v, y) :$
$dv := vx \cdot dx + vy \cdot dy$

$$-\frac{(-x+u)\,dx}{u-v} - \frac{(-y+u)\,dy}{u-v}$$

$uxx := implicitdiff(\{f, g\}, \{u(x, y), v(x, y)\}, u, x\$2) :$
$uxy := implicitdiff(\{f, g\}, \{u(x, y), v(x, y)\}, u, x, y) :$
$uyy := implicitdiff(\{f, g\}, \{u(x, y), v(x, y)\}, u, y\$2) :$
$du2 := collect(collect(factor(uxx \cdot dx^2 + 2 \cdot uxy \cdot dx \cdot dy + uyy \cdot dy^2), dx), dy)$

$$\frac{2\left(u^2 - uv + v^2 + y^2 - yv - yu\right)dy^2}{(u-v)^3} - \frac{2\left(2xy - yv - xv + v^2 - xu + u^2 - yu\right)dx\,dy}{(u-v)^3}$$
$$- \frac{2\left(u^2 - uv + v^2 + x^2 - xv - xu\right)dx^2}{(u-v)^3}$$

$vxx := implicitdiff(\{f, g\}, \{u(x, y), v(x, y)\}, v, x\$2) :$
$vxy := implicitdiff(\{f, g\}, \{u(x, y), v(x, y)\}, v, x, y) :$
$vyy := implicitdiff(\{f, g\}, \{u(x, v), v(x, y)\}, v, y\$2) :$
$dv2 := collect(collect(factor(vxx \cdot dx^2 + 2 \cdot vxy \cdot dx \cdot dy + vyy \cdot dy^2), dx), dy)$

$$\frac{2\left(u^2 - uv + v^2 + y^2 - yv - yu\right)dy^2}{(u-v)^3} + \frac{2\left(2xy - yv - xv + v^2 - xu + u^2 - yu\right)dx\,dy}{(u-v)^3}$$
$$+ \frac{2\left(u^2 - uv + v^2 + x^2 - xv - xu\right)dx^2}{(u-v)^3}$$

**Example 5.11.** Calculate the first and the second differentials for the implicit function $z = f(x, y)$, defined by:

$$F(x + z, y + z) = 0. \tag{5.4}$$

**Solution.**
We denote:

$$\begin{cases} x + z = u \\ y + z = v. \end{cases}$$

By differentiating with respect to $x$ and $y$ we have:

$$\frac{\partial F}{\partial u} \cdot \frac{\partial u}{\partial x} + \frac{\partial F}{\partial v} \cdot \frac{\partial v}{\partial x} = 0, \tag{5.5}$$

$$\frac{\partial F}{\partial u} \cdot \frac{\partial u}{\partial y} + \frac{\partial F}{\partial v} \cdot \frac{\partial v}{\partial y} = 0. \tag{5.6}$$

As

$$\begin{cases} \frac{\partial u}{\partial x} = 1 + \frac{\partial z}{\partial x} \\ \frac{\partial v}{\partial x} = \frac{\partial z}{\partial x} \end{cases}$$

the relation (5.5) becomes

$$\frac{\partial F}{\partial u}\left(1 + \frac{\partial z}{\partial x}\right) + \frac{\partial F}{\partial v} \cdot \frac{\partial z}{\partial x} = 0 \Longleftrightarrow \frac{\partial F}{\partial u} + \frac{\partial z}{\partial x}\left(\frac{\partial F}{\partial u} + \frac{\partial F}{\partial v}\right) = 0,$$

therefore

$$\frac{\partial z}{\partial x} = \frac{-\frac{\partial F}{\partial u}}{\frac{\partial F}{\partial u} + \frac{\partial F}{\partial v}} = -\frac{F_u'}{F_u' + F_v'}.$$

As

$$\begin{cases} \frac{\partial u}{\partial y} = \frac{\partial z}{\partial y} \\ \frac{\partial v}{\partial y} = 1 + \frac{\partial z}{\partial y} \end{cases}$$

the relation (5.6) becomes

$$\frac{\partial F}{\partial u} \cdot \frac{\partial z}{\partial y} + \frac{\partial F}{\partial v}\left(1 + \frac{\partial z}{\partial y}\right) = 0 \Longleftrightarrow \frac{\partial z}{\partial y}\left(\frac{\partial F}{\partial u} + \frac{\partial F}{\partial v}\right) + \frac{\partial F}{\partial v} = 0,$$

such that

$$\frac{\partial z}{\partial y} = \frac{-\frac{\partial F}{\partial v}}{\frac{\partial F}{\partial u} + \frac{\partial F}{\partial v}} = -\frac{F_v'}{F_u' + F_v'}.$$

Therefore

$$dz = -\frac{F'_u}{F'_u + F'_v}dx - \frac{F'_v}{F'_u + F'_v}dy = -\frac{F'_u dx + F'_v dy}{F'_u + F'_v}.$$

We shall derive the relation (5.5) with respect to $x$:

$$\frac{\partial}{\partial x}\left(\underbrace{\frac{\partial F}{\partial u}\cdot\frac{\partial u}{\partial x}}_{G} + \underbrace{\frac{\partial F}{\partial v}\cdot\frac{\partial v}{\partial x}}_{H}\right) = 0;$$

it results that

$$\frac{\partial F}{\partial u}\cdot\frac{\partial^2 u}{\partial x^2} + \frac{\partial G}{\partial x}\cdot\frac{\partial u}{\partial x} + \frac{\partial F}{\partial v}\cdot\frac{\partial^2 v}{\partial x^2} + \frac{\partial H}{\partial x}\cdot\frac{\partial v}{\partial x} = 0 \Longleftrightarrow$$

$$\frac{\partial F}{\partial u}\frac{\partial^2 u}{\partial x^2} + \left(\frac{\partial G}{\partial u}\cdot\frac{\partial u}{\partial x} + \frac{\partial G}{\partial v}\cdot\frac{\partial v}{\partial x}\right)\frac{\partial u}{\partial x} + \frac{\partial F}{\partial v}\frac{\partial^2 v}{\partial x^2} + \left(\frac{\partial H}{\partial u}\cdot\frac{\partial u}{\partial x} + \frac{\partial H}{\partial v}\cdot\frac{\partial v}{\partial x}\right)\cdot\frac{\partial v}{\partial x} = 0 \Longleftrightarrow$$

$$\frac{\partial F}{\partial u}\frac{\partial^2 u}{\partial x^2} + \left(\frac{\partial^2 F}{\partial u^2}\cdot\frac{\partial u}{\partial x} + \frac{\partial^2 F}{\partial u\partial v}\cdot\frac{\partial v}{\partial x}\right)\frac{\partial u}{\partial x} + \frac{\partial F}{\partial v}\frac{\partial^2 v}{\partial x^2} + \left(\frac{\partial^2 F}{\partial u\partial v}\cdot\frac{\partial u}{\partial x} + \frac{\partial^2 F}{\partial v^2}\cdot\frac{\partial v}{\partial x}\right)\cdot\frac{\partial v}{\partial x} = 0$$

$$\Longleftrightarrow \frac{\partial F}{\partial u}\cdot\frac{\partial^2 u}{\partial x^2} + \frac{\partial^2 F}{\partial u^2}\cdot\left(\frac{\partial u}{\partial x}\right)^2 + 2\frac{\partial^2 F}{\partial u\partial v}\cdot\frac{\partial u}{\partial x}\cdot\frac{\partial v}{\partial x} + \frac{\partial F}{\partial v}\cdot\frac{\partial^2 v}{\partial x^2} + \frac{\partial^2 F}{\partial v^2}\cdot\left(\frac{\partial v}{\partial x}\right)^2 = 0.$$
(5.7)

As

$$\begin{cases} \frac{\partial^2 u}{\partial x^2} = \frac{\partial^2 z}{\partial x^2} \\ \frac{\partial^2 v}{\partial x^2} = \frac{\partial^2 z}{\partial x^2} \end{cases}$$

the relation (5.7) becomes:

$$F'_u\cdot\frac{\partial^2 z}{\partial x^2} + F''_{u^2}\cdot\left(1 + \frac{\partial z}{\partial x}\right)^2 + 2F''_{uv}\cdot\left(1 + \frac{\partial z}{\partial x}\right)\cdot\frac{\partial z}{\partial x} + F'_v\cdot\frac{\partial^2 z}{\partial x^2} + F''_{v^2}\cdot\left(\frac{\partial z}{\partial x}\right)^2 = 0 \Longleftrightarrow$$

$$\left(F'_u + F'_v\right)\cdot\frac{\partial^2 z}{\partial x^2} + F''_{u^2}\cdot\left(1 - \frac{F'_u}{F'_u + F'_v}\right)^2 - 2F''_{uv}\cdot\left(1 - \frac{F'_u}{F'_u + F'_v}\right)\cdot\frac{F'_u}{F'_u + F'_v} + F''_{v^2}\cdot\left(-\frac{F'_u}{F'_u + F'_v}\right)^2 = 0;$$

therefore

$$\left(F'_u + F'_v\right)\cdot\frac{\partial^2 z}{\partial x^2} + F''_{u^2}\cdot\left(\frac{F'_v}{F'_u + F'_v}\right)^2 - 2F''_{uv}\cdot\frac{F'_u\cdot F'_v}{\left(F'_u + F'_v\right)^2} + F''_{v^2}\cdot\left(-\frac{F'_u}{F'_u + F'_v}\right)^2 = 0$$

i.e.

$$(F'_u + F'_v) \cdot \frac{\partial^2 z}{\partial x^2} + \frac{F''_{u^2} \cdot (F'_v)^2 - 2F''_{uv} F'_u F'_v + F''_{v^2} \cdot (F'_u)^2}{(F'_u + F'_v)^2} = 0,$$

such that

$$\frac{\partial^2 z}{\partial x^2} = -\frac{F''_{u^2} \cdot (F'_v)^2 - 2F''_{uv} F'_u F'_v + F''_{v^2} \cdot (F'_u)^2}{(F'_u + F'_v)^3}. \tag{5.8}$$

Similarly, we shall derive the relation (5.6) with respect to $y$:

$$\frac{\partial}{\partial y} \left( \underbrace{\frac{\partial F}{\partial u} \cdot \frac{\partial u}{\partial y}}_{G} + \underbrace{\frac{\partial F}{\partial v} \cdot \frac{\partial v}{\partial y}}_{H} \right) = 0;$$

it results that

$$\frac{\partial F}{\partial u} \cdot \frac{\partial^2 u}{\partial y^2} + \frac{\partial G}{\partial y} \cdot \frac{\partial u}{\partial y} + \frac{\partial F}{\partial v} \cdot \frac{\partial^2 v}{\partial y^2} + \frac{\partial H}{\partial y} \cdot \frac{\partial v}{\partial y} = 0 \Longleftrightarrow$$

$$\frac{\partial F}{\partial u} \cdot \frac{\partial^2 u}{\partial y^2} + \left( \frac{\partial G}{\partial u} \cdot \frac{\partial u}{\partial y} + \frac{\partial G}{\partial v} \cdot \frac{\partial v}{\partial y} \right) \cdot \frac{\partial u}{\partial y} + \frac{\partial F}{\partial v} \cdot \frac{\partial^2 v}{\partial y^2} + \left( \frac{\partial H}{\partial u} \cdot \frac{\partial u}{\partial y} + \frac{\partial H}{\partial v} \cdot \frac{\partial v}{\partial y} \right) \cdot \frac{\partial v}{\partial y} = 0 \Longleftrightarrow$$

$$\frac{\partial F}{\partial u} \cdot \frac{\partial^2 u}{\partial y^2} + \frac{\partial^2 F}{\partial u^2} \cdot \left( \frac{\partial u}{\partial y} \right)^2 + 2\frac{\partial^2 F}{\partial u \partial v} \cdot \frac{\partial u}{\partial y} \cdot \frac{\partial v}{\partial y} + \frac{\partial F}{\partial v} \cdot \frac{\partial^2 v}{\partial y^2} + \frac{\partial^2 F}{\partial v^2} \cdot \left( \frac{\partial v}{\partial y} \right)^2 = 0. \tag{5.9}$$

As

$$\begin{cases} \frac{\partial^2 u}{\partial y^2} = \frac{\partial^2 z}{\partial y^2} \\ \frac{\partial^2 v}{\partial y^2} = \frac{\partial^2 z}{\partial y^2} \end{cases}$$

the relation (5.9) becomes:

$$F'_u \cdot \frac{\partial^2 z}{\partial y^2} + F''_{u^2} \cdot \left( \frac{\partial z}{\partial y} \right)^2 + 2F''_{uv} \cdot \left( 1 + \frac{\partial z}{\partial y} \right) \cdot \frac{\partial z}{\partial y} + F'_v \cdot \frac{\partial^2 z}{\partial y^2} + F''_{v^2} \cdot \left( 1 + \frac{\partial z}{\partial y} \right)^2 = 0$$

therefore

$$(F'_u + F'_v) \cdot \frac{\partial^2 z}{\partial y^2} + F''_{u^2} \cdot \left( -\frac{F'_v}{F'_u + F'_v} \right)^2 - 2F''_{uv} \cdot \frac{F'_u \cdot F'_v}{(F'_u + F'_v)^2} + F''_{v^2} \cdot \left( \frac{F'_u}{F'_u + F'_v} \right)^2 = 0 \Longleftrightarrow$$

$$(F'_u + F'_v) \cdot \frac{\partial^2 z}{\partial y^2} + \frac{F''_{u^2} \cdot (F'_v)^2 - 2F''_{uv} F'_u F'_v + F''_{v^2} \cdot (F'_u)^2}{(F'_u + F'_v)^2} = 0,$$

such that

$$\frac{\partial^2 z}{\partial y^2} = -\frac{F''_{u^2} \cdot (F'_v)^2 - 2F''_{uv}F'_uF'_v + F''_{v^2} \cdot (F'_u)^2}{(F'_u + F'_v)^3}. \tag{5.10}$$

By deriving the relation (5.5) with respect to $y$, we have:

$$\frac{\partial}{\partial y}\left(\underbrace{\frac{\partial F}{\partial u} \cdot \frac{\partial u}{\partial x}}_{G} + \underbrace{\frac{\partial F}{\partial v} \cdot \frac{\partial v}{\partial x}}_{H}\right) = 0;$$

one obtains

$$\frac{\partial F}{\partial u} \cdot \frac{\partial^2 u}{\partial x \partial y} + \frac{\partial G}{\partial y} \cdot \frac{\partial u}{\partial x} + \frac{\partial F}{\partial v} \cdot \frac{\partial^2 v}{\partial x \partial y} + \frac{\partial H}{\partial y} \cdot \frac{\partial v}{\partial x} = 0 \iff$$

$$\frac{\partial F}{\partial u} \frac{\partial^2 u}{\partial x \partial y} + \left(\frac{\partial G}{\partial u} \cdot \frac{\partial u}{\partial y} + \frac{\partial G}{\partial v} \cdot \frac{\partial v}{\partial y}\right)\frac{\partial u}{\partial x} + \frac{\partial F}{\partial v} \cdot \frac{\partial^2 v}{\partial x \partial y} + \left(\frac{\partial H}{\partial u} \cdot \frac{\partial u}{\partial y} + \frac{\partial H}{\partial v} \cdot \frac{\partial v}{\partial y}\right) \cdot \frac{\partial v}{\partial x} = 0 \iff$$

$$\frac{\partial F}{\partial u} \cdot \frac{\partial^2 u}{\partial x \partial y} + \frac{\partial^2 F}{\partial u^2} \cdot \frac{\partial u}{\partial x} \cdot \frac{\partial u}{\partial y} + \frac{\partial^2 F}{\partial u \partial v} \cdot \frac{\partial u}{\partial x} \cdot \frac{\partial v}{\partial y} + \frac{\partial F}{\partial v} \cdot \frac{\partial^2 v}{\partial x \partial y} \tag{5.11}$$

$$+ \frac{\partial^2 F}{\partial u \partial v} \cdot \frac{\partial v}{\partial x} \cdot \frac{\partial u}{\partial y} + \frac{\partial^2 F}{\partial v^2} \cdot \frac{\partial v}{\partial x} \cdot \frac{\partial v}{\partial y} = 0.$$

As

$$\begin{cases} \frac{\partial^2 u}{\partial x \partial y} = \frac{\partial^2 z}{\partial x \partial y} \\ \frac{\partial^2 v}{\partial x \partial y} = \frac{\partial^2 z}{\partial x \partial y} \end{cases}$$

the relation (5.11) becomes:

$$(F'_u + F'_v)\frac{\partial^2 z}{\partial x \partial y} + F''_{u^2}\left(1 + \frac{\partial z}{\partial x}\right) \cdot \frac{\partial z}{\partial y} + F''_{uv}\left(1 + \frac{\partial z}{\partial x}\right)\left(1 + \frac{\partial z}{\partial y}\right)$$

$$+ F''_{uv}\frac{\partial z}{\partial x}\frac{\partial z}{\partial y} + F''_{v^2}\left(1 + \frac{\partial z}{\partial y}\right)\frac{\partial z}{\partial x} = 0 \iff$$

$$\left(F'_u + F'_v\right)\frac{\partial^2 z}{\partial x \partial y} - F''_{u^2}\left(\frac{F'_v}{F'_u + F'_v}\right)^2 + F''_{uv}\frac{F'_uF'_v}{(F'_u + F'_v)^2} + F''_{uv} \cdot \frac{F'_uF'_v}{(F'_u + F'_v)^2} - F''_{v^2}\left(\frac{F'_u}{F'_u + F'_v}\right)^2 = 0$$

namely

$$(F'_u + F'_v) \cdot \frac{\partial^2 z}{\partial x \partial y} - \frac{F''_{u^2}(F'_v)^2 - 2F''_{uv}F'_uF'_v + F''_{v^2}(F'_u)^2}{(F'_u + F'_v)^2} = 0;$$

therefore

$$\frac{\partial^2 z}{\partial x \partial y} = \frac{F''_{u^2}(F'_v)^2 - 2F''_{uv}F'_uF'_v + F''_{v^2} \cdot (F'_u)^2}{(F'_u + F'_v)^3}. \tag{5.12}$$

Taking into account the relations (5.8), (5.10) and (5.12) one deduces

$$d^2 z = \frac{\partial^2 z}{\partial x^2}dx^2 + 2\frac{\partial^2 z}{\partial x \partial y}dxdy + \frac{\partial^2 z}{\partial y^2}dy^2$$

$$= -\frac{F''_{u^2}(F'_v)^2 - 2F''_{uv}F'_uF'_v + F''_{v^2}(F'_u)^2}{(F'_u + F'_v)^3}\left(dx^2 - 2dxdy + dy^2\right).$$

We can prove the previous results using Maple 15:

$f := F(x + z, y + z) = 0:$
$dz := simplify(implicitdiff(f, z, x) \; dx + implicitdiff(f, z, y) \cdot dy)$

$$-\frac{D_1(F)(x + z, y + z)\, dx + D_2(F)(x + z, y + z)\, dy}{D_1(F)(x + z, y + z) + D_2(F)(x + z, y + z)}$$

$d := factor(implicitdiff(f, z, x\$2)dx^2 + 2 \cdot implicitdiff(f, z, x, y) \cdot dxdy + implicitdiff(f, z, y\$2)dy^2)$

$$-\frac{1}{\left(D_1(F)(x + z, y + z) + D_2(F)(x + z, y + z)\right)^3}\Big(\big(D_{1,1}(F)(x + z, y + z)\, D_2(F)(x + z, y + z)^2$$

$$- 2D_2(F)(x + z, y + z)\, D_{1,2}(F)(x + z, y + z)\, D_1(F)(x + z, y + z) + D_{2,2}(F)(x + z, y$$

$$+ z)\, D_1(F)(x + z, y + z)^2\big)\left(dx^2 - 2\, dxdy + dy^2\right)\Big)$$

## 5.3   Systems of Implicit Functions

**Theorem 5.12** (see [41], p. 200). Let $(a_1, a_2, \ldots, a_m; b_1, b_2, \ldots, b_n) \in E \subset \mathbb{R}^{n+m}$ be a solution of the system

$$\begin{cases} F_1(x_1, x_2, \ldots, x_m; y_1, y_2, \ldots, y_n) = 0 \\ F_2(x_1, x_2, \ldots, x_m; y_1, y_2, \ldots, y_n) = 0 \\ \quad\vdots \\ F_n(x_1, x_2, \ldots, x_m; y_1, y_2, \ldots, y_n) = 0, \end{cases}$$

where the functions $F_i : E \to \mathbb{R}$, $i = \overline{1, n}$ perform the conditions:

i) their first order partial derivatives are continuous on a neighborhood of
   the point $(a_1, a_2, \ldots, a_m; b_1, b_2, \ldots, b_n)$;

ii) $\frac{D(F_1, F_2, \ldots, F_n)}{D(y_1, y_2, \ldots, y_n)} \neq 0$ in the point $(a_1, a_2, \ldots, a_m; b_1, b_2, \ldots, b_n)$.

Under these conditions there exists one and only one system of $n$ implicit
functions, with $m$ variables:

$$\begin{cases} y_1 = f_1(x_1, x_2, \ldots, x_m) \\ y_2 = f_2(x_1, x_2, \ldots, x_m) \\ \qquad\vdots \\ y_n = f_n(x_1, x_2, \ldots, x_m), \end{cases}$$

defined in some neighborhood of $(a_1, a_2, \ldots, a_m; b_1, b_2, \ldots, b_n)$ and which
satisfy the above equations, for $f_i : E \subset \mathbb{R}^m \to \mathbb{R}$, $i = \overline{1, n}$ having the partial
derivatives:

$$\begin{cases} \dfrac{\partial f_1}{\partial x_i} = -\dfrac{\frac{D(F_1, F_2, \ldots, F_n)}{D(x_i, y_2, \ldots, y_n)}}{\frac{D(F_1, F_2, \ldots, F_n)}{D(y_1, y_2, \ldots, y_n)}} \\ \qquad\vdots \\ \dfrac{\partial f_n}{\partial x_i} = -\dfrac{\frac{D(F_1, F_2, \ldots, F_n)}{D(y_1, y_2, \ldots, y_{n-1}, x_i)}}{\frac{D(F_1, F_2, \ldots, F_n)}{D(y_1, y_2, \ldots, y_n)}}, \end{cases}$$

**Example 5.13.** Compute $\frac{\partial z}{\partial x}$ and $\frac{\partial z}{\partial y}$ in the point $u = 1, v = 1$ if the
equations

$$\begin{cases} x = u + \ln v \\ y = v - \ln u \\ z = 2u + v \end{cases}$$

define $u(x, y)$, $v(x, y)$, $z(x, y)$ as some implicit functions.
**Solution.**
We shall denote:

$$\begin{cases} F_1(x, y; u, v, z) = x - u - \ln v \\ F_2(x, y; u, v, z) = y - v + \ln u \\ F_3(x, y; u, v, z) = z - 2u - v. \end{cases}$$

We shall have:

$$\begin{cases} \dfrac{\partial z}{\partial x} = -\dfrac{\frac{D(F_1, F_2, F_3)}{D(u, v, x)}}{\frac{D(F_1, F_2, F_3)}{D(u, v, z)}} \\ \dfrac{\partial z}{\partial y} = -\dfrac{\frac{D(F_1, F_2, F_3)}{D(u, v, y)}}{\frac{D(F_1, F_2, F_3)}{D(u, v, z)}}. \end{cases}$$

As

$$\frac{D\left(F_1, F_2, F_3\right)}{D\left(u, v, z\right)} = \begin{vmatrix} -1 & -\frac{1}{v} & 0 \\ \frac{1}{u} & -1 & 0 \\ -2 & -1 & 1 \end{vmatrix} = 1 + \frac{1}{uv} \xrightarrow[u=1,v=1]{} 2,$$

$$\frac{D\left(F_1, F_2, F_3\right)}{D\left(u, v, x\right)} = \begin{vmatrix} -1 & -\frac{1}{v} & 1 \\ \frac{1}{u} & -1 & 0 \\ -2 & -1 & 0 \end{vmatrix} = -\frac{1}{u} - 2 \xrightarrow[u=1,v=1]{} -3,$$

$$\frac{D\left(F_1, F_2, F_3\right)}{D\left(u, v, y\right)} = \begin{vmatrix} -1 & -\frac{1}{v} & 0 \\ \frac{1}{u} & -1 & 1 \\ -2 & -1 & 0 \end{vmatrix} = -\left(1 - \frac{2}{v}\right) \xrightarrow[u=1,v=1]{} 1,$$

it results that

$$\begin{cases} \left.\dfrac{\partial z}{\partial x}\right|_{u=1,v=1} = \dfrac{3}{2} \\ \left.\dfrac{\partial z}{\partial y}\right|_{u=1,v=1} = -\dfrac{1}{2}. \end{cases}$$

We can give a computer solution only in Maple 15:

$f := u + \ln(v) = x:$
$g := v - \ln(u) = y:$
$h := 2 \cdot u + v = z:$
$d1 := implicitdiff(\{f, g, h\}, \{u(x, y), v(x, y), z(x, y)\}, z, x):$
$d1\big|_{u=1,\,v=1}$

$$\frac{3}{2}$$

$d2 := implicitdiff(\{f, g, h\}, \{u(x, y), v(x, y), z(x, y)\}, z, y):$
$d2\big|_{u=1,\,v=1}$

$$-\frac{1}{2}$$

**Example 5.14.** The system

$$\begin{cases} xe^{u+v} + 2uv = 1 \\ ye^{u-v} - \dfrac{u}{1+v} = 2x \end{cases}$$

define the implicit functions $u(x, y)$ and $v(x, y)$ such that $u(1, 2) = 0$ and $v(1, 2) = 0$. Compute the differentials $du(1, 2)$ and $dv(1, 2)$.

**Solution.**
We shall denote:

$$\begin{cases} F(x, y; u, v) = xe^{u+v} + 2uv - 1 \\ G(x, y; u, v) = ye^{u-v} - \dfrac{u}{1+v} - 2x. \end{cases}$$

We have:

$$\begin{cases} \mathrm{d}u = u'_x \mathrm{d}x + u'_y \mathrm{d}y \\ \mathrm{d}v = v'_x \mathrm{d}x + v'_y \mathrm{d}y; \end{cases}$$

therefore

$$\begin{cases} \mathrm{d}u\,(1,2) = u'_x\,(1,2)\,\mathrm{d}x + u'_y\,(1,2)\,\mathrm{d}y \\ \mathrm{d}v\,(1,2) = v'_x\,(1,2)\,\mathrm{d}x + v'_y\,(1,2)\,\mathrm{d}y, \end{cases}$$

where

$$\begin{cases} u'_x = \dfrac{\partial u}{\partial x} = -\dfrac{\frac{D(F,G)}{D(x,v)}}{\frac{D(F,G)}{D(u,v)}} \\[3em] u'_y = \dfrac{\partial u}{\partial y} = -\dfrac{\frac{D(F,G)}{D(y,v)}}{\frac{D(F,G)}{D(u,v)}} \\[3em] v'_x = \dfrac{\partial v}{\partial x} = -\dfrac{\frac{D(F,G)}{D(u,x)}}{\frac{D(F,G)}{D(u,v)}} \\[3em] v'_y = \dfrac{\partial v}{\partial y} = -\dfrac{\frac{D(F,G)}{D(u,y)}}{\frac{D(F,G)}{D(u,v)}} \end{cases}$$

and

$$\frac{D\,(F,G)}{D\,(u,v)} = \begin{vmatrix} xe^{u+v} + 2v & xe^{u+v} + 2u \\ ye^{u-v} - \frac{1}{1+v} & -ye^{u-v} + \frac{u}{(1+v)^2} \end{vmatrix}$$
$$= (xe^{u+v} + 2v)\left(-ye^{u-v} + \frac{u}{(1+v)^2}\right) - (xe^{u+v} + 2u)\left(ye^{u-v} - \frac{1}{1+v}\right)$$

$$\frac{D\,(F,G)}{D\,(x,v)} = \begin{vmatrix} e^{u+v} & xe^{u+v} + 2u \\ -2 & -ye^{u-v} + \frac{u}{(1+v)^2} \end{vmatrix} = e^{u+v}\left(-ye^{u-v} + \frac{u}{(1+v)^2}\right) + 2\,(xe^{u+v} + 2u)$$

$$\frac{D\,(F,G)}{D\,(y,v)} = \begin{vmatrix} 0 & xe^{u+v} + 2u \\ e^{u-v} & -ye^{u-v} + \frac{u}{(1+v)^2} \end{vmatrix} = -e^{u-v}\,(xe^{u+v} + 2u)$$

$$\frac{D\,(F,G)}{D\,(u,x)} = \begin{vmatrix} xe^{u+v} + 2v & e^{u+v} \\ ye^{u-v} - \frac{1}{1+v} & -2 \end{vmatrix} = -2\,(xe^{u+v} + 2v) - e^{u+v}\left(ye^{u-v} - \frac{1}{1+v}\right)$$

$$\frac{D\,(F,G)}{D\,(u,y)} = \begin{vmatrix} xe^{u+v} + 2v & 0 \\ ye^{u-v} - \frac{1}{1+v} & e^{u-v} \end{vmatrix} = e^{u-v}\,(xe^{u+v} + 2v).$$

Hence

$$\begin{cases} u'_x(1,2) = 0 \\ u'_y(1,2) = -\frac{1}{3} \\ v'_x(1,2) = -1 \\ v'_y(1,2) = \frac{1}{3} \end{cases}$$

and

$$\begin{cases} du(1,2) = -\frac{1}{3}dy \\ dv(1,2) = -dx + \frac{1}{3}dy. \end{cases}$$

Solving this problem in Maple 15 we can see the same result:

$f := x \cdot e^{u+v} + 2 \cdot u \cdot v = 1:$

$g := y \cdot e^{u-v} - \dfrac{u}{1+v} = 2 \cdot x:$

$ux := implicitdiff(\{f, g\}, \{u(x, y), v(x, y)\}, u, x)\Big|_{x=1, y=2, u=0, v=0}$  :

$uy := implicitdiff(\{f, g\}, \{u(x, y), v(x, y)\}, u, y)\Big|_{x=1, y=2, u=0, v=0}$  :

$du := ux \cdot dx + uy \cdot dy$

$$-\frac{1}{3}dy$$

$vx := implicitdiff(\{f, g\}, \{u(x, y), v(x, y)\}, v, x)\Big|_{x=1, y=2, u=0, v=0}$  :

$vy := implicitdiff(\{f, g\}, \{u(x, y), v(x, y)\}, v, y)\Big|_{x=1, y=2, u=0, v=0}$  :

$dv := vx \cdot dx + vy \cdot dy$

$$-dx + \frac{1}{3}dy$$

**Example 5.15.** Compute $\dfrac{\partial^2 z}{\partial x \partial y}$ in the point $u = 2, v = 1$ if the equations

$$\begin{cases} x = u + v^2 \\ y = u^2 - v^3 \\ z = 2uv \end{cases}$$

define $u(x, y)$, $v(x, y)$, $z(x, y)$ as some implicit functions.
**Solution.**
By differentiating we achieve:

$$\begin{cases} dx = du + 2vdv \\ dy = 2udu - 3v^2dv \\ dz = 2vdu + 2udv \end{cases}$$

i.e. it results the following system, having the unknowns $du$, $dv$, $dz$:

$$\begin{cases} du + 2vdv = dx \\ 2udu - 3v^2dv = dy \\ 2vdu + 2udv - dz = 0. \end{cases}$$

The determinant of the associated system matrix is

$$\det = \begin{vmatrix} 1 & 2v & 0 \\ 2u & -3v^2 & 0 \\ 2v & 2u & -1 \end{vmatrix} = 3v^2 + 4uv.$$

Therefore

$$du = \frac{\begin{vmatrix} dx & 2v & 0 \\ dy & -3v^2 & 0 \\ 0 & 2u & -1 \end{vmatrix}}{\det} = \frac{3v^2dx + 2vdy}{3v^2 + 4uv} \tag{5.13}$$

$$dv = \frac{\begin{vmatrix} 1 & dx & 0 \\ 2u & dy & 0 \\ 2v & 0 & -1 \end{vmatrix}}{\det} = \frac{2udx - dy}{3v^2 + 4uv} \tag{5.14}$$

$$dz = \frac{\begin{vmatrix} 1 & 2v & dx \\ 2u & -3v^2 & dy \\ 2v & 2u & 0 \end{vmatrix}}{\det} = \frac{\left(4u^2 + 6v^3\right) dx - \left(2u - 4v^2\right) dy}{3v^2 + 4uv}. \tag{5.15}$$

Taking into account that

$$\begin{cases} du = u'_x dx + u'_y dy \\ dv = v'_x dx + v'_y dy \\ dz = z'_x dx + z'_y dy \end{cases}$$

from (5.13), (5.14) and (5.15) it results:

$$\begin{aligned} u'_x &= \frac{3v^2}{3v^2+4uv} &\Longrightarrow\; u'_x(2,1) = \tfrac{3}{11} \\ u'_y &= \frac{2v}{3v^2+4uv} &\Longrightarrow\; u'_y(2,1) = \tfrac{2}{11} \\ v'_x &= \frac{2u}{3v^2+4uv} &\Longrightarrow\; v'_x(2,1) = \tfrac{4}{11} \\ v'_y &= -\frac{1}{3v^2+4uv} &\Longrightarrow\; v'_y(2,1) = -\tfrac{1}{11} \end{aligned}$$

$$z'_x = \frac{4u^2 + 6v^3}{3v^2 + 4uv}$$

$$z'_y = \frac{-2u + 4v^2}{3v^2 + 4uv}.$$

We shall deduce

$$\frac{\partial^2 z}{\partial y \partial x} = (z'_x)'_y = \frac{\left[(8uu'_y + 18v^2 v'_y)(3v^2 + 4uv) - (4u^2 + 6v^3)(6vv'_y + 4u'_y v + 4uv'_y)\right]}{(3v^2 + 4uv)^2}$$

and

$$\frac{\partial^2 z}{\partial y \partial x}(2,1) = \frac{(16 \cdot \frac{2}{11} - 18 \cdot \frac{1}{11}) \cdot 11 - (16 + 6) \cdot (-\frac{6}{11} + 4 \cdot \frac{2}{11} - \frac{4}{11} \cdot 2)}{121} = \frac{26}{121}.$$

A computer solution will be given using Maple 15:

$$f := u + v^2 = x :$$
$$g := u^2 - v^3 = y :$$
$$h := 2 \cdot u \cdot v = z :$$
$$d := implicitdiff(\{f, g, h\}, \{u(x,y), v(x,y), z(x,y)\}, z, x, y) :$$
$$d \Big|_{u=2,\, v=1}$$

$$\frac{26}{121}$$

## 5.4   Functional Dependence

**Definition 5.16** (see [15], p. 161). The real valued functions $y_k = f_k(x_1, x_2, \ldots, x_m)$, $k = \overline{1,n}$ of $m$ variables are **functionally dependent** if they satisfy a relation of the form $\Phi(y_1, y_2, \ldots, y_n) = 0$ identically.

**Proposition 5.17** (see [37], p. 172). Assuming the functions $f_k$, $k = \overline{1,n}$ to be continuously differentiable, it is claimed that the functions are functionally dependent if and only if the rank of their Jacobian

$$J_f(y_1, y_2, \ldots, y_n) = \begin{pmatrix} (y_1)'_{x_1} & (y_1)'_{x_2} & \cdots & (y_1)'_{x_m} \\ (y_2)'_{x_1} & (y_2)'_{x_2} & \cdots & (y_2)'_{x_m} \\ \cdots & \cdots & \cdots & \cdots \\ (y_n)'_{x_1} & (y_n)'_{x_2} & \cdots & (y_n)'_{x_m} \end{pmatrix}$$

is everywhere less than $n$.

**Example 5.18.** Establish the functional dependence of the functions:

$$\begin{cases} y_1 = x - y \\ y_2 = x^2 + y^2 + 2z \\ y_3 = -xy - z \end{cases}$$

defined on $\mathbb{R}^3$.

**Solution**.

As the Jacobian of our functions is

$$|J_f(y_1, y_2, y_3)| = \begin{vmatrix} (y_1)'_x & (y_1)'_y & (y_1)'_z \\ (y_2)'_x & (y_2)'_y & (y_2)'_z \\ (y_3)'_x & (y_3)'_y & (y_3)'_z \end{vmatrix} = \begin{vmatrix} 1 & -1 & 0 \\ 2x & 2y & 2 \\ -y & -x & -1 \end{vmatrix} = 0$$

it results that the three functions $y_1, y_2, y_3$ are functionally dependent in $\mathbb{R}^3$.

Having

$$\begin{vmatrix} -1 & 0 \\ 2y & 2 \end{vmatrix} \neq 0$$

one deduces that

$$\text{rank} (J_f(y_1, y_2, y_3)) = 2;$$

therefore $y_1$ and $y_2$ are functionally independent, while $y_3$ is functionally dependent by them.

One can notice that

$$y_1^2 - y_2 = 2y_3.$$

We can prove this result using Matlab 7.9:

```
>> syms x y z
>> f=[x-y;x^2+y^2+2*z;-x*y-z];
>> t=[x y z];
>> rank(jacobian(f,t))
ans =
2
>> y1=f(1);y2=f(2);y3=f(3);
>> simplify(y1^2-y2)==2*y3
ans =
1
```

and in Mathcad 14:

$$f(x) := \begin{bmatrix} x_0 - x_1 \\ \left(x_0\right)^2 + \left(x_1\right)^2 + 2 \cdot x_2 \\ -x_0 \cdot x_1 - x_2 \end{bmatrix}$$

rank$(\text{Jacob}(f(x),x)) \to 2$

$y1(x) := f(x)_0 \qquad y2(x) := f(x)_1 \qquad y3(x) := f(x)_2$

$y1(x)^2 - y2(x) = 2 \cdot y3(x)$ simplify $\to 1$

and with Mathematica 8:

In[75]:= $f := \{x - y, x^2 + y^2 + 2*z, -x*y - z\}$

In[76]:= $\text{MatrixRank}[D[f, \{\{x, y, z\}\}]]$

Out[76]= 2

In[77]:= $y1 := f[[1]]; y2 := f[[2]]; y3 := f[[3]];$

In[78]:= $\text{Simplify}[y1^2 - y2] == 2*y3$

Out[78]= $-2 (x y + z) == 2 (-x y - z)$

and using Maple 15:

$f := [x - y, x^2 + y^2 + 2 \cdot z, -x \cdot y - z]$

$$[x - y, x^2 + y^2 + 2z, -xy - z]$$

with( VectorCalculus ) :
with( LinearAlgebra ) :
Rank( Jacobian( f, [x, y, z]))

$$2$$

$y1 := f_1 : y2 := f_2 : y3 := f_3 :$
evalb( simplify( $y1^2 - y2$ ) = $2 \cdot y3$ )

$$\textit{true}$$

**Example 5.19.** Establish the functional dependence of the functions:

$$\begin{cases} f\left(x,y,z\right) = x+y+z \\ g\left(x,y,z\right) = x-y+z \\ h\left(x,y,z\right) = 4\left(xy+yz\right) \end{cases}$$

defined on $\mathbb{R}^3$. Express $h$ depending on $f$ and $g$.

**Solution.**

As the Jacobian of our functions is

$$\left|J_f\left(f,g,h\right)\right| = \begin{vmatrix} f'_x & f'_y & f'_z \\ g'_x & g'_y & g'_z \\ h'_x & h'_y & h'_z \end{vmatrix} = \begin{vmatrix} 1 & 1 & 1 \\ 1 & 1 & 1 \\ y & 4\left(x+z\right) & 4y \end{vmatrix} = 0$$

it results that the three functions $f, g, h$ are functionally dependent in $\mathbb{R}^3$.

Having

$$\operatorname{rank}\left(J_f\left(f,g,h\right)\right) = 2$$

one deduces that $f$ and $g$ are functionally independent, while $h$ is functionally dependent by them.

It is easy to check that

$$f^2 - g^2 = h.$$

We can check this result in Matlab 7.9:

```
>> syms x y z
>> f=[x+y+z;x-y+z;4*(x*y+y*z)];
>> t=[x y z];
>> rank(jacobian(f,t))
 ans =
 2
>> y1=f(1);y2=f(2);y3=f(3);
>>simple(y1^2-y2^2-y3)
ans =
 0
```

and with Mathcad 14:

$$f(x) := \begin{bmatrix} x_0 + x_1 + x_2 \\ x_0 - x_1 + x_2 \\ 4 \cdot (x_0 \cdot x_1 + x_1 \cdot x_2) \end{bmatrix}$$

$$\text{rank}(\text{Jacob}(f(x), x)) \to 2$$

$$y1(x) := f(x)_0 \qquad y2(x) := f(x)_1 \qquad y3(x) := f(x)_2$$

$$y1(x)^2 - y2(x)^2 = y3(x) \text{ simplify } \to 1$$

and using Mathematica 8:

```
f := {x + y + z, x - y + z, 4 * (x * y + y * z)}

MatrixRank[D[f, {{x, y, z}}]]

2

y1 := f[[1]]; y2 := f[[2]]; y3 := f[[3]];

Simplify[y1^2 - y2^2] == y3

4 y (x + z) == 4 (x y + y z)
```

and in Maple 15:

```
f := [x + y + z, x - y + z, 4·(x·y + y·z)] :
with(VectorCalculus) :
with(LinearAlgebra) :
Rank(Jacobian(f, [x, y, z]))
```

$$2$$

```
y1 := f_1 : y2 := f_2 : y3 := f_3 :
evalb(simplify(y1^2 - y2^2) = y3)
```

$$true$$

## 5.5   Extreme Value of a Function of Several Variables: Conditional Extremum

**Definition 5.20** (see [8]). (of an extreme value of a function). A function $f(x, y)$ has a **maximum** (**minimum**) value $f(a, b)$ at the point $P(a, b)$, if for all points $P'(x, y)$ other than $P$ in a sufficiently small neighborhood of $P$ one has the inequality $f(a, b) > f(x, y)$ (respectively, $f(a, b) < f(x, y)$). The generic term for maximum and minimum of a function is **extreme**.

**Remark 5.21** (see [8]). In a similar way, we define the extreme value of a function of three or more variables.

**Proposition 5.22.** (the necessary conditions for an extreme value, see [8]). The points at which a differentiable function $f(x, y)$ may attain an extremum (so-called *stationary points*) are found by solving the system of equations:

$$\begin{cases} f'_x(x, y) = 0 \\ f'_y(x, y) = 0 \end{cases} \tag{5.16}$$

(*necessary conditions* for an extremum).

**Proposition 5.23** (sufficient conditions for an extreme value, see [8]). Let $P(a, b)$ be a stationary point of the function $f(x, y)$, namely $f'_x(a, b) = f'_y(a, b)$. One forms the discriminant:

$$E = f''_{xy}(a, b)^2 - f''_{x^2}(a, b) \cdot f''_{y^2}(a, b). \tag{5.17}$$

Then:

A) if $E < 0$, the function has an extremum at the point $P(a, b)$, namely:

1) a minimum, if $f''_{x^2}(a, b) > 0$ (or $f''_{y^2}(a, b) > 0$);

2) a maximum, if $f''_{x^2}(a, b) < 0$ (or $f''_{y^2}(a, b) < 0$);

B) if $E > 0$, then there is no extremum at $P(a, b)$;

C) if $E = 0$, the question of an extremum of the function at $P(a, b)$ remains open (i.e., it requires a further investigation).

**Remark 5.24** (see [8]). For a function of three variables, for each stationary point $P(a, b, c)$ of the function $f(x, y, z)$ one forms the matrix

$$A = \begin{pmatrix} \dfrac{\partial^2 f}{\partial x^2} & \dfrac{\partial^2 f}{\partial x \partial y} & \dfrac{\partial^2 f}{\partial x \partial z} \\ \dfrac{\partial^2 f}{\partial x \partial y} & \dfrac{\partial^2 f}{\partial y^2} & \dfrac{\partial^2 f}{\partial y \partial z} \\ \dfrac{\partial^2 f}{\partial x \partial z} & \dfrac{\partial^2 f}{\partial y \partial z} & \dfrac{\partial^2 f}{\partial z^2} \end{pmatrix}. \tag{5.18}$$

If

- $A$ is a positive definite matrix then $P(a, b, c)$ is a mimimum point of $f$;
- $A$ is a negative definite matrix then $P(a, b, c)$ is a maximimum point of $f$;
- $A$ is neither positive nor negative definite matrix then $P(a, b, c)$ is not an extremum point;
- $A = O_3$ then we can not make any decision to the point $P(a, b, c)$.

**Definition 5.25** (see [23], p. 194). A symmetric matrix $A = (a_{ij})_{1 \le i, j \le n}$ is called **positive (negative) definite matrix** if its associated quadratic form is positive (negative) definite.

**Definition 5.26** (see [23], p. 194). Let $V$ be a real vector space.

a) The quadratic form $f : V \to \mathbb{R}$ is called **positive definite (negative definite)** if $f(\overline{x}) > 0$ ( respectively, $f(\overline{x}) < 0$), $(\forall)\, \overline{x} \in V,\, \overline{x} \ne \overline{0}$;
b) The quadratic form $f : V \to \mathbb{R}$ is called **nondefinite** if there are $\overline{a}, \overline{b} \in V$ such that $f(\overline{a}) > 0$ and $f(\overline{b}) < 0$.

**Definition 5.27** (see [8]). The **conditional extremum** of a function $f(x, y)$ is a maximum or a minimum of this function, which is attained on the condition that its arguments are related by the equation $\varphi(x, y) = 0$ (*coupling equation*).

**Proposition 5.28** (see [8]). In order to find the conditional extremum of a function $f(x, y)$, given the relationship $\varphi(x, y) = 0$, we form the so-called *Lagrange function*

$$F(x, y, \lambda) = f(x, y) + \lambda \varphi(x, y), \tag{5.19}$$

where $\lambda$ is an undetermined multiplier, and we seek the ordinary extremum of this auxiliary function.

**Remark 5.29** (see [8]). The necessary conditions for the extremum are reduced to the system of three equations

$$\begin{cases} \frac{\partial F}{\partial x} \equiv \frac{\partial f}{\partial x} + \lambda \frac{\partial \varphi}{\partial x} = 0, \\ \frac{\partial F}{\partial y} \equiv \frac{\partial f}{\partial y} + \lambda \frac{\partial \varphi}{\partial y} = 0 \\ \varphi(x, y) = 0 \end{cases} \tag{5.20}$$

with the three unknowns $x, y, \lambda$.

The question of the existence and character of a conditional extremum is solved on the basis of a study of the sign of the discriminant

$$E = F''_{xy}(a, b, \lambda)^2 - F''_{x^2}(a, b, \lambda) \cdot F''_{y^2}(a, b, \lambda) \tag{5.21}$$

of the Lagrange function $F(x, y, \lambda)$ at a stationary point $(a, b)$.

If the discriminant $E$ of the function $F(x, y, \lambda)$ at a stationary point is positive, then there is at this point a conditional maximum of the function $f(x, y)$, if $F''_{x^2}(a, b, \lambda) > 0$ (or $F''_{y^2}(a, b, \lambda) > 0$ ), a conditional minimum, if $F''_{x^2}(a, b, \lambda) < 0$ (or $F''_{y^2}(a, b, \lambda) < 0$).

**Example 5.30.** Test for an extremum the function

$$f(x, y) = (x^2 + y^2)\, e^{2x+3y}, \quad x \geq 0, \ y \geq 0.$$

**Solution.**
We shall find the first partial derivatives:

$$\begin{cases} f'_x = 2xe^{2x+3y} + 2(x^2 + y^2)\, e^{2x+3y} = 2e^{2x+3y}(x + x^2 + y^2) \\ f'_y = 2ye^{2x+3y} + 3(x^2 + y^2)\, e^{2x+3y} = e^{2x+3y}(2y + 3x^2 + 3y^2). \end{cases}$$

Then we shall form the system of equations from (5.16):

$$\begin{cases} \frac{\partial f}{\partial x} = 0 \\ \frac{\partial f}{\partial y} = 0 \end{cases} \iff \begin{cases} x + x^2 + y^2 = 0 \\ 2y + 3x^2 + 3y^2 = 0. \end{cases}$$

Solving the system and considering that $x \geq 0$, $y \geq 0$, we get the following stationary point: $P(0, 0)$.

We shall find now the second derivatives:

$$\begin{cases} f''_{x^2} = 4e^{2x+3y}(x + x^2 + y^2) + 2e^{2x+3y}(1 + 2x) \\ \qquad = 2e^{2x+3y}(1 + 4x + 2x^2 + 2y^2) \\ f''_{y^2} = 3e^{2x+3y}(2y + 3x^2 + 3y^2) + e^{2x+3y}(2 + 6y) \\ \qquad = e^{2x+3y}(2 + 12y + 9x^2 + 9y^2) \\ f''_{xy} = 2e^{2x+3y}(2y + 3x^2 + 3y^2) + e^{2x+3y} \cdot 6x = 2e^{2x+3y}(3x + 2y + 3x^2 + 3y^2) \end{cases}$$

and form the discriminant from (5.17) for our stationary point:

$$E = f''_{xy}(0, 0)^2 - f''_{x^2}(0, 0) \cdot f''_{y^2}(0, 0) = -4 < 0,$$

where

$$f''_{x^2}(0, 0) = 2 > 0.$$

Hence, the function has a minimum which is equal to the value of the function for $x = 0$, $y = 0$:

$$f_{\min} = (0 + 0) \cdot e^0 = 1.$$

We can also solve this problem using Matlab 7.9:
*Step 1.* Find the stationary points.
```
>>syms x y
>>f=(x^2+y^2)*exp(2*x+3*y);
```

```
>> u=diff(f,x);
>>v=diff(f,y);
>> [x,y]=solve(u,v,x,y)
x =
 0
 -4/13
 y =
 0
 -6/13
>>a=x(1); b=y(1);
>>aa=x(2); bb=y(2);
```

*Step 2*. Form the discriminant from (5.17) for our stationary point:

```
function [E,q]=local(a,b,u,v);
syms x y
uv=diff(u,y)^2;
uu=diff(u,x);
vv=diff(v,y);
p=subs(uv,{x,y},{a,b});
q=subs(uu,{x,y},{a,b});
r=subs(vv,{x,y},{a,b});
E=p-q*r;
end
```

Save the file with the name local.m then one writes in the command line:

```
>> [E,q]=local(a,b,u,v)
 E =
 -4
 q =
 2
```

Therefore, is a minimum local point, and the minimum value of the function is 0 and one obtains from the Matlab 7.9 command:

```
>>subs(f,{x,y},{0,0})
ans=
 0
```

and with Mathcad 14:

$$f(x,y) := \left(x^2 + y^2\right) \cdot e^{2 \cdot x + 3 \cdot y}$$

$$x := 1 \qquad y := 1 \qquad\qquad h(u,v) := \frac{\partial}{\partial u} f(u,v) \qquad m(u,v) := \frac{\partial}{\partial v} f(u,v)$$

Given

$$h(x,y) = 0 \qquad m(x,y) = 0$$

$$a := \text{Find}(x,y) \rightarrow \begin{pmatrix} 0 \\ 0 \end{pmatrix}$$

$$\text{local}(a) := \begin{vmatrix} x \leftarrow a_0 \\[4pt] y \leftarrow a_1 \\[8pt] p \leftarrow \left(\dfrac{d}{dx}\dfrac{d}{dy} f(x,y)\right)^2 \\[12pt] q \leftarrow \dfrac{d^2}{dx^2} f(x,y) \\[12pt] r \leftarrow \dfrac{d^2}{dy^2} f(x,y) \\[8pt] E \leftarrow p - q \cdot r \\[4pt] d_0 \leftarrow E \\[4pt] d_1 \leftarrow q \\[4pt] d \end{vmatrix}$$

$$\text{local}(a) = \begin{pmatrix} -4 \\ 2 \end{pmatrix}$$

$$f(0,0) = 0$$

and in Mathematica 8:

```
In[9]:= f[x_, y_] := (x^2 + y^2) * Exp[2 * x + 3 * y]

In[10]:= FindMinimum[{f[x, y], x ≥ 0, y ≥ 0}, {x, y}]

Out[10]= {1.935 × 10^-6, {x → 0.00107021, y → 0.000883406}}
```

and using Maple 15:

$f := (x, y) \rightarrow \left(x^2 + y^2\right) \cdot e^{2 \cdot x + 3 \cdot y}$:

$solve\left(\left\{\dfrac{\partial}{\partial x} f(x, y), \dfrac{\partial}{\partial y} f(x, y)\right\}\right)$

$$\{x = 0, y = 0\}, \left\{x = -\dfrac{4}{13}, y = -\dfrac{6}{13}\right\}$$

$p := \left. \dfrac{\partial}{\partial x} \dfrac{\partial}{\partial y} f(x, y) \right|_{x = 0, y = 0}$ :

$q := \left. \dfrac{\partial^2}{\partial x^2} f(x, y) \right|_{x = 0, y = 0}$

2

$r := \left. \dfrac{\partial^2}{\partial y^2} f(x, y) \right|_{x = 0, y = 0}$ :

$E := p^2 - q \cdot r$

-4

$f(0, 0)$

0

**Example 5.31.** Find extremum of the function

$$f(x, y) = 6 - 4x - 3y,$$

provided the variables $x$ and $y$ satisfy the equation:

$$x^2 + y^2 = 1.$$

**Solution.**

Geometrically speaking, the problem reduces to finding the largest and smallest values of the z-coordinate of the plane

$$z = 6 - 4x - 3y$$

for points of its intersection with the cylinder $x^2 + y^2 = 1$.

We shall form the Lagrange function

$$F(x, y, \lambda) = 6 - 4x - 3y + \lambda\left(x^2 + y^2 - 1\right).$$

We can define this function in Matlab 7.9, in the file F.m:

```
function r=F(x,y,la)
r=6-4*x-3*y+la*(x^2+y^2-1);
```

**end**

The necessary conditions yield the system of equations:

$$\begin{cases} \frac{\partial F}{\partial x} \equiv -4 + 2\lambda x = 0 \\ \frac{\partial F}{\partial y} \equiv -3 + 2\lambda y = 0 \\ \frac{\partial F}{\partial \lambda} \equiv x^2 + y^2 - 1 = 0. \end{cases}$$

Solving this system, we shall find

$$\begin{cases} \lambda_1 = -\frac{5}{2}, \ x_1 = -\frac{4}{5}, \ y_1 = -\frac{3}{5} \\ \lambda_2 = \frac{5}{2}, \ x_2 = \frac{4}{5}, \ y_2 = \frac{3}{5}. \end{cases}$$

We can also solve this system in Matlab 7.9:

```
>> syms x y la
>> u=diff(F(x,y,la),x);
>> v=diff(F(x,y,la),y);
>> q=diff(F(x,y,la),la);
>> [xx,yy]=solve(u,v,x,y);
>> g=subs(q,{x,y},{xx(1),yy(1)});
>> p=solve(g,la)
p =
-5/2
5/2
>> for i=1:2
xs(i)=subs(xx,la,p(i));
end
>>xs
xs =
[ -4/5, 4/5]
>> for i=1:2
ys(i)=subs(yy,la,p(i));
end
>>ys
ys =
[ -3/5, 3/5]
```

Since

$$\begin{cases} F''_{x^2} = 2\lambda \\ F''_{y^2} = 2\lambda \\ F''_{xy} = 0 \end{cases}$$

it follows that the discriminant $E$ of the function $F(x, y, \lambda)$ at a stationary point is

$$E = 0 - 2\lambda \cdot 2\lambda = -4\lambda^2.$$

If $\lambda_1 = -\frac{5}{2}, \ x_1 = -\frac{4}{5}, \ y_1 = -\frac{3}{5}$, then

$$E = -25 < 0$$

and
$$F''_{x^2}\left(-\frac{4}{5}, -\frac{3}{5}, -\frac{5}{2}\right) = -5 < 0,$$

consequently, the function has a conditional maximum at the point $\left(-\frac{4}{5}, -\frac{3}{5}\right)$.

If $\lambda_2 = \frac{5}{2}$, $x_2 = \frac{4}{5}$, $y_2 = \frac{3}{5}$, then

$$E = -25 < 0$$

and

$$F''_{x^2}\left(\frac{4}{5}, \frac{3}{5}, \frac{5}{2}\right) = 5 > 0,$$

consequently, the function has a conditional minimum at the point $\left(\frac{4}{5}, \frac{3}{5}\right)$.
Hence

$$\begin{cases} f_{max} = 6 - 4 \cdot \left(-\frac{4}{5}\right) 4 \cdot \left(-\frac{4}{5}\right) = 11 \\ f_{min} = 6 - 4 \cdot \frac{4}{5} - 3 \cdot \frac{3}{5} = 1. \end{cases}$$

We can determine the conditional maximum and minimum of the function in Matlab 7.9:

```
function [A t1]=local(x1,y1,la1,u,v)
syms x y z la
t1=subs(diff(u,x),{x,y,la},{x1,y1,la1}));
t2=subs(diff(u,y),{x,y,la},{x1,y1,la1});
t3=subs(diff(v,y),{x,y,la},{x1,y1,la1});
A=t2^2-t1*t3;
end
>> [A t]=local(xs(1),ys(1),p(1),u,v)
A =
-25
t =
-5
>> F(xs(1),ys(1),p(1))
ans =
11
>> [A t]=local(xs(2),ys(2),p(2),u,v)
A =
-25
t =
5
>> F(xs(2),ys(2),p(2))
ans =
```

1

and in Mathcad 14:

$$F(x,y,\lambda) := 6 - 4 \cdot x - 3 \cdot y + \lambda \cdot \left(x^2 + y^2 - 1\right)$$

$x := 1 \qquad y := 1 \qquad \lambda := 1$

Given

$$\frac{\partial}{\partial x} F(x,y,\lambda) = 0 \qquad \frac{\partial}{\partial y} F(x,y,\lambda) = 0 \qquad \frac{\partial}{\partial \lambda} F(x,y,\lambda) = 0$$

$$a := \text{Find}(x,y,\lambda) \rightarrow \begin{pmatrix} \dfrac{4}{5} & -\dfrac{4}{5} \\[2mm] \dfrac{3}{5} & -\dfrac{3}{5} \\[2mm] \dfrac{5}{2} & -\dfrac{5}{2} \end{pmatrix}$$

$$\text{local}(a,i) := \begin{vmatrix} x \leftarrow a_{0,i} \\[2mm] y \leftarrow a_{1,i} \\[2mm] \lambda \leftarrow a_{2,i} \\[2mm] p \leftarrow \left(\dfrac{d}{dx}\dfrac{d}{dy} F(x,y,\lambda)\right)^2 \\[3mm] q \leftarrow \dfrac{d^2}{dx^2} F(x,y,\lambda) \\[3mm] r \leftarrow \dfrac{d^2}{dy^2} F(x,y,\lambda) \\[3mm] E \leftarrow p - q \cdot r \\[2mm] d_0 \leftarrow E \\[2mm] d_1 \leftarrow q \\[2mm] d \end{vmatrix}$$

$$\text{local}(a,0) = \begin{pmatrix} -25 \\ 5 \end{pmatrix} \qquad F\left(a_{0,0}, a_{1,0}, a_{2,0}\right) = 1$$

$$\text{local}(a,1) = \begin{pmatrix} -25 \\ -5 \end{pmatrix} \qquad F\left(a_{0,1}, a_{1,1}, a_{2,1}\right) = 11$$

and using Mathematica 8:

```
In[74]:= F[x_, y_, λ_] := 6 - 4*x - 3*y + λ*(x^2 + y^2 - 1)
```

```
In[75]:= Solve[D[F[x, y, λ], x] == 0 && D[F[x, y, λ], y] == 0 && D[F[x, y, λ], λ] == 0,
         {x, y, λ}]
```

$$\text{Out[75]= } \left\{ \left\{ x \to -\frac{4}{5}, \ y \to -\frac{3}{5}, \ \lambda \to -\frac{5}{2} \right\}, \ \left\{ x \to \frac{4}{5}, \ y \to \frac{3}{5}, \ \lambda \to \frac{5}{2} \right\} \right\}$$

```
In[76]:= a := {{-4/5, 4/5}, {-3/5, 3/5}, {-5/2, 5/2}}
```

```
In[77]:= lc[a_, i_] :=
         Module[{d},
           p := D[F[x, y, λ], x, y] /. {x → a[[1, i]], y → a[[2, i]], λ → a[[3, i]]};
           q := D[F[x, y, λ], {x, 2}] /. {x → a[[1, i]], y → a[[2, i]], λ → a[[3, i]]};
           r := D[F[x, y, λ], {y, 2}] /. {x → a[[1, i]], y → a[[2, i]], λ → a[[3, i]]};
           Ex := p^2 - q*r;
           d := {Ex, q, F[a[[1, i]], a[[2, i]], a[[3, i]]]};
           d
         ]
```

```
In[78]:= lc[a, 1]
```

```
Out[78]= {-25, -5, 11}
```

```
In[79]:= lc[a, 2]
```

```
Out[79]= {-25, 5, 1}
```

and with Maple 15:

$$F := (x, y, \lambda) \rightarrow 6 - 4 \cdot x - 3 \cdot y + \lambda \cdot (x^2 + y^2 - 1):$$

$$a := solve\left(\left\{\frac{\partial}{\partial x}F(x, y, \lambda), \frac{\partial}{\partial y}F(x, y, \lambda), \frac{\partial}{\partial \lambda}F(x, y, \lambda)\right\}\right)$$

$$\left\{\lambda = \frac{5}{2}, x = \frac{4}{5}, y = \frac{3}{5}\right\}, \left\{\lambda = -\frac{5}{2}, x = -\frac{4}{5}, y = -\frac{3}{5}\right\}$$

$$a := Matrix\left(3, 2, \left[-\frac{4}{5}, \frac{4}{5}, -\frac{3}{5}, \frac{3}{5}, -\frac{5}{2}, \frac{5}{2}\right]\right):$$

$lc := proc(a, i)$

$$p := \left.\frac{\partial}{\partial x}\frac{\partial}{\partial y} F(x, y, \lambda)\right|_{x = a_{1,i}, y = a_{2,i}, \lambda = a_{3,i}} ;$$

$$q := \left.\frac{\partial^2}{\partial x^2} F(x, y, \lambda)\right|_{x = a_{1,i}, y = a_{2,i}, \lambda = a_{3,i}} ;$$

$$r := \left.\frac{\partial^2}{\partial y^2} F(x, y, \lambda)\right|_{x = a_{1,i}, y = a_{2,i}, \lambda = a_{3,i}} ;$$

$E := p^2 - q \cdot r,$
$Matrix\left(1, 3, \left[E, q, F\left(a_{1,i}, a_{2,i}, a_{3,i}\right)\right]\right))$
**end proc;**

$lc(a, 1)$

$$\begin{bmatrix} -25 & -5 & 11 \end{bmatrix}$$

$lc(a, 2)$

$$\begin{bmatrix} -25 & 5 & 1 \end{bmatrix}$$

**Example 5.32.** Find the extremum of the function

$$f(x, y, z) = x - 2y + 2z$$

provided the variables $x, y$ and $z$ satisfy the equation:

$$x^2 + y^2 + z^2 = 1.$$

**Solution.** *Step 1.* Determine the stationary points, i.e. the system solutions:

$$\begin{cases} F'_x(x, y, z, \lambda) = 0 \\ F'_y(x, y, z, \lambda) = 0 \\ F'_z(x, y, z, \lambda) = 0 \\ F'_\lambda(x, y, z, \lambda) = 0 \end{cases}$$

where

$$F\left(x, y, z, \lambda\right) = x - 2y + 2z + \lambda\left(x^2 + y^2 + z^2 - 1\right).$$

We have to solve the system

$$\left\{ \begin{array}{l} 1 + 2\lambda x = 0 \\ -2 + 2\lambda y = 0 \\ 2 + 2\lambda z = 0 \\ x^2 + y^2 + z^2 - 1 = 0 \end{array} \right. \Longleftrightarrow \left\{ \begin{array}{l} x = -\frac{1}{2\lambda} \\ y = \frac{1}{\lambda} \\ z = -\frac{1}{\lambda} \\ (-\frac{1}{2\lambda})^2 + \frac{1}{\lambda^2} + (-\frac{1}{\lambda})^2 - 1 = 0 \end{array} \right. \Longleftrightarrow \left\{ \begin{array}{l} \lambda_1 = -\frac{3}{2}, \lambda_2 = \frac{3}{2} \\ x_1 = \frac{1}{3}, x_2 = -\frac{1}{3} \\ y_1 = -\frac{2}{3}, y_2 = \frac{2}{3} \\ z_1 = \frac{2}{3}, z_2 = -\frac{2}{3}. \end{array} \right.$$

Therefore there are the stationary points: $\left(\frac{1}{3}, -\frac{2}{3}, \frac{2}{3}\right)$ and $\left(-\frac{1}{3}, \frac{2}{3}, -\frac{2}{3}\right)$.
We can also compute them in Matlab 7.9:

```
function r=F(x,y,z,la)
r=x-2*y+2*z+la*(x^2+y^2+z^2-1);
end
>> syms x y z la
>> u=diff(F(x,y,z,la),x);
>>v=diff(F(x,y,z,la),y);
>>w=diff(F(x,y,z,la),z);
>>q=diff(F(x,y,z,la),la);
>>[xx,yy,zz]=solve(u,v,w,x,y,z);
>>g=subs(q,{x,y,z},{xx(1),yy(1),zz(1)});
>>p=solve(g,la)
p =
 -3/2
 3/2
>> for l=1:2
xs(i)=subs(xx,la,p(i));
end
>> xs
xs =
[ 1/3, -1/3]
>> for i=1:2
ys(i)=subs(yy,la,p(i));
end
>> ys
ys =
[ -2/3, 2/3]
>> for i=1:2
zs(i)=subs(zz,la,p(i));
end
zs =
```

[ **2/3, -2/3**]

*Step 2.* For each stationary point $(a, b, c)$ of the function $F(x, y, z, \lambda)$ we build the matrix by the form (5.18).

We shall have

$$\begin{cases} \frac{\partial^2 F}{\partial x^2} = 2\lambda \\ \frac{\partial^2 F}{\partial y^2} = 2\lambda \\ \frac{\partial^2 F}{\partial z^2} = 2\lambda \\ \frac{\partial^2 F}{\partial x \partial y} = 0 \\ \frac{\partial^2 F}{\partial x \partial z} = 0 \\ \frac{\partial^2 F}{\partial y \partial z} = 0. \end{cases}$$

For the stationary point:

- $\left(\frac{1}{3}, -\frac{2}{3}, \frac{2}{3}\right)$ achieved for $\lambda_1 = -\frac{3}{2}$ it results the matrix $A_1 = \begin{pmatrix} -3 & 0 & 0 \\ 0 & -3 & 0 \\ 0 & 0 & -3 \end{pmatrix}$,

  which is negative definite; hence $\left(\frac{1}{3}, -\frac{2}{3}, \frac{2}{3}\right)$ is a maximum point;

- $\left(-\frac{1}{3}, \frac{2}{3}, -\frac{2}{3}\right)$ achieved for $\lambda_2 = \frac{3}{2}$ it results the matrix $A_2 = \begin{pmatrix} 3 & 0 & 0 \\ 0 & 3 & 0 \\ 0 & 0 & 3 \end{pmatrix}$, which

  is positive definite; hence $\left(-\frac{1}{3}, \frac{2}{3}, -\frac{2}{3}\right)$ is a minimum point.

In Matlab 7.9 we need the following sequence:

```
function [A]=local(x1,y1,z1,la1,u,v,w)
syms x y z la
t11=subs(diff(u,x),{x,y,z,la},{x1,y1,z1,la1});
t12=subs(diff(u,y),{x,y,z,la},{x1,y1,z1,la1});
t13=subs(diff(u,z),{x,y,z,la},{x1,y1,z1,la1});
t22=subs(diff(v,y),{x,y,z,la},{x1,y1,z1,la1});
t23=subs(diff(v,z),{x,y,z,la},{x1,y1,z1,la1});
t33=subs(diff(w,z),{x,y,z,la},{x1,y1,z1,la1});
A=[t11 t12 t13 ;t12 t22 t23; t13 t23 t33];
end
>> [A]=local(xs(1),ys(1),zs(1),p(1),u,v,w)
A =
[-3, 0, 0]
[0, -3, 0]
[0, 0, -3]
```

Therefore, $\left(\frac{1}{3}, -\frac{2}{3}, \frac{2}{3}\right)$ is a maximum point, and the maximum value of the function $F(x, y, z, \lambda)$ is 3; it can be achieved from the Matlab 7.9 command:

```
>> F(xs(1),ys(1),zs(1),p(1))
ans =
3
```

```
>> [A]=local(xs(2),ys(2),zs(2),p(2),u,v,w)
A =
[3, 0, 0]
[0, 3, 0]
[0, 0, 3]
```

Therefore, $\left(-\frac{1}{3}, \frac{2}{3}, -\frac{2}{3}\right)$ is a minimum point, and the minimum value of the function $F(x, y, z, \lambda)$ is -3; it can be achieved from the Matlab 7.9 command:

```
>> F(xs(2),ys(2),zs(2),p(2))
ans =
-3
```

We shall also give a computer solution using Mathcad 14:

$$F(x,y,z,\lambda) := x - 2 \cdot y + 2 \cdot z + \lambda \cdot \left(x^2 + y^2 + z^2 - 1\right)$$

$$x := 1 \qquad y := 1 \qquad z := 1 \qquad \lambda := 1$$

Given

$$\frac{\partial}{\partial x} F(x,y,z,\lambda) = 0 \qquad \frac{\partial}{\partial y} F(x,y,z,\lambda) = 0 \qquad \frac{\partial}{\partial z} F(x,y,z,\lambda) = 0 \qquad \frac{\partial}{\partial \lambda} F(x,y,z,\lambda) = 0$$

$$a := \text{Find}(x,y,z,\lambda) \rightarrow \begin{pmatrix} -\dfrac{1}{3} & \dfrac{1}{3} \\[2mm] \dfrac{2}{3} & -\dfrac{2}{3} \\[2mm] -\dfrac{2}{3} & \dfrac{2}{3} \\[2mm] \dfrac{3}{2} & -\dfrac{3}{2} \end{pmatrix}$$

$$\text{local}(a,i) := \begin{vmatrix} x \leftarrow a_{0,i} \\ y \leftarrow a_{1,i} \\ z \leftarrow a_{2,i} \\ \lambda \leftarrow a_{3,i} \\ A \leftarrow \begin{pmatrix} \dfrac{d^2}{dx^2} F(x,y,z,\lambda) & \dfrac{d}{dx}\dfrac{d}{dy} F(x,y,z,\lambda) & \dfrac{d}{dx}\dfrac{d}{dz} F(x,y,z,\lambda) \\[3mm] \dfrac{d}{dx}\dfrac{d}{dy} F(x,y,z,\lambda) & \dfrac{d^2}{dy^2} F(x,y,z,\lambda) & \dfrac{d}{dy}\dfrac{d}{dz} F(x,y,z,\lambda) \\[3mm] \dfrac{d}{dx}\dfrac{d}{dz} F(x,y,z,\lambda) & \dfrac{d}{dy}\dfrac{d}{dz} F(x,y,z,\lambda) & \dfrac{d^2}{dz^2} F(x,y,z,\lambda) \end{pmatrix} \\ A \end{vmatrix}$$

$$\text{local}(a,0) = \begin{pmatrix} 3 & 0 & 0 \\ 0 & 3 & 0 \\ 0 & 0 & 3 \end{pmatrix} \qquad \text{local}(a,1) = \begin{pmatrix} -3 & 0 & -0 \\ 0 & -3 & 0 \\ -0 & 0 & -3 \end{pmatrix}$$

$$F\left(a_{0,0}, a_{1,0}, a_{2,0}, a_{3,0}\right) = -3 \qquad F\left(a_{0,1}, a_{1,1}, a_{2,1}, a_{3,1}\right) = 3$$

and in Mathematica 8:

```
In[17]:= F[x_, y_, z_, λ_] := x - 2*y + 2*z + λ*(x^2 + y^2 + z^2 - 1)

In[18]:= Solve[D[F[x, y, z, λ], x] == 0 && D[F[x, y, z, λ], y] == 0 && D[F[x, y, z, λ], z] == 0
         && D[F[x, y, z, λ], λ] == 0, {x, y, z, λ}]
```

$$\text{Out[18]}= \left\{\left\{x \to \frac{1}{3}, y \to -\frac{2}{3}, z \to \frac{2}{3}, \lambda \to -\frac{3}{2}\right\}, \left\{x \to -\frac{1}{3}, y \to \frac{2}{3}, z \to -\frac{2}{3}, \lambda \to \frac{3}{2}\right\}\right\}$$

```
In[19]:= a := {{1/3, -1/3}, {-2/3, 2/3}, {2/3, -2/3}, {-3/2, 3/2}}

In[20]:= lc[a_, i_] :=

Module[{A},
    t1 := D[F[x, y, z, λ], {x, 2}] /. {x → a[[1, i]], y → a[[2, i]], z → a[[3, i]], λ → a[[4, i]]};
    t2 := D[F[x, y, z, λ], x, y] /. {x → a[[1, i]], y → a[[2, i]], z → a[[3, i]], λ → a[[4, i]]};
    t3 := D[F[x, y, z, λ], x, z] /. {x → a[[1, i]], y → a[[2, i]], z → a[[3, i]], λ → a[[4, i]]};
    t4 := D[F[x, y, z, λ], y, z] /. {x → a[[1, i]], y → a[[2, i]], z → a[[3, i]], λ → a[[4, i]]};
    t5 := D[F[x, y, z, λ], {y, 2}] /. {x → a[[1, i]], y → a[[2, i]], z → a[[3, i]], λ → a[[4, i]]};
    t6 := D[F[x, y, z, λ], {z, 2}] /. {x → a[[1, i]], y → a[[2, i]], z → a[[3, i]], λ → a[[4, i]]};
    A := {{t1, t2, t3}, {t2, t5, t4}, {t3, t4, t6}};
    A
]

In[21]:= lc[a, 1]

Out[21]= {{-3, 0, 0}, {0, -3, 0}, {0, 0, -3}}

In[22]:= F[a[[1, 1]], a[[2, 1]], a[[3, 1]], a[[4, 1]]]

Out[22]= 3

In[23]:= lc[a, 2]

Out[23]= {{3, 0, 0}, {0, 3, 0}, {0, 0, 3}}

In[24]:= F[a[[1, 2]], a[[2, 2]], a[[3, 2]], a[[4, 2]]]

Out[24]= -3
```

and in Maple 15:

$F := (x, y, z, \lambda) \rightarrow x - 2 \cdot y + 2 \cdot z + \lambda \cdot (x^2 + y^2 + z^2 - 1):$

$solve\left(\left\{\frac{\partial}{\partial x}F(x, y, z, \lambda), \frac{\partial}{\partial y}F(x, y, z, \lambda), \frac{\partial}{\partial z}F(x, y, z, \lambda), \frac{\partial}{\partial \lambda}F(x, y, z, \lambda)\right\}\right)$

$\left\{\lambda = \frac{3}{2}, x = -\frac{1}{3}, y = \frac{2}{3}, z = -\frac{2}{3}\right\}, \left\{\lambda = -\frac{3}{2}, x = \frac{1}{3}, y = -\frac{2}{3}, z = \frac{2}{3}\right\}$

$a := Matrix\left(4, 2, \left[-\frac{1}{3}, \frac{1}{3}, \frac{2}{3}, -\frac{2}{3}, -\frac{2}{3}, \frac{2}{3}, \frac{3}{2}, -\frac{3}{2}\right]\right):$

$lc := proc(a, i)$

$t1 := \left.\frac{\partial^2}{\partial x^2} F(x, y, z, \lambda)\right|_{x = a_{1, i}, y = a_{2, i}, z = a_{3, i}, \lambda = a_{4, i}};$

$t2 := \left.\frac{\partial^2}{\partial x \partial y} F(x, y, z, \lambda)\right|_{x = a_{1, i}, y = a_{2, i}, z = a_{3, i}, \lambda = a_{4, i}};$

$t3 := \left.\frac{\partial^2}{\partial x \partial z} F(x, y, z, \lambda)\right|_{x = a_{1, i}, y = a_{2, i}, z = a_{3, i}, \lambda = a_{4, i}};$

$t4 := \left.\frac{\partial^2}{\partial y \partial z} F(x, y, z, \lambda)\right|_{x = a_{1, i}, y = a_{2, i}, z = a_{3, i}, \lambda = a_{4, i}};$

$t5 := \left.\frac{\partial^2}{\partial y^2} F(x, y, z, \lambda)\right|_{x = a_{1, i}, y = a_{2, i}, z = a_{3, i}, \lambda = a_{4, i}};$

$t6 := \left.\frac{\partial^2}{\partial z^2} F(x, y, z, \lambda)\right|_{x = a_{1, i}, y = a_{2, i}, z = a_{3, i}, \lambda = a_{4, i}};$

$Matrix(3, 3, [t1, t2, t3, t2, t5, t4, t3, t4, t6])$

**end proc;**

$[lc(a, 1), F(a_{1, 1}, a_{2, 1}, a_{3, 1}, a_{4, 1})]$

$$\left[\begin{bmatrix} 3 & 0 & 0 \\ 0 & 3 & 0 \\ 0 & 0 & 3 \end{bmatrix}, -3\right]$$

$[lc(a, 2), F(a_{1, 2}, a_{2, 2}, a_{3, 2}, a_{4, 2})]$

$$\left[\begin{bmatrix} -3 & 0 & 0 \\ 0 & -3 & 0 \\ 0 & 0 & -3 \end{bmatrix}, 3\right]$$

## 5.6  Problems

1. Find the derivative $y'$ if

$$x^3 + y^3 - 3axy = 0, \ a \in \mathbb{R}.$$

**Computer solution.**

We shall compute $y'$ in Maple 15:

$f := x^3 + y^3 - 3 \cdot a \cdot x \cdot y = 0$

$$x^3 + y^3 - 3\,axy = 0$$

$y1 := implicitdiff(f, y, x)$

$$\frac{-x^2 + ay}{y^2 - ax}$$

2. Find the second derivative of the function

$$y = \ln(1 - x).$$

3. The equations

$$\begin{cases} u + v = x + y \\ xu + yv = 1 \end{cases}$$

define $u$ and $v$ as functions of $x$ and $y$; find:

$$\frac{\partial u}{\partial x}, \frac{\partial u}{\partial y}, \frac{\partial v}{\partial x}, \frac{\partial v}{\partial y}.$$

4. The function $z$ of the arguments $x$ and $y$ is defined by the equations

$$\begin{cases} x = u + v \\ y = u^2 + v^2 \\ z = u^3 + v^3 \end{cases}, u \neq v.$$

Find:

$$\frac{\partial z}{\partial x}, \frac{\partial z}{\partial y}.$$

**Computer solution.**
Using Maple 15 we shall have:

$f := u + v = x :$
$g := u^2 + v^2 = y :$
$h := u^3 + v^3 = z :$
$d1 := implicitdiff(\{f, g, h\}, \{u(x, y), v(x, y), z(x, y)\}, z, x)$

$$-3\,vu$$

$d2 := implicitdiff(\{f, g, h\}, \{u(x, y), v(x, y), z(x, y)\}, z, y)$

$$\frac{3}{2}u + \frac{3}{2}v$$

5. The functions $u$ and $v$ of the independent variables $x$ and $y$ are defined implicitly by the system of equations

$$\begin{cases} u + v = x \\ u - yv = 0. \end{cases}$$

Find

$$du, dv, d^2u, d^2v.$$

6. Prove that the function $z(x, y)$ defined by $F(x + z/y, \ y + z/x) = 0$ satisfies the equation:

$$x\frac{\partial z}{\partial x} + y\frac{\partial z}{\partial y} = z - xy.$$

**Computer solution.**
We shall prove that relation in Maple 15:

$f := F\left(x + \dfrac{z}{y}, y + \dfrac{z}{x}\right) = 0 :$
$simplify(x \cdot implicitdiff(f, z, x) + y \cdot implicitdiff(f, z, y))$

$$-xy + z$$

7. If the function $f$ is implicitly defined by the relation $f(x, y, z) = 0$ show that

$$\frac{\partial x}{\partial y}\frac{\partial y}{\partial z}\frac{\partial z}{\partial x} = -1$$

**Computer solution.**
We shall prove this relation in Maple 15:

$F := f(x, y, z) = 0 :$
$implicitdiff(F, z, x) \cdot implicitdiff(F, y, z) \cdot implicitdiff(F, x, y)$

$$-1$$

8. Let be $D \subset \mathbb{R}^2$, $f \in C^1(D)$ and the functions $z$ and $u$ of the variables $x$ and $y$, implicitly defined by the system:

$$\begin{cases} f(x - y, z + u) = 1 \\ xyz + \ln(x^2 + y^2 + z^2) = 2. \end{cases}$$

Compute:

$$\frac{\partial z}{\partial x}, \frac{\partial z}{\partial y}, \frac{\partial u}{\partial x}, \frac{\partial u}{\partial y}.$$

**Computer solution.**
We shall give a computer solution in Maple 15:

$F := f(x - y, z + u) = 1 :$
$G := x \cdot y \cdot z + \ln(x^2 + y^2 + z^2) = 2 :$
$d1 := implicitdiff(\{F, G\}, \{z(x, y), u(x, y)\}, z, x)$

$$-\frac{yzx^2 + y^3 z + yz^3 + 2x}{x^3 y + xy^3 + xyz^2 + 2z}$$

$d2 := implicitdiff(\{F, G\}, \{z(x, y), u(x, y)\}, z, y)$

$$-\frac{x^3 z + xzy^2 + xz^3 + 2y}{x^3 y + xy^3 + xyz^2 + 2z}$$

$d3 := simplify(implicitdiff(\{F, G\}, \{z(x, y), u(x, y)\}, u, x)) :$
$d4 := simplify(implicitdiff(\{F, G\}, \{z(x, y), u(x, y)\}, u, y)) :$

**9. If**

$$z = F(r, \varphi),$$

where $r$ and $\varphi$ are functions of the variables $x$ and $y$, defined by the system of equations:

$$\begin{cases} x = r \cos \varphi \\ y = r \sin \varphi, \end{cases}$$

find $\frac{\partial z}{\partial x}$ and $\frac{\partial z}{\partial y}$.

**10.** Regarding $z$ as a function of $x$ and $y$, find $\frac{\partial z}{\partial x}$ and $\frac{\partial z}{\partial y}$ if:

$$\begin{cases} x = a \cos \varphi \cos \psi \\ y = b \sin \varphi \cos \psi \\ z = \sin \psi. \end{cases}$$

**Computer solution.**
Using Maple 15 we shall find:

$f := a \cdot \cos(\varphi) \cdot \cos(\psi) = x:$
$g := b \cdot \sin(\varphi) \cdot \cos(\psi) = y:$
$h := c \cdot \sin(\psi) = z:$
$simplify(implicitdiff(\{f, g, h\}, \{\varphi, \psi, z\}, z, x))$

$$-\frac{c \cos(\psi) \cos(\varphi)}{a \sin(\psi)}$$

$simplify(implicitdiff(\{f, g, h\}, \{\varphi, \psi, z\}, z, y))$

$$-\frac{c \cos(\psi) \sin(\varphi)}{b \sin(\psi)}$$

11.  Let $z$ be a function of the variables $x$ and $y$, defined by the equation:

$$2x^2 + 2y^2 + z^2 - 8xz - z + 8 = 0.$$

Find $dz$ and $d^2z$ for the values $x = 2$, $y = 0$, $z = 1$.

12.  Study the functional dependence of the functions:

$$\begin{cases} f(x,y) = \dfrac{ax}{\sqrt{x^2+y^2}} \\[2mm] g(x,y) = \dfrac{by}{\sqrt{x^2+y^2}}, \end{cases}$$

defined on $\mathbb{R}^2 \backslash \{(0,0)\}$.
   **Computer solution.**
   We shall solve this problem using Matlab 7.9:

```
>> syms x y a b
>> f=[x+y+z;x-y+z;4*(x*y+y*z)];
>> t=[x y];
>> rank(jacobian(f,t))
ans =
1
>> y1=f(1);y2=f(2);
>> simplify((y1/a)^2+(y2/b)^2)
ans =
1
```

and in Mathcad 14:

$$f(x,a,b) := \begin{bmatrix} a \cdot \dfrac{x_0}{\sqrt{\left(x_0\right)^2 + \left(x_1\right)^2}} \\ b \cdot \dfrac{x_1}{\sqrt{\left(x_0\right)^2 + \left(x_1\right)^2}} \end{bmatrix}$$

$$\mathrm{rank}\left(\mathrm{Jacob}(f(x,a,b),x)\right) \to 1$$

$$y1(x,a,b) := f(x,a,b)_0 \quad y2(x,a,b) := f(x,a,b)_1$$

$$\left(\frac{y1(x,a,b)}{a}\right)^2 + \left(\frac{y2(x,a,b)}{b}\right)^2 \quad \text{simplify} \to 1$$

and with Mathematica 8:

```
In[1]:= f := {a * x / Sqrt[x^2 + y^2], b * y / Sqrt[x^2 + y^2]}

In[2]:= MatrixRank[D[f, {{x, y}}]]

Out[2]= 1

In[3]:= y1 := f[[1]]; y2 := f[[2]];

In[4]:= Simplify[(y1 / a)^2 + (y2 / b)^2]

Out[4]= 1
```

and in Maple 15:

$$f := \left[ \frac{a \cdot x}{\sqrt{x^2 + y^2}}, \frac{b \cdot y}{\sqrt{x^2 + y^2}} \right] :$$

*with*( *VectorCalculus* ) :
*with*( *LinearAlgebra* ) :
*Rank*( *Jacobian*( *f*, [*x, y*]))

1

$$y1 := f_1 : y2 := f_2 :$$

$$simplify\left( \left( \frac{y1}{a} \right)^2 + \left( \frac{y2}{b} \right)^2 \right)$$

1

13. Test for an extremum the function

$$y = x^3 + 3xy^2 - 15x - 12y.$$

14. Test for an extremum the implicitly represented function $y = f(x)$, defined by:

$$x^3 + y^3 - 3x^2y - 3 = 0.$$

**Computer solution.**
Using Maple 15, we shall have:

$f := x^3 + y^3 - 3 \cdot x^2 \cdot y - 3 = 0$ :
$y1 := implicitdiff(f, y, x)$ :
$u := solve(\{y1 = 0, x^3 + y^3 - 3 \cdot x^2 \cdot y - 3 = 0\}, \{x, y\})$ :
$convert([u_1, u_2, u_3], radical)$

$$\left[ \{x = 0, y = 3^{1/3}\}, \{x = -2, y = -1\}, \left\{ x = 1 + I\sqrt{3}, y = \frac{1}{2} + \frac{1}{2} I\sqrt{3} \right\} \right]$$

$y2 := implicitdiff(f, y, x\$2)$ :
$y2\big|_{x=0, y=3^{1/3}}$

$$\frac{2}{3} 3^{2/3}$$

$y2\big|_{x=-2, y=-1}$

$$-\frac{2}{3}$$

We can deduce:

- $y''(0) = \frac{2}{\sqrt[3]{3}} > 0$, hence

$$x = 0$$

  is a minimum point;

- $y''(-2) = -\frac{2}{3} < 0$, hence

$$x = -2$$

  is a maximum point.

15. Test for an extremum the implicitly represented function:

$$x^3 - y^2 - 3x + 4y + z^2 + z - 8 = 0.$$

16. Find the extrema of the following implicitly represented function:

$$x^2 + y^2 + z^2 - 2x + 4y - 6z - 11 = 0.$$

17. Determine the conditional extrema of the following function:

$$\begin{cases} z = xy \\ x + y = 1. \end{cases}$$

**Computer soluttion.**
We shall achieve in Matlab 7.9:
function r = F(x,y,la)
r=x*y+la*(x+y-1);
end
>> syms x y la
>> u=diff(F(x,y,la),x);
>> v=diff(F(x,y,la),y);
>> q=diff(F(x,y,la),la);
>> [xx yy]=solve(u,v,x,y);
>> g=subs(q,{x,y},{xx,yy});
>> p=solve(g,la)
 p =
 -1/2
>> xs=subs(xx,la,p);
>> ys=subs(yy,la,p);
function [A t1] = local(x1,y1,la1,u,v)
syms x y la
t1=subs(diff(u,x),{x,y,la},{x1,y1,la1});
t2=subs(diff(u,y),{x,y,la},{x1,y1,la1});
t3=subs(diff(v,y),{x,y,la},{x1,y1,la1});
A=t2^2-t1*t3;
end
>> [A,t]=local(xs,ys,p,u,v)
A =
 1
t =
 0
and with Mathcad 14:

$F(x,y,\lambda) := x \cdot y + \lambda \cdot (x + y - 1)$

$x := 1 \qquad y := 1 \qquad \lambda := 1$

Given

$$\frac{\partial}{\partial x} F(x,y,\lambda) = 0 \qquad \frac{\partial}{\partial y} F(x,y,\lambda) = 0 \qquad \frac{\partial}{\partial \lambda} F(x,y,\lambda) = 0$$

$$a := \text{Find}(x,y,\lambda) \rightarrow \begin{pmatrix} \frac{1}{2} \\ \frac{1}{2} \\ -\frac{1}{2} \end{pmatrix}$$

$$\text{local}(a,i) := \begin{vmatrix} x \leftarrow a_{0,i} \\ y \leftarrow a_{1,i} \\ \lambda \leftarrow a_{2,i} \\ p \leftarrow \left( \frac{d}{dx}\frac{d}{dy} F(x,y,\lambda) \right)^2 \\ q \leftarrow \frac{d^2}{dx^2} F(x,y,\lambda) \\ r \leftarrow \frac{d^2}{dy^2} F(x,y,\lambda) \\ E \leftarrow p - q \cdot r \\ d_0 \leftarrow E \\ d_1 \leftarrow q \\ d \end{vmatrix}$$

$$\text{local}(a,0) = \begin{pmatrix} 1 \\ -2.87 \times 10^{-15} \end{pmatrix}$$

and using Mathematica 8:

```
In[1]:= F[x_, y_, λ_] := x * y + λ * (x + y - 1)
```

```
In[2]:= Solve[D[F[x, y, λ], x] == 0 && D[F[x, y, λ], y] == 0 && D[F[x, y, λ], λ] == 0,
        {x, y, λ}]
```

$$Out[2]= \left\{\left\{x \to \frac{1}{2}, y \to \frac{1}{2}, λ \to -\frac{1}{2}\right\}\right\}$$

```
In[3]:= a := {1/2, 1/2, -1/2}
```

```
In[4]:= lc[a_] :=
        Module[{d},
            p := D[F[x, y, λ], x, y] /. {x → a[[1]], y → a[[2]], λ → a[[3]]};
            q := D[F[x, y, λ], {x, 2}] /. {x → a[[1]], y → a[[2]], λ → a[[3]]};
            r := D[F[x, y, λ], {y, 2}] /. {x → a[[1]], y → a[[2]], λ → a[[3]]};
            Ex := p^2 - q * r;
            d := {Ex, q};
            d
        ]
```

```
In[5]:= lc[a]
```

```
Out[5]= {1, 0}
```

and in Maple 15:

$F := (x, y, \lambda) \rightarrow x \cdot y + \lambda \cdot (x + y - 1) :$

$a := solve\left(\left[\dfrac{\partial}{\partial x}F(x, y, \lambda), \dfrac{\partial}{\partial y}F(x, y, \lambda), \dfrac{\partial}{\partial \lambda}F(x, y, \lambda)\right]\right)$

$$\left\{\lambda = -\dfrac{1}{2}, x = \dfrac{1}{2}, y = \dfrac{1}{2}\right\}$$

$a := Vector\left(\left[\dfrac{1}{2}, \dfrac{1}{2}, -\dfrac{1}{2}\right]\right) :$

$lc := \mathbf{proc}(a)$

$p := \dfrac{\partial}{\partial x} \dfrac{\partial}{\partial y} F(x, y, \lambda)\Big|_{x = a_1, y = a_2, \lambda = a_3}$ ;

$q := \dfrac{\partial^2}{\partial x^2} F(x, y, \lambda)\Big|_{x = a_1, y = a_2, \lambda = a_3}$ ;

$r := \dfrac{\partial^2}{\partial y^2} F(x, y, \lambda)\Big|_{x = a_1, y = a_2, \lambda = a_3}$ ;

$E := p^2 - q \cdot r,$

$Matrix(1, 2, [E, q])$

**end proc**;

$lc(a)$

$$\begin{bmatrix} 1 & 0 \end{bmatrix}$$

Hence there is no extremum at $P(\frac{1}{2}; \frac{1}{2})$

18.  Does the function $z = x^2 + y^2$ have any conditional extremum for

$$\frac{x}{2} + \frac{y}{3} = 1?$$

19.  Check if the function

$$u = x^2 + y^2 + z^2$$

has a conditional extremum, provided

$$2x + 3y - z = 1.$$

**Computer solution.**

We shall build a computer solution using Matlab 7.9:

```
function r = F(x,y,z,la)
r=x^2+y^2+z^2+la*(2*x+3*y-z-1);
end
>> syms x y z la
>> u=diff(F(x,y,z,la),x);
>> v=diff(F(x,y,z,la),y);
>> w=diff(F(x,y,z,la),z);
>> q=diff(F(x,y,z,la),la);
>> [xx yy zz]=solve(u,v,w,x,y,z);
>> g=subs(q,{x,y,z},{xx,yy,zz}) ;
>> p=solve(g,la)
 p =
 -1/7
>> xs=subs(xx,la,p);
>> ys=subs(yy,la,p);
>> zs=subs(zz,la,p);
function [A] = local(x1,y1,z1,la1,u,v,w)
syms x y z la
t11=subs(diff(u,x),{x,y,z,la},{x1,y1,z1,la1});
t12=subs(diff(u,y),{x,y,z,la},{x1,y1,z1,la1});
t13=subs(diff(u,z),{x,y,z,la},{x1,y1,z1,la1});
t22=subs(diff(v,y),{x,y,z,la},{x1,y1,z1,la1});
t23=subs(diff(v,z),{x,y,z,la},{x1,y1,z1,la1});
t33=subs(diff(w,z),{x,y,z,la},{x1,y1,z1,la1});
A=[t11 t12 t13;t12 t22 t23;t13 t23 t33];
end
>> A=local(xs,ys,zs,p,u,v,w)
A =
 2 0 0
 0 2 0
 0 0 2
>> F(xs,ys,zs,p)
 ans =
 1/14
```

and in Mathcad 14:

$$\underset{\sim}{F}(x,y,z,\lambda) := x^2 + y^2 + z^2 + \lambda \cdot (2 \cdot x + 3 \cdot y - z - 1)$$

$$x := 1 \qquad y := 1 \qquad z := 1 \qquad \lambda := 1$$

Given

$$\frac{\partial}{\partial x} F(x,y,z,\lambda) = 0 \qquad \frac{\partial}{\partial y} F(x,y,z,\lambda) = 0 \qquad \frac{\partial}{\partial z} F(x,y,z,\lambda) = 0 \qquad \frac{\partial}{\partial \lambda} F(x,y,z,\lambda) = 0$$

$$a := Find(x,y,z,\lambda) \rightarrow \begin{pmatrix} \dfrac{1}{7} \\[2mm] \dfrac{3}{14} \\[2mm] -\dfrac{1}{14} \\[2mm] -\dfrac{1}{7} \end{pmatrix}$$

$$local(a,i) := \begin{vmatrix} x \leftarrow a_{0,i} \\[1mm] y \leftarrow a_{1,i} \\[1mm] z \leftarrow a_{2,i} \\[1mm] \lambda \leftarrow a_{3,i} \\[2mm] A \leftarrow \begin{pmatrix} \dfrac{d^2}{dx^2}F(x,y,z,\lambda) & \dfrac{d}{dx}\dfrac{d}{dy}F(x,y,z,\lambda) & \dfrac{d}{dx}\dfrac{d}{dz}F(x,y,z,\lambda) \\[3mm] \dfrac{d}{dx}\dfrac{d}{dy}F(x,y,z,\lambda) & \dfrac{d^2}{dy^2}F(x,y,z,\lambda) & \dfrac{d}{dy}\dfrac{d}{dz}F(x,y,z,\lambda) \\[3mm] \dfrac{d}{dx}\dfrac{d}{dz}F(x,y,z,\lambda) & \dfrac{d}{dy}\dfrac{d}{dz}F(x,y,z,\lambda) & \dfrac{d^2}{dz^2}F(x,y,z,\lambda) \end{pmatrix} \\[3mm] A \end{vmatrix}$$

$$local(a,0) = \begin{pmatrix} 2 & -6.95 \times 10^{-15} & 1.2 \times 10^{-14} \\ -6.95 \times 10^{-15} & 2 & -2.33 \times 10^{-14} \\ 1.2 \times 10^{-14} & -2.33 \times 10^{-14} & 2 \end{pmatrix}$$

$$F\left(a_{0,0}, a_{1,0}, a_{2,0}, a_{3,0}\right) \rightarrow \frac{1}{14}$$

and in Mathematica 8:

In[11]:= F[x_ , y_ , z_ , λ_] := x^2 + y^2 + z^2 + λ * (2 * x + 3 * y - z - 1)

In[12]:= Solve[D[F[x, y, z, λ], x] == 0 && D[F[x, y, z, λ], y] == 0 && D[F[x, y, z, λ], z] == 0 &&
        D[F[x, y, z, λ], λ] == 0, {x, y, z, λ}]

Out[12]= $\left\{\left\{x \to \frac{1}{7}, y \to \frac{3}{14}, z \to -\frac{1}{14}, λ \to -\frac{1}{7}\right\}\right\}$

In[14]:= a := {1/7, 3/14, -1/14, -1/7}

In[15]:= lc[a_] :=
        Module[ {A},
          t1 := D[F[x, y, z, λ], {x, 2}] /. {x → a[[1]], y → a[[2]], z → a[[3]], λ → a[[4]]};
          t2 := D[F[x, y, z, λ], x, y] /. {x → a[[1]], y → a[[2]], z → a[[3]], λ → a[[4]]};
          t3 := D[F[x, y, z, λ], x, z] /. {x → a[[1]], y → a[[2]], z → a[[3]], λ → a[[4]]};
          t4 := D[F[x, y, z, λ], y, z] /. {x → a[[1]], y → a[[2]], z → a[[3]], λ → a[[4]]};
          t5 := D[F[x, y, z, λ], {y, 2}] /. {x → a[[1]], y → a[[2]], z → a[[3]], λ → a[[4]]};
          t6 := D[F[x, y, z, λ], {z, 2}] /. {x → a[[1]], y → a[[2]], z → a[[3]], λ → a[[4]]};
          A := {{t1, t2, t3}, {t2, t5, t4}, {t3, t4, t6}};
          A
        ]

In[16]:= lc[a]

Out[16]= {{2, 0, 0}, {0, 2, 0}, {0, 0, 2}}

In[17]:= F[a[[1]], a[[2]], a[[3]], a[[4]]]

Out[17]= $\frac{1}{14}$

and with Maple 15:

$F := (x, y, z, \lambda) \rightarrow x^2 + y^2 + z^2 + \lambda \cdot (2 \cdot x + 3 \cdot y - z - 1)$ :

$solve\left( \left\{ \frac{\partial}{\partial x} F(x, y, z, \lambda), \frac{\partial}{\partial y} F(x, y, z, \lambda), \frac{\partial}{\partial z} F(x, y, z, \lambda), \frac{\partial}{\partial \lambda} F(x, y, z, \lambda) \right\} \right)$

$$\left\{ \lambda = -\frac{1}{7}, x = \frac{1}{7}, y = \frac{3}{14}, z = -\frac{1}{14} \right\}$$

$a := Vector\left( \left[ \frac{1}{7}, \frac{3}{14}, -\frac{1}{14}, -\frac{1}{7} \right] \right)$ :

$lc := \mathbf{proc}(a)$

$t1 := \left. \frac{\partial^2}{\partial x^2} F(x, y, z, \lambda) \right|_{x = a_1, y = a_2, z = a_3, \lambda = a_4}$ ;

$t2 := \left. \frac{\partial^2}{\partial x \partial y} F(x, y, z, \lambda) \right|_{x = a_1, y = a_2, z = a_3, \lambda = a_4}$ ;

$t3 := \left. \frac{\partial^2}{\partial x \partial z} F(x, y, z, \lambda) \right|_{x = a_1, y = a_2, z = a_3, \lambda = a_4}$ ;

$t4 := \left. \frac{\partial^2}{\partial y \partial z} F(x, y, z, \lambda) \right|_{x = a_1, y = a_2, z = a_3, \lambda = a_4}$ ;

$t5 := \left. \frac{\partial^2}{\partial y^2} F(x, y, z, \lambda) \right|_{x = a_1, y = a_2, z = a_3, \lambda = a_4}$ ;

$t6 := \left. \frac{\partial^2}{\partial z^2} F(x, y, z, \lambda) \right|_{x = a_1, y = a_2, z = a_3, \lambda = a_4}$ ;

$Matrix(3, 3, [t1, t2, t3, t2, t5, t4, t3, t4, t6])$
**end proc**;
$[lc(a), F(a_1, a_2, a_3, a_4)]$

$$\left[ \begin{bmatrix} 2 & 0 & 0 \\ 0 & 2 & 0 \\ 0 & 0 & 2 \end{bmatrix}, \frac{1}{14} \right]$$

Therefore,

$$\left( \frac{1}{7}, \frac{3}{14}, -\frac{1}{14} \right)$$

is a minimum point, and the minimum value of the function $F(x, y, z, \lambda)$ is $\frac{1}{14}$.

20.  Find the conditional extrema for the function:

$$u = xy^2 z^3$$

provided

$$x + y + z = 12, x > 0, y > 0, z > 0.$$

# 6

# Terminology about Integral Calculus

## 6.1 Indefinite Integrals

### 6.1.1 Integrals of Rational Functions

Integration of a rational function reduces to integration of the rational fractions

$$\frac{P(x)}{Q(x)}, \tag{6.1}$$

where $P(x)$ and $Q(x)$ are polynomials and the degree of the numerator is lower than that of the denominator.

In order to achieve a rational fractions by the form (6.1) we need to use the partial fraction expansion techniques. There are two methods of partial fraction expansions:

A) the method through which the coefficients from the partial fraction expansion may be determined by setting (in (6.1) or an equivalent equation) $x$ equal to suitably chosen numbers;

B) the method of undetermined coefficients, which implies one of the following cases (see [59], p. 130):

**Case I**. If the rational function can be written as a sum of a nonrepetead linear factors, i.e. if it has simple distinct roots, then

G.A. Anastassiou and I.F. Iatan: Intelligent Routines, ISRL 39, pp. 245–316.
springerlink.com                    © Springer-Verlag Berlin Heidelberg 2013

$$\frac{P\left(x\right)}{Q\left(x\right)} = \sum_{k=1}^{n} \frac{A_k}{x + x_k} \tag{6.2}$$

where

$$A_k = \left(\frac{\left(x + x_k\right) P\left(x\right)}{Q\left(x\right)}\right)_{x=-x_k}. \tag{6.3}$$

**Case II.** If the rational function can be written as a sum of a repetead linear factors $\left(x + x_k\right)^t$, then there corresponds the partial fractions

$$\frac{P\left(x\right)}{Q\left(x\right)} = \frac{A_1}{x + x_k} + \frac{A_2}{\left(x + x_k\right)^2} + \cdots + \frac{A_m}{\left(x + x_k\right)^m}, \tag{6.4}$$

where the coefficient $A_m$ can be determined by the formula

$$A_m = \left(\frac{\left(x + x_k\right)^m P\left(x\right)}{Q\left(x\right)}\right)_{x=-x_k}. \tag{6.5}$$

**Case III.** In the case of an unrepeated irreducible second degree polynomial, i.e. when the denominator has complex roots, we set up fractions of the form

$$\frac{Ax + B}{x^2 + ax + b} \quad \text{or} \quad \frac{Ax + B}{\left(x + p\right)^2 + q^2}. \tag{6.6}$$

In this case, the best way to find the coefficients is to equate the coefficients of different powers of $x$.

**Case IV.** For repeated irreducible second degree factors, we have fractions of the form

$$\frac{P\left(x\right)}{Q\left(x\right)} = \frac{A_1 x + B_1}{x^2 + ax + b} + \frac{A_2 x + B_2}{\left(x^2 + ax + b\right)^2} + \cdots + \frac{A_r x + B_r}{\left(x^2 + ax + b\right)^r}. \tag{6.7}$$

Again, the best way to find the coefficients is to equate the coefficients of different powers of $x$.

**Example 6.1.** Find the integrals in the following problems:

a) $\displaystyle\int \frac{dx}{x^2 + 4x + 29}$

b) $\displaystyle\int \frac{x}{x^4 + 0.25} dx$

c) $\displaystyle\int \frac{x}{x^4 + x^2 + 1} dx$

d) $I = \displaystyle\int \frac{x^2}{(x+2)^2 (x+4)^2} dx$

e) $I_1 = \displaystyle\int \left(\frac{x+2}{x-2}\right)^2 \cdot \frac{1}{x} dx$

f) $\displaystyle\int \frac{x^2}{(1+x^2)^3} dx.$

**Solutions.**

a) One can write

$$\int \frac{dx}{x^2 + 4x + 29} = \int \frac{dx}{(x+2)^2 + 5^2} = \frac{1}{5} \arctan \frac{x+2}{5} + C.$$

We can also compute this integral using Matlab 7.9:

>> **syms x**
>> **int(1/(x^2+4*x+29))**
ans =
1/5*atan(1/5*x+2/5)

and in Mathcad 14:

$$\int \frac{1}{x^2 + 4 \cdot x + 29} \, dx \rightarrow \frac{\mathrm{atan}\left(\dfrac{x}{5} + \dfrac{2}{5}\right)}{5}$$

and with Mathematica 8:

In[2]:= **Integrate[1 / (x^2 + 4 * x + 29), x]**

Out[2]= $\dfrac{1}{5}$ ArcTan$\left[\dfrac{2+x}{5}\right]$

and in Maple 15:

$$\int \frac{1}{x^2 + 4 \cdot x + 29} \, dx$$

$$\frac{1}{5} \arctan\left(\frac{1}{5} x + \frac{2}{5}\right)$$

b) We shall have

$$\int \frac{x}{x^4 + 0.25}\,dx = \int \frac{x}{x^4 + \frac{1}{4}}\,dx = \int \frac{4x}{4x^4 + 1}\,dx = \int \left(\arctan 2x^2\right)'\,dx = \arctan 2x^2 + C.$$

c) Here we use the substitution

$$x^2 = t$$

from which we compute

$$2x\,dx = dt.$$

Making this substitution then gives

$$\int \frac{x}{x^4 + x^2 + 1}\,dx = \frac{1}{2}\int \frac{dt}{t^2 + t + 1} = \frac{1}{2}\int \frac{dt}{\left(t + \frac{1}{2}\right)^2 + \frac{3}{4}}$$

$$= \frac{1}{2}\cdot\frac{2}{\sqrt{3}}\arctan \frac{t + \frac{1}{2}}{\frac{\sqrt{3}}{2}} + C = \frac{1}{\sqrt{3}}\arctan \frac{2t + 1}{\sqrt{3}} + C$$

$$= \frac{1}{\sqrt{3}}\arctan \frac{2x^2 + 1}{\sqrt{3}} + C.$$

d) We shall use the idea of partial fraction expansion which consists to take a proper rational function and express it as the sum of simpler rational functions.

Our expansion takes the form

$$\frac{x^2}{(x + 2)^2 (x + 4)^2} = \frac{A}{x + 2} + \frac{B}{(x + 2)^2} + \frac{C}{x + 4} + \frac{D}{(x + 4)^2},$$

such that

$$A (x + 2) (x + 4)^2 + B (x + 4)^2 + C (x + 2)^2 (x + 4) + D (x + 2)^2 = x^2,$$
$$(6.8)$$

where $A, B, C, D$ are constants which are yet to be determined.

Making $x = -2$, $x = -4$, $x = 0$, $x = -1$ in (6.8) one obtains:

$$x = -2 \Longrightarrow 4B = 4 \Longrightarrow B = 1$$
$$x = -4 \Longrightarrow 4D = 16 \Longrightarrow D = 4$$

$$\left.\begin{array}{l} x = 0 \Longrightarrow 32A + 16B + 16C + 4D = 0 \Longrightarrow 2A + C = -2 \\ x = -1 \Longrightarrow 9A + 9B + 3C + D = 1 \Longrightarrow 3A + C = -4 \end{array}\right\} \Longrightarrow A = -2, C = 2.$$

It results

$$I = (-2) \int \frac{2}{x+2} dx + \int \frac{1}{(x+2)^2} dx + 2 \int \frac{1}{x+4} dx + 4 \int \frac{1}{(x+4)^2} dx$$

$$= -2 \ln|x+2| - \frac{1}{x+2} + 2 \ln|x+4| - \frac{4}{x+4} + C$$

$$= \ln \left( \frac{x+4}{x+2} \right)^2 - \frac{1}{x+2} - \frac{4}{x+4} + C$$

$$= \ln \left( \frac{x+4}{x+2} \right)^2 - \frac{5x+12}{(x+2)(x+4)} + C.$$

e) One can notice that:

$$I_1 = \int \left( \frac{x+2}{x-2} \right)^2 \cdot \frac{1}{x} dx = \int \frac{x^2 + 4x + 4}{x(x^2 - 2x + 1)} dx$$

$$= \int \frac{x^2 - 2x + 1}{x(x^2 - 2x + 1)} dx + \int \frac{6x + 3}{x(x^2 - 2x + 1)} dx,$$

namely

$$I_1 = \int \frac{1}{x} dx + 3 \int \frac{2x+1}{x(x-1)^2} dx.$$

Using the method of partial fraction expansion we can write

$$\frac{2x+1}{x(x-1)^2} = \frac{A}{x} + \frac{B}{x-1} + \frac{C}{(x-1)^2};$$

therefore

$$A(x-1)^2 + Bx(x-1) + Cx = 2x + 1. \qquad (6.9)$$

Making $x = 1$, $x = 0$, $x = 2$ in (6.9) we shall obtain:

$$x = 1 \Longrightarrow C = 3$$
$$x = 0 \Longrightarrow A = 1$$

$$x = 2 \Longrightarrow A + 2B + 2C = 5 \Longrightarrow B = \frac{5 - 2C - A}{2} = -\frac{2}{2} = -1.$$

Therefore

$$I_1 = \ln|x| + 3\int \frac{1}{x}dx - 3\int \frac{1}{x-1}dx + 3\int \frac{3}{(x-1)^2}dx$$

$$= \ln|x| + 3\ln|x| - 3\ln|x-1| - \frac{9}{x-1} + C;$$

finally, one obtains

$$I_1 = \ln\frac{x^4}{|x-1|^3} - \frac{9}{x-1} + C.$$

f) One can notice that:

$$\int \frac{x^2}{(1+x^2)^3}dx = -\frac{1}{4}\int x\left(\frac{1}{(1+x^2)^2}\right)' dx = -\frac{1}{4}\cdot x\cdot\frac{1}{(1+x^2)^2} + \frac{1}{4}\int \frac{1}{(1+x^2)^2}dx$$

$$= -\frac{1}{4}\cdot\frac{x}{(1+x^2)^2} + \frac{1}{4}\int \frac{x^2+1-x^2}{(1+x^2)^2}dx$$

$$= -\frac{1}{4}\cdot\frac{x}{(1+x^2)^2} + \frac{1}{4}\int \frac{1}{1+x^2}dx - \frac{1}{4}\int \frac{x^2}{(1+x^2)^2}dx$$

$$= -\frac{1}{4}\cdot\frac{x}{(1+x^2)^2} + \frac{1}{4}\arctan x + \frac{1}{8}\int x\left(\frac{1}{1+x^2}\right)' dx$$

$$= -\frac{1}{4}\cdot\frac{x}{(1+x^2)^2} + \frac{1}{4}\arctan x + \frac{1}{8}\cdot\frac{x}{1+x^2} - \frac{1}{8}\int \frac{1}{1+x^2}dx.$$

Finally, one obtains

$$\int \frac{x^2}{(1+x^2)^3}dx = -\frac{1}{4}\cdot\frac{x}{(1+x^2)^2} + \frac{1}{8}\arctan x + \frac{1}{8}\cdot\frac{x}{1+x^2} + C$$

We shall have in Matlab 7.9:
>>**syms x**
>> **int(x^2/((x^2+1)^3))**
    ans =
-1/4*x/(x^2+1)^2+1/8*x/(x^2+1)+1/8*atan(x)
and in Mathematica 8:

In[3]:= **Integrate[x^2 / (1 + x^2) ^3, x]**

Out[3]=  $-\dfrac{x}{4\left(1+x^2\right)^2} + \dfrac{x}{8\left(1+x^2\right)} + \dfrac{\text{ArcTan}[x]}{8}$

and with Maple 15:

$$\int \frac{x^2}{\left(1+x^2\right)^3}\, dx$$

$$-\frac{1}{4}\frac{x}{\left(1+x^2\right)^2}+\frac{1}{8}\frac{x}{1+x^2}+\frac{1}{8}\arctan(x)$$

We can not obtain the same result using Mathcad 14.

## 6.1.2 Reducible Integrals to Integrals of Rational Functions

### 6.1.2.1 Integrating Trigonometric Functions

We shall analyse the following cases (see [38, p. 12]):
**Case 1.** When the integrand is of the form

$$R\left(\sin x, \cos x\right) \tag{6.10}$$

one uses the substitution

$$\tan \frac{x}{2} = t \Longrightarrow x = 2\arctan t, \tag{6.11}$$

which implies

$$dx = \frac{2}{1+t^2}dt. \tag{6.12}$$

In this case we have to use the formulas

$$\begin{cases} \sin x = \frac{2\tan\frac{x}{2}}{1+\tan^2\frac{x}{2}} \\ \cos x = \frac{1-\tan^2\frac{x}{2}}{1+\tan^2\frac{x}{2}}. \end{cases} \tag{6.13}$$

**Case 2.** When the integrand is of the form

$$R\left(-\sin x, \cos x\right) = -R\left(\sin x, \cos x\right) \tag{6.14}$$

one uses the substitution

$$\cos x = t, \tag{6.15}$$

which implies

$$-\sin x\, dx = dt. \tag{6.16}$$

**Case 3.** When the integrand is of the form

$$R\left(\sin x, -\cos x\right) = -R\left(\sin x, \cos x\right) \tag{6.17}$$

one uses the substitution

$$\sin x = t, \tag{6.18}$$

which implies

$$\cos x \mathrm{d}x = \mathrm{d}t. \tag{6.19}$$

**Case 4.** When the integrand is of the form

$$R\left(-\sin x, -\cos x\right) = R\left(\sin x, \cos x\right) \tag{6.20}$$

one uses the substitution

$$\tan x = t \Longrightarrow x = \arctan t, \tag{6.21}$$

which implies

$$\mathrm{d}x = \frac{1}{1+t^2}\mathrm{d}t. \tag{6.22}$$

In this case we have to use the formulas

$$\begin{cases} \sin^2 x = \frac{\tan^2 x}{1+\tan^2 x} \\ \cos^2 x = \frac{1}{1+\tan^2 x}. \end{cases} \tag{6.23}$$

### 6.1.2.2    Integrating Certain Irrational Functions

**Case 1.** When the integrand is a function of $k+1$ variables, having the form

$$P\left(x, \left(\frac{ax+b}{cx+d}\right)^{\frac{p_1}{q_1}}, \left(\frac{ax+b}{cx+d}\right)^{\frac{p_2}{q_2}}, \ldots, \left(\frac{ax+b}{cx+d}\right)^{\frac{p_k}{q_k}}\right), \ p_i, \ q_i \in \mathbb{N}, \ i = \overline{1,k}, \ k \in \mathbb{N} \tag{6.24}$$

one uses the substitution (see [8])

$$\frac{ax+b}{cx+d} = t^n, n = \text{least common multiple } (q_1, q_2, \ldots, q_k). \tag{6.25}$$

**Case 2.** We have a binomial integral of the form

$$\int x^m \left(ax^n + b\right)^p \mathrm{d}x, \ a, b \neq 0, \ m, n, p \in \mathbb{Q}. \tag{6.26}$$

The integrals here spoken of can be reduced to integrals of rational functions in the following three situations (see [38], p. 16):

a) When in the integrand we have $p \in \mathbb{Z}$, one uses the substitution

$$x = t^s \,, \tag{6.27}$$

where $s$ is the common denominator of $m$ and $n$.

Choosing this substitution implies

$$\mathrm{d}x = st^{s-1} \, \mathrm{d}t. \tag{6.28}$$

b) It is the case when we have

$$\begin{cases} p \notin \mathbb{Z} \\ \frac{m+1}{n} \in \mathbb{Z}; \end{cases} \tag{6.29}$$

one uses the substitution

$$ax^n + b = t^s \,, \tag{6.30}$$

where $s$ is the denominator of $p$.

c) When we have

$$\begin{cases} p \notin \mathbb{Z} \\ \frac{m+1}{n} \notin \mathbb{Z} \\ \frac{m+1}{n} + p \in \mathbb{Z} \end{cases} \tag{6.31}$$

one makes the substitution

$$a + bx^{-n} = t^s \,, \tag{6.32}$$

where $s$ is the denominator of $p$.

**Case 3.** When there is an integral of the form

$$\int R\left(x, \sqrt{ax^2 + bx + c}\right) \mathrm{d}x \,, \tag{6.33}$$

where $R$ is a rational function we can to apply the following two methods of computing (see [44], p. 570 and [42], p. 36):

1) Depending on the values of coefficients $a, b$ and $c$ may use different substitutions that transform the function $R$ in a rational function of a new variable $z$:

a) if $a$ and $ax^2 + bx + c$ are positive and $b^2 - 4ac \neq 0$, one makes the substitution

$$\sqrt{ax^2 + bx + c} = x\sqrt{a} + z;$$

b) if $c$ is not negative, one makes the substitution

$$\sqrt{ax^2 + bx + c} = xz + \sqrt{c};$$

(6.35)

c) if $a$ is negative and the equation $ax^2 + bx + c = 0$ has the real and distinct roots $x_1$ and $x_2$, one makes the substitution

$$\sqrt{ax^2 + bx + c} = z\left(x - x_1\right).$$

(6.36)

2) Transforming the quadratic $ax^2 + bx + c$ into a sum or difference of squares, the integral becomes reducible to one of the types of integrals:

$$a) \int R\left(z, \sqrt{m^2 - z^2}\right) dz,$$

(6.37)

$$b) \int R\left(z, \sqrt{m^2 + z^2}\right) dz,$$

(6.38)

$$c) \int R\left(z, \sqrt{z^2 - m^2}\right) dz,$$

(6.39)

which are evaluated using the substitutions:

$$a) \; z = m \sin t,$$

(6.40)

$$b) \; z = m \tan t,$$

(6.41)

$$c) \; z = \frac{m}{\cos t}.$$

(6.42)

**Example 6.2.** Compute the following integrals by reducing them to integrals of rational functions:

$$a) \; I = \int \frac{e^{x/2}}{e^{x/6}\left(e^{x/3} + 1\right)} dx$$

$$b) \int \frac{1}{\sin^3 x} dx$$

$$c) \int \frac{\cos^5 x}{\sin^4 x} dx$$

$$d) \int \frac{\sqrt{x+1}}{1 + \sqrt{x+1}} dx, \; x \in (-1, \infty)$$

e) $\displaystyle\int x^5 \cdot \sqrt[3]{\left(1+x^3\right)^2}\,dx$

f) $\displaystyle\int \frac{\sqrt[3]{1+\sqrt[4]{x}}}{\sqrt{x}}\,dx, \quad x \in (0, \infty)$.

**Solutions.**

a) One can notice that

$$I = \int \frac{e^{x/2}}{e^{x/6}\left(e^{x/3}+1\right)}\,dx = \int \frac{e^{x/2}}{e^{x/6}\cdot e^{x/3}\left(1+e^{-x/3}\right)}\,dx = \int \frac{dx}{1+e^{-x/3}}.$$

Therefore, we substitute

$$e^{-x/3} = y, \ y > 0,$$

which implies

$$-\frac{1}{3}e^{-x/3}dx = dy \implies dx = -3^{x/3}dy;$$

therefore

$$dx = -\frac{3}{y}dy.$$

Making this substitution gives

$$I = -3\int \frac{dy}{y\left(1+y\right)}.$$

Using the method of partial fraction expansion we can write

$$\frac{1}{y\left(1+y\right)} = \frac{A}{y} + \frac{B}{y+1} \iff A\left(y+1\right) + By = 1$$

$$\iff \left(A+B\right)y + A = 1 \implies \begin{cases} A = 1 \\ A+B = 0 \implies B = -A - 1. \end{cases}$$

Therefore

$$I = (-3) \cdot \left(\int \frac{dy}{y} - \int \frac{dy}{y+1}\right) = -3\ln y + 3\ln\left(y+1\right) + C = 3\ln\frac{y+1}{y} + C$$

$$= 3\ln\frac{e^{-x/3}+1}{e^{-x/3}} + C = 3\ln\left(1+e^{x/3}\right) + C.$$

We can use the following Matlab 7.9 sequence to calculate the integral:

```
>>syms x
>> int(exp(x/2)/(exp(x/6)*(exp(x/3)+1)))
    ans =
    3*log(exp(1/3*x)+1)
```

and the Mathcad 14:

$$\int \frac{e^{\frac{x}{2}}}{e^{\frac{x}{6}}\left(e^{\frac{x}{3}}+1\right)}\, dx \rightarrow 3\cdot\ln\left(e^{\frac{x}{3}}+1\right)$$

and the Mathematica 8:

In[1]:= **Integrate[Exp[x/2] / (Exp[x/6] * (Exp[x/3] + 1)), x]**

Out[1]= $3\,\text{Log}\left[1 + e^{x/3}\right]$

and the Maple 15:

$$simplify\left(\int \frac{e^{\frac{x}{2}}}{e^{\frac{x}{6}}\left(e^{\frac{x}{3}}+1\right)}\, dx\right)$$

$$\frac{3\ln\left(e^{\frac{1}{3}x}+1\right)}{\ln(e)}$$

b) It is an integral involving trigonometric functions. We are in the **Case 1** with the integral from the point b). Therefore, one chooses the substitution

$$\tan\frac{x}{2} = z \implies x = 2\arctan z,$$

which implies

$$dx = \frac{2}{1+z^2}dz.$$

We shall obtain

$$\sin x = \frac{2\tan\frac{x}{2}}{1+\tan^2\frac{x}{2}} = \frac{2z}{1+z^2}.$$

Choosing this substitution gives

$$\int \frac{1}{\sin^3 x}\,dx = \int \frac{\left(1+z^2\right)^3}{8z^3}\cdot\frac{2}{1+z^2}\,dz = \frac{1}{4}\int \frac{1+2z^2+z^4}{z^3}\,dz$$

$$= \frac{1}{4}\int z^{-3}dz + \frac{1}{2}\int z^{-1}dz + \frac{1}{4}\int z\,dz = \frac{1}{4}\cdot\frac{z^{-3+1}}{-3+1} + \frac{1}{2}\ln|z| + \frac{1}{4}\cdot\frac{z^2}{2} + C$$

$$= -\frac{1}{8}z^{-2} + \frac{1}{2}\ln|z| + \frac{1}{8}z^2 + C$$

$$= \left(-\frac{1}{8}\right)\cdot\frac{1}{\tan^2\frac{x}{2}} + \frac{1}{2}\ln\left|\tan\frac{x}{2}\right| + \frac{1}{8}\tan^2\frac{x}{2} + C$$

$$= \frac{1}{8}\left(\tan^2\frac{x}{2} - \frac{1}{\tan^2\frac{x}{2}}\right) + \ln\left(\tan^2\frac{x}{2}\right) + C.$$

c) It is also an integral which involves trigonometric functions. We are in the **Case 3**.

One uses the substitution

$$\sin x = t,$$

which implies

$$\cos x\,dx = dt.$$

Choosing this substitution makes that

$$\int \frac{\cos^5 x}{\sin^4 x}\,dx = \int \frac{\left(1-\sin^2 x\right)^2\cdot\cos x}{\sin^4 x}\,dx = \int \frac{\left(1-t^2\right)^2}{t^4}\,dt$$

$$= \int \frac{dt}{t^4} - \int \frac{2t^2}{t^4}\,dt + \int \frac{t^4}{t^4}\,dt = -\frac{1}{3t^3} + \frac{2}{t} + t + C$$

$$= -\frac{1}{3\sin^3 x} + \frac{2}{\sin x} + \sin x + C.$$

d) It is an integral which involves irrational functions.
For the integral from d) we have:

$$\begin{cases} a = 1, b = 1, c = 0, d = 1 \\ \qquad n = 2. \end{cases}$$

One chooses the substitution

$$\sqrt{x+1} = t \implies x + 1 = t^2,$$

which implies

$$dx = 2t\,dt.$$

Making this substitution gives

$$\int \frac{\sqrt{x+1}}{1+\sqrt{x+1}}dx = \int \frac{t}{1+t}\cdot 2t\,dt = \int 2t\,dt - \int \frac{2t}{1+t}dt = 2\cdot\frac{t^2}{2} - 2\int \frac{t+1-1}{1+t}dt$$

$$= t^2 - 2\int dt + 2\int \frac{dt}{1+t} = t^2 - 2t + 2\ln|t+1|$$

$$= x+1 - 2\sqrt{x+1} + 2\ln\left|\sqrt{x+1}+1\right| + C.$$

e) We have a binomial integral.
We are in the **Case 2** with integral from e) because

$$p = \frac{2}{3} \notin \mathbb{Z}$$

$$m = 5, n = 3, \frac{m+1}{n} = 2 \in \mathbb{Z}.$$

One uses the substitution

$$x^3 + 1 = t^3 \implies x = \sqrt[3]{t^3 - 1},$$

which implies

$$dx = \frac{t^2}{(t^3 - 1)^{2/3}}dt.$$

We shall deduce that

$$\int x^5 \sqrt[3]{(1+x^3)^2}\,dx = \int (t^3 - 1)^{5/3}\cdot t^2 \cdot \frac{t^2}{(t^3 - 1)^{2/3}}dt = \int t^4 (t^3 - 1)\,dt$$

$$= \int t^7\,dt - \int t^4\,dt = \frac{t^8}{8} - \frac{t^5}{5} + C$$

$$= \frac{1}{8}(x^3 + 1)^{8/3} - \frac{1}{5}(x^3 + 1)^{5/3} + C$$

$$= (x^3 + 1)^{5/3}\cdot \frac{5x^3 - 3}{40} + C$$

$$= \frac{1}{40}(x^3 + 1)^{2/3}\cdot(x^3 + 1)\cdot(5x^3 - 3) + C,$$

namely

$$\int x^5 \sqrt[3]{(1+x^3)^2}\,dx = \frac{1}{40}(x^3 + 1)^{2/3}\cdot(5x^6 + 2x^3 - 3) + C.$$

We shall solves the integral using Matlab 7.9:
>>int(x^5*((1+x^3)^(2/3)),x)
ans =
1/40*(-3+2*x^3+5*x^6)*(1+x^3)^(2/3)
and Mathcad 14:

$$\int x^5 \cdot \sqrt[3]{\left(1+x^3\right)^2}\, dx \rightarrow \left(\frac{x^3}{8} - \frac{3}{40}\right) \left[\left(x^3+1\right)^2\right]^{\frac{1}{3}} \cdot \left(x^3+1\right)$$

and Mathematica 8:

In[5]:= **Integrate[x^5 * (1 + x^3) ^ (2 / 3) , x]**

Out[5]= $\dfrac{1}{40}\left(1+x^3\right)^{5/3}\left(-3+5x^3\right)$

and Maple 15:

$$\int x^5 \cdot \sqrt[3]{\left(1+x^3\right)^2}\, dx$$

$$\frac{1}{40}\left(1+x\right)\left(x^2-x+1\right)\left(-3+5x^3\right)\left(\left(1+x^3\right)^2\right)^{1/3}$$

f) We can notice that

$$\int \frac{\sqrt[3]{1+\sqrt[4]{x}}}{\sqrt{x}}\, dx = \int x^{-1/2}\left(1+x^{1/4}\right)^{1/3}\, dx.$$

Therefore, one chooses the substitution (we are in the third case of the binomial integral)

$$x^{1/4} + 1 = t^3 \implies x = \left(t^3-1\right)^4,$$

which implies

$$dx = 4 \cdot 3t^2 \left(t^3-1\right)^3\, dt.$$

Making this substitution gives

$$\int \left(t^3-1\right)^{-2} t \cdot 12t^2 \left(t^3-1\right)^3\, dt = 12\int t^3\left(t^3-1\right)\, dt = 12\left(\int t^6 dt - \int t^3 dt\right)$$

$$= 12\left(\frac{t^7}{7}-\frac{t^4}{4}\right) + C.$$

Finally, one obtains

$$\int \frac{\sqrt[3]{1 + \sqrt[4]{x}}}{\sqrt{x}} dx = \frac{12}{7} \left(1 + \sqrt[4]{x}\right)^{7/3} - 3 \left(1 + \sqrt[4]{x}\right)^{4/3} + C.$$

We shall check this result in Matlab 7.9:

`>> int(((1+x^(1/4))^(1/3))/(x^0.5),x)`

ans =

`12/7*(1+x^(1/4))^(7/3)-3*(1+x^(1/4))^(4/3)`

and in Mathcad 14:

$$\int \frac{\left(1 + x^{\frac{1}{4}}\right)^{\frac{1}{3}}}{x^{\frac{1}{2}}} dx \rightarrow \left(\frac{12 \cdot x^{\frac{1}{4}}}{7} - \frac{9}{7}\right) \left(x^{\frac{1}{4}} + 1\right)^{\frac{4}{3}}$$

and Mathematica 8:

In[15]:= `Integrate[ (1 + x^ (1 / 4))^(1 / 3) / Sqrt[x], x]`

Out[15]= $\frac{3}{7} \left(1 + x^{1/4}\right)^{4/3} \left(-3 + 4 x^{1/4}\right)$

and Maple 15:

$$\int \frac{\sqrt[3]{1 + \sqrt[4]{x}}}{\sqrt{x}} dx$$

$$\frac{12}{7} \left(1 + x^{1/4}\right)^{7/3} - 3 \left(1 + x^{1/4}\right)^{4/3}$$

## 6.2    Some Applications of the Definite Integrals in Geometry and Physics

### 6.2.1    The Area under a Curve

The definite integral represents the area under the curve $y = f(x)$ from to $x = a$ to $x = b$. There are several situations (see the Figure 6.1-6.4, [11]):

**Case 1.** Curves which are entirely above the $Ox$-axis.

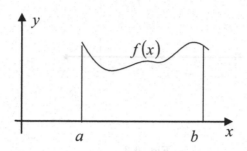

Fig. 6.1

In this case, we find the area by simply finding the integral:

$$\text{Area} = \int_a^b f(x)\,dx. \tag{6.43}$$

**Case 2.** Curves which are entirely below the $Ox$-axis.

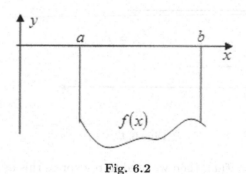

Fig. 6.2

In this case, the integral gives a negative number. We need to take the absolute value of this to find our area:

$$\text{Area} = \left| \int_a^b f(x)\,dx \right|. \tag{6.44}$$

**Case 3.** Part of the curve is below the $Ox$-axis and part of the curve is above the $Ox$-axis.

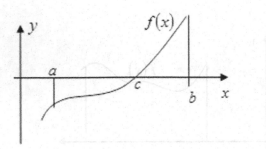

**Fig. 6.3**

In this case, we have to sum the individual parts, taking the absolute value for the section where the curve is below the $Ox$-axis (from $x = a$ to $x = c$):

$$\text{Area} = \left| \int_a^c f(x)\, dx \right| + \int_c^b f(x)\, dx. \tag{6.45}$$

**Case 4.** Certain curves are much easier to sum vertically.

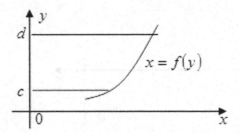

**Fig. 6.4**

If we are given $y = f(x)$, then we need to re-express this as $x = f(y)$. In this case we have:

$$\text{Area} = \int_c^d f(y)\, dy. \tag{6.46}$$

If the curve is defined by equations in parametric form: $x = x(t)$, $y = y(t)$, $t \in [a, b]$, the area under the curve $y = f(x)$ from to $x = a$ to $x = b$ is given by the integral:

$$\text{Area} = \int_a^b y(t) \cdot x'(t)\, dt. \tag{6.47}$$

For a plane curve given in polar coordinates: $\rho = \rho(\theta)$, $\theta \in [a, b]$, the area is:

$$\text{Area} = \frac{1}{2} \int_a^b \rho(\theta)^2 \, d\theta. \tag{6.48}$$

**Example 6.3.** Compute the area of an ellipse of semi-axes $a$ and $b$.
**Solution.**
One knows that an ellipse is the set

$$\Gamma = \left\{ (x, y) \in \mathbb{R}^2 \,\big|\, y = \pm \frac{b}{a} \sqrt{a^2 - x^2}, \ x \in [-a, a] \right\}$$

and one can be represented like in the Figure 6.5.

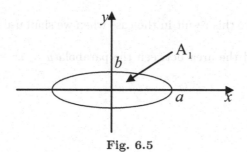

**Fig. 6.5**

One deduces that

$$A_1 = \frac{1}{4} \cdot \text{Area}(\Gamma) = \frac{b}{a} \int_0^a \sqrt{a^2 - x^2} \, dx.$$

One makes the substitution

$$x = a \cos \theta,$$

which implies

$$dx = -a \sin \theta \, d\theta.$$

Using this substitution gives

$$A_1 = ab \int_0^{\pi/2} \sin^2 \theta \, d\theta = ab \cdot \frac{\pi}{4}.$$

Therefore, one deduces that the area of the ellipse of semi-axes $a$ and $b$ is:

$$A = 4A_1 = \pi ab.$$

We can easier to compute this area in Matlab 7.9:
**>>symx x a b**

```
>> area=4*simple(int(b/a*sqrt(a^2-x^2),x,0,a))
area =
b*a*pi
```

and using Mathematica 8:

In[33]:= **1 := Integrate[b / a \* Sqrt[a^2 - x^2], {x, 0, a}]**

In[34]:= **Area = 4 \* 1**

Out[34]= $\sqrt{a^2}\; b\,\pi$

We can not achieve this result in the case when we shall use Mathcad 14 and Maple 15.

**Example 6.4.** Find the area between the parabola $y = 4x - x^2$ and the $Ox$- axis.

**Solution.**

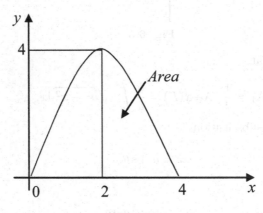

**Fig. 6.6**

One deduces that

$$\text{Area} = \int_0^4 \left(4x - x^2\right) \mathrm{d}x = 4 \cdot \left.\frac{x^2}{2}\right|_0^4 - \left.\frac{x^3}{3}\right|_0^4 = \frac{32}{3}.$$

We can prove that using Matlab 7.9:

```
>>syms x
>>Area=int(4*x-x^2,x,0,4)
Area=
32/2
```

and Mathcad 14:

$$\int_0^4 \left(4 \cdot x - x^2\right) dx \rightarrow \frac{32}{3}$$

and Mathematica 8:

In[37]:= **Area = Integrate[4 * x - x^2, {x, 0, 4}]**

Out[37]= $\dfrac{32}{3}$

and Maple 15:

$Area = \displaystyle\int_0^4 \left(4 \cdot x - x^2\right) dx$

$$Area = \frac{32}{3}$$

## 6.2.2    The Area between by Two Curves

We need the following steps to compute the area bounded by the two curves $f(x)$ and $g(x)$, which one intersects in the two points $(x_1, y_1)$ and $(x_2, y_2)$ (see the Figure 6.7 , [44], p. 557):

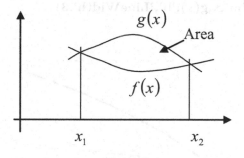

**Fig. 6.7**

**Step 1.** Graph both functions in your calculator or on graph paper to see what the area looks like.

**Step 2.** Find the intersection of the two functions to find the limits of integration.

**Step 3.** Compute the area bounded by the two curves using the formula

$$\int_{x_1}^{x_2} |f(x) - g(x)|\,dx. \tag{6.49}$$

**Example 6.5.** Compute the area bounded by the two curves: $y = \ln x$, $y = \ln^2 x$.

**Solution.**

We shall obtain:

$$\text{Area} = \int_1^e \left(\ln x - \ln^2 x\right) dx = \underbrace{\int_1^e \ln x\,dx}_{I_1} - \underbrace{\int_1^e \ln^2 x\,dx}_{I_2} = I_1 - I_2,$$

where

$$I_1 = \int_1^e \ln x\,dx = \int_1^e x' \cdot \ln x\,dx = x\ln x\big|_1^e - \int_1^e x \cdot \frac{1}{x}dx = \text{e-}\, x\big|_1^e = 1,$$

and

$$I_2 = \int_1^e \ln^2 x\,dx = \int_1^e x' \cdot \ln^2 x\,dx = x\ln^2 x\big|_1^e - 2\int_1^e x \cdot \frac{1}{x}\cdot \ln x\,dx = \text{e-}2I_1 = \text{e-}2;$$

hence

$$\text{Area} = 1 - \text{e} + 2 = 3 - \text{e}.$$

In Matlab 7.9 one obtains:

```
>> x=0.1:0.01:20;
>> f=@(x) log(x);
>> g=@(x) log(x)^2;
>> plot(x,f(x),'m',x,g(x),'b','LineWidth',3)
```

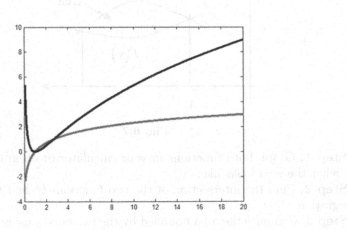

**Fig. 6.8**

```
>> syms y
>> u=solve(log(y)-log(y)^2,y)
u =
1
exp(1)
>> A=eval(int(f(y)-g(y),y,u(1),u(2)))
A =
0.2817
```

We can also give a computer solution using Mathcad 14:

$$x := 0.1, 1.01 .. 20$$

$$f(x) := \ln(x) \qquad g(x) := \ln(x)^2$$

$$u := f(y) - g(y) \text{ solve}, y \rightarrow \begin{pmatrix} 1 \\ e \end{pmatrix}$$

$$A := \int_{u_0}^{u_1} f(y) - g(y) \, dy \qquad A = 0.2817$$

and with Mathematica 8:

In[40]:= **f[x_] := Log[x]**

In[41]:= **g[x_] := Log[x]^2**

In[42]:= **Plot[{f[x], g[x]}, {x, 0.1, 20}]**

Out[42]=

In[53]:= **u := y /. Solve[f[y] - g[y] == 0, y]**

In[55]:= **Integrate[f[y] - g[y], {y, u[[1]], u[[2]]}]**

Out[55]= **3 - e**

and using Maple 15:

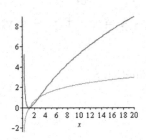

$f := x \rightarrow \ln(x)$ :
$g := x \rightarrow \ln(x)^2$ :
$plot(\{f(x), g(x)\}, x = 0.1 .. 20)$

$u := solve(f(x) - g(x), x)$

$$Area = \int_{u_1}^{u_2} (f(x) - g(x)) \, dx$$

1, e

$Area = 3 - e$

## 6.2.3  Arc Length of a Curve

We can have the following situations (see [8] and [42], p. 43):

**Case 1.** The arc length $L$ of a curve $y = f(x)$, contained between two points with abscissac $x = a$ and $x = b$ is

$$L = \int_a^b \sqrt{1 + f'^2(x)}dx, \ f : [a, b] \to \mathbb{R}. \tag{6.50}$$

**Case 2.** For a curve in plane, given in the parametric form

$$\begin{cases} x = x(t) \\ y = y(t) \end{cases}, \ t \in [a, b], \tag{6.51}$$

the arc length of the curve is

$$L = \int_a^b \sqrt{x'^2(t) + y'^2(t)}dt. \tag{6.52}$$

**Case 3.** For a plane curve given in polar coordinates: $\rho = \rho(\theta)$, $\theta \in [a, b]$, the arc length of the curve is:

$$L = \int_a^b \sqrt{\rho(\theta) + \rho'^2(\theta)}d\theta. \tag{6.53}$$

**Case 4.** For a curve in space, given in the parametric form

$$\begin{cases} x = x(t) \\ y = y(t), \ t \in [a, b], \\ z = z(t) \end{cases} \tag{6.54}$$

the arc length of the curve is

$$L = \int_a^b \sqrt{x'^2(t) + y'^2(t) + z'^2(t)}dt. \tag{6.55}$$

**Example 6.6.** Compute the arc length of the curve:

$$\text{a) } y = \ln \sin x, x \in \left[\frac{\pi}{3}, \frac{\pi}{2}\right]$$

$$\text{b) } \rho = \sin^3 \theta, \ \theta \in \left[0, \frac{\pi}{2}\right]$$

$$\text{c) } \begin{cases} x = 3 \cos t \\ y = 3 \sin t, \ t \in \left[0, \frac{\pi}{2}\right]. \\ z = 4t \end{cases}$$

**Solutions.**
a) Using the formula (6.50) it results that

$$L = \int_{\pi/3}^{\pi/2} \sqrt{1 + \left(\frac{\partial}{\partial x}\left(\ln\left(\sin x\right)\right)\right)^2}\, dx = \int_{\pi/3}^{\pi/2} \sqrt{1 + \left(\frac{\cos x}{\sin x}\right)^2}\, dx = \int_{\pi/3}^{\pi/2} \frac{dx}{\sin x}.$$

Choosing the substitution

$$\tan\frac{x}{2} = z,$$

which implies

$$dx = \frac{2}{1 + z^2} dz.$$

Since

$$\sin x = \frac{2z}{1 + z^2},$$

this substitution gives

$$L = \int_{1/\sqrt{3}}^{1} \frac{2}{1 + z^2} \cdot \frac{1 + z^2}{2z} dz = \int_{1/\sqrt{3}}^{1} \frac{1}{z} dz = \ln z\big|_{\frac{1}{\sqrt{3}}}^{1} = \ln\sqrt{3} = \frac{\ln 3}{2}.$$

We can also compute the arc length in Matlab 7.9:

```
>> simplify(int(sqrt(1+diff(log(sin(x)))^2),pi/3,pi/2))
ans =
1/2*log(3)
```

and in Mathcad 14:

$$f(x) := \ln(\sin(x))$$

$$L := \int_{\frac{\pi}{3}}^{\frac{\pi}{2}} \sqrt{1 + \left(\frac{\partial}{\partial x} f(x)\right)^2}\, dx \qquad L = 0.5493$$

and in Mathematica 8:

```
In[16]:= f[x_] := Log[Sin[x]]

In[17]:= L = Integrate[Sqrt[1 + D[f[x], x]^2], {x, Pi/3, Pi/2}]

Out[17]=  Log[3]
         ───────
            2
```

and with Maple 15:

$$f := x \rightarrow \ln(\sin(x)) :$$

$$L := \int_{\frac{\pi}{3}}^{\frac{\pi}{2}} \sqrt{1 + \left(\frac{\partial}{\partial x} f(x)\right)^2} \, dx$$

$$\frac{1}{2} \ln(3)$$

b) With (6.53) one obtains

$$L = \int_0^{\pi/2} \sqrt{\sin^6 \frac{\theta}{3} + \left(\frac{3}{3} \sin^2 \frac{\theta}{3} \cos \frac{\theta}{3}\right)^2} \, d\theta = \int_0^{\pi/2} \sqrt{\sin^6 \frac{\theta}{3} + \sin^4 \frac{\theta}{3} \cos^2 \frac{\theta}{3}} \, d\theta,$$

namely

$$L = \int_0^{\pi/2} \sin^2 \frac{\theta}{3} d\theta = -3 \int_0^{\pi/2} \sin \frac{\theta}{3} \left(\cos \frac{\theta}{3}\right)' d\theta =$$

$$- 3 \sin \frac{\theta}{3} \cos \frac{\theta}{3} \Big|_0^{\pi/2} + 3 \cdot \frac{1}{3} \int_0^{\pi/2} \cos^2 \frac{\theta}{3} d\theta;$$

therefore

$$L = -3 \sin \frac{\pi}{6} \cos \frac{\pi}{6} + \int_0^{\pi/2} d\theta - \int_0^{\pi/2} \sin^2 \frac{\theta}{3} d\theta = (-3) \cdot \frac{1}{2} \cdot \frac{\sqrt{3}}{2} + \frac{\pi}{2} - L.$$

One deduces

$$2L = (-3) \cdot \frac{\sqrt{3}}{4} + \frac{\pi}{2} \implies L = -\frac{3\sqrt{3}}{8} + \frac{\pi}{4} = 0.1359.$$

The sequence of Matlab 7.9 commands which allow us to compute the arc length of the curve is:

```
>> syms t
>> ro=sin(t/3)^3
ro =
sin(1/3*t)^3
>> L=eval(int(sqrt(ro^2+diff(ro)^2),0,pi/2))
L =
 0.1359
```

We shall also prove this result using Mathcad 14:

$$\rho(\theta) := \sin\left(\frac{\theta}{3}\right)^3$$

$$L := \int_0^{\frac{\pi}{2}} \sqrt{\rho(\theta)^2 + \left(\frac{\partial}{\partial\theta}\rho(\theta)\right)^2}\, d\theta \qquad L = 0.1359$$

and in Mathematica 8:

In[5]:= $\rho[\theta\_] := \text{Sin}[\theta/3]\wedge 3$

In[6]:= $L = \text{Integrate}[\text{Sqrt}[\rho[\theta]\wedge 2 + D[\rho[\theta], \theta]\wedge 2], \{\theta, 0, \text{Pi}/2\}]$

Out[6]= $\dfrac{1}{8}\left(-3\sqrt{3} + 2\pi\right)$

and with Maple 15:

$$\rho := \theta \rightarrow \sin\left(\frac{\theta}{3}\right)^3 :$$

$$L := \int_0^{\frac{\pi}{2}} \sqrt{\rho(\theta)^2 + \left(\frac{\partial}{\partial\theta}\rho(\theta)\right)^2}\, d\theta$$

$$\frac{1}{4}\pi - \frac{3}{8}\sqrt{3}$$

c) Using (6.55) it results:

$$L = \int_0^{\pi/2} \sqrt{(-3\sin t)^2 + (3\cos t)^2 + 4^2}\, dt = 5t\big|_0^{\pi/2} = \frac{5\pi}{2} = 0.7854.$$

The sequence of Matlab 7.9 commands requisite to compute the length arc of the curve is:

```
>> syms t
>> x=diff(3*cos(t))
x =
-3*sin(t)
>> y=diff(3*sin(t))
```

```
y =
3*cos(t)
>> z=diff(4*t)
z =
4
>> L=int(sqrt(x^2+y^2+z^2),0,pi/2)
L=
5/2*pi
```

We shall give a computer solution using Mathcad 14:

$$x(t) := 3 \cdot \cos(t) \qquad y(t) := 3 \cdot \sin(t) \qquad z(t) := 4 \cdot t$$

$$\underset{\sim}{L} := \int_{0}^{\frac{\pi}{2}} \sqrt{\left(\frac{\partial}{\partial t} x(t)\right)^2 + \left(\frac{\partial}{\partial t} y(t)\right)^2 + \left(\frac{\partial}{\partial t} z(t)\right)^2} \, dt \qquad\qquad L = 7.854$$

and in Mathematica 8:

In[10]:= x[t_] := 3 * Cos[t]

In[11]:= y[t_] := 3 * Sin[t]

In[12]:= z[t_] := 4 * t

In[13]:= L = Integrate[Sqrt[D[x[t], t]^2 + D[y[t], t]^2 + D[z[t], t]^2], {t, 0, Pi/2}]

Out[13]= $\dfrac{5\pi}{2}$

and with Maple 15:

$x := t \to 3 \cdot \cos(t) :$
$y := t \to 3 \cdot \sin(t) :$
$z := t \to 4 \cdot t :$

$$L := \int_0^{\frac{\pi}{2}} \sqrt{\left(\frac{\partial}{\partial t} x(t)\right)^2 + \left(\frac{\partial}{\partial t} y(t)\right)^2 + \left(\frac{\partial}{\partial t} z(t)\right)^2}\ dt$$

$$\frac{5}{2}\pi$$

## 6.2.4  Area of a Surface of Revolution

The area of a surface

$$\Sigma = \left\{(x, y, z) \in \mathbb{R}^3 \mid \sqrt{y^2 + z^2} = f(x)\right\} \tag{6.56}$$

formed by rotation about the $x$ axis of an arc of the curve $y = f(x)$ between the points $x = a$ and $x = b$, is expressed by (see [35], p. 485):

$$A = 2\pi \int_a^b f(x) \sqrt{1 + f'^2(x)}\,dx. \tag{6.57}$$

**Remark 6.7.** If the equation of the curve is represented differently, the area of the surface is obtained from (6.57) by an appropriate change of variables.

**Example 6.8.** Find the area of the surface formed by rotation about the $x$-axis of a parabola

$$y^2 = \sqrt{2px},\ p > 0,\ 1 \le x \le 3.$$

**Solution.**
We shall deduce

$$A = 2\pi \int_1^3 \sqrt{2px} \cdot \sqrt{1 + \frac{p}{2x}}\,dx,$$

i.e.

$$A = 2\pi\sqrt{p} \int_1^3 \sqrt{2x + p}\,dx.$$

We shall deduce

$$I = \int_1^3 \sqrt{2x+p}\,dx = \int_1^3 \frac{2x}{\sqrt{2x+p}}\,dx + \int_1^3 \frac{p}{\sqrt{2x+p}}\,dx$$

$$= 2\int_1^3 x\left(\sqrt{2x+p}\right)'dx + p\sqrt{2x+p}\,\Big|_{x=1}^{x=3}$$

$$= 2x\sqrt{2x+p}\,\Big|_{x=1}^{x=3} - 2\int_1^3 \sqrt{2x+p}\,dx + p\left(\sqrt{6+p}-\sqrt{2+p}\right)$$

$$= 6\sqrt{6+p} - 2\sqrt{2+p} - 2I + p\left(\sqrt{6+p}-\sqrt{2+p}\right)$$

$$= (6+p)\sqrt{6+p} - (2+p)\sqrt{2+p} - 2I;$$

hence

$$I = \frac{1}{3}\left[(6+p)\sqrt{6+p} - (2+p)\sqrt{2+p}\right]$$

and

$$A = \frac{2\pi}{3}\sqrt{p}\left[(6+p)^{3/2} - (2+p)^{3/2}\right].$$

We need the following sequence in Matlab 7.9:

```
>>symx x p
>> A=2*pi*int(sqrt(p)*sqrt(2*x+p),x,1,3);
>>simple(A)
ans =
2*pi*(1/3*(6+p)^(3/2)-1/3*(2+p)^(3/2))*p^(1/2)
```

or in Mathcad 14:

$$A(p) := 2 \cdot \pi \cdot \sqrt{p} \cdot \int_1^3 \sqrt{2 \cdot x + p}\ dx \qquad A(p) \to -2 \cdot \pi \cdot \sqrt{p} \cdot \left[\frac{(p+2)^{\frac{3}{2}}}{3} - \frac{(p+6)^{\frac{3}{2}}}{3}\right]$$

or in Mathematica 8:

```
In[6]:= u := Integrate[Sqrt[2 * x + p], {x, 1, 3}]

In[7]:= A = 2 * Pi * Sqrt[p] * u

Out[7]= 2/3 Sqrt[p] (-(2 + p)^(3/2) + (6 + p)^(3/2)) π
```

and in Maple 15:

$$A = 2 \cdot \pi \cdot \sqrt{p} \cdot \int_1^3 \sqrt{2 \cdot x + p} \ dx$$

$$A = 2\pi\sqrt{p}\left(-\frac{1}{3}(p+2)^{3/2} + \frac{1}{3}(p+6)^{3/2}\right)$$

## 6.2.5  Volumes of Solids

The volume formed by the revolution of a curvilinear trapezoid, bounded by the curve $y = f(x)$, the $x$-axis and two vertical lines $x = a$ and $x = b$ about the $x$- and $y$-axes are expressed by (see [35], p. 474 and [42], p. 41):

$$V = \pi \int_a^b f^2(x) \, dx. \tag{6.58}$$

**Example 6.9.** One supposes the plane domain

$$F = \{(x, y) \mid 0 \leq x \leq \pi, \ 0 \leq y \leq \sin x\}.$$

Compute the volume of the solid generated by rotating the region $F$ around the $Ox$-axis.

**Solution.**

Using the formula (6.58) we can find the volume of the solid generated by rotating the region $F$ around the $Ox$-axis.

Therefore, it results

$$V = \pi \int_0^\pi \sin^2 x \, dx = \pi \cdot \frac{\pi}{2} = \frac{\pi^2}{2}.$$

We shall compute this volume in Matlab 7.9:

```
>> syms x
>> V=pi*int(sin(x)^2,x,0,pi)
V =
pi^2/2
```

and using Mathcad 14:

$$V := \pi \cdot \int_0^\pi \sin(x)^2 \, dx \qquad V = 4.935$$

and in Mathematica 8:

In[1]:= V = Pi * Integrate[Sin[x]^2, {x, 0, Pi}]

Out[1]= $\dfrac{\pi^2}{2}$

and with Maple 15:

$$V := \pi \cdot \int_0^\pi \sin(x)^2 \, dx$$

$$\frac{1}{2}\pi^2$$

## 6.2.6  Centre of Gravity

The coordinates for the centre of gravity $G(x_G, y_G)$ corresponding to a homogeneous plate having the form of a plane domain

$$F = \{(x, y) \mid a \le x \le b,\ 0 \le y \le f(x)\}.$$

one determines using the formulas (see [42], p. 45):

$$\begin{cases} x_G = \dfrac{\int_a^b x f(x)\, dx}{\int_a^b f(x)\, dx} \\[2mm] y_G = \dfrac{\frac{1}{2}\int_a^b f^2(x)\, dx}{\int_a^b f(x)\, dx}. \end{cases} \qquad (6.59)$$

**Example 6.10.** Find the coordinates for the centre of gravity corresponding to a homogeneous plate, which has the form of

$$F = \{(x, y) \mid 0 \le x \le \pi,\ 0 \le y \le \sin x\}.$$

**Solution.**
One achieves:

$$\begin{cases} x_G = \dfrac{\int_0^\pi x \sin x\, dx}{\int_0^\pi \sin x\, dx} \\[2mm] y_G = \dfrac{\frac{1}{2}\int_0^\pi \sin^2 x\, dx}{\int_0^\pi \sin x\, dx}. \end{cases}$$

Computing

$$I_1 = \int_0^\pi x \sin x \ dx = - \int_0^\pi x \, (\cos x)' \ dx = - x \cos x |_0^\pi + \int_0^\pi \cos x \ dx$$

$$= \pi + \sin x |_0^\pi = \pi,$$

$$I_2 = \int_0^\pi \sin x \ dx = - \cos x |_0^\pi = 1 + 1 = 2,$$

$$I_3 = \int_0^\pi \sin^2 x \ dx = - \int_0^\pi \sin x \, (\cos x)' \ dx$$

$$= - \sin x \cos x |_0^\pi + \int_0^\pi \cos^2 x \ dx = \int_0^\pi \left( 1 - \sin^2 x \right) \ dx$$

$$= \pi - \int_0^\pi \sin^2 x \ dx = \pi - I_3 \Longrightarrow 2I_3 = \pi \Longrightarrow I_3 = \frac{\pi}{2};$$

one deduces

$$\begin{cases} x_G = \frac{\pi}{2} \\ y_G = \frac{\pi}{8} \end{cases}$$

i.e.

$$G \left( \frac{\pi}{2}, \frac{\pi}{8} \right).$$

We can also determine the coordinates for the centre of gravity corresponding to a homogeneous plate in Matlab 7.9:

```
>> syms x
>> I1=int(x*sin(x),0,pi);
>> I2=int(sin(x),0,pi);
>> I3=int(sin(x)^2,0,pi)/2;
>> xg=I1/I2
xg =
1/2*pi
>> yg=I3/I2
yg =
1/8*pi
```

or with Mathcad 14:

$$f(x) := \sin(x)$$

$$xg := \frac{\displaystyle\int_0^\pi x \cdot f(x)\, dx}{\displaystyle\int_0^\pi f(x)\, dx} \qquad\qquad xg = 1.571$$

$$yg := \frac{\displaystyle\frac{1}{2}\cdot\int_0^\pi f(x)^2\, dx}{\displaystyle\int_0^\pi f(x)\, dx} \qquad\qquad yg = 0.393$$

and using Mathematica 8:

In[4]:= f[x_] := Sin[x]

In[5]:= xg = Integrate[x * f[x], {x, 0, Pi}] / Integrate[f[x], {x, 0, Pi}]

Out[5]= $\dfrac{\pi}{2}$

In[6]:= yg = 1/2 * Integrate[f[x]^2, {x, 0, Pi}] / Integrate[f[x], {x, 0, Pi}]

Out[6]= $\dfrac{\pi}{8}$

and in Maple 15:

$f := x \rightarrow \sin(x)$ :

$$xg = \frac{\displaystyle\int_0^{\pi} x \cdot f(x)\,dx}{\displaystyle\int_0^{\pi} f(x)\,dx}$$

$$xg = \frac{1}{2}\,\pi$$

$$yg = \frac{\dfrac{1}{2} \cdot \displaystyle\int_0^{\pi} f(x)^2\,dx}{\displaystyle\int_0^{\pi} f(x)\,dx}$$

$$yg = \frac{1}{8}\,\pi$$

## 6.3  Improper Integrals

### 6.3.1  Integrals of Unbounded Functions

If a function $f(x)$ is unbounded in any neighbourhood of a point $c$ of an interval $[a, b]$ and is continuous for $a \leq x \leq c$ and $c < x \leq b$, then, by definition, we write (see [8] and [61])

$$\int_a^b f(x)\;dx = \lim_{\varepsilon \to 0} \int_a^{c-\varepsilon} f(x)\;dx + \lim_{\varepsilon \to 0} \int_{c+\varepsilon}^b f(x)\;dx. \qquad (6.60)$$

If the limits on the right hand side of (6.60) exist and are finite, the *improper integral* is said to *converge*, otherwise it *diverges*. When $c = a$ or $c = b$, the definition is correspondingly simplified, as in the following two cases:
**Case a).** If

$$\lim_{\substack{x \to b \\ x < b}} f(x) = \pm\infty$$

then $\int_a^b f(x)\,dx$ is convergent if there is the following limit:

$$\lim_{\substack{\varepsilon \to 0 \\ \varepsilon > 0}} \int_a^{b-\varepsilon} f(x) \, dx$$

and it is finite.

In this case one can be written:

$$\int_a^b f(x) \, dx = \lim_{\substack{\varepsilon \to 0 \\ \varepsilon > 0}} \int_a^{b-\varepsilon} f(x) \, dx. \tag{6.61}$$

**Case b).** If

$$\lim_{\substack{x \to a \\ x > a}} f(x) = \pm\infty$$

then $\int_a^b f(x) \, dx$ is convergent if there is the following limit:

$$\lim_{\substack{\varepsilon \to 0 \\ \varepsilon > 0}} \int_{a+\varepsilon}^b f(x) \, dx$$

exists and it is finite.

In this case one can be written:

$$\int_a^b f(x) \, dx = \lim_{\substack{\varepsilon \to 0 \\ \varepsilon > 0}} \int_{a+\varepsilon}^b f(x) \, dx. \tag{6.62}$$

**Proposition 6.11 (The criterion for absolute convergence, see [42], p. 62).** One calculates in the:

**Case a)**

$$L = \lim_{\substack{x \to b \\ x < b}} (b - x)^p \, f(x) \, dx; \tag{6.63}$$

**Case b)**

$$L = \lim_{\substack{x \to a \\ x > a}} (x - a)^p \, f(x) \, dx. \tag{6.64}$$

If:

- $p < 1, 0 \leq L < \infty$ it results that $\int_a^b f(x)\, dx$ is absolute convergent;
- $p \geq 1, 0 < L \leq \infty$, it results that $\int_a^b f(x)\, dx$ is divergent.

**Example 6.12.** Establish if the following improper integrals are convergent or not and in the affirmative case calculate their values:

$$a)\ I_1 = \int_0^1 \frac{dx}{\sqrt{1-x^2}}$$

$$b)\ I_2 = \int_{-1}^1 \frac{dx}{(x+1)^{2/3}}$$

$$c)\ I_3 = \int_0^1 \frac{dx}{\sqrt{x(1-x)}}.$$

**Solutions.**
a) Since

$$\lim_{\substack{x \to 1 \\ x < 1}} \frac{1}{\sqrt{1-x^2}} = \infty$$

it results that the integral $\int_0^1 \frac{dx}{\sqrt{1-x^2}}$ is improper.
Using the formula (6.61) one obtains

$$\lim_{\substack{\varepsilon \to 0 \\ \varepsilon > 0}} \int_0^{1-\varepsilon} \frac{1}{\sqrt{1-x^2}}\, dx = \lim_{\substack{\varepsilon \to 0 \\ \varepsilon > 0}} \arcsin x \big|_0^{1-\varepsilon} = \lim_{\substack{\varepsilon \to 0 \\ \varepsilon > 0}} \arcsin(1-\varepsilon) = \frac{\pi}{2} < \infty;$$

therefore $I_1$ is convergent and

$$\int_0^1 \frac{dx}{\sqrt{1-x^2}} = \frac{\pi}{2}.$$

Using Matlab 7.9 we shall have:
>>**syms x**
>> **int(1/sqrt(1-x^2),0,1)**
        ans =
**1/2*pi**
while in Mathcad 14:

$$\int_0^1 \frac{1}{\sqrt{1-x^2}}\, dx = 1.571$$

and with Mathematica 8:

In[1]:= **Integrate[1/Sqrt[1 - x^2], {x, 0, 1}]**

Out[1]= $\dfrac{\pi}{2}$

and using Maple 15:

$$\int_0^1 \frac{1}{\sqrt{1-x^2}}\, dx$$

$$\frac{1}{2}\pi$$

b) We have

$$\lim_{\substack{x \to -1 \\ x > -1}} \frac{1}{(x+1)^{2/3}} = \infty;$$

it results that $I_2$ is improper integral.
Using the formula (6.62) one obtains

$$\lim_{\substack{\varepsilon \to 0 \\ \varepsilon > 0}} \int_{-1+\varepsilon}^1 \frac{dx}{(x+1)^{2/3}} - \lim_{\substack{\varepsilon \to 0 \\ \varepsilon > 0}} \left. \frac{(x+1)^{-\frac{2}{3}+1}}{-\frac{2}{3}+1} \right|_{-1+\varepsilon}^1$$

$$= 3 \lim_{\substack{\varepsilon \to 0 \\ \varepsilon > 0}} (x+1)^{1/3} \Big|_{-1+\varepsilon}^1 = 3\sqrt[3]{2};$$

therefore $I_2$ is convergent and

$$\int_{-1}^1 \frac{dx}{(x+1)^{2/3}} = 3\sqrt[3]{2}.$$

One can calculate this integral using Matlab 7.9:
>>**syms x**
>> **int(1/((x+1)^(2/3)),-1,1)**
ans =
**3*2^(1/3)**
and in Mathcad 14:

$$\int_{-1}^{1} \frac{1}{(x+1)^{\frac{2}{3}}} \, dx = 3.78$$

and with Mathematica 8:

In[2]:= **Integrate[1 / (x + 1) ^ (2 / 3), {x, -1, 1}]**

Out[2]= $3 \times 2^{1/3}$

and using Maple 15:

$$\int_{-1}^{1} \frac{1}{(x+1)^{\frac{2}{3}}} \, dx$$

$$3\,2^{1/3}$$

c) Since

$$\left.\begin{array}{c} \lim\limits_{\substack{x \to 1 \\ x < 1}} \dfrac{1}{\sqrt{x(1-x)}} = \infty \\[2mm] \lim\limits_{\substack{x \to 0 \\ x > 0}} \dfrac{1}{\sqrt{x(1-x)}} = \infty \end{array}\right\} \implies I_3 \text{ is an improper integral.}$$

We shall write

$$I_3 = \underbrace{\int_{0}^{1/2} \frac{dx}{\sqrt{x(1-x)}}}_{I} + \underbrace{\int_{1/2}^{1} \frac{dx}{\sqrt{x(1-x)}}}_{J}.$$

Using the formula (6.62) one obtains

$$\lim_{\substack{\varepsilon \to 0 \\ \varepsilon > 0}} \int_{\varepsilon}^{1/2} \frac{dx}{\sqrt{x(1-x)}} = \lim_{\substack{\varepsilon \to 0 \\ \varepsilon > 0}} \int_{\varepsilon}^{1/2} \frac{2}{\sqrt{1-(2x-1)^2}} dx$$

$$= \lim_{\substack{\varepsilon \to 0 \\ \varepsilon > 0}} \arcsin(2x-1)\big|_{\varepsilon}^{1/2}$$

$$= \lim_{\substack{\varepsilon \to 0 \\ \varepsilon > 0}} [\arcsin 0 - \arcsin(2\varepsilon - 1)]$$

$$= -\arcsin(-1) = \arcsin 1 = \frac{\pi}{2};$$

therefore $I$ is convergent and

$$I = \int_0^{1/2} \frac{dx}{\sqrt{x(1-x)}} = \frac{\pi}{2}.$$

Using the formula (6.61) one obtains

$$\lim_{\substack{\varepsilon \to 0 \\ \varepsilon > 0}} \int_{1/2}^{1-\varepsilon} \frac{dx}{\sqrt{x(1-x)}} = \lim_{\substack{\varepsilon \to 0 \\ \varepsilon > 0}} \arcsin(2x-1)\big|_{1/2}^{1-\varepsilon}$$

$$= \lim_{\substack{\varepsilon \to 0 \\ \varepsilon > 0}} \arcsin(2 - 2\varepsilon - 1) = \arcsin 1 = \frac{\pi}{2};$$

therefore $J$ is convergent and

$$J = \int_{1/2}^{1} \frac{dx}{\sqrt{x(1-x)}} = \frac{\pi}{2}.$$

Such that $I_3$ is convergent and

$$I_3 = I + J = \frac{\pi}{2} + \frac{\pi}{2} = \pi.$$

We can simplify the previous calculus in Matlab 7.9:
```
>>syms x
>> int(1/((x*(1-x))^(1/2)),0,1)
ans =
pi
```
and in Mathcad 14:

$$\int_0^1 \frac{1}{\sqrt{x \cdot (1-x)}}\, dx = 3.142$$

and with Mathematica 8:

```
In[3]:= Integrate[1 / Sqrt[x * (1 - x)], {x, 0, 1}]

Out[3]= π
```

and using Maple 15:

$$\int_0^1 \frac{1}{\sqrt{x \cdot (1-x)}}\, dx$$

$$\pi$$

**Example 6.13.** Use the criterion for absolute convergence to establish if the following integrals are convergent or not:

$$\text{a) } I_1 = \int_{-1}^1 \frac{e^x}{x+1}\, dx$$

$$\text{b) } I_2 = \int_0^\pi \frac{dx}{\sqrt{\sin x}}.$$

**Solutions.**
a) As

$$\lim_{\substack{x \to -1 \\ x > -1}} \frac{e^x}{x+1} = \infty$$

one deduces that $I_1$ is an improper integral.
    Using (6.64) we have

$$L = \lim_{\substack{x \to -1 \\ x > -1}} (x+1)^p \frac{e^x}{x+1} \overset{p=1}{=} e^{-1},$$

namely

$$I_1 = \int_{-1}^1 \frac{e^x}{x+1}\, dx$$

is divergent.

We shall compute the limit in Matlab 7.9:

```
>>syms x p
>> limit(subs((x+1)^p*exp(x)/(x+1),p,1),x,-1,'right')
ans =
1/exp(1)
```

and in Mathcad 14:

$$\lim_{x \to -1^+}\left[(x+1)^p \cdot \frac{e^x}{x+1} \quad \text{substitute}, p=1 \to e^x\right] \to e^{-1}$$

and with Mathematica 8:

```
In[15]:= Limit[(x+1)^p*Exp[x] / (x+1) /. p→1, x→-1, Direction→-1]
```

$$\text{Out[15]=} \quad \frac{1}{e}$$

and using Maple 15:

$$\lim_{x \to -1^+} (x+1)^p \cdot \left.\frac{e^x}{x+1}\right|_{p=1}$$

$$e^{-1}$$

b) We can notice that

$$I_2 = \int_0^\pi \frac{dx}{\sqrt{\sin x}} = \underbrace{\int_0^{\pi/2} \frac{dx}{\sqrt{\sin x}}}_{I} + \underbrace{\int_{\pi/2}^\pi \frac{dx}{\sqrt{\sin x}}}_{J}.$$

One achieves:

$$\lim_{\substack{x \to 0 \\ x > 0}} x^p \cdot \frac{1}{\sqrt{\sin x}} \overset{p=1/2}{=} 1 \Longrightarrow I \text{ is absolute convergent;}$$

$$\lim_{\substack{x \to \pi \\ x > \pi}} (\pi - x)^p \cdot \frac{1}{\sqrt{\sin x}} = \lim_{\substack{x \to \pi \\ x > \pi}} (\pi - x)^p \cdot \frac{1}{\sqrt{\sin (\pi - x)}}$$

$$\overset{p=1/2}{=} 1 \implies J \text{ is absolute convergent;}$$

therefore

$$I_2 = I + J = \int_0^\pi \frac{dx}{\sqrt{\sin x}}$$

is absolute convergent.

## 6.3.2   Integrals with Infinite Limits

**Case a).** If

$$\lim_{b \to \infty} \int_a^b f(x)$$

exists and it is finite then

$$\int_a^\infty f(x)$$

is convergent and:

$$\int_a^\infty f(x) = \lim_{b \to \infty} \int_a^b f(x). \tag{6.65}$$

**Case b).** If

$$\lim_{a \to -\infty} \int_a^b f(x)$$

exists and it is finite then

$$\int_{-\infty}^b f(x)$$

is convergent and:

$$\int_{-\infty}^b f(x) = \lim_{a \to -\infty} \int_a^b f(x). \tag{6.66}$$

**Proposition 6.14 (The criterion for absolute convergence, see [42], p. 68).** One computes:

$$L = \lim_{x \to \infty} x^p f(x). \tag{6.67}$$

If:

- $p > 1, 0 \le L < \infty$ it results that $\int_a^\infty f(x)\,dx$ is convergent;
- $p \le 1, 0 < L \le \infty$, it results that $\int_a^\infty f(x)\,dx$ is divergent.

**Example 6.15.** Establish if the following improper integrals on a unbounded interval are convergent or not and in the affirmative case calculate their values:

$$\text{a) } I_1 = \int_1^\infty \frac{dx}{1 + x^2}$$

$$\text{b) } I_2 = \int_0^\infty \frac{dx}{\sqrt{1 + x}}$$

$$\text{c) } I_3 = \int_0^\infty xe^{-x^2}\,dx$$

$$\text{d) } I_4 = \int_{-\infty}^\infty \frac{dx}{1 + x^2}$$

$$\text{e) } I_5 = \int_3^\infty \frac{dx}{\sqrt{x}\,(x + 1)}.$$

**Solutions.**
a) We can write

$$\lim_{b \to \infty} \int_1^b \frac{dx}{1 + x^2} = \lim_{b \to \infty} \arctan x \big|_1^b = \lim_{b \to \infty} (\arctan b - \arctan 1) = \frac{\pi}{2} - \frac{\pi}{4} = \frac{\pi}{4};$$

therefore $I_1$ is convergent and

$$\int_1^\infty \frac{dx}{1 + x^2} = \frac{\pi}{4}.$$

b) We shall have

$$\lim_{b \to \infty} \int_0^\infty \frac{dx}{\sqrt{1 + x}} = \lim_{b \to \infty} \frac{(1 + x)^{-\frac{1}{2}+1}}{-\frac{1}{2} + 1} \bigg|_0^b = \infty,$$

namely

$$\int_0^\infty \frac{dx}{\sqrt{1 + x}}$$

is divergent.

c) One obtains

$$\lim_{b\to\infty}\int_0^b xe^{-x^2}\,dx = \lim_{b\to\infty} -\frac{1}{2}e^{-x^2}\Big|_0^b = -\frac{1}{2}\lim_{b\to\infty}\left(e^{-b}-1\right) = \frac{1}{2};$$

it results that $I_3$ is convergent and

$$I_3 = \int_0^\infty xe^{-x^2}\,dx = \frac{1}{2}.$$

d) One can notice that

$$\int_{-\infty}^\infty \frac{dx}{1+x^2} = \underbrace{\int_{-\infty}^1 \frac{dx}{1+x^2}}_{I} + \underbrace{\int_1^\infty \frac{dx}{1+x^2}}_{J}.$$

Calculating

$$\lim_{a\to-\infty}\int_a^1 \frac{dx}{1+x^2} = \lim_{a\to-\infty} \arctan x\big|_a^1 = \lim_{a\to-\infty}\left(-\arctan a + \frac{\pi}{4}\right) = \frac{\pi}{2} + \frac{\pi}{4} = \frac{3\pi}{4}$$

one deduces that $I$ is convergent and

$$I = \int_{-\infty}^1 \frac{dx}{1+x^2} = \frac{3\pi}{4}.$$

Similarly, one obtains that $J = I_1$ is convergent and

$$J = \int_1^\infty \frac{dx}{1+x^2} = \frac{\pi}{4}.$$

Therefore $I_4$ is convergent and

$$I_4 = \int_{-\infty}^\infty \frac{dx}{1+x^2} = I + J = \frac{3\pi}{4} + \frac{\pi}{4} = \pi.$$

In Matlab 7.9 we shall have:
```
>>syms x
>> int(1/(1+x^2),-inf,inf)
ans =
pi
```
while in Mathcad 14:

$$\int_{-\infty}^\infty \frac{1}{1+x^2}\,dx = 3.142$$

and in Mathematica 8:

$\text{In[16]:=}$ **Integrate[1 / (1 + x^2), {x, -∞, ∞}]**

$\text{Out[16]=}$ $\pi$

and with Maple 15:

$$\int_{-\infty}^{\infty} \frac{1}{1+x^2}\, dx$$

$$\pi$$

e) In the case of this integral

$$\lim_{b \to \infty} \int_3^b \frac{dx}{\sqrt{x}\,(x+1)} = 2 \lim_{b \to \infty} \arctan \sqrt{x}\Big|_3^b$$

$$= 2 \lim_{b \to \infty} \left( \arctan \sqrt{b} - \arctan \sqrt{3} \right) = 2 \left( \frac{\pi}{2} - \arctan \sqrt{3} \right),$$

namely $I_5$ is convergent and

$$I_5 = \int_3^{\infty} \frac{dx}{\sqrt{x}\,(x+1)} = \pi - 2 \arctan \sqrt{3}.$$

**Example 6.16.** Use the criterion for absolute convergence to establish if the following integrals are convergent or not:

a) $\int_0^{\infty} \frac{x^2}{x^4 - x^2 + 1}\, dx$

b) $\int_0^{\infty} \frac{x\sqrt{x}}{1 + x^2}\, dx$

c) $\int_1^{\infty} \frac{1}{x\sqrt[3]{x^2 + 1}}\, dx.$

**Solutions.**
a) We can notice that

$$L = \lim_{x \to \infty} x^p \cdot \frac{x^2}{x^4 - x^2 + 1} \overset{p=2}{=} 1,$$

therefore

$$\int_0^\infty \frac{x^2}{x^4 - x^2 + 1}\,dx$$

is convergent.

We shall verify that in Matlab 7.9:

```
>> syms x p
>> limit(subs(x^p*x^2/(x^4-x^2+1),p,2),x,inf)
ans =
1
```

and in Mathcad 14:

$$\lim_{x \to \infty}\left(x^p \cdot \frac{x^2}{x^4 - x^2 + 1} \quad \text{substitute}, p = 2 \to \frac{x^4}{x^4 - x^2 + 1}\right) \to 1$$

and using Mathematica 8:

In[17]:= **Limit[x^p * x^2 / (x^4 - x^2 + 1) /. p → 2, x → ∞]**

Out[17]= **1**

and with Maple 15:

$$\lim_{x \to \infty} x^p \cdot \left.\frac{x^2}{x^4 - x^2 + 1}\right|_{p=2}$$

$$1$$

b) One obtains

$$L = \lim_{x \to \infty} x^p \cdot \frac{x\sqrt{x}}{1 + x^2} \overset{p=1/2}{=} 1,$$

namely

$$\int_0^\infty \frac{x\sqrt{x}}{1 + x^2}\,dx$$

is divergent.

We can deduce this conclusion using Matlab 7.9:

```
>> clear all
>> syms x p
```

```
>> limit(subs(x^p*x*sqrt(x)/(1+x^2),p,1/2),x,inf)
ans =
1
```
and Mathcad 14:

$$\lim_{x \to \infty} \left( x^p \cdot \frac{x \cdot \sqrt{x}}{1 + x^2} \quad \text{substitute}, p = \frac{1}{2} \to \frac{x^2}{x^2 + 1} \right) \to 1$$

and with Mathematica 8:

In[18]:= **Limit[x^p * x * Sqrt[x] / (1 + x^2) /. p → 1/2, x → ∞]**

Out[18]= **1**

and using Maple 15:

$$\lim_{x \to \infty} x^p \cdot \left. \frac{x \cdot \sqrt{x}}{1 + x^2} \right|_{p = \frac{1}{2}}$$

1

c) We shall have

$$L = \lim_{x \to \infty} x^p \cdot \frac{1}{x \sqrt[3]{x^2 + 1}} \overset{p = 4/3}{=} \lim_{x \to \infty} \sqrt[3]{\frac{x}{x^2 + 1}} = 0,$$

therefore

$$\int_1^\infty \frac{1}{x \sqrt[3]{x^2 + 1}} dx$$

is convergent.

## 6.3.3  The Comparison Criterion for the Integrals

**Proposition 6.17 (The comparison criterion for the integrals,** see [42], p. 69). One compares

$$\int_a^\infty f(x) \, dx \le \int_a^\infty \varphi(x) \, dx.$$

If :

- $\int_a^\infty \varphi(x)dx$ is convergent then $\int_a^\infty f(x)dx$ is convergent;
- $\int_a^\infty f(x)dx$ is divergent then $\int_a^\infty \varphi(x)dx$ is divergent.

**Example 6.18.** Use the comparison criterion to establish if the following integrals are convergent or not and in the affirmative case find a majorant of them:

$$a) \int_1^\infty \frac{e^{-x^2}}{x^2}dx$$

$$b) \int_2^\infty \frac{1}{\ln x}dx.$$

**Solutions.**
a) We can see that

$$\int_1^\infty \frac{e^{-x^2}}{x^2}dx \le \int_1^\infty xe^{-x^2}dx.$$

Since

$$\int_1^\infty xe^{-x^2}dx$$

is convergent (see $I_3$ from the Example 6.15), using the comparison criterion for the integrals one deduces that

$$\int_1^\infty \frac{e^{-x^2}}{x^2}dx$$

is convergent.

We can see that, using Matlab 7.9:
>>**syms x**
>> **eval(int(exp(-x^2)/(x^2),1,inf))**
ans =
        **0.0891**
and with Mathematica 8:

In[3]:= **SetPrecision[Integrate[Exp[-x^2] / (x^2), {x, 1, Infinity}], 6]**

Out[3]= 0.0891

and in Maple 15:

*Digits* := 5 :

$$\text{evalf}\left(\int_{1}^{\infty}\frac{e^{-x^2}}{x^2}\,dx\right)$$

0.0891

We can't do that in Mathcad 14.

b) One can notice that

$$\int_{2}^{\infty}\frac{2}{x\ln x}\,dx \le \int_{2}^{\infty}\frac{1}{\ln x}\,dx.$$

However,

$$\lim_{b\to\infty}\int_{2}^{b}\frac{2}{x\ln x}\,dx = 2\lim_{b\to\infty}\int_{2}^{\infty}\frac{(\ln x)'}{\ln x}\,dx = 2\lim_{b\to\infty}\ln(\ln x)|_{2}^{b}$$
$$= 2\lim_{b\to\infty}(\ln(\ln b) - \ln(\ln 2)) = \infty;$$

therefore

$$\int_{2}^{\infty}\frac{2}{x\ln x}\,dx$$

is divergent. Using the comparison criterion for the integrals it results that

$$\int_{2}^{\infty}\frac{1}{\ln x}\,dx.$$

is divergent.

The same conclusion can be deduced in Matlab 7.9:

>>**syms x**
>> **int(1/log(x),2,inf)**
ans =
Inf

and in Mathcad 14:

$$\int_{2}^{\infty}\frac{1}{\ln(x)}\,dx \to \infty$$

and with Mathematica 8:

In[23]:= **Integrate[1 / Log[x] , {x, 2, ∞}];**

Integrate::idiv : Integral of $\dfrac{1}{\text{Log}[x]}$ does not converge on {2, ∞}. »

and using Maple 15:

$$\int_2^\infty \frac{1}{\ln(x)}\,dx$$

∞

## 6.4  Parameter Integrals

**Proposition 6.19** (see [42], p. 76). If $\alpha(t)$ and $\beta(t)$ are two derivable functions and there is $\frac{\partial f(x,t)}{\partial t}$ and it is continuous, then

$$F(t) = \int_{\alpha(t)}^{\beta(t)} f(x,t)\,dx$$

is derivable and

$$F'(t) = \int_{\alpha(t)}^{\beta(t)} \frac{\partial f(x,t)}{\partial t}\,dx + f(\beta(t),t)\cdot\beta'(t) + f(\alpha(t),t)\cdot\alpha'(t). \quad (6.68)$$

**Proposition 6.20** (see [42], p. 80). Let $f(x,t)$ and $\frac{\partial f(x,t)}{\partial t}$ be continuous functions. If

$$F(t) = \int_a^\infty f(x,t)\,dx = \lim_{b\to\infty} \int_a^b f(x,t)\,dx$$

converges and $\int_a^\infty \frac{\partial f(x,t)}{\partial t}\,dx$ converges uniformly then $F(t)$ is derivable and

$$F'(t) = \int_a^\infty \frac{\partial f(x,t)}{\partial t}\,dx. \quad (6.69)$$

**Example 6.21.** Use the derivation formula relative to the respective parameter to compute the following parameter integral:

$$\text{a) } F(t) = \int_0^1 \frac{x^t - 1}{\ln x}\,dx \quad (6.70)$$

b) $\int_0^1 \dfrac{x^\alpha - x^\beta}{\ln x} dx$ \hfill (6.71)

c) $F(t) = \int_0^\infty \dfrac{e^{-tx^2} - e^{-3x^2}}{x} dx$ \hfill (6.72)

d) $F(\alpha) = \int_0^\infty \dfrac{\arctan \alpha x}{x(1+x^2)} dx, \ \alpha > 0.$ \hfill (6.73)

**Solutions.**

a) Using the formula (6.68) one obtains

$$F'(t) = \int_0^1 \frac{x^t \cdot \ln x}{\ln x} dx = \int_0^1 x^t dx = \frac{1}{t+1}.$$

Therefore,

$$F(t) = \int \frac{1}{t+1} dt = \ln|t+1| + C. \tag{6.74}$$

From (6.74) it results $F(0) = 0$ but from (6.70) one knows $F(0) = C$. One deduces $C = 0$ and

$$F(t) = \int \frac{1}{t+1} dt = \ln|t+1|.$$

So that

$$F(t) = \int_0^1 \frac{x^t - 1}{\ln x} dx = \ln|t+1|. \tag{6.75}$$

A computer solution can be given only using Mathematica 8:

In[9]:= **Integrate[(x^t - 1) / Log[x], {x, 0, 1}]**

Out[9]= **ConditionalExpression[Log[1 + t], Re[t] > -1]**

b) It will result

$$F(\alpha, \beta) = \int_0^1 \frac{x^\alpha - x^\beta}{\ln x} dx = \int_0^1 \frac{x^\alpha - 1}{\ln x} dx - \int_0^1 \frac{x^\beta - 1}{\ln x} dx$$

$$\overset{(6.70)}{=} F(\alpha) - F(\beta) = \ln|\alpha+1| - \ln|\beta+1| = \ln \frac{|\alpha+1|}{|\beta+1|}.$$

We can not compute this parameter integral using Matlab 7.9, Mathcad 14 or Maple 15; however in Mathematica 8 we shall have:

In[1]:= **Integrate[ (x^α - x^β) / Log[x], {x, 0, 1}]**

Out[1]= ConditionalExpression[Log[1 + α] - Log[1 + β], Re[β] > -1 && Re[α] > -1]

c) Using the formula (6.69) it results

$$F'(t) = \int_0^\infty \frac{-x^2 e^{-tx^2}}{x} dx = -\int_0^\infty x e^{-tx^2} dx = \frac{1}{2t} \int_0^\infty \left( e^{-tx^2} \right)' dx$$

$$= \frac{1}{2t}(0 - 1) = -\frac{1}{2t}$$

and

$$F(t) = -\int \frac{1}{2t} dt = -\frac{1}{2} \ln |t| + C. \tag{6.76}$$

By calculating $F(3)$ in two ways, namely using (6.72) and respectively (6.76) one obtains

$$\left.\begin{array}{l} F(3) \stackrel{(6.72)}{=} 0 \\ F(3) \stackrel{(6.76)}{=} -\frac{1}{2} \ln 3 + C \end{array}\right\} \Longrightarrow C = \frac{1}{2} \ln 3.$$

Introducing the value of $C$ in the formula (6.76) one obtains

$$F(t) = -\int \frac{1}{2t} dt = -\frac{1}{2} \ln |t| + \frac{1}{2} \ln 3 = \frac{1}{2} \ln \frac{3}{|t|}.$$

We can also achieve this result only using Mathematica 8:

In[2]:= **Integrate[ (Exp[-t * x^2] - Exp[-3 * x^2]) / x, {x, 0, ∞}]**

Out[2]= ConditionalExpression$\left[-\frac{1}{2} \text{Log}\left[\frac{t}{3}\right], \text{Re}[t] > 0\right]$

d) We shall have

$$F'(\alpha) = \int_0^\infty \frac{x}{1 + \alpha^2 x^2} \cdot \frac{1}{x(1 + x^2)} dx = \int_0^\infty \frac{1}{(1 + x^2)(1 + \alpha^2 x^2)} dx.$$

Using the method of partial fraction expansions, we can write

$$\frac{1}{(1 + x^2)(1 + \alpha^2 x^2)} = \frac{Ax + B}{1 + x^2} + \frac{Cx + D}{1 + \alpha^2 x^2},$$

namely

$$(Ax + B)\left(1 + \alpha^2 x^2\right) + (Cx + D)\left(1 + x^2\right) \;=\; 1 \Longleftrightarrow$$

$$\begin{cases} \alpha^2 A + C = 0 \\ \alpha^2 B + D = 0 \\ A + C = 0 \Longrightarrow C = -A \\ B + D = 1 \Longrightarrow D = 1 - B \end{cases} \Longrightarrow \begin{cases} A = 0 \\ B = -\frac{1}{\alpha^2 - 1} \\ C = 0 \\ D = \frac{\alpha^2}{\alpha^2 - 1}. \end{cases}$$

Finally, one deduces

$$\frac{1}{\left(1 + x^2\right)\left(1 + \alpha^2 x^2\right)} = \left(-\frac{1}{\alpha^2 - 1}\right) \cdot \frac{1}{1 + x^2} + \frac{\alpha^2}{\alpha^2 - 1} \cdot \frac{1}{1 + \alpha^2 x^2}$$

and

$$F'(\alpha) = -\frac{1}{\alpha^2 - 1} \int_0^\infty \frac{1}{1 + x^2}\, dx + \frac{\alpha^2}{\alpha^2 - 1} \int_0^\infty \frac{1}{1 + \alpha^2 x^2}\, dx$$

$$= -\frac{1}{\alpha^2 - 1}\, \arctan x \Big|_0^\infty + \frac{\alpha^2}{\alpha^2 - 1} \cdot \frac{1}{\alpha} \cdot \arctan \alpha x \Big|_0^\infty.$$

Therefore,

$$F'(\alpha) = -\frac{1}{\alpha^2 - 1} \cdot \frac{\pi}{2} + \frac{\alpha}{\alpha^2 - 1} \cdot \frac{\pi}{2},$$

namely

$$F(\alpha) = -\frac{\pi}{2} \cdot \frac{1}{2} \ln \left|\frac{\alpha - 1}{\alpha + 1}\right| + \frac{\pi}{4} \ln \left|\alpha^2 - 1\right|$$

$$\overset{\alpha \geq 0}{=} \frac{\pi}{4}\left[ -\ln|\alpha - 1| + \ln(\alpha + 1) + \ln(\alpha + 1) + \ln|\alpha - 1| \right]$$

$$= \frac{\pi}{2} \ln(\alpha + 1).$$

**Example 6.22.** Use the parameter integral

$$F(t) = \int_0^1 \frac{\ln(1 + tx)}{1 + x^2}\, dx \qquad\qquad (6.77)$$

to calculate

$$\int_0^1 \frac{\ln(1 + x)}{1 + x^2}\, dx.$$

**Solution.**
Using the formula (6.68) it results

$$F'(t) = \int_0^t \frac{x}{1+tx} \cdot \frac{1}{1+x^2} \, dx + \frac{\ln\left(1+t^2\right)}{1+t^2}.$$

We can write

$$\frac{x}{(1+tx)(1+x^2)} = \frac{A}{1+tx} + \frac{Bx+C}{1+x^2},$$

namely

$$A\left(1+x^2\right) + (Bx+C)(1+tx) = x \Longleftrightarrow$$

$$\begin{cases} Bt + A = 0 \\ B + Ct = 1 \\ A + C = 0 \end{cases} \Longrightarrow \begin{cases} A = -\frac{t}{1+t^2} \\ B = \frac{1}{1+t^2} \\ C = \frac{t}{1+t^2}. \end{cases}$$

One obtains

$$F'(t) = -\frac{t}{1+t^2} \int_0^t \frac{dx}{1+tx} + \int_0^t \frac{\frac{1}{1+t^2}x + \frac{t}{1+t^2}}{1+x^2} \, dx + \frac{\ln\left(1+t^2\right)}{1+t^2},$$

i.e.

$$F'(t) = -\frac{t}{1+t^2} \cdot \frac{1}{t} \ln\left(1+tx\right) \Big|_0^t + \frac{1}{1+t^2} \int_0^t \frac{x}{1+x^2} \, dx$$

$$+ \frac{t}{1+t^2} \int_0^t \frac{dx}{1+x^2} + \frac{\ln\left(1+t^2\right)}{1+t^2}$$

and further,

$$F'(t) = -\frac{\ln\left(1+t^2\right)}{1+t^2} + \frac{1}{1+t^2} \cdot \frac{1}{2} \ln\left(1+x^2\right)\Big|_0^t + \frac{t}{1+t^2} \arctan x \Big|_0^t + \frac{\ln\left(1+t^2\right)}{1+t^2}.$$

Finally,

$$F'(t) = \frac{1}{2} \cdot \frac{1}{1+t^2} \ln\left(1+t^2\right) + \frac{t}{1+t^2} \arctan t.$$

Therefore

$$F(t) = \frac{1}{2} \int \frac{\ln\left(1+t^2\right)}{1+t^2} \, dt + \int \frac{t}{1+t^2} \arctan t \, dt$$

$$= \frac{1}{2} \int \frac{\ln\left(1+t^2\right)}{1+t^2} \, dt + \frac{1}{2} \int \left(\ln\left(1+t^2\right)\right)' \arctan t \, dt$$

$$= \frac{1}{2} \int \frac{\ln\left(1+t^2\right)}{1+t^2} \, dt + \frac{1}{2} \arctan t \cdot \ln\left(1+t^2\right) - \frac{1}{2} \int \frac{\ln\left(1+t^2\right)}{1+t^2} + C$$

namely

$$F(t) = \frac{1}{2} \arctan t \cdot \ln \left(1 + t^2\right) + C.$$

Since $F(0) = 0$ one deduces $C = 0$ and

$$F(t) = \frac{1}{2} \arctan t \cdot \ln \left(1 + t^2\right).$$

From (6.77) it results

$$F(1) = \int_0^1 \frac{\ln(1+x)}{1+x^2} dx = \frac{1}{2} \cdot \frac{\pi}{4} \ln 2 = \frac{\pi}{8} \ln 2.$$

We can give a computer solution of the problem using Mathematica 8:

In[43]:= **v := Integrate[Log[1 + t * x] / (1 + x^2), {x, 0, 1}];**

In[44]:= **u := v /. t -> 1**

In[46]:= **SetPrecision[u, 6]**

Out[46]= **0.2722 + 0. × 10$^{-5}$ i**

We shall check this result with Matlab 7.9:
```
>> syms x
>> u=int(log(1+x)/(1+x^2),x,0,1)
u =
(pi*log(2))/8
>> vpa(u,6)
ans =
0.272198
```

and in Mathcad 14:

$$\int_0^1 \frac{\ln(1+x)}{1+x^2} \, dx \to \frac{\pi \cdot \ln(2)}{8} = 0.272198$$

and using Mathematica 8:

In[49]:= **Integrate[Log[1 + x] / (1 + x^2), {x, 0, 1}]**

Out[49]= $\dfrac{1}{8} \pi \, \text{Log}[2]$

In[50]:= **SetPrecision[%, 6]**

Out[50]= 0.272198

We can not do the same in Maple 15.

## 6.5  Problems

1. Find the integral:

$$\int \frac{\ln x}{x\sqrt{1 - 4\ln x - \ln^2 x}}dx.$$

**Computer solution.**
We shall compute this indefinite integral in Mathematica 8:

In[4]:= **Simplify[Integrate[Log[x] / (x * Sqrt[1 - 4 * Log[x] - Log[x]^2]), x]]**

Out[4]= $-2\,\text{ArcSin}\left[\dfrac{2 + \text{Log}[x]}{\sqrt{5}}\right] - \sqrt{1 - 4\,\text{Log}[x] - \text{Log}[x]^2}$

and with Maple 15:

$$\int \frac{\ln(x)}{x\sqrt{1 - 4\cdot\ln(x) - \ln(x)^2}}\,dx$$

$$-\sqrt{1 - 4\ln(x) - \ln(x)^2} - 2\arcsin\left(\frac{1}{5}\sqrt{5}\,(2 + \ln(x))\right)$$

We shall not achieve this result using Maple 7.9 or Mathcad 14.

2. Evaluate the integral:

$$\int_{\frac{\pi}{6}}^{\frac{\pi}{3}} \cot^4 \varphi\, d\varphi.$$

3. Find the area enclosed by the straight line $x+y+2 = 0$ and the parabola $y = -x^2$.

**Computer solution.**
We shall compute this area in Matlab 7.9:
```
>> x=-2:0.1:2;
>> f=@(x)-x-2;
>> g=@(x)-x.^2;
>> plot(x,f(x),'m',x,g(x),'b','LineWidth',3)
```

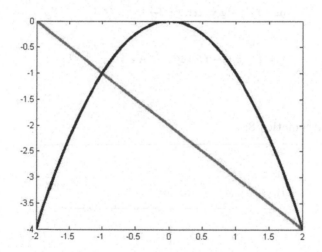

```
>> syms y
>> u=solve(f(y)-g(y),y)
u =
 -1
  2
>> Area=int(g(y)-f(y),y,u(1),u(2))
Area =
9/2
```
and with Mathcad 14:

$$x := -2, -1.99 .. 2$$

$$f(x) := -x - 2 \qquad g(x) := -x^2$$

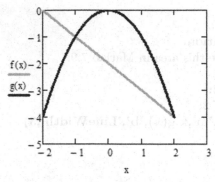

$$u := f(y) - g(y) \text{ solve}, y \rightarrow \begin{pmatrix} -1 \\ 2 \end{pmatrix}$$

$$A := \int_{u_0}^{u_1} g(y) - f(y) \, dy \qquad A = 4.5$$

and using Mathematica 8:

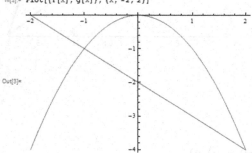

In[1]:= f[x_] := -x - 2

In[2]:= g[x_] := -x^2

In[3]:= Plot[{f[x], g[x]}, {x, -2, 2}]

Out[3]=

In[4]:= u := y /. Solve[f[y] - g[y] == 0, y]

In[5]:= Integrate[g[y] - f[y], {y, u[[1]], u[[2]]}]

Out[5]= $\dfrac{9}{2}$

and  with Maple 15:

$f := x \to -x - 2:$
$g := x \to -x^2:$
plot( {f(x), g(x)}, x = -2..2)

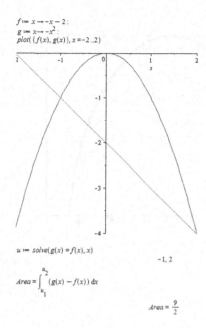

$u := solve(g(x) = f(x), x)$

$-1, 2$

$Area = \int_{u_1}^{u_2} (g(x) - f(x)) \, dx$

$Area = \dfrac{9}{2}$

4. Compute thea area contained between the parabolas

$$\{ y = x^2,$$

and the straight line

$$y = 2x.$$

5. Find the entire area bounded by the astroid:

$$x^{\frac{2}{3}} + y^{\frac{2}{3}} = a^{\frac{2}{3}}.$$

6. Find the area bounded by the $Ox$- axis and one arc of the cycloid:

$$\begin{cases} x(t) = a(t - \sin t) \\ y(t) = a(1 - \cos t). \end{cases}, \quad t \in [0, 2\pi].$$

**Computer solution.**
We shall compute this area using Matlab 7.9:
```
>> x=@(t,a)a*(t-sin(t));
>> y=@(t,a)a*(1-cos(t));
>> syms t a
>> int(y*diff(x,t),t,0,2*pi)
ans =
3*pi*a^2
>> t=0:0.01*pi:2*pi;
>> a=1;
>> plot(x(t,a),y(t,a),'m','LineWidth',4)
```

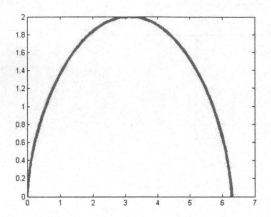

and with Mathcad 14:

$$x(t,a) := a \cdot (t - \sin(t)) \qquad y(t,a) := a \cdot (1 - \cos(t))$$

$$\int_0^{2 \cdot \pi} y(t,a) \cdot \frac{d}{dt} x(t,a) \ dt \rightarrow 3 \cdot \pi \cdot a^2$$

$$t := 0, 0.1 \cdot \pi .. 2 \cdot \pi$$

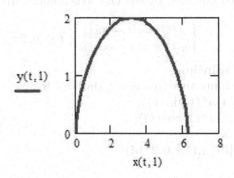

and in Mathematica 8:

In[9]:= **x[t_, a_] := a \* (t - Sin[t])**

In[10]:= **y[t_, a_] := a \* (1 - Cos[t])**

In[11]:= **Integrate[y[t, a] \*D[x[t, a], t], {t, 0, 2\*Pi}]**

Out[11]= $3 a^2 \pi$

In[13]:= **ParametricPlot[{x[t, 1], y[t, 1]}, {t, 0, 2 Pi}]**

Out[13]=

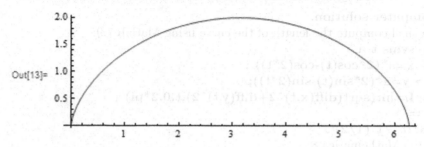

and using Maple 15:

$$x := (t, a) \rightarrow a \cdot (t - \sin(t))$$

$$(t, a) \rightarrow a (t - \sin(t))$$

$$y := (t, a) \rightarrow a \cdot (1 - \cos(t))$$

$$(t, a) \rightarrow a (1 - \cos(t))$$

$$\int_0^{2 \cdot \pi} y(t, a) \cdot \frac{d}{dt} x(t, a) dt$$

$$3 \pi a^2$$

$$plot \left( \left[ a \cdot (t - \sin(t)) \Big|_{a=1}, a \cdot (1 - \cos(t)) \Big|_{a=1}, t = 0 .. 2 \pi \right] \right)$$

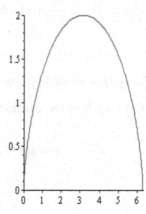

7. Find the entire area of the cardiod

$$\rho = a\left(1 + \cos\theta\right).$$

8. Find the length of the curve:

$$\begin{cases} x(t) = a(2\cos t - \cos 2t) \\ y(t) = a(2\sin t - \sin 2t). \end{cases}, \quad t \in [0, 2\pi].$$

**Computer solution.**
We shall compute the length of the curve using Matlab 7.9:
>> **syms t a**
>> **x=a\*(2\*cos(t)-cos(2\*t)) ;**
>> **y=a\*(2\*sin(t)-sin(2\*t));**
>> **L=int(sqrt(diff(x,t)^2+diff(y,t)^2),t,0,2\*pi)**
**L =**
**16\*(a^2)^(1/2)**
and in Mathematica 8:

In[11]:= **x[t_] := a \* (2 \* Cos[t] - Cos[2 \* t])**

In[12]:= **y[t_] := a \* (2 \* Sin[t] - Sin[2 \* t])**

In[13]:= **L = Integrate[Sqrt[D[x[t], t]^2 + D[y[t], t]^2], {t, 0, 2 \* Pi}]**

Out[13]= $16\sqrt{a^2}$

and with Maple 15:

$x := t \rightarrow a \cdot (2 \cdot \cos(t) - \cos(2 \cdot t)) :$
$y := t \rightarrow a \cdot (2 \cdot \sin(t) - \sin(2 \cdot t)) :$

$$L := \int_0^{2\cdot\pi} \sqrt{\left(\frac{\partial}{\partial t}x(t)\right)^2 + \left(\frac{\partial}{\partial t}y(t)\right)^2}\, dt$$

$16\, a\,\mathrm{csgn}(a)$

We can't achieve this result in Mathcad 14.

9. Find the entire length of the cardiod

$$r = a(1 + \cos\varphi).$$

10. Find the area of the surface formed by the rotation of a part of the curve

$$y = \tan x$$

from $x = 0$ to $x = \frac{\pi}{4}$, about the $x$- axis.

11. Find the volume of a torus formed by rotation of the circle

$$x^2 + (y - y_0)^2 = r^2, \quad y_0 \geq r.$$

about the $Ox$- axis.

12. Find the volume of an ellipsoid formed by the rotation of the ellipse

$$\frac{x^2}{a^2} + \frac{y^2}{b^2} = 1$$

about the $Ox$- axis.

**Computer solution.**
We shall compute this volume using Matlab 7.9:
>> syms x a b
>> V=pi*b^2/a^2*int(a^2-x^2,x,-a,a)
V =
(4*pi*a*b^2)/3
and in Mathcad 14:

$$\pi \cdot \frac{b^2}{a^2} \cdot \int_{-a}^{a} \left(a^2 - x^2\right) dx \rightarrow \frac{4 \cdot \pi \cdot a \cdot b^2}{3}$$

and with Mathematica 8:

In[2]:= V = Pi * b^2 / a^2 * Integrate[ (a^2 - x^2), {x, -a, a}]

Out[2]= $\frac{4}{3}$ a b$^2$ $\pi$

and in Maple 15:

$$V = \pi \cdot \frac{b^2}{a^2} \cdot \int_{-a}^{a} \left(a^2 - x^2\right) dx$$

$$V = \frac{4}{3} \pi b^2 a$$

13.Find the centre of gravity of an arc of the semi-circle

$$x^2 + y^2 = r^2, \ r \geq 0.$$

14.Find the coordinates of the centre of gravity of an area bounded by the curves

$$\begin{cases} y = x^2, \\ y = \sqrt{x}. \end{cases}$$

**Computer solution.**
We shall solve this problem using Matlab 7.9:
```
>> x=0:0.01:2;
>> y1=@(x)x.^2;
>> y2=@(x)sqrt(x);
>> plot(x,y1(x),'m',x,y2(x),'b','LineWidth',3)
```

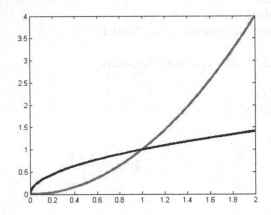

```
>> syms v
>> u=solve(y1(v)-y2(v),v);
>> xg=int(v*(y2(v)-y1(v)),v,u(1),u(2))/int(y2(v)-y1(v),
v,u(1),u(2))
   xg =
   9/20
>> yg=int((y2(v)^2-y1(v)^2),v,u(1),u(2))/int(y2(v)-y1(v),
v,u(1),u(2))
   yg =
   9/10
```
and in Mathcad 14:

$$y1(x) := x^2 \qquad y2(x) := \sqrt{x}$$

$$u := y2(x) - y1(x) \ \text{solve}, x \ \rightarrow \begin{pmatrix} 0 \\ 1 \end{pmatrix}$$

$$t := 0, 0.1 .. 2$$

$$xg := \dfrac{\displaystyle\int_{u_0}^{u_1} x \cdot (y2(x) - y1(x)) \, dx}{\displaystyle\int_{u_0}^{u_1} y2(x) - y1(x) \, dx} \qquad xg \rightarrow \dfrac{9}{20} = 0.45$$

$$yg := \dfrac{\dfrac{1}{2} \cdot \displaystyle\int_{u_0}^{u_1} \left( y2(x)^2 - y1(x)^2 \right) dx}{\displaystyle\int_{u_0}^{u_1} y2(x) - y1(x) \, dx} \qquad yg \rightarrow \dfrac{9}{20} = 0.45$$

and in Mathematica 8:

```
In[1]:= y1[x_] := x^2

In[2]:= y2[x_] := Sqrt[x]

In[3]:= Plot[{y1[x], y2[x]}, {x, 0, 2}]
```

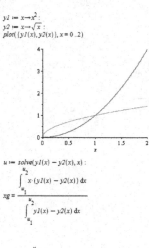

Out[3]=

```
In[4]:= u := x /. Solve[y1[x] - y2[x] == 0, x]

In[5]:= xg = Integrate[x * (y2[x] - y1[x]), {x, u[[1]], u[[2]]}] /
           Integrate[y2[x] - y1[x], {x, u[[1]], u[[2]]}]
```

Out[5]= $\dfrac{9}{20}$

```
In[6]:= yg = 1/2 * Integrate[y2[x]^2 - y1[x]^2, {x, u[[1]], u[[2]]}] /
           Integrate[y2[x] - y1[x], {x, u[[1]], u[[2]]}]
```

Out[6]= $\dfrac{9}{20}$

and with Maple 15:

```
y1 := x → x^2 :
y2 := x → √x :
plot({y1(x), y2(x)}, x = 0..2)
```

$$xg = \dfrac{\displaystyle\int_{u_1}^{u_2} x \cdot (y1(x) - y2(x))\, dx}{\displaystyle\int_{u_1}^{u_2} y1(x) - y2(x)\, dx}$$

```
u := solve(y1(x) - y2(x), x) :
```

$$xg = \dfrac{9}{20}$$

$$yg = \dfrac{\dfrac{1}{2}\displaystyle\int_{u_1}^{u_2} y1(x)^2 - y2(x)^2\, dx}{\displaystyle\int_{u_1}^{u_2} y1(x) - y2(x)\, dx}$$

$$yg = \dfrac{9}{20}$$

15.Evaluate the improper integral or establish their divergence:

$$\int_{-\infty}^{\infty} \frac{dx}{x^2 + 4x + 9}.$$

16.Find the convergence of the improper integral:

$$\int_{0}^{100} \frac{dx}{\sqrt[3]{x} + 2\sqrt[4]{x} + \sqrt[5]{x}}.$$

**Computer solution.**

We shall give a computer solution in Matlab 7.9:

>> **syms x p**

>> **limit(subs(x^p/(x^(1/3)+2*x^(1/4)+x^(1/5)),p,1/3),x,0,**
'right')

ans =

0

and with Mathcad 14:

$$\lim_{x \to 0^+} \left( x^p \cdot \frac{1}{\sqrt[3]{x} + 2 \cdot \sqrt[4]{x} + \sqrt[5]{x}} \quad \text{substitute}, p = \frac{1}{3} \to \frac{x^{\frac{1}{3}}}{\sqrt[3]{x} + 2 \cdot \sqrt[4]{x} + \sqrt[5]{x}} \right) \to 0$$

and using Mathematica 8:

In[1]:= **Limit[x^p/ (x^(1/3) + 2*x^(1/4) + x^(1/5)) /. p → 1/3, x → 0, Direction → -1]**

Out[1]= 0

and in Maple 15:

$$\lim_{x \to 0^+} x^p \cdot \frac{1}{\sqrt[3]{x} + 2 \cdot \sqrt[4]{x} + \sqrt[5]{x}} \bigg|_{p = \frac{1}{3}}$$

0

Using the Proposition 6.11, we shall deduce that

$$\int_{0}^{100} \frac{dx}{\sqrt[3]{x} + 2\sqrt[4]{x} + \sqrt[5]{x}}$$

is absolute convergent.

17.Test the convergence of the integral

$$\int_0^\infty \frac{\arctan x}{x^2+1}\,dx$$

and evaluate it in the case of convergence.
   **Computer solution.**
   We shall solve this problem in Matlab 7.9:
   >> **syms x p**
   >> **limit(subs(x^p*atan(x)/(1+x^2),p,2),x,0)**
   ans =
   **0**
   >> **int(atan(x)/(1+x^2),x,0,inf)**
   ans =
   **pi^2/8**
   and with Mathcad 14:

$$\lim_{x\to 0}\left(x^p\cdot\frac{\operatorname{atan}(x)}{x^2+1}\quad\text{substitute},p=2\ \to\ \frac{x^2\cdot\operatorname{atan}(x)}{x^2+1}\right)\to 0$$

$$\int_0^\infty \frac{\operatorname{atan}(x)}{x^2+1}\,dx\to\frac{\pi^2}{8}$$

and in Mathematica 8:

In[2]:= **Limit[x^p \* ArcTan[x] / (1 + x^2) /. p → 2, x → 0]**

Out[2]= 0

In[3]:= **Integrate[ArcTan[x] / (1 + x^2), {x, 0, ∞}]**

Out[3]= $\dfrac{\pi^2}{8}$

and using Maple 15:

$$\lim_{x \to 0} x^p \cdot \left. \frac{\arctan(x)}{x^2 + 1} \right|_{p=2}$$

0

$$\int_0^\infty \frac{\arctan(x)}{x^2 + 1} \, dx$$

$$\frac{1}{8} \pi^2$$

18. Use the formula for differentiation with respect to the respective parameter to compute the following parameter integral:

$$\int_0^\infty \frac{x^m}{1 + x^n} dx, \ m, n \in \mathbb{R}, \ n < m + 1 < 0.$$

19. By differentiating with respect to a parameter, evaluate

$$\int_0^1 \frac{\ln\left(1 - \alpha^2 x^2\right)}{\sqrt{1 - x^2}} dx, \ |\alpha| < 1.$$

**Computer solution.**
Using Mathematica 8 we shall have:

In[2]:= **Integrate[Log[1 - α^2 * x^2] / Sqrt[1 - x^2], {x, 0, 1}]**

Out[2]= **ConditionalExpression$\left[ \pi \, \text{Log}\left[\frac{1}{2} \left(1 + \sqrt{1 - \alpha^2}\right)\right], \ \text{Im}[\alpha] \ne 0 \ || \ -1 < \text{Re}[\alpha] < 1\right]$**

We can't compute this integral with Matlab 7.9 or Mathcad 14 or Maple15.

20. Applying differentiation with respect to a parameter, evaluate the following integral:

$$\int_0^\infty e^{-\alpha x} \cdot \frac{\sin \beta x}{x} dx, \ \alpha > 0.$$

**Computer solution.**
We can compute this integral only with Mathematica 8:

In[11]:= **Integrate[Exp[-α\*x] \* Sin[β\*x] / x, {x, 0, ∞}]**

Out[11]= ConditionalExpression$\left[\text{ArcTan}\left[\frac{\beta}{\alpha}\right], \text{Abs[Im[}\beta\text{]]} \le \text{Re[}\alpha\text{]}\ \&\&\ \text{Re[}\alpha\text{]} > 0\right]$

# 7

# Equations and Systems of Linear Ordinary Differential Equations

## 7.1 Successive Approximation Method

**Definition 7.1** (see [15], p. 292). A **Cauchy problem** consists in a differential equation

$$y' = f(x, y), \ f : D \to \mathbb{R}, \ D \subseteq \mathbb{R}^2$$

and an initial condition:

$$y(x_0) = y_0.$$

Let be the Cauchy problem:

$$\begin{cases} y' = f(x, y) \\ y(x_0) = y_0. \end{cases} \tag{7.1}$$

The solution of the problem (7.1) can be found as a limit of the sequence of functions defined by the successive approximation method, (see [45], p. 77):

$$y_n(x_0) = y_0 + \int_{x_0}^{x} f(t, y_{n-1}(t)) \, dt, \ n = 1, 2 \ldots \tag{7.2}$$

**Example 7.2.** Solve the Cauchy problem:

$$\begin{cases} y' = xy \\ y(0) = 1 \end{cases}$$

G.A. Anastassiou and I.F. Iatan: Intelligent Routines, ISRL 39, pp. 317–394.
springerlink.com                    © Springer-Verlag Berlin Heidelberg 2013

using the successive approximation method.

**Solution.**

The sequence of the successive approximation method (7.2) is:

$$y_0(x) = 1$$

$$y_1(x) = 1 + \int_0^x t\,dt = 1 + \frac{t^2}{2}\Big|_0^x = 1 + \frac{x^2}{2} = 1 + \frac{\frac{x^2}{2}}{1!}$$

$$y_2(x) = 1 + \int_0^x t\left(1 + \frac{t^2}{2}\right)dt = 1 + \int_0^x t\,dt + \int_0^x \frac{t^3}{2}\,dt$$

$$= 1 + \frac{t^2}{2}\Big|_0^x + \frac{t^4}{8}\Big|_0^x = 1 + \frac{x^2}{2} + \frac{x^4}{8} = 1 + \frac{\frac{x^2}{2}}{1!} + \frac{\left(\frac{x^2}{2}\right)^2}{2!}$$

$$y_3(x) = 1 + \int_0^x t\left(1 + \frac{t^2}{2} + \frac{t^4}{8}\right)dt = 1 + \int_0^x t\,dt + \int_0^x \frac{t^3}{2}\,dt + \int_0^x \frac{t^5}{8}\,dt$$

$$= 1 + \frac{t^2}{2}\Big|_0^x + \frac{t^4}{8}\Big|_0^x + \frac{t^6}{48}\Big|_0^x = 1 + \frac{x^2}{2} + \frac{x^4}{8} + \frac{x^6}{48}$$

$$= 1 + \frac{\frac{x^2}{2}}{1!} + \frac{\left(\frac{x^2}{2}\right)^2}{2!} + \frac{\left(\frac{x^2}{2}\right)^3}{3!}$$

$$\vdots$$

$$y_n(x) = \sum_{k=0}^n \frac{\left(\frac{x^2}{2}\right)^k}{k!}.$$

The solution of our Cauchy problem will be

$$y(x) = \lim_{n\to\infty} y_n(x) = \sum_{k=0}^\infty \frac{\left(\frac{x^2}{2}\right)^k}{k!} = e^{\frac{x^2}{2}}.$$

We shall also solve this problem in Matlab 7.9:

```
function w=f(u,v)
w= u*v;
end
function r=y(n,t,x,y0)
r=y0;
for k=1:n
r1=int(f(t,r),t,0,x);
r=subs(r1,x,t)+y0;
end
end
```

In the command line we shall write:

```
>>syms x t
>> y1=subs(y(1,t,x,1),t,x)
y1 =
1+1/2*x^2
>> y2=subs(y(2,t,x,1),t,x)
y2 =
1+1/8*x^4+1/2*x^2
>> y3=subs(y(3,t,x,1),t,x)
y3 =
1+1/48*x^6+1/8*x^4+1/2*x^2
```

and using Mathcad 14:

$$f(x,y) := x \cdot y$$

$$y(n,t,x,y0) := \begin{vmatrix} r \leftarrow y0 \\ \text{for } k \in 1..n \\ \quad \begin{vmatrix} x \leftarrow t \\ r1 \leftarrow \int_0^x f(t,r)\, dt \\ r \leftarrow r1 + y0 \end{vmatrix} \\ r \end{vmatrix}$$

$$y(1,t,x,1) \text{ substitute}, t = x \;\rightarrow\; \frac{x^2}{2} + 1$$

$$y(2,t,x,1) \text{ substitute}, t = x \;\rightarrow\; \frac{x^4}{8} + \frac{x^2}{2} + 1$$

$$y(3,t,x,1) \text{ substitute}, t = x \;\rightarrow\; \frac{x^6}{48} + \frac{x^4}{8} + \frac{x^2}{2} + 1$$

and in Mathematica 8:

```
In[66]:= f[u_, v_] := u*v

In[67]:= y[n_, t_, x_, y0_] :=
         Module[{r = y0},
          For[k = 1, k <= n, k++, x = t; r1 = Integrate[f[t, r], {t, 0, x}]; r = y0 + r1];
          r
         ]

In[68]:= y[1, x, x, 1]

Out[68]= 1 + x^2/2

In[69]:= y[2, x, x, 1]

Out[69]= 1 + x^2/2 + x^4/8

In[70]:= y[3, x, x, 1]

Out[70]= 1 + x^2/2 + x^4/8 + x^6/48
```

and with Maple 15:

$$f := (x, y) \rightarrow x \cdot y:$$
$$y := \mathbf{proc}(n, t, x, y0)$$
$$\mathbf{local}\, r, r1, k;$$
$$r := y0;$$
$$\mathbf{for}\, k\, \mathbf{from}\, 1\, \mathbf{to}\, n\, \mathbf{do}$$
$$r1 := \int_0^x f(t, r)\, dt;$$
$$r := y0 + r1 \Big|_{x=t}$$
$$\mathbf{end\, do};$$
$$\mathbf{end\, proc};$$
$$y(1, t, x, 1)\Big|_{t=x}$$

$$1 + \frac{1}{2}x^2$$

$$y(2, t, x, 1)\Big|_{t=x}$$

$$1 + \frac{1}{8}x^4 + \frac{1}{2}x^2$$

$$y(3, t, x, 1)\Big|_{t=x}$$

$$1 + \frac{1}{48}x^6 + \frac{1}{8}x^4 + \frac{1}{2}x^2$$

## 7.2  First Order Differential Equations Solvable by Quadratures

**Definition 7.3** (see [3]). A equality of the form

$$F(x, y, y') = 0, \tag{7.3}$$

where $F : D \rightarrow \mathbb{R}, D \subseteq \mathbb{R}^3$ it is called a **first order differential equation**.
**Definition 7.4** (see [3]). A relation of the form

$$F(x, y, \lambda) = 0 \tag{7.4}$$

in which $\lambda$ is a real parameter, represents a **family of curves** in the plane $xOy$, each curve from the family being determined by the value of the respective parameter $\lambda$.

**Definition 7.5** (see [3]). The **general solution** of a differential equation means a family of functions that depend on a constant $C$ with the following two properties:

a) for each fixed $C$, each function is a solution on an interval, which depends on $C$;

b) in every point of the domain $D$ it crosses a solution of the family and only one.

**Definition 7.6** (see [3]). We call a **particular solution** of a differential equation any solution which is obtained from the general one, through the particularization of the constant $C$.

**Definition 7.7** (see [3]). The **singular solution** is a solution which has the property that through every point of its, at least one solution of the equation crosses. The singular solution can not be obtained from the general solution, for any value of the constant $C$.

**Remark 7.8.** From the point of geometrical view, the singular solution is the envelope of the family of those curves that define the general solution.

**Definition 7.9** (see [23], p. 263). The **envelope of a family of curves** is the curve, which is tangent at every point of its, of a curve from the respective family (see Fig 7.1).

**Fig. 7.1** The envelope of a family of curves

To get the equation of the considered family envelope it must that the parameter to be eliminated between the equations

$$F(x, y, \lambda) = 0$$

and

$$F'(x, y, \lambda) = 0.$$

## 7.2.1  *First Order Differential Equations with Separable Variables*

**Definition 7.10 (see [3]).** A **differential equation with separable variables** is a first-order equation of the type

$$y' = p(x)\, q(y), \tag{7.5}$$

where $p, q : (a, b) \to \mathbb{R}$ are continuous, $q \neq 0$.

Formal, if we write

$$y' = \frac{dy}{dx}$$

then the equation (7.5) becomes

$$\frac{dy}{q(y)} = p(x)\, dx$$

and it admits the unique solution, defined implicitly by the equality:

$$\int \frac{dy}{q(y)} = \int p(x)\, dx + C . \tag{7.6}$$

**Remark 7.11.** If $a$ is a finite number and $q(a) = 0$ then $y = a$ is a solution of the equation (7.5). This solution is often singular.

**Example 7.12.** Find the solution of the equation:

$$\text{a) } y' = 1 + \frac{1}{x} - \frac{1}{y^2 + 2} - \frac{1}{x(y^2 + 2)}$$

$$\text{b) } 2yy' = \frac{e^x}{e^x + 1}.$$

**Solutions.**

a) The equation can be written:

$$y' = 1 + \frac{1}{x} - \frac{1}{y^2 + 2}\left(1 + \frac{1}{x}\right)$$

i.e.

$$y' = \left(1 + \frac{1}{x}\right)\left(1 - \frac{1}{y^2 + 2}\right)$$

or

$$y' = \left(1 + \frac{1}{x}\right) \cdot \frac{y^2 + 1}{y^2 + 2}$$

i.e.

$$\frac{y^2 + 2}{y^2 + 1} dy = \left(1 + \frac{1}{x}\right) dx;$$

therefore

$$\int \frac{y^2 + 2}{y^2 + 1} dy = \int \left(1 + \frac{1}{x}\right) dx;$$

hence

$$y + \arctan y = x + \ln|x| + C.$$

We can solve this equation only using Matlab 7.9:
>>y= dsolve('Dy=1+1/x-(1/(y^2+2))*(1+1/x)','x')
 y =
  i
 -i
   solve(y + atan(y) = C3 + x + log(x), y)
We shall not achieve any solution of this equation if we use Mathcad 14, Mathematica 8 or Maple 15.
   b) We shall have:

$$2y dy = \frac{e^x}{e^x + 1} dx;$$

hence

$$2\int y dy = \int \frac{e^x}{e^x + 1} dx,$$

i.e.

$$y^2 = \ln(e^x + 1) + C.$$

We can also solve this equation in Matlab 7.9:
>> y=dsolve('Dy=exp(x)/(2*y*(exp(x)+1))','x')
 y =
(log(exp(x)+1)+C1)^(1/2)
-(log(exp(x)+1)+C1)^(1/2)
and with Mathematica 8.0:

In[2]:= DSolve[D[y[x], x] == Exp[x] / (2 * y[x] * (Exp[x] + 1)), y[x], x]

Out[2]= $\left\{\left\{y[x] \rightarrow -\sqrt{2 C[1] + Log[1 + e^x]}\right\}, \left\{y[x] \rightarrow \sqrt{2 C[1] + Log[1 + e^x]}\right\}\right\}$

and using Maple 15:

$$ode := \text{diff}(y(x), x) = \frac{e^x}{2 \cdot y(x) \cdot (e^x + 1)}$$

$$\frac{d}{dx} y(x) = \frac{1}{2} \frac{e^x}{y(x) (e^x + 1)}$$

$$dsolve(ode)$$

$$y(x) = \sqrt{\ln(e^x + 1)} + \_C1 , y(x) = -\sqrt{\ln(e^x + 1)} + \_C1$$

## 7.2.2  First Order Homogeneous Differential Equations

**Definition 7.13 (see [3]).** We call homogeneous differential equation, an equation of the form:

$$y' = f(x, y), \qquad (7.7)$$

$f$ being a continuous and homogeneous (of degree zero) function.
  Using the change of variable

$$u(x) = \frac{y}{x}, \qquad (7.8)$$

the homogeneous equations are reduced to some equations with separable variables.
  Substituting (7.8) in the equation (7.7) it results the equation

$$y' = u'x + u,$$

i.e.

$$u' = \frac{f(x, u) - u}{x},$$

which is an equation with separable variables.
**Remark 7.14.** The homogeneous equations also admit singular solutions, namely the roots of equation

$$f(1, u) - u = 0. \qquad (7.9)$$

As $u = C$ is a root of the equation (7.7), from the relation (7.8) it results

$$y = Cx,$$

i.e. the homogeneous equations admit the straight lines as some particular solutions.
**Example 7.15.** Integrate the equation:

$$y' = \frac{x + y}{x - y}.$$

**Solution.**
We obtain:
$$y' = \frac{1 + \frac{y}{x}}{1 - \frac{y}{x}}.$$

Making the change of variable (7.8) it results that
$$y = ux \Rightarrow y' = u'x + u;$$

hence, the equation will be
$$u'x + u = \frac{1 + u}{1 - u} \iff (u'x + u)(1 - u) = 1 + u \iff u'(1 - u)x = u^2 + 1,$$

i.e. one achieves the equation with separable variables:
$$\frac{1 - u}{u^2 + 1}du = \frac{dx}{x};$$

therefore
$$\int \frac{1 - u}{u^2 + 1}du = \int \frac{dx}{x} \iff \int \frac{1}{u^2 + 1}du - \int \frac{u}{u^2 + 1}du = \ln|x| + C,$$

i.e.
$$\arctan u - \frac{1}{2}\ln(u^2 + 1) = \ln|x| + C \iff \arctan u - \ln\left|x\sqrt{u^2 + 1}\right| = C.$$

Finally, we shall have
$$\arctan\frac{y}{x} - \ln\left|x\sqrt{\frac{y^2}{x^2} + 1}\right| = C$$

or
$$\arctan\frac{y}{x} - \ln\left|\sqrt{x^2 + y^2}\right| = C.$$

We shall have in Matlab 7.0:
```
>>y=dsolve('Dy=(x+y)/(x-y)','x')
y =
-1/2*log((y^2+x^2)/x^2)+atan(y/x)-log(x)-C1 = 0
```
and with Mathematica 8:

In[3]:= **DSolve[D[y[x], x] == (x + y[x]) / (x - y[x]), y[x], x]**

Out[3]= $\text{Solve}\left[-\text{ArcTan}\left[\frac{y[x]}{x}\right] + \frac{1}{2}\text{Log}\left[1 + \frac{y[x]^2}{x^2}\right] == C[1] - \text{Log}[x], y[x]\right]$

and in Maple 15:

$$\text{ode} := \frac{d}{dx} y(x) = \frac{x + y(x)}{x - y(x)} :$$
$$\textit{dsolve(ode)}$$

$$y(x) = \tan\left( RootOf\left( -2\_Z + \ln\left( \frac{1}{\cos(\_Z)^2} \right) + 2\ln(x) + 2\_C1 \right) \right) x$$

We can't achieve this solution using Matlab 7.9 or in Mathcad 14.

## 7.2.3  Equations with Reduce to Homogeneous Equations

**Definition 7.16** (see [3]). The equations that one reduce to homogeneous equations are some equations of the type:

$$y' = \frac{a_1 x + b_1 y + c_1}{a_2 x + b_2 y + c_2}, \ a_i, b_i, c_i \in \mathbb{R}, i = \overline{1,2}. \tag{7.10}$$

As the equations

$$\begin{cases} a_1 x + b_1 y + c_1 = 0 \\ a_2 x + b_2 y + c_2 = 0 \end{cases}$$

represents the equations of two straight lines, we shall analyze the following cases:

1. the straight lines are parallel:

$$\frac{a_1}{a_2} = \frac{b_1}{b_2} = \lambda \Longrightarrow \begin{cases} a_1 = \lambda a_2 \\ b_1 = \lambda b_2; \end{cases}$$

the equation (7.10) becomes

$$y' = \frac{\lambda (a_2 x + b_2 y) + c_1}{a_2 x + b_2 y + c_2}.$$

Denoting

$$a_2 x + b_2 y = u$$

we have

$$u' = a_2 + b_2 y';$$

we achieve

$$y' = \frac{u' - a_2}{b_2} = \frac{u'}{b_2} - \frac{a_2}{b_2}. \tag{7.11}$$

From (7.10) and (7.11) we deduce the following equation with separable variables:

$$\frac{u'}{b_2} - \frac{a_2}{b_2} = \frac{\lambda u + c_1}{u + c_2}. \tag{7.12}$$

2. If the straight lines one cross in the point $(x_0, y_0)$, then is made a translation of the axes, such that the new origin to be the point of coordinates $(x_0, y_0)$:

$$\begin{cases} u = x - x_0 \\ v = y - y_0. \end{cases} \tag{7.13}$$

Therefore

$$y' = v'.$$

We shall obtain

$$v' = \frac{a_1(u + x_0) + b_1(v + y_0) + c_1}{a_2(u + x_0) + b_2(v + y_0) + c_2} = \frac{a_1 u + b_1 v + \overbrace{a_1 x_0 + b_1 y_0 + c_1}^{=0}}{a_1 u + b_1 v + \underbrace{a_2 x_0 + b_2 y_0 + c_2}_{=0}},$$

i.e.

$$v' = \frac{a_1 u + b_1 v}{a_1 u + b_1 v}.$$

Hence, it results the homogeneous equation

$$v' = \frac{a_1 + b_1 \frac{v}{u}}{a_2 + b_2 \frac{v}{u}}. \tag{7.14}$$

**Definition 7.17** (see [3]). The same method one applies to the equations of the general form:

$$y' = f\left(\frac{a_1 x + b_1 y + c_1}{a_2 x + b_2 y + c_2}\right), \quad a_i, b_i, c_i \in \mathbb{R}, i = \overline{1, 2}. \tag{7.15}$$

**Example 7.18.** Find the general solution of the differential equation:

$$(4x + y + 1)\, dx + (2x + y - 1)\, dy = 0.$$

**Solution.**

We obtain:

$$y' = -\frac{4x + y + 1}{2x + y - 1}.$$

As

$$\left.\begin{array}{l} \frac{a_1}{a_2} = \frac{4}{2} \\ \frac{b_1}{b_2} = 1 \end{array}\right\} \implies \frac{a_1}{a_2} \neq \frac{b_1}{b_2} \implies \text{the straightlines are not parallel(we are in the second case)}.$$

Solving the system

$$\begin{cases} 4x + y + 1 = 0 \\ 2x + y - 1 = 0 \end{cases}$$

we achieve the solution

$$\begin{cases} x = -1 \\ y = 3; \end{cases}$$

therefore, the straight lines one cross in the point $(-1, 3)$.
With (7.13) we shall have:

$$\begin{cases} u = x + 1 \\ v = y - 3 \end{cases}$$

and it results the homogeneous equation

$$v' = \frac{-4 + \frac{v}{u}}{2 + \frac{v}{u}}.$$

Denoting

$$\frac{v}{u} = t,$$

we can transform the previous differential equation into the following equation with separable variables:

$$t'u + t = -\frac{4 + t}{2 + t},$$

i.e.

$$-\frac{t + 2}{t^2 + 3t + 4}dt = \frac{du}{u}.$$

We shall deduce that:

$$-\int \frac{t + 2}{t^2 + 3t + 4}dt = \int \frac{du}{u},$$

i.e.

$$-\frac{1}{2}\int\frac{2t+3}{t^2+3t+4}dt - \frac{1}{2}\int\frac{dt}{t^2+3t+4} = \ln|u| + C \iff$$

$$-\frac{1}{2}\ln|t^2+3t+4| - \frac{1}{2}\int\frac{dt}{\left(t+\frac{3}{2}\right)^2+\frac{7}{4}} = \ln|u| + C \iff$$

$$-\frac{1}{2}\ln|t^2+3t+4| - \frac{1}{\sqrt{7}}\arctan\frac{2t+3}{\sqrt{7}} = \ln|u| + C.$$

Finally, it will result:

$$-\frac{1}{2}\ln\left|\left(\frac{y-3}{x+1}\right)^2 + 3\left(\frac{y-3}{x+1}\right) + 4\right| - \frac{1}{\sqrt{7}}\arctan\frac{2\left(\frac{y-3}{x+1}\right)+3}{\sqrt{7}} = \ln|x+1| + C,$$

i.e.

$$-\frac{1}{2}\ln\left|(y-3)^2 + 3(y-3)(x+1) + 4(x+1)^2\right| - \frac{1}{\sqrt{7}}\arctan\frac{3x+2y-3}{\sqrt{7}(x+1)} = \ln|x+1|+C.$$

In Matlab 7.0 (but not in Matlab 7.9) we shall have:
>>y=dsolve('Dy=-(4*x+y+1)/(2*x+y-1)','x')
y =
-1/2*log((4*(x+1)^2-3*(-y+3)*(x+1)+(-y+3)^2)/(x+1)^2)+
1/7*7^(1/2)*atan(1/7*(-3*x+3-2*y)*7^(1/2)/(x+1))-log(x+1)-
C1 = 0
and with Mathematica 8:

In[6]:= DSolve[D[y[x], x] == - (4*x + y[x] + 1) / (2*x + y[x] - 1), y[x], x];

In[7]:= Simplify[%]

Out[7]= Solve$\left[2\sqrt{7}\ \text{ArcTan}\left[\dfrac{-1+\frac{4(1+x)}{-1+2x+y[x]}}{\sqrt{7}}\right] ==\right.$

$\left. 7\left(4\,C[1] + 2\,\text{Log}[2\,(1+x)] + \text{Log}\left[\dfrac{4 - x + 4x^2 + 3(-1+x)\,y[x] + y[x]^2}{2(1+x)^2}\right]\right), y[x]\right]$

and in Maple 15:

$$\text{ode} := \frac{d}{dx}y(x) = -\frac{4 \cdot x + y(x) + 1}{2 \cdot x + y(x) - 1} :$$
$$\text{dsolve(ode)} :$$

We can't achieve this result using Mathcad 14.

## 7.2.4  First Order Linear Differential Equations

**Definition 7.19** (see [3]). A **differential non-homogeneous equation** is of the form

$$y' = p(x) y + q(x), \tag{7.16}$$

where $p, q$ are two continuous functions.

**Remark 7.20.** As the equation (7.16) has the degree one in $y$ and $y'$ is also called *linear*.

**Proposition 7.21** (see [3]). The general solution non-homogeneous equation (7.16) is equal to the general solution of the homogeneous equation

$$y' = p(x) y,$$

adding the particular solution of the non-homogeneous equation (which is obtained using the method of variation of the constants of Lagrange) and it has the analytical expression:

$$y(x) = e^{\int p(x) dx} \left( C_1 + \int q(x) e^{-\int p(x) dx} dx \right), \quad C_1 \in \mathbb{R}. \tag{7.17}$$

**Example 7.22.** Solve the equation:

$$xy' - y = x^2 \cos x, \quad x > 0.$$

**Solution.**

*First Method.* Writing the equation of the form:

$$y' = \frac{y}{x} + x \cos x,$$

one notices that

$$\begin{cases} p(x) = \frac{1}{x} \\ q(x) = x \cos x; \end{cases}$$

therefore, using (7.17) it results that:

$$y(x) = e^{\int \frac{1}{x} dx} \left( C_1 + \int x \cos x \cdot e^{-\int \frac{1}{x} dx} dx \right) = e^{\ln x} \left( C_1 + \int x \cos x \cdot e^{-\ln x} dx \right)$$

and, further:

$$y(x) = x \left( C_1 + \int x \cdot \frac{1}{x} \cos x \, dx \right),$$

i.e.

$$y\left(x\right) = x\left(C_1 + \sin x\right),$$

*Second Method.* The equation could also be solved as follows:
*Step 1.* We solve the homogeneous equation:

$$y' = \frac{y}{x};$$

denoting

$$\frac{y}{x} = u$$

it will result the equation with separable variables:

$$u' = 0,$$

which has the solution

$$u = C;$$

hence the solution of the homogeneous equation will be

$$y\left(x\right) = Cx.$$

*Step 2.* We apply the method of variation of the constants:

$$y_p\left(x\right) = C\left(x\right)x \Longrightarrow y_p'\left(x\right) = xC'\left(x\right) + C\left(x\right).$$

From the condition that the particular solution to satisfy the non- homogeneous equation, we deduce

$$xC'\left(x\right) + C\left(x\right) = \frac{C\left(x\right)\cdot x}{x} + x\cos x,$$

i.e.

$$xC'\left(x\right) + C\left(x\right) = C\left(x\right) + x\cos x.$$

Finally, it results that

$$C'\left(x\right) = \cos x;$$

therefore

$$C\left(x\right) = \int \cos x \; dx \Longrightarrow C\left(x\right) = \sin x + C_2,$$

$$y_p\left(x\right) = C\left(x\right)\cdot x = x\left(\sin x + C_2\right)$$

and

$$y(x) = y_o(x) + y_p(x) = Cx + x\sin x + xC_2 = x(C + C_2) + x\sin x,$$

i.e.

$$y(x) = x(C_1 + \sin x).$$

Using Matlab 7.9, we shall achieve:

>> y=dsolve('x*Dy-y=x^2*cos(x)','x')

y =

x*sin(x)+x*C1

We can also solve this equation using Mathematica 8:

In[21]:= DSolve[x*D[y[x], x] == y[x] + x^2*Cos[x], y[x], x]

Out[21]= {{y[x] → x C[1] + x Sin[x]}}

and with Maple 15:

$$\text{ode} := x \cdot \text{diff}(y(x), x) - y(x) = x^2 \cdot \cos(x)$$

$$\left(\frac{d}{dx} y(x)\right) x - y(x) = x^2 \cos(x)$$

dsolve(ode)

$$y(x) = x\sin(x) + x\_C1$$

## 7.2.5  Exact Differential Equations

**Definition 7.23** (see [3]). A **total differential equation** is of the form:

$$f(x, y)\, dx + g(x, y)\, dy = 0, \quad f, g : D \subseteq \mathbb{R}^2 \to \mathbb{R}. \tag{7.18}$$

If the left side of the equation (7.18) is the total differential of a function $\Phi : D \to \mathbb{R}$, i.e.

$$d\Phi = f(x, y)\, dx + g(x, y)\, dy, \tag{7.19}$$

then the differential equation is called **exact differential equation**.

**Proposition 7.24** (see [3]). The necessary and sufficient condition that the equation (7.18) to be *exact differential* is that

$$\frac{\partial f}{\partial y} = \frac{\partial g}{\partial x}. \tag{7.20}$$

**Proposition 7.25** (see [3]). The general solution of the exact differential equation is:

$$\Phi(x, y) = C, \tag{7.21}$$

where

$$\Phi(x, y) = \int_{x_0}^{x} f(t, y_0)\, dt + \int_{y_0}^{y} g(x, t)\, dt, \quad (x_0, y_0) \in D. \tag{7.22}$$

If the condition (7.20) is not accomplished, then the differential equation (7.18) must be multiplied by an integrating factor $\mu(x, y)$ such that the equation to become an exact differential equation.

There are two cases:

Case 1. If $\mu = \mu(x)$, then the condition (7.20) becomes

$$\frac{\partial}{\partial y}(f \cdot \mu) = \frac{\partial}{\partial x}(g \cdot \mu) \Longleftrightarrow \mu\frac{\partial f}{\partial y} = \mu\frac{\partial g}{\partial x} + g\mu' \Longrightarrow$$

$$\frac{\mu'}{\mu} = \frac{\frac{\partial f}{\partial y} - \frac{\partial g}{\partial x}}{g} = \varphi(x)$$

and

$$\mu(x) = e^{\int \varphi(x)\,dx}. \tag{7.23}$$

Case 2. If $\mu = \mu(y)$, then reasoning as in the case 1, the condition (7.20) becomes:

$$\frac{\mu'}{\mu} = \frac{\frac{\partial f}{\partial y} - \frac{\partial g}{\partial x}}{-f} = \varphi(y)$$

and

$$\mu(y) = e^{\int \varphi(y)\,dy}. \tag{7.24}$$

**Example 7.26.** Solve the equation:

$$\text{a) } (ye^{xy} - 4xy)\, dx + (xe^{xy} - 2x^2)\, dy = 0$$

$$\text{b) } y(1 + xy)\, dx - x dy = 0, \quad y \neq 0.$$

**Solutions.**
Denoting

$$\begin{cases} f(x, y) = ye^{xy} - 4xy \\ g(x, y) = xe^{xy} - 2x^2 \end{cases}$$

it will result that:

$$\left.\begin{array}{l}\frac{\partial f}{\partial y} = e^{xy} + xye^{xy} - 4x \\ \frac{\partial g}{\partial x} = e^{xy} + xye^{xy} - 4x\end{array}\right\} \implies \frac{\partial f}{\partial y} = \frac{\partial g}{\partial x},$$

i.e. the given equation is an exact differential equation.

Using (7.21) and (7.22) we deduce that:

$$\int_{x_0}^{x} \left(y_0 e^{ty_0} - 4ty_0\right) dt + \int_{y_0}^{y} \left(xe^{xt} - 2x^2\right) dt = C \iff$$

$$e^{ty_0}\Big|_{x_0}^{x} - 4y_0 \cdot \frac{t^2}{2}\Big|_{x_0}^{x} + e^{xt}\Big|_{y_0}^{y} - 2x^2 t\Big|_{y_0}^{y} = C \iff$$

$$e^{xy} - 2x^2 y = C_1,$$

where

$$C_1 = C + e^{x_0 y_0} + 2x_0^2 y_0.$$

Solving in Matlab 7.9 the proposed differential equation, we distinguish the following steps:

*Step 1.* We check if the equation is an exact differential equation.

```
>> syms x y t y0 x0
>>f=y*exp(x*y)-4*x*y;
>> g=x*exp(x*y)-2*x^2;
>> d1=diff(f,y);
>> d2=diff(g,x);
>> d1==d2
ans =
1
```

*Step 2.* As the equation is an exact differential equation, we can apply the formula (7.22) to determine its solution.

```
>> Phi=int(subs(f,{x,y},{t,y0}),t,x0,x)+int(subs(g,y,t),t,y0,y)
Phi =
exp(x*y) - exp(x0*y0) - 2*x^2*y + 2*x0^2*y0
```

We shall also solve this problem using Mathcad 14:

$$f(x,y) := y \cdot e^{x \cdot y} - 4 \cdot x \cdot y \qquad g(x,y) := x \cdot e^{x \cdot y} - 2 \cdot x^2$$

$$\frac{d}{dy} f(x,y) = \frac{d}{dx} g(x,y) \rightarrow 1$$

$$\phi(x,y,x0,y0) := \int_{x0}^{x} f(t,y0) \, dt + \int_{y0}^{y} g(x,t) \, dt$$

$$\phi(x,y,x0,y0) \rightarrow e^{x \cdot y} - e^{x0 \cdot y0} - 2 \cdot x^2 \cdot y + 2 \cdot x0^2 \cdot y0$$

and with Mathematica 8:

In[15]:= f[x_, y_] := y * Exp[x * y] - 4 * x * y

In[16]:= g[x_, y_] := x * Exp[x * y] - 2 * x^2

In[17]:= D[f[x, y], y] == D[g[x, y], x]

Out[17]= True

In[18]:= φ = Simplify[Integrate[f[t, y0], {t, x0, x}] + Integrate[g[x, t], {t, y0, y}]]

Out[18]= e^{xy} - e^{x0 y0} - 2 x^2 y + 2 x0^2 y0

and in Maple 15:

$$f := (x, y) \rightarrow y \cdot e^{x \cdot y} - 4 \cdot x \cdot y :$$
$$g := (x, y) \rightarrow x \cdot e^{x \cdot y} - 2 \cdot x^2 :$$
$$evalb\left( \frac{\partial}{\partial y} f(x, y) = \frac{\partial}{\partial x} g(x, y) \right)$$

$$true$$

$$\Phi := \int_{x0}^{x} f(t, y0) \, dt + \int_{y0}^{y} g(x, t) \, dt$$

$$-e^{x0 \, y0} + 2 \, x0^2 \, y0 + e^{xy} - 2 \, x^2 \, y$$

b) Denoting

$$\begin{cases} f(x,y) = y(1 + xy) \\ g(x,y) = -x \end{cases}$$

it will result that:

$$\left.\begin{array}{l} \frac{\partial f}{\partial y} = 1 + 2xy \\ \frac{\partial g}{\partial x} = -1 \end{array}\right\} \implies \frac{\partial f}{\partial y} \neq \frac{\partial g}{\partial x},$$

i.e. the given equation is not an exact differential equation.

We can notice that

$$\frac{\frac{\partial f}{\partial y} - \frac{\partial g}{\partial x}}{-f} = \frac{1 + 2xy + 1}{-y(1 + xy)} = \frac{2(1 + xy)}{-y(1 + xy)} = -\frac{2}{y} = \varphi(y);$$

hence

$$\mu(y) = e^{\int -\frac{2}{y} dy} = e^{-2 \ln|y|} = e^{-\ln y^2} = \frac{1}{y^2}$$

is the integrating factor.

Multiplying the given equation with $\frac{1}{y^2}$ we achieve:

$$\frac{1 + xy}{y} dx - \frac{x}{y^2} dy = 0,$$

which is an exact differential equation.

Using (7.21) and (7.22) we deduce that:

$$\int_{x_0}^x \frac{1 + ty_0}{y_0} dt + \int_{y_0}^y \frac{x}{t^2} dt = C \iff$$

$$\frac{1}{y_0}(x - x_0) + \left.\frac{t^2}{2}\right|_{x_0}^x - x \cdot \left.\frac{t^{-2+1}}{-2 + 1}\right|_{y_0}^y = C \iff$$

$$\frac{x^2}{2} + \frac{x}{y} = C_1,$$

where

$$C_1 = C + \frac{x_0^2}{2} + \frac{x_0}{y_0}.$$

We can also solve this equation in Matlab 7.9:

*Step 1.* We check if the equation is an exact differential equation.

```
>> syms x y t y0 x0
>> f=y*(1+x*y);
>> g=-x;
>> d1=diff(f,y);
>> d2=diff(g,x);
>> d1==d2
ans =
0
```

*Step 2.* Since the equation is not an exact total differential equation, we must determine the integrating factor (7.24):
>> phi=simple((d1-d2)/(-f))
phi =
-2/y
>> miu=exp(int(phi,y))
miu =
1/y^2
*Step 3.* We can apply the formula (7.22) in order to determine the solution of the equation.
>>Phi=int(subs(f*miu,{x,y},{t,y0}),t,x0,x)+int
(subs(g*miu,y,t),t,y0,y);
>> Phi=expand(Phi)
Phi =
piecewise([0 <= y and y0 <= 0, x/y0 - x*Inf - x0/y0 + x^2/2
- x0^2/2], [y < 0 or 0 < y0, x/y - x0/y0 + x^2/2 - x0^2/2])
We shall also solve this problem using Mathcad 14:

$$f(x,y) := y \cdot (1 + x \cdot y) \qquad g(x,y) := -x$$

$$\frac{d}{dy} f(x,y) = \frac{d}{dx} g(x,y) \rightarrow 2 \cdot x \cdot y + 1 = -1$$

$$\mu(y) := e^{\displaystyle\int \frac{\frac{d}{dy} f(x,y) - \frac{d}{dx} g(x,y)}{-f(x,y)} \, dy} \qquad \text{simplify} \rightarrow \frac{1}{y^2}$$

$$f1(x,y) := f(x,y) \cdot \mu(y) \qquad g1(x,y) := g(x,y) \cdot \mu(y)$$

$$\phi(x,y,x0,y0) := \int_{x0}^{x} f1(t,y0) \, dt + \int_{y0}^{y} g1(x,t) \, dt$$

and with Mathematica 8:

In[130]:= `f[x_, y_] := y * (1 + x * y)`

In[131]:= `g[x_, y_] := -x`

In[132]:= `D[f[x, y], y] == D[g[x, y], x]`

Out[132]= $1 + 2 x y == -1$

In[133]:= `φ := (D[f[x, y], y] - D[g[x, y], x]) / (-f[x, y])`

In[134]:= `μ[y_] := Simplify[Exp[Integrate[φ, y]]]`

In[135]:= `f1[x_, y_] = f[x, y] * μ[y];`

In[136]:= `g1[x_, y_] = g[x, y] * μ[y];`

In[137]:= `φ = Simplify[Integrate[f1[t, y0], {t, x0, x}] + Integrate[g1[x, t], {t, y0, y}]]`

Out[137]= $\text{ConditionalExpression}\left[\frac{1}{2}\left(x^2 + \frac{2x}{y} - \frac{x0\,(2 + x0\,y0)}{y0}\right),\right.$

$\left.\left(y\,y0 + y0^2\,\&\&\,\text{Re}\left[\frac{y0}{y-y0}\right] \geq 0\right) \,||\, \text{Re}\left[\frac{y0}{y-y0}\right] \leq -1 \,||\, \frac{y0}{y-y0} \in \text{Reals}\right]$

and in Maple 15:

$f := (x, y) \rightarrow y \cdot (1 + x \cdot y):$
$g := (x, y) \rightarrow -x:$
$evalb\left(\frac{\partial}{\partial y}f(x, y) = \frac{\partial}{\partial x}g(x, y)\right)$

$\qquad\qquad\qquad\qquad\qquad\qquad\qquad\qquad\qquad\qquad\qquad\qquad$ *false*

$\mu := (y) \rightarrow e^{\displaystyle\int \frac{\frac{\partial}{\partial y}f(x,y) - \frac{\partial}{\partial x}g(x,y)}{-f(x,y)}\,dy} :$

$f1 := (x, y) \rightarrow f(x, y) \cdot \mu(y):$
$g1 := (x, y) \rightarrow g(x, y) \cdot \mu(y):$
$assume(0 < y0 < y)$

$\Phi := \int_{x0}^{x} f1(t, y0)\,dt + \int_{y0}^{y} g1(x, t)\,dt$

$\qquad\qquad\qquad\qquad \frac{1}{2}x^2 - \frac{1}{2}x0^2 + \frac{x - x0}{y0\sim} + \frac{x\,(y0\sim - y\sim)}{y0\sim\,y\sim}$

## 7.2.6   Bernoulli's Equation

**Definition 7.27** (see [3]). The differential equation of the form

$$y' = p(x)y + q(x)y^\alpha \tag{7.25}$$

constitutes the **Bernoulli's equation**, $p, q$ being two continuous functions.
**Proposition 7.28** (see [3]). If

- $\alpha = 0$ then the equation (7.25) becomes an non-homogeneous linear differential equation;
- $\alpha = 1$ then the equation (7.25) becomes a differential equation with separable variables.

Otherwise, i.e. for $\alpha \in \mathbb{R} \setminus \{0, 1\}$, using the change of function

$$y = z^{\frac{1}{1-\alpha}}, \tag{7.26}$$

the equation (7.25) one reduces to an non-homogeneous linear differential equation.
**Example 7.29.** Solve the equation:

$$y' - 4\frac{y}{x} - x\sqrt{y} = 0, \ x > 0, \ y \geq 0.$$

**Solution.**
We can notice that

$$\alpha = \frac{1}{2},$$

hence, using the change of function

$$y = z^2,$$

from the fact that

$$y' = 2zz'$$

it will result the linear differential equation:

$$z' = 2\frac{z}{x} + \frac{x}{2}.$$

From the relation (7.17) it results that the solution of this equation will be

$$z(x) = e^{\int \frac{2}{x} dx} \left( C_1 + \int \frac{x}{2} \cdot e^{-\int \frac{2}{x} dx} dx \right) = x^2 \left( C_1 + \frac{1}{2} \ln x \right);$$

therefore

$$y(x) = x^4 \left( C_1 + \frac{1}{2} \ln x \right)^2.$$

We shall achieve this explicit solution in Matlab 7.9:
```
>> y=simplify(dsolve('Dy-4*y/x-x*sqrt(y)','x'))
y =
0
(x^4*(2*C3 + log(x))^2)/4
```
and using Mathematica 8:

In[8]:= **DSolve[y'[x] == 4 * y[x] /x + x * Sqrt[y[x]], y[x], x];**

In[9]:= **Simplify[%]**

Out[9]= $\left\{\left\{ y[x] \rightarrow \frac{1}{4} x^4 (2 C[1] + Log[x])^2 \right\}\right\}$

We shall find only an implicit solution using Maple 15:

$$ode := \text{diff}(y(x), x) - \frac{4 \cdot y(x)}{x} - x \cdot \sqrt{y(x)} = 0$$

$$\frac{d}{dx} y(x) - \frac{4 y(x)}{x} - x\sqrt{y(x)} = 0$$

$$dsolve(ode)$$

$$\sqrt{y(x)} - \left( \frac{1}{2} \ln(x) + \_C1 \right) x^2 = 0$$

We can not solve the equation in Mathcad 14.

## 7.2.7  Riccati's Equation

**Definition 7.30** (see [3]). A differential equation, which is of the form

$$y' = p(x) y^2 + q(x) y + r(x) \tag{7.27}$$

represents the **Riccati's equation**, $p, q, r$ being three continuous functions.
**Proposition 7.31 (see [3]).** If you know a particular solution $y_p(x)$ of the equation (7.27), then using the substitution

$$y = y_p + \frac{1}{z} \tag{7.28}$$

the equation (7.27) becomes a non-homogeneous differential equation.
**Example 7.32.** Solve the equation:

$$xy' = y^2 - (2x+1)y + x^2 + 2x, \; y_1(x) = x, \; x > 0.$$

**Solution.**

This is a Riccati equation.

We can write the equation on the form:

$$y' = \frac{y^2}{x} - \frac{2x+1}{y}y + x + 2.$$

Using the substitution from (7.28), i.e.:

$$y = x + \frac{1}{z} \Longrightarrow y' = 1 - \frac{z'}{z^2},$$

the equation becomes:

$$1 - \frac{z'}{z^2} = \frac{1}{x}\left(x + \frac{1}{z}\right)^2 - \frac{2x+1}{x}\left(x + \frac{1}{z}\right) + x + 2;$$

after some calculus it will result the equation:

$$-z' = 2z + \frac{1}{x} - \frac{2x+1}{x}z,$$

i.e. the linear differential equation:

$$z' = \frac{1}{x}z - \frac{1}{x}.$$

The solution of this equation will be:

$$z(x) = e^{\int \frac{1}{x}dx}\left(C_1 + \int \left(-\frac{1}{x}\right) \cdot e^{-\int \frac{1}{x}dx}dx\right) = x\left(C_1 - \int \frac{1}{x} \cdot \frac{1}{x}dx\right)$$

$$= x\left(C_1 + \frac{1}{x}\right)$$

i.e.

$$z(x) = C_1 x + 1$$

and

$$y(x) = \frac{C_1 x^2 + x + 1}{C_1 x + 1}.$$

We shall obtain using Matlab 7.9:
>> y=dsolve('x*Dy=y^2-(2*x+1)*y+x^2+2*x','x')
y =
x
x + 1/(C2*x + 1)
and with Mathematica 8:

In[4]:= **DSolve[x\*y'[x] == y[x]^2 - (2\*x + 1) \*y[x] + x^2 + 2\*x, y[x], x];**

In[5]:= **Simplify[%]**

Out[5]= $\left\{\left\{y[x] \rightarrow \dfrac{x^2 - C[1] - x\,C[1]}{x - C[1]}\right\}\right\}$

and in Maple 15:

$$\text{ode} := x \cdot \text{diff}(y(x), x) = y(x)^2 - (2 \cdot x + 1) \cdot y(x) + x^2 + 2 \cdot x$$

$$x\left(\frac{d}{dx} y(x)\right) = y(x)^2 - (2x + 1) y(x) + x^2 + 2x$$

*dsolve(ode)*

$$y(x) = \frac{-x - 1 + \_C1 x^2}{-1 + \_C1 x}$$

## 7.2.8  Lagrange's Equation

**Definition 7.33** (see [3]). A differential equation of the form

$$y = xA(y') + B(y'),  \tag{7.29}$$

in which $A, B$ are two continuous functions means the **Lagrange/s equation**.

Denoting

$$y' = p \Longleftrightarrow \frac{dy}{dx} = p \Longleftrightarrow dy = p\, dx  \tag{7.30}$$

the equation (7.29) becomes

$$y = xA(p) + B(p);  \tag{7.31}$$

by differentiation we get:

$$p\, dx = A(p)\, dx + [xA'(p) + B'(p)]\, dp,$$

i.e.

$$(p - A(p))\, dx = [xA'(p) + B'(p)]\, dp.$$

If

1. $p - A(p) \neq 0$

then it results

$$\frac{\mathrm{d}x}{\mathrm{d}p} = \frac{A'(p)}{p - A(p)}x + \frac{B'(p)}{p - A(p)},$$

a non-homogeneous differential equation, having the general solution:

$$x = e^{\int \frac{A'(p)}{p-A(p)}\mathrm{d}p}\left(C + \int e^{-\int \frac{A'(p)}{p-A(p)}\mathrm{d}p} \cdot \frac{B'(p)}{p - A(p)}\mathrm{d}p\right). \tag{7.32}$$

Thereby, from (7.31) and (7.32) we deduce that the general solution of the Lagrange equation is:

$$y = A(p) e^{\int \frac{A'(p)}{p-A(p)}\mathrm{d}p}\left(C + \int e^{-\int \frac{A'(p)}{p-A(p)}\mathrm{d}p} \cdot \frac{B'(p)}{p - A(p)}\mathrm{d}p\right) + B(p). \tag{7.33}$$

2. $p - A(p)$ one cancels on the common interval of definition of the functions $A$ and $B$, then we shall denote by $p_1$ the solution of the equation:

$$p - A(p) = 0 \tag{7.34}$$

whose corresponds a solution of the Lagrange equation, i.e.

$$y = xA(p_1) + B(p_1). \tag{7.35}$$

The Lagrange's equation admits the straight lines of the form (7.35) as singular solutions, given by the roots of the equation (7.34).

**Example 7.34.** Solve the equation:

$$x - y = \frac{4}{9}y'^2 - \frac{8}{27}y'^3.$$

**Solution.**
This is a Lagrange equation.
Denoting

$$y' = p$$

the equation will become:

$$y = x - \frac{4}{9}p^2 + \frac{8}{27}p^3. \tag{7.36}$$

By differentiation with respect to $x$ we obtain:

$$\mathrm{d}y = \mathrm{d}x - \frac{8}{9}p\,\mathrm{d}p + \frac{8}{9}p^2\mathrm{d}p;$$

using (7.30) it results:

$$p\,dx = dx - \frac{8}{9}p\,dp + \frac{8}{9}p^2 dp,$$

i.e.

$$(p-1)\,dx = \frac{8}{9}(p-1)\,p\,dp$$

and, finally (if $p \neq 1$):

$$dx = \frac{8}{9}p\,dp,$$

which is a differential equation with separable variables; therefore

$$x = \frac{4}{9}p^2 + C. \tag{7.37}$$

From (7.36) we deduce:

$$y = \frac{4}{9}p^2 + C - \frac{4}{9}p^2 + \frac{8}{27}p^3,$$

i.e.

$$y = \frac{8}{27}p^3 + C. \tag{7.38}$$

If $p \neq 1$, the general solution of the differential equation can be obtained from (7.37) and (7.38) in the following way:

- from (7.37) it results that

$$p = \pm\frac{3}{2}\sqrt{x - C};$$

- substituting $p$ into (7.38) we shall have:

$$y(x) = \pm(x - C)^{3/2} + C.$$

If $p = 1$, the singular solution will be:

$$y(x) = -\frac{4}{27} + x.$$

We shall find the solutions of this equation and we shall plot them in Matlab 7.9:

```
>> y=simplify(dsolve('x-y=(4/9)*Dy^2-(8/27)*Dy^3','x'))
y =
x - 4/27
C2 + (-(C2 - x)^3)^(1/2)
```

```
   C2 - (-(C2 - x)^3)^(1/2)
>> t=-12:0.1:12;
>> C1=12;
>>x=(t.^2).^(1/3)+C1;
>>y=t+C1;
>> C2=11;
>>x1=(t.^2).^(1/3)+C2;
>>y1=t+C2;
>> C3=8;
>> x2=(t.^2).^(1/3)+C3;
>>y2=t+C3;
>> C4=2;
>> x4=(t.^2).^(1/3)+C4;
>> y4=t+C4;
>> C5=10;
>> x5=(t.^2).^(1/3)+C5;
>>y5=t+C5;
>> x3=-15:15;
>>y3=-4/27+x3;
>> plot(x,y,'b',x2,y2,'b',x4,y4,'b',x5,y5,'b',x3,y3,'m',
'LineWidth',2)
```

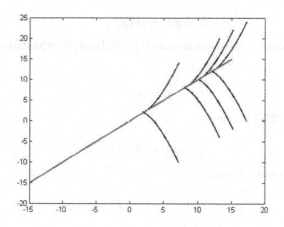

and using Maple 15:

$\text{ode} := x - y(x) = \frac{4}{9} \cdot \text{diff}(y(x), x)^2 - \frac{8}{27} \cdot \text{diff}(y(x), x)^3 :$

$\text{dsolve(ode)}$

$$y(x) = x - \frac{4}{27}, y(x) = \_C1 - (x - \_C1)^{3/2}, y(x) = \_C1 + (x - \_C1)^{3/2} \tag{1}$$

$C1 := 12 : C2 := 11 : C3 := 8 : C4 := 2 : C5 := 10 :$

$\text{plot}\left(\left[x3, \left[(t^2)^{\frac{1}{3}} + C1, t + C1, t = -12..12\right], \left[(t^2)^{\frac{1}{3}} + C2, t + C2, t = -12..12\right], \left[(t^2)^{\frac{1}{3}} + C3, t + C3, t = -12..12\right],\right.$
$\left.\left[(t^2)^{\frac{1}{3}} + C4, t + C4, t = -12..12\right], \left[(t^2)^{\frac{1}{3}} + C5, t + C5, t = -12..12\right]\right], x3 = -15..15\right)$

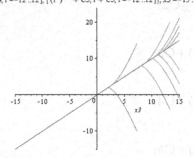

## 7.2.9 Clairaut's Equation

**Definition 7.35** (see [3]). The differential equation of the form

$$y = xy' + B(y'), \tag{7.39}$$

$B$ being a continuous function means the **Clairaut's equation**.
  Denoting

$$y' = p$$

the equation (7.39) becomes

$$y = xp + B(p); \tag{7.40}$$

by differentiation we obtain:

$$p\,dx = p\,dx + x\,dp + B'(p)\,dp,$$

i.e.

$$[x + B'(p)]\,dp = 0.$$

  If

1. $dp = 0 \Longrightarrow p = C;$

  then from (7.40) we achieve

$$y = Cx + B(C),$$ (7.41)

equation which represents a family of straight lines in plane, which is the general solution of the Clairaut equation.

2. $x + B'(p) = 0 \Longrightarrow x = -B'(p);$

we get the singular solution of the Clairaut equation:

$$y = -pB'(p) + B(p).$$ (7.42)

**Example 7.35.** Solve the equation:

$$y = xy' - 4y'^2.$$

**Solution.**
This is a Clairaut equation.
Denoting

$$y' = p$$

the equation becomes:

$$y = px - 4p^2,$$

i.e.

$$dy = p dx + x dp - 8p dp.$$

As

$$\frac{dy}{dx} = p \Longrightarrow dy = p \, dx$$

we shall deduce that:

$$p dx = p dx + x \, dp - 8p dp,$$

i.e.

$$(x - 8p) \, dp = 0.$$

We shall analyse the following two cases:

Case 1.    $dp = 0 \Longrightarrow p = C.$

The solution of the differential equation is

$$y = Cx - 4C^2$$

and means a family of straight lines.

Case 2.  $x - 8p = 0$;

it results that:

$$\begin{cases} x = 8p \\ y = 8p^2 - 4p^2 = 4p^2 \end{cases}$$

or

$$y = \frac{x^2}{16}$$

(one achieves by eliminating $p$ between the two equations) constitutes a singular solution.

We shall solve this problem in Matlab 7.9:

```
>> y=dsolve('y=x*Dy-4*Dy^2','x')
y =
x^2/16
C*x - 4*C^2
>> t=-2:0.1:2;
>> x=t;
>> y=t.^2/16;
>>C=0.01;
>>y1=C*x-4*C^2
>>C1=0.04;
>>y2=C1*x-4*C1^2 ;
>> C2=0.09;
>> y3=C2*x-4*C2^2;
>> plot(x,y,'b',x,y1,x,y2,x,y3,'LineWidth',2)
```

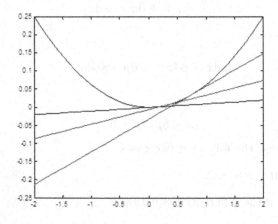

and with Maple 15:

$ode := y(x) = x \cdot \text{diff}(y(x), x) - 4 \cdot \text{diff}(y(x), x)^2 :$
$dsolve(ode)$

$$y(x) = \frac{1}{16} x^2, y(x) = x\_C1 - 4\_C1^2 \tag{1}$$

$C := 0.01 : C1 := 0.04 : C2 := 0.09 :$

$$plot\left(\left[\left[\frac{t^2}{16}, t, t = -2 \ldots 2\right], [C \cdot x - 4 C^2, x, x = -2 \ldots 2], [C1 \cdot x - 4 C1^2, x, \right.\right.$$

$$\left.\left. x = -2 \ldots 2], [C2 \cdot x - 4 C2^2, x, x = -2 \ldots 2]\right]\right)$$

We can't solve this problem in Mathcad 14.

Mathematica 8 finds only that solution of the equation, which represents a family of straight lines; it can't determine the singular solution of this Clairaut equation.

## 7.3 Higher Order Differential Equations

### 7.3.1 Homogeneous Linear Differential Equations with Constant Coefficients

**Definition 7.36** (see [25]). A differential equation of the form

$$a_0 y^{(n)} + a_1 y^{(n-1)} + \ldots + a_{n-1} y' + a_n y = 0, \tag{7.43}$$

where $a_0, a_1, \ldots, a_n$ are some real constants, $a_0 \neq 0$ is called **homogeneous differential linear equation, of order** $n$,**with constant coefficients**.

**Definition 7.37** (see [25]). The polynomial

$$P(\lambda) = a_0 \lambda^n + a_1 \lambda^{n-1} + \ldots + a_{n-1} \lambda^n + a_n \tag{7.44}$$

represents the **attached characteristic polynomial of the homogeneous differential equation of order** $n$**, with constant coefficients** from (7.43) and the equation

$$P(\lambda) = 0 \tag{7.45}$$

constitutes the **attached characteristic of the homogeneous differential equation of order $n$, with constant coefficients** from (7.43).

The method by which we obtain the differential equation solutions (7.43) was introduced of L. Euler, who tried to satisfy this equation with some functions of the form $e^{\lambda t}$.

We can notice that the solutions of differential equation (7.43) depend on the characteristic equation roots (see [25]).

Case 1. Firstly, we consider the case when the roots of the characteristic equation are real and in turn, we analyze, the subcase when the roots are distinct and then the case when the characteristic equation has also multiple roots.

a) Assuming that the characteristic equation has all the roots distinct and real, we can write the solution of the homogeneous linear differential equation (7.43) of the form

$$y(x) = C_1 e^{\lambda_1 x} + C_2 e^{\lambda_2 x} + \ldots + C_n e^{\lambda_n x}. \tag{7.46}$$

b) If the characteristic equation has the real root $\lambda = \lambda_1$, multiple of the order $p$, $p \leq n$ we can write the general solution of the homogeneous linear differential (7.43) of the form

$$y(x) = C_1 e^{\lambda_1 x} + C_2 x e^{\lambda_1 x} + \ldots + C_p x^{p-1} e^{\lambda_1 x}; \tag{7.47}$$

this expression of $y(x)$ is also called the *contribution* of the real multiple root, of the order $p$, $\lambda = \lambda_1$ of the characteristic equation, to the general solution of the homogeneous equation.

c) The characteristic equation has real $k$ roots $\lambda_1, \lambda_2 \ldots, \lambda_k$, with the multiplicities $p_1, p_2 \ldots, p_k$, $p_1 + p_2 + \ldots + p_k = n$, then the general solution of the homogeneous linear differential (7.43) is

$$y(x) = Q_{p_1-1}(x) e^{\lambda_1 x} + Q_{p_2-1}(x) e^{\lambda_2 x} + \ldots + Q_{p_k-1}(x) e^{\lambda_k x}, \tag{7.48}$$

where

$$Q_{p_i-1}(x) = C_1 + C_2 x + \ldots + C_{p_i} x^{p_i-1} \tag{7.49}$$

is a polynomial, by at most $p_i - 1$ degree.

Case 2. We suppose that the roots of the characteristic equation are complex and in turn, we analyze the subcase when the roots are distinct and then the case when the characteristic equation has also multiple roots.

a) We assume that all the characteristic equation has distinct complex roots; it results that they are two by two complex-conjugate. The general solution of the homogeneous differential equation (7.43) will be:

$$y\left(x\right) = C_1 e^{\alpha_1 x} \cos \beta_1 x + C_2 e^{\alpha_2 x} \cos \beta_2 x + \ldots + \tag{7.50}$$
$$C_k e^{\alpha_k x} \cos \beta_k x + C_1^* e^{\alpha_1 x} \sin \beta_1 x + C_2^* e^{\alpha_2 x} \sin \beta_2 x + \ldots + C_k^* e^{\alpha_k x} \sin \beta_k x,$$

where $C_i, C_i^*, i = \overline{1,k}$ are some arbitrary constants.

b) Supposing that the characteristic equation has the complex multiple root $\lambda_1 = \alpha_1 + i\beta_1$ of the order $p_1$ it results that the general solution of the homogeneous differential equation will be:

$$y\left(x\right) = C_1 e^{\alpha_1 x} \cos \beta_1 x + C_2 x e^{\alpha_1 x} \cos \beta_1 x + \ldots + \tag{7.51}$$
$$C_{p_1} x^{p_1-1} e^{\alpha_1 x} \cos \beta_1 x + C_1^* e^{\alpha_1 x} \sin \beta_1 x + C_2^* x e^{\alpha_1 x} \sin \beta_1 x + \ldots + C_{p_1}^* x^{p_1-1} e^{\alpha_1 x} \sin \beta_1 x.$$

c) The characteristic equation has the complex roots

$$\begin{cases} \lambda_1 = \alpha_1 + i\beta_1 \\ \lambda_2 = \alpha_2 + i\beta_2 \\ \quad \vdots \\ \lambda_j = \alpha_j + i\beta_j \end{cases}$$

of multiplicities $p_1, p_2, \ldots, p_j$, where $2\left(p_1 + p_2 + \ldots + p_j\right) = n$.

The general solution of the homogeneous differential equation (7.43) will be

$$y\left(x\right) = \left[R_1\left(x\right) \cos \beta_1 x + S_1\left(x\right) \sin \beta_1 x\right] e^{\alpha_1 x} \tag{7.52}$$
$$+ \ldots + \left[R_{p_j-1}\left(x\right) \cos \beta_j x + S_{p_j-1}\left(x\right) \sin \beta_j x\right] e^{\alpha_j x},$$

where

■ $R_{p_j-1}\left(x\right) = C_1 + C_2 x + \ldots + C_{p_j} x^{p_j-1}$ is a polynomial by at most $p_j - 1$ degree,

■ $S_{p_j-1}\left(x\right) = C_1^* + C_2^* x + \ldots + C_{p_j}^* x^{p_j-1}$ is a polynomial by at most $p_j - 1$ degree.

Case 3. We suppose that the characteristic equation has

■ the real roots $\lambda_1, \lambda_2 \ldots, \lambda_j$, with the multiplicities $p_1, p_2 \ldots, p_j$

and

■ the complex roots

$$\begin{cases} \lambda_{j+1} = \alpha_1 + i\beta_1 \\ \lambda_{j+2} = \alpha_2 + i\beta_2 \\ \quad\vdots \\ \lambda_{j+l} = \alpha_l + i\beta_l \end{cases}$$

with the multiplicities $p_{j+1}, p_{j+2} \ldots, p_{j+l}$ , where

$$p_1 + p_2 + \ldots + p_j + 2\,(p_{j+1} + p_{j+2} + \ldots + p_{j+l}) = n.$$

The general solution of the homogeneous differential equation (7.43) will be:

$$y\,(x) = \sum_{i=1}^{j} Q_{p_i-1}\,(x)\,e^{\lambda_1 x} + \sum_{k=1}^{l} e^{\alpha_k x} \left[ R_{p_j+k-1}\,(x) \cos \beta_k x + S_{p_j+k-1}\,(x) \sin \beta_k x \right],$$

(7.53)

where:

■ $Q_{p_i-1}\,(x)$ is a polynomial by at most $p_i - 1$ degree and it has the expression (7.49),

■ $R_{p_j+k-1}\,(x) = c_1 + c_2 x + \ldots + c_{p_k} x^{p_k-1}$ is a polynomial by at most $p_k - 1$ degree,

■ $S_{p_j+k-1}\,(x) = c_1^* + c_2^* x + \ldots + c_{p_k}^* x^{p_k-1}$ is a polynomial by at most $p_k - 1$ degree.

**Example 7.38.** Find the general solution of the homogeneous differential equations, with constant coefficients:

$$\text{a) } y'' - y = 0$$

$$\text{b) } y^{(4)} - 5y'' + 4 = 0$$

$$\text{c) } y^{(7)} + 3y^{(6)} + 3y^{(5)} + y^{(4)} = 0$$

$$\text{d) } y^{(4)} + 5y'' + 4y = 0$$

$$\text{e) } y^{(4)} + 2y''' + 3y'' + 2y' + y = 0$$

$$\text{f) } 64y^{(8)} - 48y^{(6)} + 12y^{(4)} + y'' = 0.$$

**Solutions.**

a) We obtain the characteristic polynomial

$$P(\lambda) = \lambda^2 - 1.$$

The roots of the characteristic equation $P(\lambda) = 0$ are $\lambda_{1,2} = \pm 1$. Using (7.46), we shall have

$$\begin{cases} y_1(x) = e^x \\ y_2(x) = e^{-x} \end{cases}$$

and the general solution will be

$$y(x) = C_1 e^x + C_2 e^{-x}.$$

In Matlab 7.9 we need the following sequence:
>> y=dsolve('D2y=y','x')
y =
C1*exp(x)+C2/exp(x)
We can also solve this equation using Mathematica 8:

In[3]:= **DSolve[y''[x] == y[x], y[x], x]**

Out[3]= $\{\{y[x] \rightarrow e^x C[1] + e^{-x} C[2]\}\}$

and with Maple 15:

$$\text{ode} := \frac{d^2}{dx^2} y(x) = y(x)$$

$$\frac{d^2}{dx^2} y(x) = y(x)$$

dsolve(ode)

$$y(x) = \_C1\, e^x + \_C2\, e^{-x}$$

b) We have

$$P(\lambda) = 0 \Leftrightarrow \lambda^4 - 5\lambda^2 + 4 = 0 \Leftrightarrow (\lambda^2 - 4)(\lambda^2 - 1) = 0$$
$$\Leftrightarrow \begin{cases} \lambda_{1,2} = \pm 1 \\ \lambda_{3,4} = \pm 2. \end{cases}$$

We obtain

$$\begin{cases} y_1\left(x\right) = e^x \\ y_2\left(x\right) = e^{-x} \\ y_3\left(x\right) = e^{2x} \\ y_4\left(x\right) = e^{-2x}. \end{cases}$$

The general solution of the homogeneous linear differential equation will be

$$y\left(x\right) = C_1 e^x + C_2 e^{-x} + C_3 e^{2x} + C_4 e^{-2x}.$$

c) We obtain the characteristic polynomial

$$P\left(\lambda\right) = \lambda^7 + 3\lambda^6 + 3\lambda^5 + \lambda^4 = \lambda^4\left(\lambda + 1\right)^3.$$

The roots of the characteristic equation $P\left(\lambda\right) = 0$ are:

♦ $\lambda = 0$ a multiple root of the order 4,
♦ $\lambda = -1$ a multiple root of the order 3.

Using (7.48), the general solution will be:

$$y\left(x\right) = C_1 + C_2 x + C_3 x^2 + C_4 x^3 + \left(C_5 + C_6 x + C_7 x^2\right) e^{-x}.$$

d) We obtain the characteristic polynomial

$$P\left(\lambda\right) = \lambda^4 + 5\lambda^2 + 4.$$

The roots of the characteristic equation $P\left(\lambda\right) = 0$ are:

$$\begin{cases} \lambda_{1,2} = \pm i \\ \lambda_{3,4} = \pm 2i. \end{cases}$$

Using (7.52), the general solution will be:

$$y\left(x\right) = C_1 \cos x + C_2 \sin x + C_3 \cos 2x + C_4 \sin 2x.$$

We can also achieve this solution using Matlab 7.9:
>> y=dsolve('D4y+5*D2y+4*y=0','x')
y =
C1*cos(x)+C2*sin(x)+C3*cos(2*x)+C4*sin(2*x)
and in Mathematica 8:

In[1]:= DSolve[y''''[x] + 5 * y''[x] + 4 * y[x] == 0, y[x], x]

Out[1]= {{y[x] → C[3] Cos[x] + C[1] Cos[2 x] + C[4] Sin[x] + C[2] Sin[2 x]}}

or

In[3]:= `DSolve[D[y[x], {x, 4}] + 5*D[y[x], {x, 2}] + 4*y[x] == 0, y[x], x]`

Out[3]= `{{y[x] → C[3] Cos[x] + C[1] Cos[2 x] + C[4] Sin[x] + C[2] Sin[2 x]}}`

and with Maple 15:

$$\text{ode} := \frac{d^4}{dx^4} y(x) + 5 \cdot \frac{d^2}{dx^2} y(x) + 4 \cdot y(x) = 0$$

$$\frac{d^4}{dx^4} y(x) + 5 \left( \frac{d^2}{dx^2} y(x) \right) + 4 y(x) = 0$$

$dsolve(ode)$

$y(x) = \_C1 \sin(x) + \_C2 \cos(x) + \_C3 \sin(2x) + \_C4 \cos(2x)$

e) We have

$$P(\lambda) = 0 \Leftrightarrow (\lambda^2 + \lambda + 1)^2 = 0$$

$$\Leftrightarrow \lambda_{1,2} = \frac{-1 \pm i\sqrt{3}}{2}.$$

Using (7.50), we achieve:

$$\begin{cases} y_1(x) = e^{-\frac{1}{2}x} \cos \frac{\sqrt{3}}{2} x \\ y_2(x) = e^{-\frac{1}{2}x} \sin \frac{\sqrt{3}}{2} x \end{cases}$$

The general solution of the homogeneous linear differential equation will be:

$$y(x) = \left( C_1 \cos \frac{\sqrt{3}}{2} x + C_2 \sin \frac{\sqrt{3}}{2} x \right) e^{-\frac{1}{2}x} + \left( C_3 \cos \frac{\sqrt{3}}{2} x + C_4 \sin \frac{\sqrt{3}}{2} x \right) x e^{-\frac{1}{2}x}.$$

We shall have in Matlab 7.9:
>> y=simplify(dsolve('D4y+2*D3y+3*D2y+2*Dy+y=0','x'))
y =
(C2*cos((3^(1/2)*x)/2) + C4*sin((3^(1/2)
x)/2) + C3*x*cos((3^(1/2)*x)/2) + C5*x*sin((3^(1/2)*x)/2))/exp(x/2)
and using Mathematica 8:

```
Simplify[
DSolve[
D[y[x], {x, 4}] + 2*D[y[x], {x, 3}] + 3*D[y[x], {x, 2}] +
2*D[y[x], x] + y[x] == 0, y[x], x]]
```

Out[5]= $\left\{\left\{y[x] \rightarrow e^{-x/2}\left((C[3] + xC[4]) \text{Cos}\left[\frac{\sqrt{3} \ x}{2}\right] + \right.\right.\right.$

$\left.\left.\left.(C[1] + xC[2]) \text{Sin}\left[\frac{\sqrt{3} \ x}{2}\right]\right)\right\}\right\}$

and with Maple 15:

$$\text{ode} := \frac{d^4}{dx^4}y(x) + 2 \cdot \frac{d^3}{dx^3}y(x) + 3 \cdot \frac{d^2}{dx^2}y(x) + 2 \cdot \frac{d}{dx}y(x) + y(x)$$
$$= 0$$

$$\frac{d^4}{dx^4}y(x) + 2\left(\frac{d^3}{dx^3}y(x)\right) + 3\left(\frac{d^2}{dx^2}y(x)\right) + 2\left(\frac{d}{dx}y(x)\right) \qquad (1)$$
$$+ y(x) = 0$$

$\text{dsolve}(\text{ode})$

$$y(x) = \_C1\,e^{-\frac{1}{2}x}\sin\left(\frac{1}{2}\sqrt{3}\,x\right) + \_C2\,e^{-\frac{1}{2}x}\cos\left(\frac{1}{2}\sqrt{3}\,x\right) \qquad (2)$$

$$+ \_C3\,e^{-\frac{1}{2}x}\sin\left(\frac{1}{2}\sqrt{3}\,x\right)x$$

$$+ \_C4\,e^{-\frac{1}{2}x}\cos\left(\frac{1}{2}\sqrt{3}\,x\right)x$$

f) We achieve the characteristic polynomial

$$P(\lambda) = 64\lambda^8 + 48\lambda^6 + 12\lambda^4 + \lambda^2 = \lambda^2\left(4\lambda^2 + 1\right)^2.$$

The roots of the characteristic equation $P(\lambda) = 0$ are:

- $\lambda_1 = 0$ a double root,
- $\lambda_{2,3} = \pm\frac{1}{2}i$ some triple roots.

Using (7.53), the general solution will be

$$y\left(x\right) = C_1 + C_2 x + \left(C_3 + C_4 x + C_5 x^2\right)\cos\frac{x}{2} + \left(C_6 + C_7 x + C_8 x^2\right)\sin\frac{x}{2}.$$

## 7.3.2 Non-homogeneous Linear Differential Equations with Constant Coefficients

**Definition 7.39** (see [25]). A differential equation of the form

$$a_0 y^{(n)} + a_1 y^{(n-1)} + \ldots + a_{n-1}y' + a_n y = f\left(x\right), \qquad (7.54)$$

where $a_0, a_1, \ldots, a_{n-1}, a_n$ are some real constants, $a_0 \neq 0$ and $f : C^{(0)}\left(I\right) \to \mathbb{R}$ is a continuous function on an interval $I \subseteq \mathbb{R}$ is called **non-homogeneous linear differential equation of order $n$, with constant coefficients.**

The general solution of this equation is the sum of the general solution of the associated homogeneous equation and a particular solution of the non-homogeneous equation; therefore:

$$y\left(x\right) = y_o\left(x\right) + y_p\left(x\right). \qquad (7.55)$$

In the case when $f$ is an arbitrary function, for the determination of a particular solution of the non-homogeneous equation is used:

A) the *method of variation of constants*, developed by Joseph Louis Lagrange;

B) the *method of the undetermined coefficients*.

### 7.3.2.1 The Method of Variation of Constants

The particular solution of the non-homogeneous equation can be found in the form (see [3]):

$$y_p\left(x\right) = C_1\left(x\right)y_1\left(x\right) + C_2\left(x\right)y_2\left(x\right) + \cdots + C_n\left(x\right)y_n\left(x\right), \qquad (7.56)$$

where $\{C_1'\left(x\right), C_2'\left(x\right), \ldots, C_n'\left(x\right)\}$ represents the solution of the non-homogeneous linear algebraic system, of $n$ equations, with $n$ unknowns:

$$\begin{cases} C_1'\left(x\right)y_1\left(x\right) + C_2'\left(x\right)y_2\left(x\right) + \cdots + C_n'\left(x\right)y_n\left(x\right) = 0 \\ C_1'\left(x\right)y_1'\left(x\right) + C_2'\left(x\right)y_2'\left(x\right) + \cdots + C_n'\left(x\right)y_n'\left(x\right) = 0 \\ \qquad\qquad\qquad \vdots \\ C_1'\left(x\right)y_1^{(n-2)}\left(x\right) + C_2'\left(x\right)y_2^{(n-2)}\left(x\right) + \cdots + C_n'\left(x\right)y_n^{(n-2)}\left(x\right) = 0 \\ C_1'\left(x\right)y_1^{(n-1)}\left(x\right) + C_2'\left(x\right)y_2^{(n-1)}\left(x\right) + \cdots + C_n'\left(x\right)y_n^{(n-1)}\left(x\right) = \frac{f(x)}{a_0}. \end{cases}$$
$$(7.57)$$

**Remark 7.40.** If the order of the nonhomogeneous differential equation is high, then the calculus for determining the particular solution become laborious, since the system which results by applying the method of variation of the constants has $n$ equations, and $n$ unknown functions.

**Example 7.41.** Determine the general solution of the following non-homogeneous linear differential equation with constant coefficients:

$$y'' + 3y' + 2y = \frac{1}{1 + e^x}.$$

**Solution.**
First of all, we determine the general solution of the associated homogeneous equation:

$$y'' + 3y' + 2y = 0.$$

We have:

$$P(\lambda) = \lambda^2 + 3\lambda + 2 = (\lambda + 1)(\lambda + 2).$$

The characteristic equation $P(\lambda) = 0$ has the roots $\lambda_1 = -1$, $\lambda_2 = -2$. We achieve

$$y_o(x) = C_1 e^{-x} + C_2 e^{-2x}.$$

We seek

$$y_p(x) = C_1(x) e^{-x} + C_2(x) e^{-2x},$$

where

$$\begin{cases} C_1'(x) e^{-x} + C_2'(x) e^{-2x} = 0 \\ -C_1'(x) e^{-x} - 2C_2'(x) e^{-2x} = \frac{1}{1+e^x}. \end{cases} \tag{7.58}$$

We notice that the determinant of the system is

$$\Delta = \begin{vmatrix} e^{-x} & e^{-2x} \\ -e^{-x} & -2e^{-2x} \end{vmatrix} = W[x_1, x_2] \neq 0.$$

If we add the two equations of the system (7.58) we deduce

$$-C_2'(x) e^{-2x} = \frac{1}{1 + e^x};$$

therefore

$$C_2'(x) = -\frac{e^{2x}}{1+e^x} = -\frac{e^{2x} + e^x - e^x}{1+e^x} = -\frac{e^x(e^x + 1) - e^x}{1+e^x} = -e^x + \frac{e^x}{1+e^x},$$

i.e.

$$C_2\left(x\right) = \int\left(-e^x + \frac{e^x}{1+e^x}\right)dx = -e^x + \ln\left(1+e^x\right).$$

We shall have

$$C_1'\left(x\right) = -C_2'\left(x\right)\frac{e^{-2x}}{e^{-x}} = \frac{e^{2x}}{1+e^x}\cdot e^{-2x}\cdot e^x = \frac{e^x}{1+e^x}$$

and

$$C_1\left(x\right) = \int\frac{e^x}{1+e^x}dx = \ln\left(1+e^x\right).$$

It results that

$$\begin{aligned}
y_p\left(x\right) &= e^{-x}\ln\left(1+e^x\right) + \left[-e^x + \ln\left(1+e^x\right)\right]e^{-2x}\\
&= \left(e^{-x} + e^{-2x}\right)\ln\left(1+e^x\right) - e^{-x}.
\end{aligned}$$

The general solution of the nonhomogeneous equation will be

$$y\left(x\right) = C_1e^{-x} + C_2e^{-2x} + \left(e^{-x} + e^{-2x}\right)\ln\left(1+e^x\right) - e^{-x}.$$

We shall solve this equation in Matlab 7.9:
>> y=dsolve('D2y+3*Dy+2*y=1/(1+exp(x))','x')
y =
exp(-2*x)*log(1+exp(x))+exp(-x)*log(1+exp(x))-exp(-2*x)
C1+exp(-x)*C2
and with Mathematica 8:

In[16]:= **Simplify[**
     **DSolve[D[y[x], {x, 2}] + 3 \* D[y[x], x] + 2 \* y[x] ==**
     **1 / (1 + Exp[x]), y[x], x]]**

Out[16]= $\left\{\left\{y[x] \rightarrow e^{-2x}\left(C[1] + e^x\left(-1 + C[2]\right) + \left(1 + e^x\right)\text{Log}\left[1+e^x\right]\right)\right\}\right\}$

and in Maple 15:

$$\text{ode} := \frac{d^2}{dx^2}y(x) + 3\cdot\frac{d}{dx}y(x) + 2\cdot y(x) = \frac{1}{1+e^x}:$$

dsolve(ode)

$$\begin{aligned}
y(x) &= e^{-x}\ln\left(1 + e^x\right) + \ln\left(1 + e^x\right)e^{-2x} - e^{-2x}\_C1\\
&\quad + e^{-x}\_C2
\end{aligned}$$

## 7.3.2.2    The Method of the Undetermined Coefficients

The physical problems contribute to some equations of the form (7.54), where $f(x)$ has a particular form and in such cases the particular solution of the nonhomogeneous equation can be determined by the *method of the undetermined coefficients*.

We can distinguish the following situations (see [25]):

*Situation 1.* The right side of the differential equation (7.54) has the form

$$f(x) = C = const.$$

a) If $\lambda = 0$ is not a root of the characteristic equation, then the differential equation (7.54) has a particular solution of the form

$$y_p(x) = \frac{C}{a_n}. \qquad (7.59)$$

b) If $\lambda = 0$ is a multiple root of the order $m$ of the characteristic equation, then the differential equation (7.54) has a particular solution of the form

$$y_p(x) = \frac{C \cdot x^m}{m! \cdot a_{n-m}}. \qquad (7.60)$$

*Situation 2.* The right side of the differential equation (7.54) has the form

$$f(x) = Ce^{\alpha x},$$

where $\alpha$ is a constant.

a) If $\lambda = \alpha$ is not a root of the characteristic equation, then the differential equation (7.54) has a particular solution of the form

$$y_p(x) = \frac{C \cdot e^{\alpha x}}{P(\alpha)}. \qquad (7.61)$$

b) If $\lambda = \alpha$ is a multiple root of the order $m$ of the characteristic equation, then the differential equation (7.54) has a particular solution of the form

$$y_p(x) = \frac{C \cdot x^m \cdot e^{\alpha x}}{P^{(m)}(\alpha)}. \qquad (7.62)$$

*Situation 3.* The right side of the differential equation (7.54) has the form

$$f(x) = P_m(x),$$

where $P_m(x)$ is a polynomial by the degree $m$.

a) If $\lambda = 0$ is not a root of the characteristic equation, then the differential equation (7.54) has a particular solution of the form

$$y_p(x) = Q_m(x), \qquad (7.63)$$

where $Q_m(x)$ is a polynomial of the same degree as $P_m(x)$, whose coefficients are determined by identifying, with the condition that $y_p(x)$ checks the non-homogeneous equation.

b) If $\lambda = 0$ is a multiple root of the order $r$ of the characteristic equation, then the differential equation (7.54) has a particular solution of the form

$$y_p(x) = x^r Q_m(x), \qquad (7.64)$$

where $Q_m(x)$ is a polynomial of the same degree as $P_m(x)$.

*Situation 4.* The right side of the differential equation (7.54) has the form

$$f(x) = e^{\alpha x} P_m(x).$$

a) If $\lambda = \alpha$ is not a root of the characteristic equation, then the differential equation (7.54) has a particular solution of the form

$$y_p(x) = e^{\alpha x} Q_m(x), \qquad (7.65)$$

where $Q_m(x)$ is a polynomial of the same degree as $P_m(x)$, whose coefficients are determined by identifying, with the condition that $y_p(x)$ from (7.65) checks the non-homogeneous equation.

b) If $\lambda = \alpha$ is a multiple root of the order $r$ of the characteristic equation, then the differential equation (7.54) has a particular solution of the form

$$y_p(x) = e^{\alpha x} x^r Q_m(x). \qquad (7.66)$$

*Situation 5.* The right side of the differential equation (7.54) has the form

$$f(x) = M \cos \beta x + N \sin \beta x.$$

a) If $\lambda = \beta i$ is not a root of the characteristic equation, then the differential equation (7.54) has a particular solution of the form

$$y_p(x) = A \cos \beta x + B \sin \beta x. \qquad (7.67)$$

b) If $\lambda = \beta i$ is a multiple root of the order $m$ of the characteristic equation, then the differential equation (7.54) has a particular solution of the form

$$y_p(x) = x^m (A \cos \beta x + B \sin \beta x). \qquad (7.68)$$

*Situation 6.* The right side of the differential equation (7.54) has the form

$$f(x) = e^{\alpha x} (P_m(x) \cos \beta x + Q_m(x) \sin \beta x).$$

a) If $\lambda = \alpha + \beta i$ is not a root of the characteristic equation, then the differential equation (7.54) has a particular solution of the form

$$y_p(x) = e^{\alpha x} (R_m(x) \cos \beta x + S_m(x) \sin \beta x). \qquad (7.69)$$

b) If $\lambda = \alpha + \beta i$ is a multiple root of the order $r$ of the characteristic equation, then the differential equation (7.54) has a particular solution of the form

$$y_p(x) = x^r e^{\alpha x} (R_m(x) \cos \beta x + S_m(x) \sin \beta x). \qquad (7.70)$$

*Situation 7.* The right side of the differential equation (7.54) has the form

$$f(x) = f_1(x) + \cdots + f_k(x),$$

with $f_i(x)$ of the form from the situations 1- 6.

In this case, the differential equation (7.54) has a particular solution of the form

$$y_p(x) = y_{p1}(x) + \cdots + y_{pk}(x), \qquad (7.71)$$

with $y_{pi}(x)$ corresponding to $f_i(x)$.

**Example 7.42.** Find the general solution of the non-homogeneous differential equations, with constant coefficients:

a) $y'' + y' = 3$

b) $y''' + 4y'' + 5y' = 1 + 4e^{-x}$

c) $y'' - 5y' + 6y = 6x^2 - 10x + 2$

d) $y^{(4)} + 2y'' + y = \sin x.$

**Solutions.**

a) The associated characteristic polynomial of the homogeneous equation is

$$P(\lambda) = \lambda^2 + \lambda.$$

The characteristic equation has the roots $\lambda_1 = 0$, $\lambda_2 = -1$.
The solution of the homogeneous equation is

$$y_o(x) = C_1 + C_2 e^{-x}.$$

Using (7.60), the particular solution of the nonhomogeneous equation is

$$y_p(x) = \frac{3 \cdot x^1}{1! \cdot a_{2-1}} = \frac{3x}{a_1} = 3x$$

and the general solution of the differential equation will be

$$y(x) = C_1 + C_2 e^{-x} + 3x.$$

b) The associated characteristic polynomial of the homogeneous equation is

$$P(\lambda) = \lambda^3 + 4\lambda^2 + 5\lambda = \lambda \left( \lambda^2 + 4\lambda + 5 \right).$$

The characteristic equation has the roots

$$\begin{cases} \lambda_1 = 0 \\ \lambda_2 = -2 + i \\ \overline{\lambda}_2 = -2 - i. \end{cases}$$

The solution of the homogeneous equation is

$$y_o(x) = C_1 + C_2 e^{-2x} \sin x + C_3 e^{-2x} \cos x.$$

The particular solution of the non-homogeneous equation is

$$y_p(x) = y_{p1}(x) + y_{p2}(x),$$

where

$$y_{p1}(x) = \frac{1 \cdot x^1}{1! \cdot a_{3-1}} = \frac{x}{a_2} = \frac{x}{5}$$

and

$$y_{p2}(x) = \frac{4 \cdot e^{-x}}{P(-1)} = \frac{4 \cdot e^{-x}}{-2} = -2 \cdot e^{-x};$$

therefore

$$y_p(x) = \frac{x}{5} - 2 \cdot e^{-x}.$$

The general solution of the differential equation will be

$$y(x) = C_1 + C_2 e^{-2x} \sin x + C_3 e^{-2x} \cos x + \frac{x}{5} - 2 \cdot e^{-x}.$$

c) The associated characteristic polynomial of the homogeneous equation is

$$P(\lambda) = \lambda^2 - 5\lambda + 6 = (\lambda - 2)(\lambda - 3).$$

The characteristic equation has the roots $\lambda_1 = 2$, $\lambda_2 = 3$.
The solution of the homogeneous equation is

$$y_o(x) = C_1 e^{2x} + C_2 e^{3x}.$$

The particular solution of the non-homogeneous equation is

$$y_p(x) = x^2 + ax + b.$$

We have

$$\begin{cases} y_p'(x) = 2x + a \\ y_p''(x) = 2. \end{cases}$$

We achieve

$$2 - 5(2x + a) + 6(x^2 + ax + b) = 6x^2 - 10x + 2;$$

hence

$$\begin{cases} -10 + 6a = -10 \Longrightarrow a = 0 \\ 2 - 5a + 6b = 2 \Longrightarrow b = 0 \end{cases}$$

i.e.

$$y_p(x) = x^2.$$

The general solution of the differential equation will be

$$y(x) = C_1 e^{2x} + C_2 e^{3x} + x^2.$$

We shall have in Matlab 7.9:
>> y=dsolve('D2y-5*Dy+6*y=6*x^2-10*x+2','x')
y =
x^2+ C1*exp(3*x)+C2*exp(2*x)
and with Mathematica 8:

In[2]:= DSolve[D[y[x], {x, 2}] - 5*D[y[x], x] + 6*y[x] ==
        6*x^2 - 10*x + 2, y[x], x]

Out[2]= $\left\{\left\{y[x] \to x^2 + e^{2x} C[1] + e^{3x} C[2]\right\}\right\}$

and in Maple 15:

$$\text{ode} := \frac{d^2}{dx^2} y(x) - 5 \cdot \frac{d}{dx} y(x) + 6 \cdot y(x) = 6 \cdot x^2 - 10 \cdot x + 2:$$

$$\text{dsolve(ode)}$$

$$y(x) = e^{3x} \_C2 + e^{2x} \_C1 + x^2$$

d) The associated characteristic equation of the homogeneous differential equation

$$y^{(4)} + 2y'' + y = 0$$

is

$$\left(\lambda^2 + 1\right)^2 = 0 \Leftrightarrow$$

$\lambda_1 = i$, $\overline{\lambda}_1 = -i$ are some double roots.
We obtain

$$y_o(x) = (C_1 + C_2 x) \cos x + (C_1 + C_2 x) \sin x.$$

We seek

$$y_p(x) = x^2 (A \cos x + B \sin x).$$

We shall have

$$\begin{cases} y_p'(x) = 2x\left(A\cos x + B\sin x\right) + x^2\left(-A\sin x + B\cos x\right) \\ y_p''(x) = 2\left(A\cos x + B\sin x\right) + 4x\left(-A\sin x + B\cos x\right) \\ \qquad + x^2\left(-A\cos x - B\sin x\right) \\ y_p'''(x) = 6\left(-A\sin x + B\cos x\right) + 6x\left(-A\cos x - B\sin x\right) \\ \qquad + x^2\left(A\sin x - B\cos x\right) \\ y_p^{(4)}(x) = 12\left(-A\cos x - B\sin x\right) + 8x\left(A\sin x - B\cos x\right) \\ \qquad + x^2\left(A\cos x + B\sin x\right). \end{cases}$$

By emphasizing the condition that the particular solution to check the non-homogeneous equation

$$y^{(4)} + 2y'' + y = \sin x$$

we deduce

$$12\left(-A\cos x - B\sin x\right) + 8x\left(A\sin x - B\cos x\right) + x^2\left(A\cos x + B\sin x\right)$$
$$+4\left(A\cos x + B\sin x\right) + 8x\left(-A\sin x + B\cos x\right) + 2x^2\left(-A\cos x - B\sin x\right)$$

$$+x^2\left(A\cos x + B\sin x\right) = \sin x;$$

hence

$$-8A\cos x - 8B\sin x = \sin x \Longrightarrow A = 0, B = -\frac{1}{8}.$$

It results that

$$y_p(x) = -\frac{x^2}{8}\sin x$$

and

$$y(x) = (C_1 + C_2 x)\cos x + (C_1 + C_2 x)\sin x - \frac{x^2}{8}\sin x.$$

We shall check this result in Matlab 7.9:
>> y=simplify(dsolve('D4y+2*D2y+y=sin(x)','x'))
y =
  C2*cos(x) - (x^2*sin(x))/8 + C4*sin(x) - (x*cos(x))/8 +
C5*x*sin(x) + C3*x*cos(x)
and using Mathematica 8:

In[15]:= DSolve[D[y[x], {x, 4}] + 2*D[y[x], {x, 2}] + y[x] == Sin[x],
    y[x], x];

In[16]:= Simplify[%]

Out[16]= $\left\{\left\{y[x] \rightarrow \left(C[1] + x\left(-\frac{1}{4} + C[2]\right)\right) \cos[x] + \right.\right.$

$\left.\left. \frac{1}{16}\left(1 - 2x^2 + 16 C[3] + 16 x C[4]\right) \sin[x]\right\}\right\}$

and with Maple 15:

$$\mathbf{ode} := \frac{d^4}{dx^4} y(x) + 2 \cdot \frac{d^2}{dx^2} y(x) + y(x) = \sin(x) :$$

$$simplify(dsolve(ode))$$

$$y(x) = -\frac{1}{4}\cos(x)x + \frac{1}{8}\sin(x) - \frac{1}{8}\sin(x)x^2 + \_C1\cos(x)$$

$$+ \_C2\sin(x) + \_C3\cos(x)x + \_C4\sin(x)x$$

## 7.3.3  Euler's Equation

**Definition 7.43** (see [15], p. 312). A non-homogeneous linear differential equation, of higer order, with variable coefficients, which may be reduced to an equation with constant coefficients, called Euler's equation:

$$a_n x^n y^{(n)} + a_{n-1} x^{n-1} y^{(n-1)} + \ldots + a_1 xy' + a_0 y = f(x), \qquad (7.72)$$

with $a_i \in \mathbb{R}, i = \overline{0, n}$, and $f$ is a continuous function.

In the equation (7.72) the derivatives are taken in relation to $x$.

Euler's equation is reduced to an equation with constant coefficients by changing the independent variable

$$x = e^t, \ x > 0 \qquad (7.73)$$

i.e.

$$t = \ln x.$$

It will results that:

$$y' = \frac{dy}{dx} = \frac{dy}{dt} \cdot \frac{dt}{dx} = \frac{1}{x}\frac{dy}{dt} = e^{-t}\frac{dy}{dt}, \qquad (7.74)$$

$$y'' = \frac{d^2y}{dx^2} = \frac{d}{dx}\left(\underbrace{\frac{dy}{dx}}_{F}\right) = \frac{dF}{dx} = \frac{dF}{dt}\cdot\frac{dt}{dx} = \frac{1}{x}\frac{dF}{dt}$$

$$= e^{-t}\frac{d}{dt}\left(e^{-t}\frac{dy}{dt}\right) = e^{-t}\left(-e^{-t}\frac{dy}{dt} + e^{-t}\frac{d^2y}{dt^2}\right),$$

i.e.

$$y'' = e^{-2t}\left(-\frac{dy}{dt} + \frac{d^2y}{dt^2}\right); \tag{7.75}$$

similarly,

$$y''' = \frac{d^3y}{dx^3} = \frac{d}{dx}\left(\underbrace{\frac{d^2y}{dx^2}}_{G}\right) = \frac{dG}{dx} = \frac{dG}{dt}\cdot\frac{dt}{dx} = \frac{1}{x}\frac{dG}{dt}$$

$$= e^{-t}\cdot\frac{d}{dt}\left(e^{-2t}\left(-\frac{dy}{dt} + \frac{d^2y}{dt^2}\right)\right) =$$

$$= e^{-t}\left[-2e^{-2t}\left(-\frac{dy}{dt} + \frac{d^2y}{dt^2}\right) + e^{-2t}\left(-\frac{d^2y}{dt^2} + \frac{d^3y}{dt^3}\right)\right]$$

$$= e^{-3t}\left(2\frac{dy}{dt} - 2\frac{d^2y}{dt^2} - \frac{d^2y}{dt^2} + \frac{d^3y}{dt^3}\right),$$

i.e.

$$y''' = e^{-3t}\left(2\frac{dy}{dt} - 3\frac{d^2y}{dt^2} + \frac{d^3y}{dt^3}\right). \tag{7.76}$$

**Example 7.44.** Find the general solution of the differential equation:

$$x^3y''' + 3x^2y' + xy = 1.$$

**Solution.**
Using the change of variable from (7.73) and the relations (7.74), (7.76) we have:

$$e^{3t}\cdot\left[e^{-2t}\left(-\frac{dy}{dt} + \frac{d^2y}{dt^2}\right)\right] + 3e^{2t}\cdot e^{-t}\frac{dy}{dt} + e^t y = 1$$

$$\Longleftrightarrow e^t\left(\frac{d^2y}{dt^2} - \frac{dy}{dt}\right) + 3e^t\frac{dy}{dt} + e^t y = 1$$

$$\Longleftrightarrow e^t\frac{d^2y}{dt^2} + 2e^t\frac{dy}{dt} + e^t y = 1,$$

i.e. it has resulted the non-homogeneous linear differential equation with constant coefficients:
$$\frac{d^2y}{dt^2} + 2\frac{dy}{dt} + y = e^{-t}.$$

Firstly, we shall solve the homogeneous linear differential equation with constant coefficients

$$\frac{d^2y}{dt^2} + 2\frac{dy}{dt} + y = 0.$$

We obtain the characteristic polynomial

$$P(\lambda) = \lambda^2 + 2\lambda + 1.$$

The roots of the characteristic equation $P(\lambda) = 0$ are: $\lambda_{1,2} = -1$. Using (7.47), the general solution will be:

$$y_o(t) = C_1 e^{-t} + C_2 t\, e^{-t},$$

i.e.

$$y_o(x) = \frac{C_1}{x} + \frac{C_2}{x}\ln|x|.$$

Using (7.62) we shall assume

$$y_p(t) = \frac{t^2 \cdot e^{-t}}{2},$$

i.e.

$$y_p(x) = \frac{\ln^2|x|}{2x};$$

hence the general solution of the Euler's equation is:

$$y(x) = \frac{C_1}{x} + \frac{C_2}{x}\ln|x| + \frac{\ln^2|x|}{2x}.$$

We can achieve this solution only in Matlab 7.9:
>> y=dsolve('x^3*D2y+3*x^2*Dy+x*y=1','x')
y =
C7/x + log(x)^2/(2*x) + (C8*log(x))/x

## 7.3.4 Homogeneous Systems of Differential Equations with Constant Coefficients

**Definition 7.45** (see [56]). A system of differential equations of the form

$$\begin{cases} y_1' = a_{11}y_1 + a_{12}y_2 + \cdots + a_{1n}y_n + f_1(x) \\ y_2' = a_{21}y_1 + a_{22}y_2 + \cdots + a_{2n}y_n + f_2(x) \\ \quad\vdots \\ y_n' = a_{n1}y_1 + a_{n2}y_2 + \cdots + a_{nn}y_n + f_n(x), \end{cases} \qquad (7.77)$$

where

- $y_k' = \frac{\mathrm{d}y_k}{\mathrm{d}x}$, $k = \overline{1,n}$,

- $a_{ij}, f_i \in C^{(0)}(I)$, $i,j = \overline{1,n}$, $I \subseteq \mathbb{R}$,

- $x_1, x_2, \ldots, x_n \in C^{(1)}(I)$ are some unknown functions

is called a **linear system of first order differential equations**.

The functions are called the **coefficients of the system**.

If $f_1 = f_2 = \ldots = f_n = 0$ on $I$, the system is called **homogeneous**, otherwise it is called **non-homogeneous**.

A homogeneous linear system, with constant coefficients can be written in the matriceal form:

$$Y' = A \cdot Y(x), \qquad (7.78)$$

where

$$\begin{cases} Y'(x) = \begin{pmatrix} y_1' \\ y_2' \\ \vdots \\ y_n' \end{pmatrix}, \\ Y(x) = \begin{pmatrix} y_1 \\ y_2 \\ \vdots \\ y_n \end{pmatrix}, \\ A = \begin{pmatrix} a_{11} & a_{12} & \cdots & a_{1n} \\ a_{21} & a_{22} & \cdots & a_{2n} \\ \vdots & \vdots & \vdots & \vdots \\ a_{n1} & a_{n2} & \cdots & a_{nn} \end{pmatrix}. \end{cases}$$

The system (7.78) can be solved using the following two methods:

- *method of characteristic equation,*
- *elimination method.*

## 7.3.5 Method of Characteristic Equation

**Definition 7.46** (see [56]). The equation

$$\det\left(A - \lambda I_n\right) = 0 \tag{7.79}$$

is called the **characteristic equation** of the system (7.78).

*Case 1.* If the matrix $A$ has some distinct eigenvalues, then of each eigenvalue it corresponds a eigenvector, of components $(A_1, A_2, \ldots, A_n)$.

To determine the general solution of the homogeneous system (7.78) one
proceeds as follows:

1. solve the equation (7.79);

2. achieve the eigenvalues $\lambda_i$, $i = \overline{1, n}$;

3. for each eigenvalue $\lambda = \lambda_i$, its corresponding eigenvector
   $(A_{1i}, A_{2i}, \ldots, A_{ni})$ one determines;

4. write the general solution of the homogeneous system (7.78):

$$
\begin{cases}
y_1\left(x\right) = A_{11}C_1 e^{\lambda_1 x} + A_{12}C_2 e^{\lambda_2 x} + \cdots + A_{1n}C_n e^{\lambda_n x} \\
y_2\left(x\right) = A_{21}C_1 e^{\lambda_1 x} + A_{22}C_2 e^{\lambda_2 x} + \cdots + A_{2n}C_n e^{\lambda_n x} \\
\qquad\qquad\qquad\vdots \\
y_n\left(x\right) = A_{n1}C_1 e^{\lambda_1 x} + A_{n2}C_2 e^{\lambda_2 x} + \cdots + A_{nn}C_n e^{\lambda_n x}.
\end{cases}
$$

*Case 2.* The case of the multiple eigenvalues.

We suppose that $\lambda = \lambda_0$ is an eigenvalue of multiplicity $m$ order of the
matrix $A$.

The solution $Y\left(x\right)$ will be:

$$
Y\left(x\right) = \left[
\begin{pmatrix} A_{11} \\ A_{21} \\ \vdots \\ A_{n1} \end{pmatrix} \cdot C_1 +
\begin{pmatrix} A_{12} \\ A_{22} \\ \vdots \\ A_{n2} \end{pmatrix} \cdot x C_2 + \cdots +
\begin{pmatrix} A_{1m} \\ A_{2m} \\ \vdots \\ A_{nm} \end{pmatrix} \cdot x^{m-1} C_m
\right] e^{\lambda_0 x}.
$$

$$\tag{7.80}$$

**Example 7.47.** Find the general solution of the system of differential
equations

$$
\begin{cases}
y_1' = -3y_1 - y_2 \\
y_2' = y_1 - y_2.
\end{cases}
$$

**Solution.**

We have

$$A = \begin{pmatrix} -3 & -1 \\ 1 & -1 \end{pmatrix},$$

$$P(\lambda) = \det(A - \lambda I_2) = \begin{vmatrix} -3 - \lambda & -1 \\ 1 & -1 - \lambda \end{vmatrix} = -(\lambda + 2)^2.$$

The equation $P(\lambda) = 0$ has $\lambda = -2$ as a double root.
Using (7.80), we propose a solution of the form

$$Y(x) = \begin{pmatrix} y_1(x) \\ y_2(x) \end{pmatrix} = \begin{pmatrix} A_{11} + A_{12}x \\ A_{21} + A_{22}x \end{pmatrix} e^{-2x};$$

therefore

$$\begin{cases} y_1(x) = (A_{11} + A_{12}x)\, e^{-2x} \\ y_2(x) = (A_{21} + A_{22}x)\, e^{-2x}. \end{cases} \tag{7.81}$$

We require that the solution from (7.81) to check the homogeneous system

$$\begin{cases} A_{12}e^{-2x} - 2(A_{11} + A_{12}x)\, e^{-2x} = -3(A_{11} + A_{12}x)\, e^{-2x} - (A_{21} + A_{22}x)\, e^{-2x} \\ A_{22}e^{-2x} - 2(A_{21} + A_{22}x)\, e^{-2x} = (A_{11} + A_{12}x)\, e^{-2x} - (A_{21} + A_{22}x)\, e^{-2x}. \end{cases}$$

It results that

$$\begin{cases} (A_{12} - 2A_{11})\, e^{-2x} - 2A_{12}x\, e^{-2x} = (-3A_{11} - A_{21})\, e^{-2x} - (3A_{21} + A_{22})\, xe^{-2x} \\ (A_{22} - 2A_{21})\, e^{-2x} - 2A_{22}x\, e^{-2x} = (A_{11} - A_{21})\, e^{-2x} + (A_{12} - A_{22})\, xe^{-2x} \end{cases}$$

i.e.

$$\begin{cases} A_{12} - 2A_{11} = -3A_{11} - A_{21} \\ 2A_{12} = 3A_{21} + A_{22} \\ A_{22} - 2A_{21} = A_{11} - A_{21} \\ -2A_{22} = A_{12} - A_{22} \end{cases} \Longleftrightarrow \begin{cases} A_{11} + A_{12} + A_{21} = 0 \\ -A_{12} - A_{22} = 0 \\ A_{11} + A_{21} - A_{22} = 0 \\ A_{12} + A_{22} = 0. \end{cases} \tag{7.82}$$

If we let $A_{11}, A_{12}$ arbitrary, then from (7.82) we obtain

$$\begin{cases} A_{21} = -A_{11} - A_{12} \\ A_{22} = -A_{12}. \end{cases} \tag{7.83}$$

From (7.81) and (7.83) we deduce

$$Y(x) = \begin{pmatrix} y_1(x) \\ y_2(x) \end{pmatrix} = \begin{pmatrix} A_{11} + A_{12}x \\ (-A_{11} - A_{12}) - A_{12}x \end{pmatrix} e^{-2x}$$

or

$$Y\left(x\right) = \begin{pmatrix} y_1\left(x\right) \\ y_2\left(x\right) \end{pmatrix} = \begin{pmatrix} e^{-2x} \\ -e^{-2x} \end{pmatrix} A_{11} + \begin{pmatrix} xe^{-2x} \\ \left(-1-x\right)e^{-2x} \end{pmatrix} A_{12}.$$

Solving this system, we shall have in Matlab 7.9:
>> [y1,y2]=dsolve('Dy1=-3*y1-y2','Dy2=y1-y2','x')
y1 =
C9/exp(2*x) - C10/exp(2*x) - (C9*x)/exp(2*x)
y2 =
C10/exp(2*x) + (C9*x)/exp(2*x)
and in Mathematica 8:

In[3]:= DSolve[{D[y1[x], x] == -3*y1[x] - y2[x],
        D[y2[x], x] == y1[x] - y2[x]}, {y1[x], y2[x]}, x];

In[4]:= Simplify[%]

Out[4]= {{y1[x] → e^{-2x} (C[1] - x C[1] - x C[2]),
        y2[x] → e^{-2x} (C[2] + x (C[1] + C[2]))}}

and with Maple 15:

$$sys1 := \left[ \frac{d}{dx} y1(x) = -3\,y1(x) - y2(x),\ \frac{d}{dx} y2(x) = y1(x) - y2(x) \right] :$$
$$sol1 := dsolve(sys1)$$
$$\{y1(x) = e^{-2x}(\_C1 + \_C2\,x), y2(x) = -e^{-2x}(\_C1 + \_C2\,x \tag{1}$$
$$+ \_C2)\}$$

## 7.3.6 Elimination Method

Elimination method consists in reducing the system of the differential equations to a single linear differential equation of order $n$, for one of the unknown functions of the system and then solving this equation.

In the case of a system of the differential equations of the form

$$\begin{cases} y_1' = a_{11}y_1 + a_{12}y_2 \\ y_2' = a_{21}y_1 + a_{22}y_2, \end{cases}$$

the elimination method involves the following steps (see [3] and [56]):

1. We compute

$$y_1'' = a_{11}y_1' + a_{12}y_2' = a_{11}y_1' + a_{12}\left(a_{21}y_1 + a_{22}y_2\right)$$
$$= a_{11}y_1' + a_{12}a_{21}y_1 + a_{12}a_{22}y_2$$
$$= a_{11}y_1' + a_{12}a_{21}y_1 + a_{12}a_{22} \cdot \frac{y_1' - a_{11}y_1}{a_{12}}$$
$$= \left(a_{11} + a_{22}\right)y_1' + \left(a_{12}a_{21} - a_{11}a_{22}\right)y_1.$$

2. We solve the differential equation

$$y_1'' - \left(a_{11} + a_{22}\right)y_1' - \left(a_{12}a_{21} - a_{11}a_{22}\right)y_1 = 0$$

whose solution is $y_1\left(x\right)$.

3. We substitute $y_1\left(x\right)$ in the second equation of the system to determine $y_2\left(x\right)$.

**Example 7.48.** Use the elimination method to determine the general solution for the following system of differential equations

$$\begin{cases} y_1' = 3y_2 - 4y_3 \\ y_2' = -y_3 \\ y_3' = -2y_1 + y_2. \end{cases} \tag{7.84}$$

It results that the homogeneous linear differential equation with constant coefficients:

$$y_1''' = 8y_1' - 4y_2' - 3y_3' = 8y_1' + 4y_3 + 6y_1 - 3y_2$$
$$= 8y_1' + 6y_1 - y_1' = 7y_1' + 6y_1,$$

i.e.

$$y_1''' - 7y_1' - 6y_1 = 0.$$

We shall achieve the characteristic polynomial

$$P_1\left(\lambda\right) = \lambda^3 - 7\lambda - 6 = \left(\lambda + 1\right)\left(\lambda + 2\right)\left(\lambda - 3\right).$$

The roots of the characteristic equation $P_1\left(\lambda\right) = 0$ are:

$$\begin{cases} \lambda_1 = -1 \\ \lambda_2 = -2 \\ \lambda_3 = 3. \end{cases}$$

It results that:

$$y_1\left(x\right) = C_1 e^{-x} + C_2 e^{-2x} + C_3 e^{3x}. \tag{7.85}$$

We have

$$y_3'' = -2y_1' + y_2' = -2y_1' - y_3,$$

i.e.

$$y_3'' + y_3 = -2y_1' = -2\left(-C_1 e^{-x} - 2C_2\, e^{-2x} + 3C_3\, e^{3x}\right).$$

We solve the non-homogeneous differential equation with constant coefficients:

$$y_3'' + y_3 = 2C_1 e^{-x} + 4C_2\, e^{-2x} - 6C_3\, e^{3x}.$$

First of all we determine the general solution of the associated homogeneous equation:

$$y_3'' + y_3 = 0.$$

We have:

$$P_3(\lambda) = \lambda^2 + 1.$$

The characteristic equation $P_3(\lambda) = 0$ has the roots:

$$\begin{cases} \lambda_1 = i \\ \overline{\lambda}_1 = i. \end{cases}$$

We obtain

$$y_{3o}(x) = C_4 \cos x + C_5 \sin x.$$

We have

- $P_3(-1) = 2 \neq 0,$
- $P_3(-2) = 5 \neq 0,$
- $P_3(3) = 10 \neq 0.$

Therefore

$$y_{3p}(x) = \frac{2C_1 e^{-x}}{P_3(-1)} + \frac{4C_2 e^{-2x}}{P_3(-2)} - \frac{6C_3 e^{3x}}{P_3(3)}$$

$$= C_1 e^{-x} + \frac{4}{5}C_2 e^{-2x} - \frac{3}{5}C_3 e^{3x}.$$

It results that

$$y_3(x) = C_4 \cos x + C_5 \sin x + C_1 e^{-x} + \frac{4}{5}C_2 e^{-2x} - \frac{3}{5}C_3 e^{3x}. \qquad (7.86)$$

We have

$$y_2' = -y_3 = -C_4 \cos x - C_5 \sin x - C_1 e^{-x} - \frac{4}{5} C_2 e^{-2x} + \frac{3}{5} C_3 e^{3x},$$

i.e.

$$y_2(x) = -C_4 \sin x + C_5 \cos x + C_1 e^{-x} + \frac{2}{5} C_2 e^{-2x} + \frac{1}{5} C_3 e^{3x} + C_6. \quad (7.87)$$

Substituting (7.85), (7.86) and (7.87) into (7.84) we shall achieve

$$\begin{cases} -3C_4 - 4C_5 = 0 \\ 3C_5 - 4C_4 = 0 \Longrightarrow C_4 = C_5 = C_6 = 0 \\ 3C_6 = 0. \end{cases}$$

Denoting by

$$\begin{cases} C_1 = K_1 \\ \frac{1}{5}C_2 = K_2 \\ \frac{1}{5}C_3 = K_3 \end{cases}$$

the general solution of the system (7.84) will be

$$\begin{cases} y_1(x) = K_1 e^{-x} + 5K_2 e^{-2x} + 5K_3 e^{3x} \\ y_2(x) = K_1 e^{-x} + 2K_2 e^{-2x} + K_3 e^{3x} \\ y_3(x) = K_1 e^{-x} + 4K_2 e^{-2x} - 3K_3 e^{3x}. \end{cases}$$

We can also obtain this solution in Matlab 7.9:
```
>> [y1,y2,y3]=dsolve('Dy1=3*y2-4*y3','Dy2=-y3','
   Dy3=-2*y1+y2','x')
y1 =
(5*C11)/(4*exp(2*x)) - (5*C12*exp(3*x))/3 + C13/exp(x)
y2 =
C11/(2*exp(2*x)) - (C12*exp(3*x))/3 + C13/exp(x)
y3 =
C11/exp(2*x) + C12*exp(3*x) + C13/exp(x)
```
and using Mathematica 8:

In[2]:= `DSolve[{D[y1[x], x] == 3*y2[x] - 4*y3[x], D[y2[x], x] == -y3[x],`
`D[y3[x], x] == -2*y1[x] + y2[x]}, {y1[x], y2[x], y3[x]}, x];`

In[3]:= `Simplify[%]`

Out[3]= $\left\{\left\{y1[x] \rightarrow \frac{1}{4} e^{-2x} \left(2 \left(2 - e^x + e^{5x}\right) C[1] + \right.\right.\right.$

$\left(-8 + 7 e^x + e^{5x}\right) C[2] - \left(-4 + e^x + 3 e^{5x}\right) C[3]\right),$

$y2[x] \rightarrow \frac{1}{20} e^{-2x} \left(2 \left(4 - 5 e^x + e^{5x}\right) C[1] + \right.$

$\left(-16 + 35 e^x + e^{5x}\right) C[2] - \left(-8 + 5 e^x + 3 e^{5x}\right) C[3]\right),$

$y3[x] \rightarrow \frac{1}{20} e^{-2x} \left(-2 \left(-8 + 5 e^x + 3 e^{5x}\right) C[1] + \right.$

$\left.\left.\left. \left(-32 + 35 e^x - 3 e^{5x}\right) C[2] + \left(16 - 5 e^x + 9 e^{5x}\right) C[3]\right)\right\}\right\}$

and with Maple 15:

$$sys1 := \left[ \frac{d}{dx}y1(x) = 3y2(x) - 4 \cdot y3(x), \frac{d}{dx}y2(x) = -y3(x), \frac{d}{dx}y3(x) \right.$$

$$\left. = -2y1(x) + y2(x) \right]:$$

$sol1 := dsolve(sys1)$

$$\left\{ y1(x) = \_C1\,e^{-x} + 5\_C2\,e^{3x} + \frac{5}{2}\_C3\,e^{-2x}, y2(x) = \_C1\,e^{-x} \right. \quad (1)$$

$$+ \_C2\,e^{3x} + \_C3\,e^{-2x}, y3(x) = \_C1\,e^{-x} - 3\_C2\,e^{3x}$$

$$\left. + 2\_C3\,e^{-2x} \right\}$$

## 7.4 Non-homogeneous Systems of Differential Equations with Constant Coefficients

**Theorem 7.49** (see[3] and [56]). The general solution of the non-homogeneous system (7.77) is the sum of the general solution of the homogeneous system and a particular solution of the non-homogeneous system:

$$Y(x) = Y^o(x) + Y^p(x) = \begin{pmatrix} y_1^o(x) \\ y_2^o(x) \\ \vdots \\ y_n^o(x) \end{pmatrix} + \begin{pmatrix} y_1^p(x) \\ y_2^p(x) \\ \vdots \\ y_n^p(x) \end{pmatrix}. \tag{7.88}$$

A particular solution of the non-homogeneous system can be determined using the method of variation of constants.

We search the particular solution of the form:

$$\begin{cases} y_1^p(x) = K_1(x) y_{11} + K_2(x) y_{12} + \cdots + K_n(x) y_{1n} \\ y_2^p(x) = K_1(x) y_{21} + K_2(x) y_{22} + \cdots + K_n(x) y_{2n} \\ \qquad\qquad \vdots \\ y_n^p(x) = K_1(x) y_{n1} + K_2(x) y_{n2} + \cdots + K_n(x) y_{nn}, \end{cases} \tag{7.89}$$

where the functions $K_i'(x)$, $i = \overline{1, n}$ are determined from system:

$$\begin{cases} K_1'(x) y_{11} + K_2'(x) y_{12} + \cdots + K_n'(x) y_{1n} = f_1(x) \\ K_1'(x) y_{21} + K_2'(x) y_{22} + \cdots + K_n'(x) y_{2n} = f_2(x) \\ \qquad\qquad \vdots \\ K_1'(x) y_{n1} + K_2'(x) y_{n2} + \cdots + K_n'(x) y_{nn} = f_n(x). \end{cases}$$

**Example 7.50.** Solve the system of non-homogeneous linear differential equations with constant coefficients:

$$\begin{cases} y_1' = y_2 \\ y_2' = y_1 + e^x + e^{-x}. \end{cases}$$

**Solution.**

The general solution of the associated homogeneous system

$$\begin{cases} y_1' = y_2 \\ y_2' = y_1 \end{cases}$$

will be of the form

$$\begin{cases} y_1^o(x) = K_1 e^x + K_2 e^{-x} \\ y_2^o(x) = K_1 e^x - K_2 e^{-x}. \end{cases}$$

We seek the particular solution non-homogeneous system of the form

$$\begin{cases} y_1^p(x) = K_1(x) e^x + K_2(x) e^{-x} \\ y_2^p(x) = K_1(x) e^x - K_2(x) e^{-x}, \end{cases}$$

in which the functions $K_1(x)$ and $K_2(x)$ check the equations

$$\begin{cases} K_1'e^x + K_2'e^{-x} = 0 \\ K_1'e^x - K_2'e^{-x} = e^x + e^{-x}. \end{cases}$$

We obtain

$$\begin{cases} K_1' = \frac{1}{2} + \frac{1}{2}e^{-2x} \\ K_2' = -\frac{1}{2} - \frac{1}{2}e^{2x}; \end{cases}$$

hence

$$\begin{cases} K_1 = \frac{x}{2} - \frac{1}{4}e^{-2x} \\ K_2 = -\frac{x}{2} - \frac{1}{4}e^{2x} \end{cases}$$

and

$$\begin{cases} y_1^p(x) = \frac{1}{2}e^x \left(x - \frac{1}{2}\right) - \frac{1}{2}e^{-x}\left(x + \frac{1}{2}\right) \\ y_2^p(x) = \frac{1}{2}e^x \left(x + \frac{1}{2}\right) + \frac{1}{2}e^{-x}\left(x - \frac{1}{2}\right). \end{cases}$$

The general solution of the system will be

$$\begin{cases} y_1(x) = K_1e^x + K_2e^{-x} + \frac{1}{2}e^x \left(x - \frac{1}{2}\right) - \frac{1}{2}e^{-x}\left(x + \frac{1}{2}\right) \\ y_2(x) = K_1e^x - K_2e^{-x} + \frac{1}{2}e^x \left(x + \frac{1}{2}\right) + \frac{1}{2}e^{-x}\left(x - \frac{1}{2}\right). \end{cases}$$

We shall obtain in Matlab 7.9:

```
>> [y1,y2]=dsolve('Dy1=y2','Dy2=y1+exp(x)+exp(-x)','x')
y1 =
C14*exp(x) - exp(x)/4 - x/(2*exp(x)) - 1/(4*exp(x)) +
(x*exp(x))/2 - C15/exp(x)
y2 =
exp(x)/4 - 1/(4*exp(x)) + x/(2*exp(x)) + C14*exp(x) +
(x*exp(x))/2 + C15/exp(x)
```

and using Mathematica 8:

```
In[5]:= DSolve[{y1'[x] == y2[x],
            y2'[x] == y1[x] + Exp[x] + Exp[-x]}, {y1[x], y2[x]},
        x];

In[6]:= Simplify[%]
```

$$\text{Out[6]= } \left\{\left\{y1[x] \to \frac{1}{4} e^{-x} \left(-1 - 2x + 2C[1] - \right.\right.\right.$$
$$\left. 2C[2] + e^{2x} \left(-1 + 2x + 2C[1] + 2C[2]\right)\right),$$
$$y2[x] \to \frac{1}{4} e^{-x} \left(-1 + 2x - 2C[1] + 2C[2] + \right.$$
$$\left.\left.\left. e^{2x} \left(1 + 2x + 2C[1] + 2C[2]\right)\right)\right\}\right\}$$

and with Maple 15:

$$sys1 := \left[ \frac{d}{dx} y1(x) = y2(x), \frac{d}{dx} y2(x) = y1(x) + e^x + e^{-x} \right]:$$

$$sol1 := dsolve(sys1)$$

$$\left\{ y1(x) = \sinh(x)\_C2 + \cosh(x)\_C1 - \frac{1}{2}\cosh(x) \right.$$

$$+ \sinh(x) x, y2(x) = \cosh(x)\_C2 + \sinh(x)\_C1$$

$$\left. + \frac{1}{2}\sinh(x) + \cosh(x) x \right\}$$

## 7.5  Problems

1. Solve the Cauchy problem:

$$\begin{cases} y' = 2x\,(1+y) \\ \quad y\,(0) = 0 \end{cases}$$

using the successive approximation method.
   **Computer solution.**
   We shall give a computer solution in Matlab 7.9:
   **function w=f(u,v)**
   **w= 2*u*(1+v);**
   **end**
   **function r=y(n,t,x,y0)**
   **r=y0;**
   **for k=1:n**
   **r1=int(f(t,r),t,0,x);**
   **r=subs(r1,x,t)+y0;**
   **end**
   **end**
   In the command line we shall write:
   **>>syms x t**
   **>> y1=subs(y(1,t,x,0),t,x)**
   **y1 =**
   **x^2**
   **>> y2=subs(y(2,t,x,0),t,x)**
   **y2 =**
   **x^4/2+x^2**
   **>> y3=subs(y(3,t,x,0),t,x)**

y3 =
x^6/6+x^4/2+x^2
and in Mathcad 14:

$f(x,y) := 2 \cdot x \cdot (1 + y)$

$$y(n,t,x,y0) := \begin{vmatrix} r \leftarrow y0 \\ \text{for } k \in 1..n \\ \begin{vmatrix} x \leftarrow t \\ r1 \leftarrow \int_0^x f(t,r)\,dt \\ r \leftarrow r1 + y0 \end{vmatrix} \\ r \end{vmatrix}$$

$y(1,t,x,0)$ substitute, $t = x \;\rightarrow\; x^2$

$y(2,t,x,0)$ substitute, $t = x \;\rightarrow\; \dfrac{x^2 \cdot (x^2 + 2)}{2}$ expand $\;\rightarrow\; \dfrac{x^4}{2} + x^2$

$y(3,t,x,0)$ substitute, $t = x \;\rightarrow\; \dfrac{x^2 \cdot (x^4 + 3 \cdot x^2 + 6)}{6}$ expand $\;\rightarrow\; \dfrac{x^6}{6} + \dfrac{x^4}{2} + x^2$

and using Mathematica 8:

```
In[1]:= f[u_, v_] := 2*u*(1 + v)

In[2]:= y[n_, t_, x_, y0_] :=
        Module[{r = y0},
         For[k = 1, k <= n, k++, x = t; r1 = Integrate[f[t, r], {t, 0, x}]; r = y0 + r1];
         r
        ]

In[3]:= y[1, x, x, 0]

Out[3]= x^2

In[4]:= y[2, x, x, 0]

Out[4]= x^2 + x^4/2

In[5]:= y[3, x, x, 0]

Out[5]= x^2 + x^4/2 + x^6/6
```

and in Maple 15:

$$f := (x, y) \rightarrow 2 \cdot x \cdot (1 + y) :$$
$$y := \mathbf{proc}(n, t, x, y0)$$
$$\mathbf{local}\, r, r1, k;$$
$$r := y0;$$
$$\mathbf{for}\, k\, \mathbf{from}\, 1\, \mathbf{to}\, n\, \mathbf{do}$$

$$r1 := \int_0^x f(t, r)\, dt;$$

$$r := y0 + r1\Big|_{x=t}$$

$$\mathbf{end\, do};$$
$$\mathbf{end\, proc};$$
$$y(1, t, x, 0)\Big|_{t=x}$$

$$x^2$$

$$y(2, t, x, 0)\Big|_{t=x}$$

$$\frac{1}{2}x^4 + x^2$$

$$y(3, t, x, 0)\Big|_{t=x}$$

$$\frac{1}{6}x^6 + \frac{1}{2}x^4 + x^2$$

2  Find the solution of the equation:

$$\left(1 - x^2 y\right) dx + x^2 \left(y - x\right) dy = 0.$$

**Computer solution.**
We can solve this equation in Matlab 7.9:
*Step 1.* We check if the equation is an exact differential equation.
>> **syms x y t y0 x0**
>> **f=1-x^2*y;**
>> **g=x^2*(y-x);**
>> **d1=diff(f,y);**
 >> **d2=diff(g,x);**
>> **d1==d2**
ans =
 0
*Step 2.* Since the equation is not an exact total differential equation, we
must determine the integrating factor (7.24):
>> **phi=simple((d1-d2)/g)**
phi =

**-2/x**
>> miu=exp(int(phi,x))
miu =
1/x^2

*Step 3.* We can apply the formula (7.22) in order to determine the solution of the equation.

>>Phi=expand(int(subs(f*miu,{x,y},{t,y0}),t,x0,x)+
int(subs(g*miu,y,t),t,y0,y))
Phi =
piecewise([0 <= x and x0 <= 0, Inf - x*y + x0*y0 + y^2/2 - y0^2/2], [x < 0 or 0 < x0, x0*y0 - x*y - 1/x + 1/x0 + y^2/2 - y0^2/2])

We shall also solve this problem using Mathcad 14:

$$f(x,y) := 1 - x^2 \cdot y \qquad\qquad g(x,y) := x^2 \cdot (y - x)$$

$$\frac{d}{dy} f(x,y) = \frac{d}{dx} g(x,y) \rightarrow -x^2 = -2 \cdot x \cdot (x - y) - x^2$$

$$\mu(x) := e^{\displaystyle\int \frac{\frac{d}{dy}f(x,y) - \frac{d}{dx}g(x,y)}{g(x,y)} \, dx} \qquad\qquad \text{simplify} \rightarrow \frac{1}{x^2}$$

$$f1(x,y) := f(x,y) \cdot \mu(x) \qquad\qquad g1(x,y) := g(x,y) \cdot \mu(x)$$

$$\phi(x,y,x0,y0) := \int_{x0}^{x} f1(t,y0) \, dt + \int_{y0}^{y} g1(x,t) \, dt$$

and with Mathematica 8:

In[122]:= `f[x_, y_] := 1 - x^2 * y`

In[123]:= `g[x_, y_] := x^2 * (y - x)`

In[124]:= `D[f[x, y], y] == D[g[x, y], x]`

Out[124]= $-x^2 == -x^2 + 2x(-x + y)$

In[125]:= `φ := (D[f[x, y], y] - D[g[x, y], x]) / g[x, y]`

In[126]:= `μ[x_] := Simplify[Exp[Integrate[φ, x]]]`

In[127]:= `f1[x_, y_] = f[x, y] * μ[x];`

In[128]:= `g1[x_, y_] = g[x, y] * μ[x];`

In[129]:= `φ = Integrate[f1[t, y0], {t, x0, x}] + Integrate[g1[x, t], {t, y0, y}]`

Out[129]= $\text{ConditionalExpression}\left[-\dfrac{1}{x} + \dfrac{1}{x0} - xy + \dfrac{y^2}{2} + x0\,y0 - \dfrac{y0^2}{2},\right.$

$\left.\left(x\,x0 \neq x0^2 \,\&\&\, \text{Re}\left[\dfrac{x0}{x - x0}\right] \geq 0\right) \,||\, \text{Re}\left[\dfrac{x0}{x - x0}\right] \leq -1 \,||\, \dfrac{x0}{x - x0} \in \text{Reals}\right]$

and in Maple 15:

$f := (x, y) \rightarrow 1 - x^2 \cdot y :$
$g := (x, y) \rightarrow x^2 \cdot (y - x) :$
$evalb\left(\dfrac{\partial}{\partial y} f(x, y) = \dfrac{\partial}{\partial x} g(x, y)\right)$

$\qquad\qquad\qquad\qquad\qquad\qquad\qquad\qquad\qquad false$

$\mu := (x) \rightarrow e^{\displaystyle \int \frac{\frac{\partial}{\partial y} f(x,y) - \frac{\partial}{\partial x} g(x,y)}{g(x,y)} \, dx} \quad :$

$f1 := (x, y) \rightarrow f(x, y) \cdot \mu(x) :$
$g1 := (x, y) \rightarrow g(x, y) \cdot \mu(x) :$
$assume(0 < x0 < x)$

$\Phi := \displaystyle\int_{x0}^{x} f1(t, y0) \, dt + \int_{y0}^{y} g1(x, t) \, dt$

$\qquad\qquad \dfrac{x\text{\textasciitilde} + x\text{\textasciitilde} y0\,x0\text{\textasciitilde}^2 - x0\text{\textasciitilde} - x0\text{\textasciitilde} y0\,x\text{\textasciitilde}^2}{x0\text{\textasciitilde}\,x\text{\textasciitilde}} + \dfrac{1}{2}y^2 - \dfrac{1}{2}y0^2 - x\text{\textasciitilde} \cdot (y - y0)$

3. Solve the equation

$$y' = \frac{y\left(x^3 + 1\right)}{x} \cdot \left(1 - y^2\right).$$

**Computer solution.**
It will result in Matlab 7.9:
```
>> y=dsolve('Dy=y*(x^3+1)/x*(1-y^2)','x')
y =
0
1
-1
1/(x^2+exp(-2/3*x^3)*C1)^(1/2)*x
-1/(x^2+exp(-2/3*x^3)*C1)^(1/2)*x
```
and with Mathematica 8:

In[20]:= DSolve[D[y[x], x] == y[x] * (x^3 + 1) / x * (1 - y[x]^2), y[x], x]

Out[20]= $\left\{\left\{y[x] \to -\dfrac{e^{\frac{x^3}{3}} x}{\sqrt{e^{2C[1]} + e^{\frac{2x^3}{3}} x^2}}\right\}, \left\{y[x] \to \dfrac{e^{\frac{x^3}{3}} x}{\sqrt{e^{2C[1]} + e^{\frac{2x^3}{3}} x^2}}\right\}\right\}$

and using Maple 15:

$$\text{ode} := \text{diff}(y(x), x) = \frac{y(x) \cdot \left(x^3 + 1\right) \cdot \left(1 - y(x)^2\right)}{x}$$

$$\frac{d}{dx} y(x) = \frac{y(x)\left(x^3 + 1\right)\left(1 - y(x)^2\right)}{x}$$

*dsolve(ode)*

$$y(x) = \frac{x}{\sqrt{x^2 + e^{-\frac{2}{3}x^3}\_C1}}, y(x) = -\frac{x}{\sqrt{x^2 + e^{-\frac{2}{3}x^3}\_C1}}$$

4. Integrate the equations:

$$\text{a) } y' = -\frac{2xy}{x^2 + y^2}$$

$$\text{b) } y' = \frac{2\left(x + y - 1\right)^2 + 3x\left(2x - y + 1\right)}{\left(x + y - 1\right)^2 - 3x\left(2x - y + 1\right)}$$

$$c)\ xy' - y = \sqrt{x^2 + y^2}.$$

5. Solve the equations:

$$a)\ y' + 4xy = xe^{-x^2}$$

$$b)\ y' - y\tan x = \cos x.$$

**Computer solution.**

b)We shall find the solution of this non-homogeneous equation both in Matlab 7.9:

>> y=dsolve('Dy-y*tan(x)=cos(x)','x')

y =

(x/2 + sin(2*x)/4)/cos(x) + C2/cos(x)

and using Mathematica 8:

In[8]:= DSolve[D[y[x], x] - y[x] * Tan[x] == Cos[x], y[x], x]

Out[8]= $\left\{\left\{y[x] \rightarrow C[1]\ \text{Sec}[x] + \text{Sec}[x]\left(\dfrac{x}{2} + \dfrac{1}{4}\ \text{Sin}[2\,x]\right)\right\}\right\}$

and with Maple 15:

ode := diff(y(x), x) − y(x)·tan(x) = cos(x) :

dsolve(ode)

$$y(x) = \frac{\dfrac{1}{4}\sin(2x) + \dfrac{1}{2}x + \_C1}{\cos(x)}$$

6. Find the general solution of the Bernoulli's equation:

$$a)\ y' - 3xy = xy^2$$

$$b)\ y' + \frac{y}{x} = \frac{1}{x^2 y^2}, x > 0, y \neq 0.$$

7. Solve the Riccati's equation:

$$a)\ y' = y^2 - x^2 + 1$$

$$b)\ y' = \frac{x}{2}y^2 - \frac{2}{x}y - \frac{1}{2x^3}.$$

8. Determine the types of differential equations and indicate methods for their solutions:

$$a)\ y = xy' + \frac{1}{y'^2}$$

$$b)\ y = xy' + \cos y'.$$

9. Find solutions to the equations:

$$a)\ y = xy' + y'^3$$

$$b)\ yy'^2 - 2xy' + y = 0.$$

**Computer solution.**
b) We shall solve this Lagrange equation with Matlab 7.9:
>> y=dsolve('y*Dy^2-2*x*Dy+y','x')
y =
0
x
-x
2*(-C7)^(1/2)*(C7 - x)^(1/2)
(-2)*(-C7)^(1/2)*(C7 - x)^(1/2)

and using Mathematica 8:

In[1]:= DSolve[y[x]*D[y[x]]^2 - 2*x*D[y[x]] + y[x] == 0, y[x], x]

Out[1]= $\left\{\{y[x] \rightarrow 0\}, \left\{y[x] \rightarrow -\sqrt{-1+2x}\right\}, \left\{y[x] \rightarrow \sqrt{-1+2x}\right\}\right\}$

and with Maple 15:

$ode := y(x) \cdot (\text{diff}(y(x), x))^2 - 2 \cdot x \cdot \text{diff}(y(x), x) + y(x) = 0:$
$dsolve(ode)$

$y(x) = x, y(x) = -x, y(x) = \sqrt{\_C1^2 - 2\,Ix\_C1}, y(x)$
$\quad = \sqrt{\_C1^2 + 2\,Ix\_C1}, y(x) = -\sqrt{\_C1^2 - 2\,Ix\_C1}, y(x) =$
$\quad -\sqrt{\_C1^2 + 2\,Ix\_C1}$

10. Solve the equation: $x - y = \dfrac{1+y'}{\sqrt{1+y'^2}}.$

11. Solve the following equations:

$$\text{a) } y'' + y' + y = 0$$

$$\text{b) } y^{(4)} - 3y''' + 5y'' - 3y' + 4y = 0.$$

12. Find the general solution of the equations:

$$\text{a) } y'' + y' - 6y = 2\cos 2x - 10\sin 2x$$

$$\text{b) } y^{(5)} + 4y''' = x^2 e^x.$$

**Computer solution.**
b) We shall achieve in Matlab 7.9:
>> y=dsolve('D5y+4*D3y=x^2*exp(x)','x');
>> simplify(y)
ans =
   C7/16 - C9/4 + (358*exp(x))/125 + (x^2*exp(x))/5 - (C8*x)/4
- (C7*x^2)/8 - (34*x*exp(x))/25 + C10*cos(2*x) + C11*sin(2*x)
and with Mathematica 8:

In[8]:= DSolve[D[y[x], {x, 5}] + 4*D[y[x], {x, 3}] == x^2*Exp[x],
         y[x], x]

Out[8]= $\left\{\left\{y[x] \rightarrow \dfrac{1}{125}\, e^x \left(358 - 170\, x + 25\, x^2\right) - \dfrac{1}{4}\, x\, C[1] + C[3] + \right.\right.$
         $\left.\left. x\, C[4] + x^2\, C[5] + \dfrac{1}{8}\, C[2]\, \text{Cos}[2\, x] - \dfrac{1}{8}\, C[1]\, \text{Sin}[2\, x]\right\}\right\}$

and using Maple 15:

$$ode := \dfrac{d^5}{dx^5} y(x) + 4 \cdot \dfrac{d^3}{dx^3} y(x) = x^2 \cdot e^x :$$
$$dsolve(ode)$$
$$y(x) = \dfrac{1}{8}\cos(2x)\_C2 - \dfrac{1}{8}\sin(2x)\_C1 + \dfrac{358}{125}e^x$$
$$- \dfrac{34}{25}e^x x + \dfrac{1}{5}x^2 e^x + \dfrac{1}{2}\_C3 x^2 + \_C4 x + \_C5$$

13. Find the general solution of the equation $y'' = x + \sin x$.

**Computer solution.**

The solution of the differential equation will be deterrmined in Matlab 7.9:

```
>> y=dsolve('D2y=x+sin(x)','x')
y =
C5 - sin(x) + C4*x + x^3/6
```

and using Mathematica 8:

In[3]:= DSolve[D[y[x], {x, 2}] == x + Sin[x], y[x], x]

Out[3]= $\left\{\left\{y[x] \to \dfrac{x^3}{6} + C[1] + x\,C[2] - Sin[x]\right\}\right\}$

and with Maple 15:

$$ode := \frac{d^2}{dx^2}y(x) = x + \sin(x) :$$

$$dsolve(ode)$$

$$y(x) = \frac{1}{6}x^3 - \sin(x) + \_C1\,x + \_C2$$

14. Solve the following equations:

$$\text{a) } x^2 y'' - xy' + x = 6x\ln x$$

$$\text{b) } xy''' + y'' = 1 + x.$$

15. Solve the equation: $(3x+2)^2\,y'' + 7\,(3x+2)\,y' = -63x + 18$.

16. Find the general solution of the equation: $x^2 y'' + xy' + x = 2\sin(\ln x)$.

17. Find the general solution of the first-order homogeneous system of linear differential equation:

$$\text{a) } \begin{cases} y_1' = y_2 + y_3 \\ y_2' = y_3 + y_1 \\ y_3' = y_1 + y_2 \end{cases}$$

$$\text{b) } \begin{cases} y_1' = -2y_1 + 2y_2 + 2y_3 \\ y_2' = -10y_1 + 6y_2 + 8y_3 \\ y_3' = 3y_1 - y_2 - 2y_3. \end{cases}$$

18.Solve the first-order homogeneous system of linear differential equation:

$$\begin{cases} \frac{dy}{dx} - 3y + 8z - 4u = 0 \\ \frac{dz}{dx} + y - 5z + 2u = 0 \\ \frac{du}{dx} + 3y - 14z + 6u = 0. \end{cases}$$

**Computer solution.**

We shall give a computer solution using Matlab 7.9:

>> [y,z,u]=dsolve('Dy-3*y+8*z-4*u','Dz+y-5*z+2*u','Du+3*y-14*z+6*u','x')

y =

C17*exp(x) + C15/exp(x) + C16*exp(2*x)

z =

- C15/(2*exp(x)) - (4*C16*exp(2*x))/5

u =

(C17*exp(x))/2 + C15/(4*exp(x)) + (2*C16*exp(2*x))/5

and in Mathematica 8:

In[19]:= DSolve[{D[y[x], x] - 3*y[x] + 8*z[x] - 4*u[x] == 0,
    D[z[x], x] + y[x] - 5*z[x] + 2*u[x] == 0,
    D[u[x], x] + 3*y[x] - 14*z[x] + 6*u[x] == 0}, {y[x], z[x], u[x]},
    x]

Out[19]= $\left\{\left\{ u[x] \rightarrow -\frac{1}{3} e^{-x} \left(-8 + 5 e^{3x}\right) C[1] - \right.\right.$

$\frac{1}{3} e^{-x} \left(-2 - 3 e^{2x} + 5 e^{3x}\right) C[2] + \frac{2}{3} e^{-x} \left(-8 + 3 e^{2x} + 5 e^{3x}\right) C[3],$

$y[x] \rightarrow \frac{4}{3} e^{-x} \left(-1 + e^{3x}\right) C[1] + \frac{1}{3} e^{-x} \left(-1 + 4 e^{3x}\right) C[2] -$

$\frac{8}{3} e^{-x} \left(-1 + e^{3x}\right) C[3], z[x] \rightarrow -\frac{2}{3} e^{-x} \left(-1 + e^{3x}\right) C[1] -$

$\left.\left.\frac{1}{6} e^{-x} \left(-1 - 3 e^{2x} + 4 e^{3x}\right) C[2] + \frac{1}{3} e^{-x} \left(-4 + 3 e^{2x} + 4 e^{3x}\right) C[3]\right\}\right\}$

and with Maple 15:

$$sys1 := \left[ \frac{d}{dx}y(x) - 3 \cdot y(x) + 8 \cdot z(x) - 4 \cdot u(x) = 0, \; \frac{d}{dx}z(x) + y(x) - 5 \cdot z(x) + 2 \right.$$

$$\left. \cdot u(x) = 0, \; \frac{d}{dx}u(x) + 3 \cdot y(x) - 14 \cdot z(x) + 6 \cdot u(x) = 0 \right] :$$

$sol1 := dsolve(sys1)$

$$\left\{ u(x) = \_C1\, e^{-x} + \_C2\, e^{x} + \_C3\, e^{2x}, y(x) = -\frac{1}{2}\_C1\, e^{-x} - \frac{4}{5}\_C3\, e^{2x}, \right. \tag{1}$$

$$\left. z(x) = \frac{1}{4}\_C1\, e^{-x} + \frac{1}{2}\_C2\, e^{x} + \frac{2}{5}\_C3\, e^{2x} \right\}$$

19. Use the method of variation of constants to determine the general solution of the following non-homogeneous system of linear differential equations, with constant coefficients:

a)
$$\begin{cases} y_1' = 2y_1 + y_2 - 2y_3 - x + 2 \\ y_2' = -y_1 + 1 \\ y_3' = y_1 + y_2 - y_3 + 1 - x \end{cases}$$

b)
$$\begin{cases} y_1' = y_2 + y_3 - e^x \\ y_2' = y_1 + y_3 - e^x \\ y_3' = y_1 + y_2 - e^x. \end{cases}$$

**Computer solution.**

b) We shall give a computer solution In Matlab 7.9:

```
>> [y1,y2,y3]=dsolve('Dy1=y2+y3-exp(x)',
'Dy2=y1+y3-exp(x)','Dy3=y1+y2-exp(x)','x')
y1 =
exp(x) + C12*exp(2*x) - C5/exp(x) - C6/exp(x)
y2 =
exp(x) + C12*exp(2*x) + C5/exp(x)
y3 =
exp(x) + C12*exp(2*x) + C6/exp(x)
```

and with Mathematica 8:

In[5]:= `DSolve[{y1'[x] == y2[x] + y3[x] - Exp[x],`
`y2'[x] == y1[x] + y3[x] - Exp[x],`
`y3'[x] == y1[x] + y2[x] - Exp[x]},`
`{y1[x], y2[x], y3[x]}, x];`

In[6]:= `Simplify[%]`

Out[6]= $\left\{\left\{y1[x] \rightarrow \frac{1}{3} e^{-x} \left(3 e^{2x} + 2C[1] - \right.\right.\right.$

$C[2] - C[3] + e^{3x} (C[1] + C[2] + C[3]))$,

$y2[x] \rightarrow \frac{1}{3} e^{-x} \left(3 e^{2x} - C[1] + 2C[2] - C[3] + \right.$

$e^{3x} (C[1] + C[2] + C[3]))$,

$y3[x] \rightarrow \frac{1}{3} e^{-x} \left(3 e^{2x} - C[1] - C[2] + 2C[3] + \right.$

$\left.\left.\left. e^{3x} (C[1] + C[2] + C[3]))\right\}\right\}\right\}$

and in Maple 15:

$sys1 := \left[ \frac{d}{dx} y1(x) = y2(x) + y3(x) - e^x, \frac{d}{dx} y2(x) = y1(x) + y3(x) - e^x, \frac{d}{dx} y3(x) = y1(x) \right.$

$\left. + y2(x) - e^x \right]:$

$sol1 := dsolve(sys1)$

$\{y1(x) = e^{2x}\_C3 + e^{-x}\_C2 + e^x, y2(x) = e^{2x}\_C3 + e^{-x}\_C2 + e^x + e^{-x}\_C1,$  **(1)**

$y3(x) = e^{2x}\_C3 - 2e^{-x}\_C2 + e^x - e^{-x}\_C1\}$

20. Applying the method of variation of constants, find the general solution of the non-homogeneous system of linear differential equations, with constant coefficients:

$$\begin{cases} \frac{dx}{dt} = z + y - 4x + 1 \\ \frac{dy}{dt} = 2y + x - z + 1 \\ \frac{dz}{dt} = y + z - 2x + 1. \end{cases}$$

**Computer solution.**
We shall solve this system using Matlab 7.9:
`>> [x,y,z]=dsolve('Dx=z+y-4*x+1','Dy=2*y+x-z+1',`
`'Dz=y+z-2*x+1','t') ;`
`>>simple(x)`

ans =
   ((4*7^(1/2))/27 + 11/27)*(- C2*exp(7^(1/2)*t - t)*
(27*7^(1/2) - 72) -
   (C3*(5*7^(1/2) − 16))/exp(t + 7^(1/2)*t) - C1*exp(t)*
(4*7^(1/2) - 11))
   >>simple(y)
ans =
   (2*C2*exp(7^(1/2)*t - t))/3 + (2*C3)/(3*exp(t + 7^(1/2)*t))
+ (2*C1*exp(t))/3 +
   (7^(1/2)*C2*exp(7^(1/2)*t - t))/3 - (7^(1/2)*C3)/(3*exp(t +
7^(1/2)*t)) - 2/3
   >>simple(z)
ans =
   C2*exp(7^(1/2)*t - t) + C3/exp(t + 7^(1/2)*t) + C1*exp(t)
- 1/3
   and with Mathematica 8:

In[23]:= DSolve[{D[x[t], t] == z[t] + y[t] - 4*x[t] + 1, D[y[t], t] == 2*y[t] + x[t] - z[t] + 1,
        D[z[t], t] == y[t] + z[t] - 2*x[t] + 1}, {x[t], y[t], z[t]}, t];

In[24]:= Simplify[%]

$$
\text{Out[24]=} \left\{\left\{x[t] \to -\frac{1}{42} e^{-\left(1+\sqrt{7}\right)t}\left(7\left(-4-\sqrt{7}+\left(-4+\sqrt{7}\right)e^{2\sqrt{7}t}+2e^{\left(2+\sqrt{7}\right)t}\right)C[1]+\right.\right.\right.
$$

$$
\left(-7+5\sqrt{7}-\left(7+5\sqrt{7}\right)e^{2\sqrt{7}t}+14 e^{\left(2+\sqrt{7}\right)t}\right)C[2]+
$$

$$
\left.\left(14-\sqrt{7}+\left(14+\sqrt{7}\right)e^{2\sqrt{7}t}-28 e^{\left(2+\sqrt{7}\right)t}\right)C[3]\right),
$$

$$
y[t] \to -\frac{1}{42} e^{-\left(1+\sqrt{7}\right)t}\left(28 e^{3+\sqrt{7}t}-7\left(-2-\sqrt{7}\right)C[1]-35 C[2]+13\sqrt{7}\ C[2]+\right.
$$

$$
28 e^{\left(2+\sqrt{7}\right)t}(C[1]+C[2]-2C[3])+28 C[3]-11\sqrt{7}\ C[3]+
$$

$$
\left.e^{2\sqrt{7}t}\left(-7\left(2+\sqrt{7}\right)C[1]-\left(35+13\sqrt{7}\right)C[2]+\left(28-11\sqrt{7}\right)C[3]\right)\right),
$$

$$
z[t] \to \frac{1}{42} e^{-\left(1+\sqrt{7}\right)t}\left(-14 e^{3+\sqrt{7}t}-42 e^{\left(2+\sqrt{7}\right)t}(C[1]+C[2]-2C[3])+\right.
$$

$$
3\left(7 C[1]+\left(7-3\sqrt{7}\right)C[2]+\left(-7-2\sqrt{7}\right)C[3]\right)+
$$

$$
\left.\left.\left.3 e^{2\sqrt{7}t}\left(7 C[1]+\left(7+3\sqrt{7}\right)C[2]-\left(7+2\sqrt{7}\right)C[3]\right)\right)\right\}\right\}
$$

and in Maple 15:

$sys1 := \left[ \frac{d}{dt}x(t) = z(t) + y(t) - 4 \cdot x(t) + 1, \frac{d}{dt}y(t) = 2 \cdot y(t) + x(t) \right.$

$\left. -z(t) + 1, \frac{d}{dt}z(t) = y(t) + z(t) - 2 \cdot x(t) + 1 \right]:$

$sol1 := dsolve(sys1)$

$\left\{ x(t) = \_C1\, e^t + \_C2\, e^{(-1+\sqrt{7})\, t} + \_C3\, e^{-(1+\sqrt{7})\, t}, y(t) \right.$

$= 2\_C1\, e^t + \frac{5}{3}\_C2\, e^{(-1+\sqrt{7})\, t} + \frac{2}{3}\_C2\, e^{(-1+\sqrt{7})\, t}\sqrt{7}$

$+ \frac{5}{3}\_C3\, e^{-(1+\sqrt{7})\, t} - \frac{2}{3}\_C3\, e^{-(1+\sqrt{7})\, t}\sqrt{7} - \frac{2}{3},$

$z(t) = 3\_C1\, e^t + \frac{4}{3}\_C2\, e^{(-1+\sqrt{7})\, t}$

$+ \frac{1}{3}\_C2\, e^{(-1+\sqrt{7})\, t}\sqrt{7} + \frac{4}{3}\_C3\, e^{-(1+\sqrt{7})\, t}$

$\left. -\frac{1}{3}\_C3\, e^{-(1+\sqrt{7})\, t}\sqrt{7} - \frac{1}{3} \right\}$

# 8

# Line and Double Integral Calculus

## 8.1 Line Integrals of the First Type

The line integral is an extension of the definite integral in the sense that, the interval of integration $[a, b]$ is replaced by an arc of curve $AB$.

**Definition 8.1** ([42], p. 89). Let be a curve in space, having the parametrical equations:

$$(C) : \begin{cases} x = x(t) \\ y = y(t) \\ z = z(t) \end{cases}, \ t \in [a, b],$$

with $x, y, z$ continuous functions and with the first order partial derivatives being continuous in an interval $[a, b]$; o such curve is called **smooth curve**.

**Definition 8.2** (see [23], p. 252). The **element of arc of a curve** $C$ is the differential d$s$ of the function $s = s(t)$, which signifies the length of the respective arc, of the curve $C$.

**Theorem 8.3** (see [15], p. 235 and [42], p. 94). If $f$ is a continuous function on the domain $D \subset \mathbb{R}^3$ and $AB \subset D$ is an arc of smooth curve then there is the line integral of the first type and it has the expression:

$$\int_{AB} f(x, y, z) \, ds = \int_a^b f(x(t), y(t), z(t)) \sqrt{x'^2(t) + y'^2(t) + z'^2(t)} \, dt.$$

$$(8.1)$$

**Proposition 8.4** (see [15], p. 234). In the case of a plane curve

G.A. Anastassiou and I.F. Iatan: Intelligent Routines, ISRL 39, pp. 395–474.
springerlink.com © Springer-Verlag Berlin Heidelberg 2013

$$(C) : \begin{cases} x = x\,(t) \\ y = y\,(t) \end{cases}, \; t \in [a, b],$$

with $y'\,(x)$ a continuous function, the formula (8.1) becomes:

$$\int_{AB} f\,(x, y)\,\mathrm{d}s = \int_a^b f\,(x\,(t), y\,(t))\,\sqrt{x'^2\,(t) + y'^2\,(t)}\,\mathrm{d}t. \qquad (8.2)$$

**Remark. 8.5** (see [8]). The line integral of the first type does not depend on the direction of the path of integration.

## 8.1.1   Applications of Line Integral of the First Type

**Definition 8.6** (see [42], p. 97). If the integrand $f$ from (8.1) is interpreted as a linear density of the integration curve $C$, then this integral represents the **mass of the curve** $C$.

The mass corresponding to a material thread is given by the formula:

$$\mathrm{Mass} = \int_C \rho\,(x, y, z)\,\mathrm{d}s, \qquad (8.3)$$

where $\rho\,(x, y, z)$ is the linear density of the material thread in the point $(x, y, z) \in C$.

**Definition 8.7** (see [42], p. 96). The length of the arc of smooth curve is

$$\mathrm{L} = \int_{AB} \mathrm{d}s. \qquad (8.4)$$

**Definition 8.8** (see [42], p. 98). The coordinates of the centre of gravity corresponding to a material thread are:

$$\begin{cases} x_G = \dfrac{\int_{AB} x\rho(x,y,z)\mathrm{d}s}{\int_{AB} \rho(x,y,z)\mathrm{d}s} \\[2mm] y_G = \dfrac{\int_{AB} y\rho(x,y,z)\mathrm{d}s}{\int_{AB} \rho(x,y,z)\mathrm{d}s} \\[2mm] z_G = \dfrac{\int_{AB} z\rho(x,y,z)\mathrm{d}s}{\int_{AB} \rho(x,y,z)\mathrm{d}s} . \end{cases} \qquad (8.5)$$

**Example 8.9.** Use the line integral of first kind to compute the mass corresponding to a material thread having the form of a cylindrical eller:

$$(C) : \begin{cases} x = a\cos t \\ y = a\sin t \\ z = bt \end{cases}, \; t \in \left[0, \frac{\pi}{2}\right],$$

with the density $\rho\,(x, y, z) = x + y + z$.

**Solution.**
Using the relations (8.1) and (8.3), the mass of our material thread will
be:

$$\text{Mass} = \int_0^{\frac{\pi}{2}} (a\cos t + a\sin t + bt) \sqrt{a^2 \sin^2 t + a^2 \cos^2 t + b^2} dt$$

$$= \sqrt{a^2 + b^2} \int_0^{\frac{\pi}{2}} (a\cos t + a\sin t + bt) \, dt$$

$$= \sqrt{a^2 + b^2} \left( a \int_0^{\frac{\pi}{2}} \cos t \, dt + a \int_0^{\frac{\pi}{2}} \sin t \, dt + b \int_0^{\frac{\pi}{2}} t \, dt \right)$$

$$= \sqrt{a^2 + b^2} \left( a\sin t \big|_0^{\frac{\pi}{2}} - a\cos t \big|_0^{\frac{\pi}{2}} + \frac{bt^2}{2} \Big|_0^{\frac{\pi}{2}} \right),$$

therefore

$$\text{Mass} = \sqrt{a^2 + b^2} \left( a + a + b \cdot \frac{\pi^2}{8} \right) = \sqrt{a^2 + b^2} \left( 2a + b \cdot \frac{\pi^2}{8} \right).$$

In Matlab 7.9 we need the following instructions:

```
>> syms a b t u v w
>> f=@(u,v,w) u+v+w;
>> x=@(t) a*cos(t) ;
>> y=@(t) a*sin(t) ;
>> z=@(t) b*t ;
>> xt=diff(x(t))
     xt =
     -a*sin(t)
>> yt=diff(y(t))
yt =
a*cos(t)
>> zt=diff(z(t))
zt =
     b
>> factor(int(f(x(t),y(t),z(t))*sqrt(xt^2+yt^2+zt^2),0,pi/2))
ans =
1/8*(a^2+b^2)^(1/2)*(16*a+b*pi^2)
```

We shall also compute this mass using Mathcad 14:

$f(u,v,w) := u + v + w$

$x(t,a) := a \cdot \cos(t)$      $y(t,a) := a \cdot \sin(t)$      $z(t,b) := b \cdot t$

$xt(t,a) := \dfrac{d}{dt} x(t,a)$      $yt(t,a) := \dfrac{d}{dt} y(t,a)$      $zt(t,b) := \dfrac{d}{dt} z(t,b)$

$$M(a,b) := \int_0^{\frac{\pi}{2}} f(x(t,a),y(t,a),z(t,b)) \cdot \sqrt{xt(t,a)^2 + yt(t,a)^2 + zt(t,b)^2}\, dt$$

$$M(a,b) \rightarrow \left( 2 \cdot a + \frac{\pi^2 \cdot b}{8} \right) \cdot \sqrt{a^2 + b^2}$$

and with Mathematica 8:

```
In[6]:= f[u_, v_, w_] := u + v + w

In[7]:= x[t_] := a * Cos[t]

In[8]:= y[t_] := a * Sin[t]

In[9]:= z[t_] := b * t

In[10]:= xt = D[x[t], t]; yt = D[y[t], t]; zt = D[z[t], t];

In[11]:= Mass = Integrate[f[x[t], y[t], z[t]] * Sqrt[xt^2 + yt^2 + zt^2], {t, 0, Pi / 2}]
```

$\text{Out[11]}= \sqrt{a^2 + b^2} \left( 2a + \dfrac{b\pi^2}{8} \right)$

and in Maple 15:

$f := (u, v, w) \rightarrow u + v + w :$
$x := a \cdot \cos(t) :$
$y := a \cdot \sin(t) :$
$z := b \cdot t :$

$$Mass = factor \left[ \int_0^{\frac{\pi}{2}} f(x, y, z) \cdot \sqrt{\left( \frac{\partial}{\partial t} x \right)^2 + \left( \frac{\partial}{\partial t} y \right)^2 + \left( \frac{\partial}{\partial t} z \right)^2} \, dt \right]$$

$$Mass = \frac{1}{8} \sqrt{a^2 + b^2} \left( 16 \, a + b \pi^2 \right)$$

**Example 8.10.** Find the centre of gravity $G(x_G, y_G)$ corresponding to the cycloid arc:

$$(AB) : \begin{cases} x = a(t - \sin t) \\ y = a(1 - \cos t) \end{cases}, \quad t \in [0, 2\pi],$$

with the density $\rho(x, y) = \sqrt{y}$, $y > 0$, $a$ being a positive constant.
**Solution.**
We shall have:

$$I_1 = \int_{AB} \rho(x, y) \, ds = \int_0^{2\pi} \sqrt{a(1 - \cos t)} \cdot \sqrt{a^2 (1 - \cos t)^2 + a^2 \sin^2 t} \, dt$$

$$= a\sqrt{a} \int_0^{2\pi} \sqrt{1 - \cos t} \cdot \sqrt{1 - 2\cos t + \cos^2 t + \sin^2 t} \, dt$$

$$= a\sqrt{2a} \int_0^{2\pi} (1 - \cos t) \, dt = a\sqrt{2} \cdot \sqrt{a} \left( \int_0^{2\pi} dt - \int_0^{2\pi} \cos t \, dt \right)$$

$$= a\sqrt{2} \cdot \sqrt{a} \left( 2\pi - \sin t \big|_0^{2\pi} \right) = 2\pi \cdot a\sqrt{2} \cdot \sqrt{a},$$

$$I_2 = \int_{AB} x\rho(x, y) \, ds = \int_0^{2\pi} a(t - \sin t) \sqrt{a(1 - \cos t)} \cdot \sqrt{a^2 (1 - \cos t)^2 + a^2 \sin^2 t} \, dt$$

$$= a^2 \sqrt{2} \cdot \sqrt{a} \int_0^{2\pi} (t - \sin t) \cdot (1 - \cos t) \, dt.$$

One makes the substitution

$$t - \sin t = u,$$

which implies:

$$(1 - \cos t) \ dt = du;$$

$$t = 0 \Longrightarrow u = 0$$
$$t = 2\pi \Longrightarrow u = 2\pi.$$

Using this substitution gives

$$I_2 = a^2\sqrt{2} \cdot \sqrt{a} \int_0^{2\pi} u \ du = a^2\sqrt{2} \cdot \sqrt{a} \cdot \left.\frac{u^2}{2}\right|_0^{2\pi} = 2\pi^2 a^2 \sqrt{2} \cdot \sqrt{a}.$$

$$I_3 = \int_{AB} y\rho\,(x,y)\,ds = \int_0^{2\pi} \sqrt{a\,(1 - \cos t)} \cdot a\,(1 - \cos t) \cdot \sqrt{a^2\,(1 - \cos t)^2 + a^2 \sin^2 t}\,dt$$

$$= a^2\sqrt{2} \cdot \sqrt{a} \int_0^{2\pi} (1 - \cos t)^2 \ dt$$

$$= a^2\sqrt{2} \cdot \sqrt{a} \left( \int_0^{2\pi} dt - 2 \int_0^{2\pi} \cos t \ dt + \int_0^{2\pi} \cos^2 t \ dt \right)$$

$$= a^2\sqrt{2} \cdot \sqrt{a} \left( 2\pi - 2\sin t\big|_0^{2\pi} + \int_0^{2\pi} (\sin t)'\cos t \ dt \right)$$

$$= a^2\sqrt{2} \cdot \sqrt{a} \left( 2\pi + \sin t\cos t\big|_0^{2\pi} + \int_0^{2\pi} \sin^2 t \ dt \right)$$

$$= a^2\sqrt{2} \cdot \sqrt{a}\,(2\pi + \pi) = 3\pi a^2 \sqrt{2a}.$$

Finally,

$$\begin{cases} x_G = \dfrac{2\pi^2 a^2 \sqrt{2} \cdot \sqrt{a}}{2\pi \cdot a\sqrt{2} \cdot \sqrt{a}} = \pi a \\[3mm] y_G = \dfrac{3\pi a^2 \sqrt{2} \cdot \sqrt{a}}{2\pi \cdot a\sqrt{2} \cdot \sqrt{a}} = \dfrac{3a}{2}. \end{cases}$$

In Matlab 7.9 we shall have:
```
>> syms t a
>> x=a*(t-sin(t));
>> y=a*(1-cos(t));
>> I1=simple(int(sqrt(y)*sqrt(diff(x,t)^2+diff(y,t)^2)
,t,0,2*pi))
 I1 =
 2*pi*a^(3/2)*2^(1/2)
>> I2=simple(int(sqrt(x^2)*sqrt(y)*sqrt(diff(x,t)^2+diff(y,t)
^2),t,0,2*pi))
 I2 =
 2*a^(5/2)*2^(1/2)*pi^2
>> I3=simple(int(y*sqrt(y)*sqrt(diff(x,t)^2+diff(y,t)^2),
t,0,2*pi))
```

**I3 =**
**3\*pi\*a^(5/2)\*2^(1/2)**
**>> xg=I1\I2**
**xg =**
**a\*pi**
**>> yg=I3/I1**
**yg =**
**3/2\*a**

We shall also solve this problem with Mathematica 8:

In[59]:= ρ[u_, v_] := Sqrt[v]

In[60]:= x[t_] := a * (t - Sin[t])

In[61]:= y[t_] := a * (1 - Cos[t])

In[62]:= xt = D[x[t], t]; yt = D[y[t], t];

In[63]:= I1 = Integrate[ρ[x[t], y[t]] * Sqrt[xt^2 + yt^2], {t, 0, 2*Pi}]

Out[63]= $2\sqrt{2}\sqrt{a}\sqrt{a^2}\,\pi$

In[64]:= I2 = Integrate[x[t] * ρ[x[t], y[t]] * Sqrt[xt^2 + yt^2], {t, 0, 2*Pi}]

Out[64]= $2\sqrt{2}\;a^{3/2}\sqrt{a^2}\,\pi^2$

In[65]:= I3 = Integrate[y[t] * ρ[x[t], y[t]] * Sqrt[xt^2 + yt^2], {t, 0, 2*Pi}]

Out[65]= $3\sqrt{2}\;a^{3/2}\sqrt{a^2}\,\pi$

In[66]:= xG = I2 / I1

Out[66]= $a\,\pi$

In[67]:= yG = I3 / I1

Out[67]= $\dfrac{3a}{2}$

and using Maple 15:

$\rho := (u, v) \to \sqrt{v} :$
$x := a \cdot (t - \sin(t)) :$
$y := a \cdot (1 - \cos(t)) :$

$$I1 := \int_0^{2 \cdot \pi} \rho(x, y) \cdot \sqrt{\left(\frac{\partial}{\partial t} x\right)^2 + \left(\frac{\partial}{\partial t} y\right)^2} \, dt$$

$$2 \pi a^{3/2} \operatorname{csgn}(a) \sqrt{2}$$

$$I2 := \text{simplify}\left(\int_0^{2 \cdot \pi} x \cdot \rho(x, y) \cdot \sqrt{\left(\frac{\partial}{\partial t} x\right)^2 + \left(\frac{\partial}{\partial t} y\right)^2} \, dt\right)$$

$$2 a^{5/2} \sqrt{2} \operatorname{csgn}(a) \pi^2$$

$$I3 := \text{simplify}\left(\int_0^{2 \cdot \pi} y \cdot \rho(x, y) \cdot \sqrt{\left(\frac{\partial}{\partial t} x\right)^2 + \left(\frac{\partial}{\partial t} y\right)^2} \, dt\right)$$

$$3 \pi a^{5/2} \operatorname{csgn}(a) \sqrt{2}$$

$$xG = \frac{I2}{I1}$$

$$xG = a \pi$$

$$yG = \frac{I3}{I1}$$

$$yG = \frac{3}{2} a$$

We can't solve this problem using Mathcad 14.
**Example 8.11.** Compute

$$\int_C (x + y) \, ds,$$

where $C$ is the contour of the triangle, which has the vertices: $O(0,0)$, $A(1,0), B(0,1)$.
**Solution.**
The next figure emphases the outline of the triangle $OAB$.

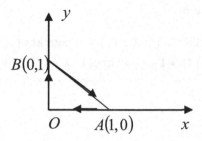

One deduces that:

$$\int_C (x+y)\,\mathrm{d}s = \int_{AO} (x+y)\,\mathrm{d}s + \int_{OB} (x+y)\,\mathrm{d}s + \int_{BA} (x+y)\,\mathrm{d}s.$$

Keeping into account the following parametrical equations corresponding to the lines $AO$, $OB$, $BA$:

$$AO: \begin{cases} x = -t+1 \\ y = 0 \end{cases}, \ t \in [0,1]$$

$$OB: \begin{cases} x = 0 \\ y = t \end{cases}, \ t \in [0,1]$$

$$BA: \begin{cases} x = t \\ y = 1-t \end{cases}, \ t \in [0,1]$$

one obtains:

$$\int_C (x+y)\,\mathrm{d}s = \int_0^1 (1-t)\,\mathrm{d}t + \int_0^1 t\,\mathrm{d}t + \int_0^1 (t+1-t)\sqrt{1^2+1^2}\mathrm{d}t$$

$$= t\Big|_0^1 - \frac{t^2}{2}\Big|_0^1 + \frac{t^2}{2}\Big|_0^1 + \sqrt{2}t\Big|_0^1 = 1 + \sqrt{2}.$$

We shall check this result in Matlab 7.9:
```
>> syms t
>> int(1-t,t,0,1)+int(t,t,0,1)+int((t+1-t)*sqrt(1^2+1^2),t,0,1)
ans =
2^(1/2) + 1
```
and with Mathcad 14:

$$\int_0^1 (1-t)\,\mathrm{d}t + \int_0^1 t\,\mathrm{d}t + \int_0^1 (t+1-t)\cdot\sqrt{1^2+1^2}\,\mathrm{d}t \to \sqrt{2}+1$$

and using Mathematica 8:

In[1]:= **Integrate[1 - t, {t, 0, 1}] + Integrate[t, {t, 0, 1}] +**
　　　　**Integrate[(t + 1 - t) \* Sqrt[1^2 + 1^2], {t, 0, 1}]**

Out[1]= $1 + \sqrt{2}$

and in Maple 15:

$$\int_0^1 (1 - t)\, dt + \int_0^1 t\, dt + \int_0^1 (t + 1 - t) \cdot \sqrt{1^2 + 1^2}\, dt$$

$$1 + \sqrt{2}$$

**Example 8.12.** Compute

$$\int_\Gamma \left(x^2 + y^2\right) \ln z\, ds,$$

where

$$\Gamma : \begin{cases} x = e^t \cos t \\ y = e^t \sin t\, , t \in [0, 1]. \\ z = e^t \end{cases}$$

**Solution.**
　　We shall compute at first the arc element:

$$ds = \sqrt{(e^t \cos t - e^t \sin t)^2 + (e^t \cos t - e^t \sin t)^2 + e^{2t}}\, dt$$
$$= \sqrt{2e^{2t} + e^{2t}}\, dt = \sqrt{3}e^t\, dt,$$

and then

$$\int_\Gamma \left(x^2 + y^2\right) \ln z\, ds = \sqrt{3} \int_0^1 \left(e^{2t} \cos^2 t + e^{2t} \sin^2 t\right) te^t\, dt = \sqrt{3} \int_0^1 te^{3t}\, dt$$

$$= \frac{\sqrt{3}}{3} \int_0^1 t \cdot \left(e^{3t}\right)'\, dt = \frac{\sqrt{3}}{3} \left( te^{3t} \Big|_0^1 - \int_0^1 e^{3t}\, dt \right)$$

$$= \frac{\sqrt{3}}{3} \left( e^3 - \frac{1}{3} e^{3t} \Big|_0^1 \right) = \frac{\sqrt{3}}{3} \left( e^3 - \frac{1}{3} e^3 + \frac{1}{3} \right)$$

$$= \frac{\sqrt{3}}{3} \left( \frac{2}{3} e^3 + \frac{1}{3} \right)$$

i.e.

$$I = \frac{\sqrt{3}}{9} \left( 2e^3 + 1 \right).$$

We can also compute this integral in Matlab 7.9:
```
>>syms t
>>x=exp(t)*cos(t);
>>y=exp(t)*sin(t);
>>z=exp(t);
>>f=@(xx,yy,zz) (xx^2+yy^2)*log(zz);
>>xt=diff(x);
>>yt=diff(y);
>>zt=diff(z);
>> I=int(f(x,y,z)*sqrt(xt^2+yt^2+zt^2),t,0,1) ;
>> simple(I)
ans =
 1/9*3^(1/2)*(2*exp(3)+1)
```
and with Mathematica 8:

In[1]:= f[u_, v_, w_] := (u^2 + v^2) Log[w]

In[2]:= x[t_] := Exp[t] * Cos[t]

In[3]:= y[t_] := Exp[t] * Sin[t]

In[4]:= z[t_] := Exp[t]

In[5]:= xt = D[x[t], t]; yt = D[y[t], t]; zt = D[z[t], t];

In[6]:= Int = Integrate[f[x[t], y[t], z[t]] * Sqrt[xt^2 + yt^2 + zt^2], {t, 0, 1}]

Out[6]= $\dfrac{1 + 2 e^3}{3 \sqrt{3}}$

and using Maple 15:

$f := (u, v, w) \to (u^2 + v^2) \cdot \ln(w) :$

$x := e^t \cdot \cos(t) :$

$y := e^t \cdot \sin(t) :$

$z := e^t :$

$$Ii := factor\left[\int_0^1 f(x, y, z) \cdot \sqrt{\left(\frac{\partial}{\partial t} x\right)^2 + \left(\frac{\partial}{\partial t} y\right)^2 + \left(\frac{\partial}{\partial t} z\right)^2}\, dt\right]$$

$$\frac{1}{9}\sqrt{3}\left(1 + 2\,e^3\right)$$

We can't compute this integral with Mathcad 14.

## 8.2   Line Integrals of the Second Type

If $P(x, y)$ and $Q(x, y)$ are continuous functions and $y = f(x)$ is a smooth curve $C$ which runs from $a$ to $b$ as $x$ varies, then the corresponding *line integral of the second type* is expressed as follows (see [8]):

$$\int_C P(x, y)\, dx + Q(x, y)\, dy = \int_a^b \left[P(x, f(x)) + f'(x)\, P(x, f(x))\right] dx.$$

(8.6)

In case when the curve $C$ is represented parametrically:

$$(C) : \begin{cases} x = x(t) \\ y = y(t) \end{cases}, \quad t \in [a, b]$$

we have (see [8]):

$$\int_C P(x, y)\, dx + Q(x, y)\, dy = \int_a^b \left[P(x(t), y(t))\, x'(t) + Q(x(t), y(t))\, y'(t)\right] dt. \quad (8.7)$$

Similar formulae hold for a line integral of the second type taken over a space curve.

A line integral of the second type changes sign when the direction of the integration path is reversed. This integral may be interpreted in terms of mechanics as the work of an appropriate variable force along the curve of integration $C$.

The mechanical work accomplished when a body on motion one moves on an arc $AB$ under the action of a variable force

$$\vec{F}(x, y, z) = P(x, y, z)\,\vec{i} + Q(x, y, z)\,\vec{j} + R(x, y, z)\,\vec{k}$$

is (see [42], p. 99):

$$L = \int_{AB} P(x, y, z)\,dx + P(x, y, z)\,dy + R(x, y, z)\,dz. \qquad (8.8)$$

When the arc $AB$ is of the form:

$$(C) : \begin{cases} x = x(t) \\ y = y(t) \\ z = z(t) \end{cases}, \quad t \in [a, b]$$

then

$$L = \int_a^b \left[ P(x(t), y(t), z(t))\,x'(t) + Q(x(t), y(t), z(t))\,y'(t) + R(x(t), y(t), z(t))\,z'(t) \right] dt.$$
$$(8.9)$$

**Definition 8.13** (see [2], p. 30). An **area** bounded by the closed contour $C$ can be computed using one of the following formulas:

$$\text{Area} = \oint_C x\,dy, \qquad (8.10)$$

$$\text{Area} = -\oint_C y\,dx, \qquad (8.11)$$

$$\text{Area} = \frac{1}{2} \oint_C x\,dy - y\,dx \qquad (8.12)$$

(the direction of circulation of the contour is chosen counter-clockwise).

**Theorem 8.14** (see [48], p. 411 and [42], p. 109). If the integral

$$I = \int_{AM} P(x, y, z)\,dx + Q(x, y, z)\,dy + R(x, y, z)\,dz \qquad (8.13)$$

doesn't depend on a path (namely it doesn't depend on the curve, which links the points $A$ and $M$ but just on these points) then

$$I = \int_a^x P(t, b, c)\,dt + \int_b^y Q(x, t, c)\,dt + \int_c^z R(x, y, t)\,dt, \qquad (8.14)$$

where $A(a, b, c)$, $M(x, y, z) \in D$, $D$ being a simple connex domain.

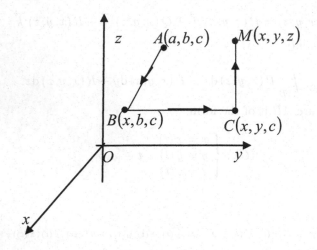

The integral from (8.13) is independent of a path if and only if

$$\begin{cases} \frac{\partial P}{\partial y} = \frac{\partial Q}{\partial x} \\ \frac{\partial Q}{\partial z} = \frac{\partial R}{\partial y} \\ \frac{\partial R}{\partial x} = \frac{\partial P}{\partial z} \end{cases} \qquad (8.15)$$

i.e. the expression from under the integral sign is a total differential.

**Example 8.15.** Use the line integral of second kind to compute the mechanical work accomplished by the force

$$\overrightarrow{F}(x,y) = \left(x^2 - 2xy\right)\overrightarrow{i} + \left(2xy + y^2\right)\overrightarrow{j}$$

in long of the parabola arc $AB : y = x^2$, which joins the points $A(1,1)$ and $B(2,4)$.

**Solution.**

The next figure represents the arc $AB$.

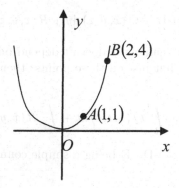

Since in case of our problem

$$(AB): \begin{cases} x = t \\ y = t^2 \end{cases}, t \in [1, 2]$$

it will result that

$$L = \int_1^2 \left[ (t^2 - 2t^3) \cdot 1 + (2t^3 + t^4) \cdot 2t \right] dt$$

$$= \int_1^2 \left( t^2 - 2t^3 + 4t^4 + 2t^5 \right) dt = \frac{1219}{30}.$$

We can also compute the mechanical work using the following Matlab 7.9 sequence:

```
>>syms x y t
>> P=@(x,y) x^2-2*x*y ;
>> Q=@(x,y) 2*x*y+y^2 ;
>> x=@(t) t ;
>> y=@(t) t^2 ;
>> xt=diff(x(t)) ;
>> yt=diff(y(t));
>> int(P(x(t),y(t))*xt+Q(x(t),y(t))*yt,1,2)
     ans =
1219/30
```

and using Mathcad 14:

$$P(x,y) := x^2 - 2 \cdot x \cdot y \qquad Q(x,y) := 2 \cdot x \cdot y + y^2$$

$$x(t) := t \qquad y(t) := t^2$$

$$xd(t) := \frac{d}{dt} x(t) \qquad yd(t) := \frac{d}{dt} y(t)$$

$$\int_1^2 P(x(t), y(t)) \cdot xd(t) + Q(x(t), y(t)) \cdot yd(t) \; dt \rightarrow \frac{1219}{30}$$

and with Mathematica 8:

In[5]:= `P[x_, y_] := x^2 - 2*x*y`

In[6]:= `Q[x_, y_] := 2*x*y + y^2`

In[7]:= `x[t_] := t`

In[8]:= `y[t_] := t^2`

In[9]:= `xt = D[x[t], t]; yt = D[y[t], t];`

In[10]:= `Int = Integrate[P[x[t], y[t]]*xt + Q[x[t], y[t]]*yt, {t, 1, 2}]`

Out[10]= $\dfrac{1219}{30}$

and in Maple 15:

$$P := (x, y) \to x^2 - 2 \cdot x \cdot y :$$
$$Q := (x, y) \to 2 \cdot x \cdot y + y^2 :$$
$$x := t :$$
$$y := t^2 :$$
$$\int_1^2 P(x, y) \cdot \frac{\partial}{\partial t} x + Q(x, y) \cdot \frac{\partial}{\partial t} y \, dt$$

$$\frac{1219}{30}$$

**Example 8.16.** Compute the following line integral of second type:

$$1)\ I = \int_\Gamma \sqrt{yz}\,dx + \sqrt{zx}\,dy + \sqrt{xy}\,dz,$$

where

$$\Gamma : \begin{cases} x\,(t) = t \\ y\,(t) = t^2, t \in [0, 1] \ ; \\ z\,(t) = t^3 \end{cases}$$

$$2)\ J = \oint_\Gamma \sqrt{y^2 - x + 2}\,(dx + dy),$$

where is the closed curve, oriented positive, with the both extremities in the point $A\,(0, 1)$, formed from the portion of the parabola $y^2 = x + 1$, placed in the dials II and III and the portion of the parabola $2y^2 = -x+2$, placed in the dials I and IV;

$$3) \ K \ = \int_C y^2 dx + x^2 dy,$$

where $C$ is the superior ellipse

$$\frac{x^2}{a^2} + \frac{y^2}{b^2} = 1, \ y \geq 0,$$

crossed in the clockwise.

**Solutions.**

1) One results that

$$I = \int_0^1 \left( \sqrt{t^5} + 2t\sqrt{t^4} + 3t^2\sqrt{t^3} \right) dt$$

$$= \int_0^1 t^{\frac{5}{2}} dt + 2 \int_0^1 t^3 dt + 3 \int_0^1 t^{\frac{7}{2}} dt = \frac{61}{42}.$$

Computing in Matlab 7.9, we shall have:

```
>> syms x y t
>> xx=t;
>> yy=t^2;
>> zz=t^3;
>> xt=diff(xx);
>> yt=diff(yy);
>> zt=diff(zz);
>> P=@(x,y,z) sqrt(y*z) ;
>> Q=@(x,y,z) sqrt(z*x);
>> R=@(x,y,z) sqrt(x*y) ;
>> I=int(P(xx,yy,zz)*xt+Q(xx,yy,zz)*yt+R(xx,yy,zz)*
zt,t,0,1)
    I =
    61/42
```

We shall also achieve this result in Mathcad 14:

$$P(x,y,z) := \sqrt{y \cdot z} \qquad\qquad Q(x,y,z) := \sqrt{z \cdot x} \qquad R(x,y,z) := \sqrt{x \cdot y}$$

$$x(t) := t \qquad\qquad y(t) := t^2 \qquad z(t) := t^3$$

$$xd(t) := \frac{d}{dt}x(t) \qquad yd(t) := \frac{d}{dt}y(t) \qquad zd(t) := \frac{d}{dt}z(t)$$

$$\int_0^1 P(x(t),y(t),z(t)) \cdot xd(t) + Q(x(t),y(t),z(t)) \cdot yd(t) + R(x(t),y(t),z(t)) \cdot zd(t) \; dt \to \frac{61}{42}$$

and using Mathematica 8:

```
In[1]:= P[x_, y_, z_] := Sqrt[y * z]

In[2]:= Q[x_, y_, z_] := Sqrt[z * x]

In[3]:= R[x_, y_, z_] := Sqrt[x * y]

In[4]:= x[t_] := t

In[5]:= y[t_] := t^2

In[6]:= z[t_] := t^3

In[7]:= xt = D[x[t], t]; yt = D[y[t], t]; zt = D[z[t], t];

In[8]:= Integrate[P[x[t], y[t], z[t]] * xt + Q[x[t], y[t], z[t]] * yt +
           R[x[t], y[t], z[t]] * zt, {t, 0, 1}]

Out[8]=  61
        ――
         42
```

and with Maple 15:

$P := (x, y, z) \rightarrow \sqrt{y \cdot z}$ :
$Q := (x, y, z) \rightarrow \sqrt{z \cdot x}$ :
$R := (x, y, z) \rightarrow \sqrt{x \cdot y}$ :
$x := t$ :
$y := t^2$ :
$z := t^3$ :

$$\int_0^1 P(x, y, z) \cdot \frac{\partial}{\partial t}x + Q(x, y, z) \cdot \frac{\partial}{\partial t}y + R(x, y, z) \cdot \frac{\partial}{\partial t}z \, dt$$

$$\frac{61}{42}$$

2) The next figure shows the curve $\Gamma$.

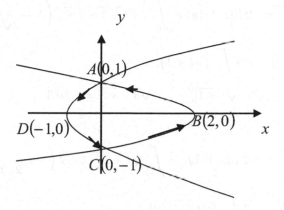

One can notice that

$$\Gamma = AD \cup DC \cup CB \cup BA$$

and therefore, one can write

$$J = \underbrace{\oint_{AD} \sqrt{y^2 - x + 2}\,(dx + dy)}_{I_1} + \underbrace{\oint_{DC} \sqrt{y^2 - x + 2}\,(dx + dy)}_{I_2}$$

$$+ \underbrace{\oint_{CB} \sqrt{y^2 - x + 2}\,(dx + dy)}_{I_3} + \underbrace{\oint_{AD} \sqrt{y^2 - x + 2}\,(dx + dy)}_{I_4}$$

$$= I_1 + I_2 + I_3 + I_4.$$

Using the parametrical equations of the arcs:

$$(AD): \begin{cases} x\,(t) = t \\ y\,(t) = \sqrt{t+1} \end{cases}, \quad t \in [0, -1]$$

$$(DC): \begin{cases} x\,(t) = t \\ y\,(t) = -\sqrt{t+1} \end{cases}, \quad t \in [-1, 0]$$

$$(CB): \begin{cases} x\,(t) = t \\ y\,(t) = -\sqrt{\frac{-t+2}{2}} \end{cases}, \quad t \in [0, 2]$$

$$(DC): \begin{cases} x\,(t) = t \\ y\,(t) = \sqrt{\frac{-t+2}{2}} \end{cases}, \quad t \in [2, 0]$$

it will result that:

$$I_1 = \oint_{AD} \sqrt{y^2 - x + 2}\,(dx + dy) = \int_0^{-1} \sqrt{t+1-t+2}\left(1 + \frac{1}{2\sqrt{t+1}}\right) dt$$

$$= -\sqrt{3} \int_{-1}^0 dt - \sqrt{3} \int_{-1}^0 \left(\sqrt{t+1}\right)' dt$$

$$= -\sqrt{3}\, t\big|_{-1}^0 - \sqrt{3} \cdot \sqrt{t+1}\big|_{-1}^0 = -\sqrt{3} - \sqrt{3} = -2\sqrt{3}$$

$$I_2 = \oint_{DC} \sqrt{y^2 - x + 2}\,(dx + dy) = \int_{-1}^0 \sqrt{t+1-t+2}\left(1 + \frac{1}{2\sqrt{t+1}}\right) dt$$

$$= \sqrt{3} \int_{-1}^0 dt - \sqrt{3} \int_{-1}^0 \left(\sqrt{t+1}\right)' dt$$

$$= \sqrt{3}\, t\big|_{-1}^0 - \sqrt{3} \cdot \sqrt{t+1}\big|_{-1}^0 = \sqrt{3} - \sqrt{3} = 0$$

$$I_3 = \oint_{CB} \sqrt{y^2 - x + 2}\,(dx + dy) = \int_0^2 \sqrt{\frac{-t+2}{2} - t + 2}\left(1 + \frac{1}{2\sqrt{2}\cdot\sqrt{-t+2}}\right) dt$$

$$= \frac{\sqrt{3}}{\sqrt{2}} \int_0^2 \left(\sqrt{-t+2} + \frac{1}{2\sqrt{2}}\right) dt = \frac{\sqrt{3}}{\sqrt{2}} \left(\underbrace{\int_0^2 \sqrt{-t+2}\,dt}_{I_5} + \frac{1}{2\sqrt{2}}\, t\big|_0^2\right)$$

$$= \frac{\sqrt{3}}{\sqrt{2}}\left(I_5 + \frac{1}{\sqrt{2}}\right),$$

where

$$I_5 = \int_0^2 \sqrt{-t+2}\, dt = \int_0^2 \frac{-t+2}{\sqrt{-t+2}}\, dt = 2\int_0^2 t\left(\sqrt{-t+2}\right)'dt + \int_0^2 \frac{2}{\sqrt{-t+2}}\, dt$$

$$= 2t\sqrt{-t+2}\Big|_0^2 - 2\int_0^2 \sqrt{-t+2}\, dt - 4\sqrt{-t+2}\Big|_0^2 = 2I_5 + 4\sqrt{2} \Longrightarrow I_5 = \frac{4\sqrt{2}}{3};$$

therefore

$$I_3 = \frac{4}{\sqrt{3}} + \frac{\sqrt{3}}{2};$$

$$I_4 = \oint_{BA} \sqrt{y^2 - x + 2}\,(dx + dy) = \int_2^0 \sqrt{\frac{-t+2}{2} - t + 2}\left(1 - \frac{1}{2\sqrt{2}\cdot\sqrt{-t+2}}\right)dt$$

$$= -\frac{\sqrt{3}}{\sqrt{2}} \int_0^2 \left(\sqrt{-t+2} - \frac{1}{2\sqrt{2}}\right)dt = -\frac{\sqrt{3}}{\sqrt{2}}\left(\underbrace{\int_0^2 \sqrt{-t+2}\,dt}_{I_5} - \frac{1}{2\sqrt{2}}\,t\Big|_0^2\right)$$

$$= -\frac{\sqrt{3}}{\sqrt{2}}\left(I_5 + \frac{1}{\sqrt{2}}\right) = -\frac{4}{\sqrt{3}} + \frac{\sqrt{3}}{2}.$$

Finally, one obtains:

$$I = I_1 + I_2 + I_3 + I_4$$
$$= -2\sqrt{3} + 0 + \frac{4}{\sqrt{3}} + \frac{\sqrt{3}}{2} - \frac{4}{\sqrt{3}} + \frac{\sqrt{3}}{2} = -\sqrt{3}.$$

3) Using the parametrical representation of the ellipse:

$$(C): \begin{cases} x = a\cos\theta \\ y = b\sin\theta \end{cases}, \ \theta \in [0, \pi]$$

we shall have:

$$K = \int_C y^2 dx + x^2 dy$$
$$= \int_0^\pi \left[b^2 \sin^2\theta \cdot (-a\sin\theta) + a^2\cos^2\theta \cdot b\cos\theta\right]d\theta$$
$$= -ab^2 \underbrace{\int_0^\pi \sin^3\theta\, d\theta}_{I_6} + a^2 b \underbrace{\int_0^\pi \cos^3\theta\, d\theta}_{I_7} = -ab^2 \cdot I_6 + a^2 b \cdot I_7,$$

where

$$I_6 = \int_0^\pi \sin^3 \theta \ d\theta = - \int_0^\pi (\cos \theta)' \sin^2 \theta \ d\theta =$$

$$= - \sin^2 \theta \cos \theta \big|_0^\pi + 2 \int_0^\pi \sin \theta \cos^2 \theta \ d\theta$$

$$= \int_0^\pi \sin \theta \left(1 - \sin^2 \theta \right) \ d\theta = 2 \int_0^\pi \sin \theta \ d\theta - 2 \int_0^\pi \sin^3 \theta \ d\theta$$

$$= -2 \cos \theta \big|_0^\pi - 2I_6 = 4 - 2I_6 \Longrightarrow I_6 = \frac{4}{3}$$

and

$$I_7 = \int_0^\pi \cos^3 \theta \ d\theta = \int_0^\pi (\sin \theta)' \cos^2 \theta \ d\theta =$$

$$= \sin \theta \cos^2 \theta \big|_0^\pi + 2 \int_0^\pi \sin^2 \theta \cos \theta \ d\theta$$

$$= 2 \int_0^\pi \left(1 - \cos^2 \theta \right) \cos \theta \ d\theta = 2 \int_0^\pi \cos \theta \ d\theta - 2 \int_0^\pi \cos^3 \theta \ d\theta$$

$$= 2 \sin \theta \big|_0^\pi - 2I_7 = 0 - 2I_7 \Longrightarrow I_7 = 0.$$

Therefore,

$$K = \int_C y^2 dx + x^2 dy = -ab^2 \cdot \frac{4}{3} + a^2 b \cdot 0 = -\frac{4ab^2}{3}.$$

We can also obtain this result in Matlab 7.9:

```
>> syms x y theta a b
>> xx=a*cos(theta);
>> yy=b*sin(theta);
>> xt=diff(xx,theta);
>> yt=diff(yy,theta);
>> P=@(x,y) y^2;
>> Q=@(x,y) x^2;
>> I=int(P(xx,yy)*xt+Q(xx,yy)*yt,theta,0,pi)
I =
-4/3*b^2*a
```

and using Mathcad 14:

$xx(a, \theta) := a \cdot \cos(\theta)$       $yy(b, \theta) := b \cdot \sin(\theta)$

$xt(a, \theta) := \dfrac{d}{d\theta} xx(a, \theta)$       $yt(b, \theta) := \dfrac{d}{d\theta} yy(b, \theta)$

$P(x, y) := y^2$       $Q(x, y) := x^2$

$\displaystyle\int_0^\pi P(xx(a, \theta), yy(b, \theta)) \cdot xt(a, \theta) + Q(xx(a, \theta), yy(b, \theta)) \cdot yt(b, \theta)\, d\theta$ simplify $\to -\dfrac{4 \cdot a \cdot b^2}{3}$

and with Mathematica 8:

```
In[1]:= P[x_, y_] := y^2

In[2]:= Q[x_, y_] := x^2

In[3]:= xx := a * Cos[θ]

In[4]:= yy := b * Sin[θ]

In[5]:= xt = D[xx, θ]; yt = D[yy, θ];

In[6]:= Integrate[P[xx, yy] * xt + Q[xx, yy] * yt, {θ, 0, Pi}]
```

$$\text{Out[6]}= -\frac{4\, a\, b^2}{3}$$

and in Maple 15:

$P := (x, y) \to y^2 :$
$Q := (x, y) \to x^2 :$
$xx := a \cdot \cos(\theta) :$
$yy := b \cdot \sin(\theta) :$

$$\int_0^\pi P(xx, yy) \cdot \frac{\partial}{\partial\theta} xx + Q(xx, yy) \cdot \frac{\partial}{\partial\theta} yy\, d\theta$$

$$-\frac{4}{3} b^2 a$$

**Example 8.17.** Compute the area bounded by an arc of cycloid:

$$(C) : \begin{cases} x = a\,(t - \sin t) \\ y = a\,(1 - \cos t) \end{cases}, \ t \in [0, 2\pi]$$

and the $Ox$ axis.

**Solution.**

The next figure represents the area which has to be computed.

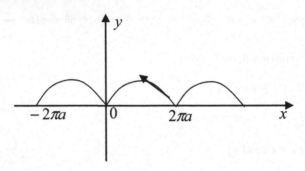

As

$$dy = a \sin t \ dt$$

it will result that

$$\text{Area} = \oint_C x \ dy = -\int_0^{2\pi} a\,(t - \sin t) \cdot a \sin t \ dt$$

$$= -a^2 \left( \int_0^{2\pi} t \sin t \ dt - \underbrace{\int_0^{2\pi} \sin^2 t \ dt}_{I} \right) = -a^2 \left( -\int_0^{2\pi} t \cdot (\cos t)' \ dt - I \right)$$

$$= -a^2 \left( -t \cos t \big|_0^{2\pi} + \int_0^{2\pi} \cos t \ dt - I \right) = -a^2 \left( -2\pi - I \right),$$

where

$$I = \int_0^{2\pi} \sin^2 t \ dt = -\int_0^{2\pi} \sin t \cdot (\cos t)' \ dt$$

$$= \sin t \cos t \big|_0^{2\pi} + \int_0^{2\pi} \cos^2 t \ dt = \int_0^{2\pi} \left( 1 - \sin^2 t \right) \ dt$$

$$= 2\pi - \int_0^{2\pi} \sin^2 t \ dt = 2\pi - I \Longrightarrow I = \pi.$$

Finally,

$$\text{Area} = -a^2 \left( -2\pi - \pi \right) = 3\pi a^2.$$

In Matlab 7.9 we shall achieve:

```
>> syms a t
>> x=a*(t-sin(t));
>> y=a*(1-cos(t));
>> yt=diff(y,t)
yt =
a*sin(t)
>> A=-int(x*yt,t,0,2*pi)
A =
3*pi*a^2
```

We shall also compute this area in Mathcad 14:

$$x(a,t) := a \cdot (t - \sin(t)) \quad y(a,t) := a \cdot (1 - \cos(t))$$

$$yt(a,t) := \frac{d}{dt} y(a,t)$$

$$A(a) := -\int_0^{2 \cdot \pi} x(a,t) \cdot yt(a,t) \, dt \rightarrow 3 \cdot \pi \cdot a^2$$

and using Mathematica 8:

```
In[5]:= x := a * (t - Sin[t])

In[6]:= y := a * (1 - Cos[t])

In[7]:= yt = D[y, t];

In[8]:= A = -Integrate[x * yt, {t, 0, 2 * Pi}]

Out[8]= 3 a² π
```

and with Maple 15:

$$x := a \cdot (t - \sin(t)) :$$
$$y := a \cdot (1 - \cos(t)) :$$
$$A = -\int_0^{2 \cdot \pi} x \cdot \frac{\partial}{\partial t} y \, dt$$

$$A = 3 \pi a^2$$

**Example 8.18.** Find that the expression from under the integral sign is a total differential to compute the line integral:

$$I = \int_{(1,1,0)}^{(2,3,4)} yz\ dx + xz\ dy + xy\ dz,$$

where only the ends of the integration curve have specified.

**Solution.**

In the case of the integral $I$ are accomplished the conditions:

$$\begin{cases} \frac{\partial P}{\partial y} = \frac{\partial Q}{\partial x} = z \\ \frac{\partial Q}{\partial z} = \frac{\partial R}{\partial y} = x \\ \frac{\partial R}{\partial x} = \frac{\partial P}{\partial z} = y \end{cases}$$

where

$$\begin{cases} P\left(x,y,z\right) = yz \\ Q\left(x,y,z\right) = xz \\ R\left(x,y,z\right) = xy; \end{cases}$$

therefore the line integral $I$ doesn't depends on a path from $\mathbb{R}^3$; its value is:

$$I = \int_1^2 P\left(t,1,0\right) dt + \int_1^3 Q\left(2,t,0\right) dt + \int_0^4 R\left(2,3,t\right) dt$$

$$= \int_1^2 0 dt + \int_1^3 0 dt + \int_0^4 6 dt = 0 + 0 + 6t\big|_0^4 = 24.$$

We can also obtain this result in Matlab 7.9:

```
>>syms x y z t
>> a=1;b=1;c=0;x=2;y=3;z=4;
>> P=@(x,y,z)y*z;
>> Q=@(x,y,z)x*z;
>> R=@(x,y,z)x*y;
>> L=int(P(t,b,c),t,a,x)+int(Q(x,t,c),t,b,y)+int(R(x,y,t),t,c,z)
L =
24
```

and with Mathcad 14:

$$a := 1 \qquad b := 1 \qquad c := 0 \qquad x := 2 \qquad y := 3 \qquad z := 4$$

$$P(x,y,z) := y \cdot z \qquad Q(x,y,z) := x \cdot z \qquad R(x,y,z) := x \cdot y$$

$$L := \int_a^x P(t,b,c)\, dt + \int_b^y Q(x,t,c)\, dt + \int_c^z R(x,y,t)\, dt$$

$$L = 24$$

and using Mathematica 8:

```
In[7]:= a := 1; b := 1; c := 0; x := 2; y := 3; z := 4;

In[8]:= P[x_, y_, z_] := y * z

In[9]:= Q[x_, y_, z_] := x * z

In[10]:= R[x_, y_, z_] := x * y

In[11]:= L = Integrate[P[t, b, c], {t, a, x}] +
            Integrate[Q[x, t, c], {t, b, y}] + Integrate[R[x, y, t], {t, c, z}]

Out[11]= 24
```

and in Maple 15:

$$a = 1 : b = 1 : c = 0 : x = 2 : y = 3 : z = 4 :$$
$$P := (x, y, z) \rightarrow y \cdot z :$$
$$Q := (x, y, z) \rightarrow x \cdot z :$$
$$R := (x, y, z) \rightarrow x \cdot y :$$
$$\int_a^x P(t, b, c)\, dt + \int_b^y Q(x, t, c)\, dt + \int_c^z R(x, y, t)\, dt$$

$$24$$

# 8.3  Calculus Way of the Double Integrals

When we have to compute the double integral

$$\iint_D f(x, y)\, dx dy, \tag{8.16}$$

the following two situations can appear (see [38, p. 106 and [8]]):

A) the domain of integration $D$ is bounded on the left and right hand sides by the straight lines $x = a$ and $x = b$ $(b > a)$, from below and from above, by the continuous curves $y = \varphi_1(x)$ and $y = \varphi_2(x)$, see Fig. 8.1; the domain $D$ is simple in the $Oy$ axis.

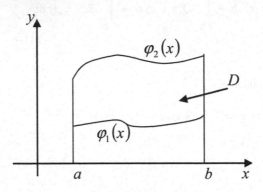

**Fig. 8.1**

From Fig. 8.1, the integral (8.16) may be computed by the formula:

$$\iint_D f(x,y)\,dxdy = \int_a^b \left( \int_{\varphi_1(x)}^{\varphi_2(x)} f(x,y)\,dy \right) dx, \qquad (8.17)$$

where

$$D = \left\{ (x,y) \in \mathbb{R}^2 \mid a \le x \le b,\ \varphi_1(x) \le y \le \varphi_2(x) \right\}. \qquad (8.18)$$

B) the domain $D$ is simple relative to the $Ox$ axis, i.e. it is bounded from below and from above by the straight lines $y = c$ and $y = d$ $(d > c)$ and from the left and the right hand side, by the continuous curves $x = \varphi_1(y)$ and $x = \varphi_2(y)$, see Fig. 8.2.

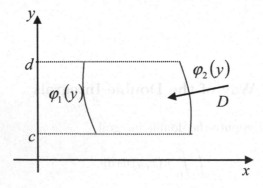

**Fig. 8.2**

From Fig. 8.2 one can notice that:

$$D = \left\{ (x, y) \in \mathbb{R}^2 \mid c \leq y \leq d,\ \varphi_1(y) \leq x \leq \varphi_2(y) \right\};\qquad (8.19)$$

hence

$$\int\int_D f(x, y)\, dxdy = \int_c^d \left( \int_{\varphi_1(y)}^{\varphi_2(y)} f(x, y)\, dx \right) dy. \qquad (8.20)$$

*Geometric interpretation of the double integral* (see [44], p. 574): The definite integral represents the area bounded by a curve. In the same way, the double integral of a function of two variables $z = f(x, y)$ can be interpreted as the volume bounded by the surface $z = f(x, y)$.

**Example 8.19.** Indicate the integration limits both in an order and in the other for the double integral from (8.16), if the domain is:

1) the trapezium of vertices $O(0, 0)$, $A(2, 0)$, $B(1, 1)$, $C(0, 1)$;

2) the circular sector $OAB$ with the centre $O(0, 0)$ and the extremities of the circle arc are: $A(1, 1)$ and $B(-1, 1)$.

Solutions
1) The next figure shows the domain $D$.

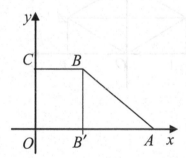

One can notice that:

$$AB : y = 2 - x \Longrightarrow x = 2 - y.$$

Supposing that the domain $D$ is as in Fig. 8.1 then

$$D = D_1 \cup D_2,$$

where

$$D_1 = \left\{ (x, y) \in \mathbb{R}^2 \mid 0 \leq x \leq 1,\ 0 \leq y \leq 1 \right\},$$

$$D_2 = \left\{ (x, y) \in \mathbb{R}^2 \mid 1 \leq x \leq 2,\ 0 \leq y \leq 2 - x \right\}.$$

It will result that

$$\iint_D f(x,y)\,dxdy = \iint_{D_1} f(x,y)\,dxdy + \iint_{D_2} f(x,y)\,dxdy$$
$$= \int_0^1 \left( \int_0^1 f(x,y)\,dy \right) dx + \int_1^2 \left( \int_0^{2-x} f(x,y)\,dy \right) dx.$$

In the case when the domain $D$ is as in Fig.8.2 one obtains that

$$D = \left\{ (x,y) \in \mathbb{R}^2 |\ 0 \le y \le 1,\ 0 \le x \le 2 - y \right\}.$$

and

$$\iint_D f(x,y)\,dxdy = \int_0^1 \left( \int_0^{2-y} f(x,y)\,dx \right) dy.$$

2) The next figure shows the domain $D$.

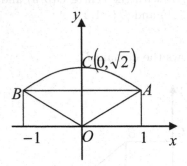

We can notice that:

$$OA : y = x \Longrightarrow x = y$$
$$OB : y = -x \Longrightarrow x = -y$$
$$AB : y = \sqrt{2 - x^2} \Longrightarrow x = \pm\sqrt{2 - y^2}.$$

Supposing that the domain is as in Fig. 8.1 then

$$D = D_1 \cup D_2,$$

where

$$D_1 = \left\{ (x,y) \in \mathbb{R}^2 |\ -1 \le x \le 0,\ -x \le y \le \sqrt{2 - x^2} \right\},$$

$$D_2 = \left\{ (x,y) \in \mathbb{R}^2 |\ 0 \le x \le 1,\ x \le y \le \sqrt{2 - x^2} \right\}.$$

It will result that

$$\iint_D f(x,y)\,dxdy = \iint_{D_1} f(x,y)\,dxdy + \iint_{D_2} f(x,y)\,dxdy$$

$$= \int_{-1}^{0} \left( \int_{-x}^{\sqrt{2-x^2}} f(x,y)\,dy \right) dx + \int_{0}^{1} \left( \int_{x}^{\sqrt{2-x^2}} f(x,y)\,dy \right) dx.$$

In the case when the domain is as in Fig. 8.2 one obtains that

$$D = D_3 \cup D_4,$$

where

$$D_3 = \left\{ (x,y) \in \mathbb{R}^2 \mid 0 \le y \le 1, \ -y \le x \le y \right\},$$

$$D_4 = \left\{ (x,y) \in \mathbb{R}^2 \mid 1 \le y \le \sqrt{2}, -\sqrt{2-y^2} \le x \le \sqrt{2-y^2} \right\}.$$

It results that

$$\iint_D f(x,y)\,dxdy = \iint_{D_3} f(x,y)\,dxdy + \iint_{D_4} f(x,y)\,dxdy$$

$$= \int_{0}^{1} \left( \int_{-y}^{y} f(x,y)\,dx \right) dy + \int_{1}^{\sqrt{2}} \left( \int_{-\sqrt{2-y^2}}^{\sqrt{2-y^2}} f(x,y)\,dx \right) dy.$$

**Example 8.20.** Compute the double integral:

1) $\displaystyle\iint_D \frac{\sin^2 x}{\cos^2 y}\,dxdy$, where the domain $D : \left\{ 0 \le x \le \dfrac{\pi}{2}, \ 0 \le y \le \dfrac{\pi}{4} \right\}$;

2) $\displaystyle\iint_D x\,dxdy$, where the domain $D$ is bounded by the parabola

$y = x^2 + 1$ and the lines $y = 2x$ and $x = 0$;

3) $\displaystyle\iint_D \sqrt{xy}\,dxdy$, where the domain $D$ is bounded by the curves

$y = x^2$ and the lines $y = \sqrt{x}$, for $x \in [0,1]$.

**Solutions.**

1) One obtains:

$$\int\int_D \frac{\sin^2 x}{\cos^2 y}\,dxdy = \int_0^{\frac{\pi}{2}}\left(\int_0^{\frac{\pi}{4}}\frac{dy}{\cos y}\right)\sin^2 x\,dx$$

$$= \int_0^{\frac{\pi}{2}}\left(\tan y|_0^{\frac{\pi}{4}}\right)\sin^2 x\,dx = \int_0^{\frac{\pi}{2}}\sin^2 x\,dx,$$

where

$$I = \int_0^{\frac{\pi}{2}}\sin^2 x\,dx = -\int_0^{\frac{\pi}{2}}\sin x\cdot(\cos x)'\,dx$$

$$= -\sin x\cos x|_0^{\frac{\pi}{2}} + \int_0^{\frac{\pi}{2}}\cos^2 x\,dx = \int_0^{\frac{\pi}{2}}\left(1 - \sin^2 x\right)\,dx$$

$$= x|_0^{\frac{\pi}{2}} - I \Longrightarrow 2I = \frac{\pi}{2} \Longrightarrow I = \frac{\pi}{4},$$

namely

$$\int\int_D \frac{\sin^2 x}{\cos^2 y}\,dxdy = \frac{\pi}{4}.$$

We can compute this double integral in Matlab 7.9, in two ways:
```
>> syms x y
>> f=@(x,y) sin(x).^2/cos(y).^2
f =
 @(x,y) sin(x).^2/cos(y).^2
>> dblquad(f,0,pi/2,0,pi/4)
ans =
 0.7854
```
or
```
>> int(int(1/cos(y)^2,0,pi/4)*sin(x)^2,0,pi/2)
ans =
 1/4*pi
```
and with Mathcad 14:

$$\int_0^{\frac{\pi}{2}}\left(\int_0^{\frac{\pi}{4}}\frac{1}{\cos(y)^2}\,dy\right)\cdot\sin(x)^2\,dx \to \frac{\pi}{4}$$

and using Mathematica 8:

```
In[1]:= Integrate[Integrate[1/ (Cos[y]^2), {y, 0, Pi/4}] * Sin[x]^2, {x, 0, Pi/2}]

Out[1]= π
        ─
        4
```

and with Maple 15:

$$\int_0^{\frac{\pi}{2}} \left( \int_0^{\frac{\pi}{4}} \frac{1}{\cos(y)^2} dy \right) \cdot \sin(x)^2 dx$$

$$\frac{1}{4}\pi$$

2) The next figure shows the domain $D$.

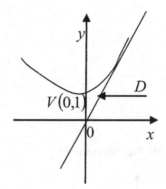

As

$$D = \{(x,y) \in \mathbb{R}^2 |\ 0 \le x \le 1,\ 2x \le y \le x^2 + 1\}$$

one deduces that

$$\iint_D x dx dy = \int_0^1 \left( \int_{2x}^{x^2+1} dy \right) x\ dx$$

$$= \int_0^1 \left( y|_{2x}^{x^2+1} \right) x\ dx = \int_0^1 \left( x^2 + 1 - 2x \right) x dx = \frac{1}{12}.$$

In Matlab 7.9 we shall have:
>>**syms x**

>> int(int(1,2*x,x^2+1)*x,0,1)
    ans =
1/12

We shall achieve the same result using Mathcad 14:

$$\int_0^1 \left( \int_{2\cdot x}^{x^2+1} 1\,dy \right) \cdot x\,dx \to \frac{1}{12}$$

and with Mathematica 8:

In[1]:= **Integrate[Integrate[1, {y, 2\*x, x^2 + 1}] \*x, {x, 0, 1}]**

Out[1]= $\dfrac{1}{12}$

and with Maple 15:

$$\int_0^1 \left( \int_{2\cdot x}^{x^2+1} 1\,dy \right) \cdot x\,dx$$

$$\frac{1}{12}$$

3) The next figure represents the domain $D$.

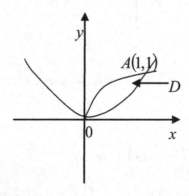

One can notice that

$$D = \left\{ (x,y) \in \mathbb{R}^2 \,\middle|\, 0 \le x \le 1,\ x^2 \le y \le \sqrt{x} \right\};$$

one obtains

$$\int\int_D \sqrt{xy}\,dx\,dy = \int_0^1 \left( \int_{x^2}^{\sqrt{x}} \sqrt{y}\,dy \right) \sqrt{x}\ dx$$

$$= \int_0^1 \left( \frac{y^{\frac{1}{2}+1}}{\frac{1}{2}+1} \bigg|_{x^2}^{\sqrt{x}} \right) \sqrt{x}\ dx = \frac{2}{3} \int_0^1 \left( x^{\frac{3}{4}} - x^3 \right) \sqrt{x}\,dx$$

$$= \frac{2}{3} \int_0^1 \left( x^{\frac{5}{4}} - x^{\frac{7}{2}} \right)\ dx = \frac{2}{3} \left( \frac{x^{\frac{5}{4}+1}}{\frac{5}{4}+1} \bigg|_0^1 - \frac{x^{\frac{7}{2}+1}}{\frac{7}{2}+1} \bigg|_0^1 \right)$$

$$= \frac{2}{3} \left( \frac{4}{9} - \frac{2}{9} \right) = \frac{4}{27}.$$

We can check this result using Matlab 7.9:
```
>> syms x y
>> int(int(sqrt(y),x^2,sqrt(x))*sqrt(x),0,1)
ans =
        4/27
```
and in Mathcad 14:

$$\int_0^1 \left( \int_{x^2}^{\sqrt{x}} \sqrt{y}\ dy \right) \cdot \sqrt{x}\ dx \rightarrow \frac{4}{27}$$

and using Mathematica 8:

In[2]:= `Integrate[Integrate[Sqrt[y], {y, x^2, Sqrt[x]}] * Sqrt[x], {x, 0, 1}]`

Out[2]= $\dfrac{4}{27}$

and with Maple 15:

$$\int_0^1 \left( \int_{x^2}^{\sqrt{x}} \sqrt{y}\ dy \right) \cdot \sqrt{x}\ dx$$

$$\frac{4}{27}$$

## 8.4   Applications of the Double Integral

### 8.4.1   Computing Areas

The area of a closed and bounded plane domain $D \subset \mathbb{R}^2$ can be computed (see [42], p. 114) using the double integral:

$$\text{Area} = \int dxdy. \tag{8.21}$$

**Example 8.21.** Use a double integral to compute the area of the domain

$$D = \left\{ (x,y) \in \mathbb{R}^2 \mid \ -y^2 \le x \le y, \ 0 \le y \le 1 \right\}.$$

**Solution.**

The next figure emphases the domain $D$.

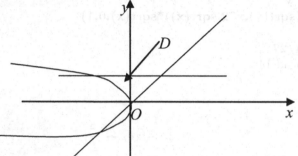

One deduces that

$$\text{Area} = \iint_D f(x,y) \, dxdy = \int_0^1 \left( \int_{-y^2}^{y} dx \right) dy = \int_0^1 (y + y^2) \, dy = \frac{5}{6}.$$

We can also compute this area in Matlab 7.9:

```
>> syms x y
>>A=int(int(1,x,-y^2,y),y,0,1)
A =
5/6
```

and using Mathcad 14:

$$\underset{\sim}{A} := \int_0^1 \left( \int_{-y^2}^{y} 1 \, dx \right) dy \to \frac{5}{6}$$

and with Mathematica 8:

In[2]:= **A = Integrate[Integrate[1, {x, -y^2, y}], {y, 0, 1}]**

Out[2]= $\dfrac{5}{6}$

and in Maple 15:

$$A = \int_0^1 \left( \int_{-y^2}^{y} 1 dx \right) dy$$

$$A = \frac{5}{6}$$

## 8.4.2  Mass of a Plane Plate

We shall use the following formula (see [42], p. 115) in order to compute the *mass* of a plane plate, having the form of a domain $D$ and the density $\rho(x, y)$:

$$\text{Mass}(D) = \int\int_D \rho(x, y) \, dxdy. \tag{8.22}$$

**Example 8.22.** Compute mass corresponding to a plane plate, having the form of the domain

$$D = \left\{ (x, y) \in \mathbb{R}^2 \mid x^2 + y^2 \leq 4, \ 3y \geq x^2 \right\}$$

and the density

$$\rho(x, y) = y.$$

**Solution.**
The following figure shows the domain $D$.

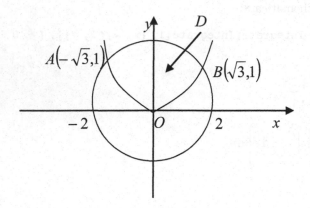

One can notice that

$$D = \left\{ (x,y) \in \mathbb{R}^2 \mid -\sqrt{3} \leq x \leq \sqrt{3}, \frac{x^2}{3} \leq y \leq \sqrt{4 - x^2} \right\}.$$

It will result

$$\text{Mass}(D) = \int_{-\sqrt{3}}^{\sqrt{3}} \left( \int_{\frac{x^2}{3}}^{\sqrt{4-x^2}} y \, dy \right) dx = \int_{-\sqrt{3}}^{\sqrt{3}} \left( y^2 \Big|_{\frac{x^2}{3}}^{\sqrt{4-x^2}} \right) dy$$

$$= \frac{1}{2} \int_{-\sqrt{3}}^{\sqrt{3}} \left( 4 - x^2 - \frac{x^4}{9} \right) dx$$

$$= \frac{1}{2} \left( 4x \Big|_{-\sqrt{3}}^{\sqrt{3}} - \frac{x^3}{3} \Big|_{-\sqrt{3}}^{\sqrt{3}} - \frac{x^5}{45} \Big|_{-\sqrt{3}}^{\sqrt{3}} \right) = \frac{14\sqrt{3}}{5}.$$

The same result can be obtained using Matlab 7.9:
```
>>syms x y
>> M=int(int(y,y,x^2/3,sqrt(4-x^2)),x,-sqrt(3),sqrt(3))
M=
14/5*3^(1/2)
```
and in Mathcad 14:

$$\text{Mass} := \int_{-\sqrt{3}}^{\sqrt{3}} \left( \int_{\frac{x^2}{3}}^{\sqrt{4-x^2}} y \, dy \right) dx \rightarrow \frac{14 \cdot \sqrt{3}}{5}$$

and using Mathematica 8:

```
In[2]:= Mass = Integrate[Integrate[y, {y, x^2/3, Sqrt[4 - x^2]}],
         {x, -Sqrt[3], Sqrt[3]}]
```

$$\text{Out[2]=} \quad \frac{14\sqrt{3}}{5}$$

and with Maple 15:

$$Mass = \int_{-\sqrt{3}}^{\sqrt{3}} \left( \int_{\frac{x^2}{3}}^{\sqrt{4-x^2}} y\,dy \right) dx$$

$$Mass = \frac{14}{5}\sqrt{3}$$

## 8.4.3 Coordinates the Centre of Gravity of a Plane Plate

The next formulae (see [42], p. 115) help us to compute the coordinates of the *centre of gravity* $G\left(x_G, y_G\right)$, corresponding to a plane plate, having the form of a domain $D$ and the density $\rho\left(x, y\right)$:

$$\begin{cases} x_G = \frac{\int\int_D x\rho(x,y)\,dxdy}{\int\int_D \rho(x,y)\,dxdy} \\[2ex] y_G = \frac{\int\int_D y\rho(x,y)\,dxdy}{\int\int_D \rho(x,y)\,dxdy}. \end{cases} \tag{8.23}$$

**Example 8.23.** Compute the area and the centre of gravity coresponding to a homogeneous plane plate, with the form of the domain bounded by the curve $y = \sin x$ and the straight line $(OA)$ which crosses through the origin and through the point $A\left(\frac{\pi}{2}, 1\right)$ from the first dial.

   **Solution.**
   The next figure shows the domain $D$.

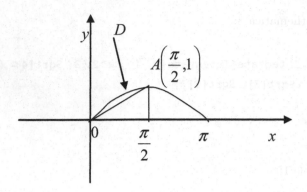

We notice that

$$D = \left\{ (x,y) \in \mathbb{R}^2 \mid 0 \le x \le \frac{\pi}{2}, \frac{2}{\pi}x \le y \le \sin x \right\}$$

and

$$(OA): \frac{x-0}{\frac{\pi}{2}} = \frac{y-0}{1-0} \implies (OA): \frac{2}{\pi}x = y.$$

Using the relation (8.22) we shall obtain that

$$\text{Area} = \int\!\!\int_D dxdy = \int_0^{\frac{\pi}{2}} \left( \int_{\frac{2}{\pi}x}^{\sin x} dy \right) dx = \int_0^{\frac{\pi}{2}} \left( y \Big|_{\frac{2}{\pi}x}^{\sin x} \right) dx$$

$$= \int_0^{\frac{\pi}{2}} \left( \sin x - \frac{2}{\pi}x \right) dx = -\cos x \Big|_0^{\frac{\pi}{2}} - \frac{2}{\pi} \cdot \frac{x^2}{2} \Big|_0^{\frac{\pi}{2}} = 1 - \frac{\pi}{4}.$$

We have to compute the following integrals:

$$\int\!\!\int_D xdxdy = \int_0^{\frac{\pi}{2}} \left( \int_{\frac{2}{\pi}x}^{\sin x} dy \right) xdx = \int_0^{\frac{\pi}{2}} \left( x \sin x - \frac{2}{\pi}x^2 \right) dx$$

$$= -\int_0^{\frac{\pi}{2}} x (\cos x)' dx - \frac{2}{\pi} \cdot \frac{x^3}{3} \Big|_0^{\frac{\pi}{2}} = -x \cos x \Big|_0^{\frac{\pi}{2}} + \int_0^{\frac{\pi}{2}} \cos x dx - \frac{\pi^2}{12} = 1 - \frac{\pi^2}{12}.$$

$$\int\!\!\int_D ydxdy = \int_0^{\frac{\pi}{2}} \left( \int_{\frac{2}{\pi}x}^{\sin x} y \, dy \right) dx = \int_0^{\frac{\pi}{2}} \left( \frac{y^2}{2} \Big|_{\frac{2}{\pi}x}^{\sin x} \right) dx = \int_0^{\frac{\pi}{2}} \frac{\sin^2 x}{2} dx - \frac{4}{\pi^2} \int_0^{\frac{\pi}{2}} \frac{x^2}{2} dx = \frac{\pi}{24};$$

such that $\begin{cases} x_G = \frac{12-\pi^2}{12-3\pi} \\ y_G = \frac{\pi}{24-6\pi}. \end{cases}$

We can also solve this problem using Matlab 7.9:

>>syms x y
>> a=int(int(1,y,(2*x)/pi,sin(x)),x,0,pi/2);
>> I1=int(int(x,y,(2*x)/pi,sin(x)),x,0,pi/2); I2=int(int(y,y,
(2*x)/pi,sin(x)),x,0,pi/2);
>> xg=simple(I1/a)
xg =
(-12+pi^2)/(3*pi-12)
>> yg=simple(I2/a)
yg =
-pi/(6*pi-24)

and in Mathcad 14:

$$I1 := \int_0^{\frac{\pi}{2}} \left( \int_{\frac{2}{\pi} \cdot x}^{\sin(x)} 1 \, dy \right) dx \qquad I2 := \int_0^{\frac{\pi}{2}} \left( \int_{\frac{2}{\pi} \cdot x}^{\sin(x)} 1 \, dy \right) \cdot x \, dx \qquad I3 := \int_0^{\frac{\pi}{2}} \left( \int_{\frac{2}{\pi} \cdot x}^{\sin(x)} y \, dy \right) dx$$

$$xG := \frac{I2}{I1} \qquad xG \text{ simplify} \rightarrow \frac{\pi^2 - 12}{3 \cdot (\pi - 4)} \qquad yG := \frac{I3}{I1} \qquad yG \rightarrow -\frac{\pi}{6 \cdot \pi - 24}$$

and using Mathematica 8:

In[6]:= I1 := Integrate[Integrate[1, {y, 2 / Pi * x, Sin[x]}], {x, 0, Pi / 2}]

In[7]:= I2 := Integrate[Integrate[1, {y, 2 / Pi * x, Sin[x]}] * x, {x, 0, Pi / 2}]

In[8]:= I3 := Integrate[Integrate[y, {y, 2 / Pi * x, Sin[x]}], {x, 0, Pi / 2}]

In[9]:= xG = I2 / I1

Out[9]= $\dfrac{1 - \frac{\pi^2}{12}}{1 - \frac{\pi}{4}}$

In[10]:= yG = I3 / I1

Out[10]= $\dfrac{\pi}{24 \left(1 - \frac{\pi}{4}\right)}$

and in Maple 15:

$$I1 := \int_0^{\frac{\pi}{2}} \left( \int_{\frac{2}{\pi} \cdot x}^{\sin(x)} 1 \, dy \right) dx :$$

$$I2 := \int_0^{\frac{\pi}{2}} \left( \int_{\frac{2}{\pi} \cdot x}^{\sin(x)} 1 \, dy \right) x dx :$$

$$I3 := \int_0^{\frac{\pi}{2}} \left( \int_{\frac{2}{\pi} \cdot x}^{\sin(x)} y \, dy \right) dx :$$

$$xG = factor\left( \frac{I2}{I1} \right)$$

$$xG = \frac{1}{3} \frac{-12 + \pi^2}{-4 + \pi}$$

$$yG = factor\left( \frac{I3}{I1} \right)$$

$$yG = -\frac{1}{6} \frac{\pi}{-4 + \pi}$$

## 8.4.4   Moments of Inertia of a Plane Plate

The *moments of inertia* of a plane plate, having the form of a domain $D$ and the density $\rho(x, y)$, relative to the $Ox$ and $Oy$- axes are, respectively (see [15], p. 244):

$$I_x = \int \int_D y^2 \rho(x, y) \, dx dy \tag{8.24}$$

$$I_y = \int \int_D x^2 \rho(x, y) \, dx dy. \tag{8.25}$$

**Example 8.24.** Calculate moments of inertia relative to the axes of coordinates, for the plate which has the form of the domain

$$D = \{(x, y) \in \mathbb{R}^2 | \ x + y \leq 1, \ x \geq 0, \ y \geq 0\}$$

if its density is

$$\rho(x, y) = xy.$$

**Solution.**
The following figure shows the domain $D$.

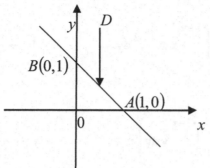

In order to compute $I_x$, we shall assume that the domain $D$ is simple relative to the $Ox$ axis:

$$D = \left\{ (x, y) \in \mathbb{R}^2 \mid 0 \leq y \leq 1, \ 0 \leq x \leq 1 - y \right\}.$$

Therefore

$$I_x = \int_0^1 \left( \int_0^{1-y} x \, dx \right) y^3 \, dy = \int_0^1 \left( \frac{x^2}{2} \Big|_0^{1-y} \right) y^3 \, dy = \frac{1}{2} \int_0^1 y^3 (1-y)^2 \, dy$$

$$= \frac{1}{2} \int_0^1 \left( y^3 - 2y^4 + y^5 \right) dy = \frac{1}{120}.$$

In order to compute $I_y$, we shall assume that the domain $D$ is simple relative to the $Oy$ axis:

$$D = \left\{ (x, y) \in \mathbb{R}^2 \mid 0 \leq x \leq 1, \ 0 \leq y \leq 1 - x \right\}.$$

Therefore

$$I_y = \int_0^1 \left( \int_0^{1-x} y \, dy \right) x^3 \, dx = \int_0^1 \left( \frac{y^2}{2} \Big|_0^{1-x} \right) x^3 \, dx = \frac{1}{2} \int_0^1 x^3 (1-x)^2 \, dx$$

$$= \frac{1}{2} \int_0^1 \left( x^3 - 2x^4 + x^5 \right) dx = \frac{1}{120}.$$

We can easier obtain this result using Matlab 7.9:

```
>> syms x y
>>Ix=int(y^3*int(x,x,0,1-y),y,0,1)
Ix =
1/120
>> Iy=int(x^3*int(y,y,0,1-x),x,0,1)
Iy =
1/120
```

or in Mathcad 14:

$$\text{Ix} := \int_0^1 \left( \int_0^{1-y} x \, dx \right) \cdot y^3 \, dy \qquad\qquad \text{Ix} \to \frac{1}{120}$$

$$\text{Iy} := \int_0^1 \left( \int_0^{1-x} y \, dy \right) \cdot x^3 \, dx \qquad\qquad \text{Iy} \to \frac{1}{120}$$

or with Mathematica 8:

In[1]:= **Ix = Integrate[y^3 \* Integrate[x, {x, 0, 1 - y}], {y, 0, 1}]**

Out[1]= $\dfrac{1}{120}$

In[2]:= **Iy = Integrate[x^3 \* Integrate[y, {y, 0, 1 - x}], {x, 0, 1}]**

Out[2]= $\dfrac{1}{120}$

or using Maple 15:

$$\text{Ix} := \int_0^1 \left( \int_0^{1-y} x \, dx \right) y^3 \, dy$$

$$\frac{1}{120}$$

$$\text{Iy} := \int_0^1 \left( \int_0^{1-x} y \, dy \right) x^3 \, dx$$

$$\frac{1}{120}$$

## 8.4.5  Computing Volumes

The *volume* $V$ of the body bounded above by a continuous surface $z = f(x, y)$, $(x, y) \in D$, below by the plane $z = 0$ and on the sides by a right cylindrical surface which cuts out of the $xOy$ -plane a region $D$ (see Fig. 8.3), is equal to (see [8] and [42], p. 114)

$$V_{body} = \int\int_D f(x, y) \, dxdy. \qquad (8.26)$$

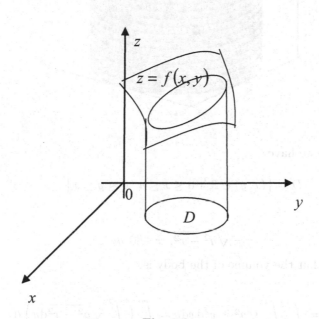

Fig. 8.3

**Example 8.25.** Use the double integral to compute the body volume from the first octant bounded by the cylinder

$$x^2 + z^2 = a^2 \ (a = \mathrm{ct} > 0)$$

and the planes

$$\begin{cases} y = 0 \\ z = 0 \\ y = x. \end{cases}$$

**Solution.**
The next figure pictures the body.

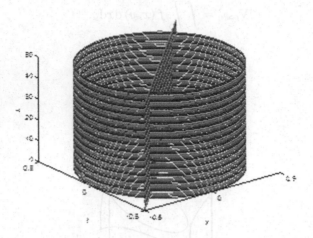

In our case we have

$$D = \left\{ (x,y) \in \mathbb{R}^2 \mid 0 \le x \le a, \ 0 \le y \le x \right\}.$$

As

$$z = \sqrt{a^2 - x^2}, \ x \in [0, a]$$

one deduces that the volume of the body is:

$$V_{body} = \int_0^a \int_0^x \sqrt{a^2 - x^2} \, dx \, dy = \int_0^a \left( \int_0^x \sqrt{a^2 - x^2} \, dy \right) dx$$
$$= \int_0^a x\sqrt{a^2 - x^2} \, dx,$$

where

$$I = \int_0^a x\sqrt{a^2 - x^2} \, dx = \int_0^a x \cdot \frac{a^2 - x^2}{\sqrt{a^2 - x^2}} \, dx$$
$$= -\int_0^a a^2 \left( \sqrt{a^2 - x^2} \right)' dx + \int_0^a x^2 \left( \sqrt{a^2 - x^2} \right)' dx$$
$$= -a^2 \sqrt{a^2 - x^2} \Big|_0^a + x^2 \sqrt{a^2 - x^2} \Big|_0^a - 2 \underbrace{\int_0^a x\sqrt{a^2 - x^2} \, dx}_{I};$$

therefore

$$3I = -a^2 (0 - a) = a^3 \implies$$
$$I = \frac{1}{3} a^3,$$

i.e.

$$V_{body} = \frac{1}{3}a^3.$$

In Matlab 7.9 we shall have:
>> **syms x y a**
>> **simple(int(int(sqrt(a^2-x^2),y,0,x),x,0,a))**
**ans =**
    **1/3*a^3**
We shall achieve the same result using Mathcad 14:

$$\int_0^a \left( \int_0^x \sqrt{a^2 - x^2} \, dy \right) dx \text{ simplify } \rightarrow \frac{\left(a^2\right)^{\frac{3}{2}}}{3}$$

or with Mathematica 8:

In[5]:= **V = Integrate[Integrate[Sqrt[a^2 - x^2], {y, 0, x}], {x, 0, a}]**

Out[5]= $\frac{1}{3} \left(a^2\right)^{3/2}$

or in Maple 15:

$$\int_0^a \left( \int_0^x \sqrt{a^2 - x^2} \, dy \right) dx$$

$$\frac{1}{3} a^3 \, csgn(a)$$

# 8.5   Change of Variables in Double Integrals

## 8.5.1   *Change of Variables in Polar Coordinates*

Using the change of variables in polar coordinates:

$$\begin{cases} x = \rho \cos \theta \\ y = \rho \sin \theta \end{cases}, \ \rho \geq 0, \ \theta \in [0, 2\pi] \tag{8.27}$$

we shall have (see [15], p. 243 and [8])

$$\int\int_D f(x,y)\,dxdy = \int\int_{D'} f(\rho\cos\theta,\rho\sin\theta)\cdot|J|\,d\rho d\theta, \qquad (8.28)$$

where

$$|J| = \frac{D(x,y)}{D(\rho,\theta)} = \rho$$

is the functional determinant (the Jacobian) of the functions $x$ and $y$.

**Example 8.26.** Use the change of variables in polar coordinates to compute the double integral:

1) $\displaystyle\int\int_D \frac{\ln(x^2+y^2)}{x^2+y^2}\,dxdy$, where $D = \{1 \le x^2 + y^2 \le e^2\}$

2) $\displaystyle\int\int_D (x^2+y^2)\,dxdy$, where $D = \{a^2 \le x^2 + y^2 \le b^2,\ 2x \le y \le 4x\}$

3) $\displaystyle\int\int_D x\,dxdy$, where $D = \{x^2 + y^2 \le 2x,\ x \le y\}$.

**Solution.**

1) Since computing this integral in rectangular coordinates is too difficult, we change to polar coordinates. After we apply the change of variables in polar coordinates, the domain $D$ will become $D'$:

$$D' = \{1 \le \rho \le e,\ 0 \le \theta \le 2\pi\}$$

and

$$\int\int_D \frac{\ln(x^2+y^2)}{x^2+y^2}\,dxdy = \int_1^e\int_0^{2\pi} \frac{\ln\rho^2}{\rho^2}\cdot\rho\,d\rho d\theta$$

$$= \int_1^e\left(\int_0^{2\pi} d\theta\right)\frac{\ln\rho^2}{\rho}\,d\rho = 2\pi\int_1^e \frac{\ln\rho^2}{\rho}\,d\rho$$

$$= \frac{2\pi}{2}\int_1^e \ln\rho^2\cdot(\ln\rho^2)'\,d\rho$$

$$= \pi\left.\frac{\ln^2\rho^2}{2}\right|_1^e = \frac{\pi}{2}\left(\ln^2 e^2 - \ln^2 1\right) = 2\pi.$$

We can also compute this double integral using Matlab 7.9, in the following two ways:

*Method 1.* Select successively File->New->M-file and write the following instructions:

```
function yt=w(r,th)
x=r*cos(th);
y=r*sin(th);
f=log(x.^2+y.^2)./(x.^2+y.^2);
yt=f.*r;
end
```
Save the file with w.m then write in the command line:
>> I=dblquad(@w,1,exp(1),0,2*pi)
I =
  6.2832
*Method 2.*
>> syms rho th
>> x=rho*cos(th);
>> y=rho*sin(th);
>> eval(int(int(log(x^2+y^2)/(x^2+y^2)*rho,rho,1,exp(1))),
th,0,2*pi))
  ans =
  6.2832
and using Mathematica 8:

In[5]:= x := ρ * Cos[θ]

In[6]:= y := ρ * Sin[θ]

In[7]:= Integrate[Integrate[1, {θ, 0, 2 * pi}] * Log[x^2 + y^2] / (x^2 + y^2) *
        ρ, {ρ, 1, Exp[1]}]

Out[7]= 2 pi

and in Maple 15:

$$x := \rho \, \cos(\theta) :$$
$$y := \rho \cdot \sin(\theta) :$$
$$\int_1^e \left( \left( \int_0^{2 \cdot \pi} 1 d\theta \right) \cdot \frac{\ln(x^2 + y^2)}{x^2 + y^2} \cdot \rho \right) d\rho$$

$$2\pi$$

We can't compute this result in Mathcad 14.
2) Using the formulae (8.27) we obtain:

$$\left. \begin{array}{c} a^2 \leq \rho^2 \leq b^2 \\ \rho \geq 0 \end{array} \right\} \Longrightarrow \rho \in [a, b],$$

$$2x \leq y \leq 4x \Longrightarrow 2 \leq \tan\theta \leq 4 \Longrightarrow \theta \in [\arctan 2, \ \arctan 4].$$

After the change of variables in polar coordinates, the domain will become :

$$D' = \{a \leq \rho \leq b, \ \arctan 2 \leq \theta \leq \arctan 4\}$$

and

$$\iint_D (x^2 + y^2) \, dxdy = \int_a^b \int_{\arctan 2}^{\arctan 4} \rho^2 \cdot \rho \, d\rho \, d\theta = \int_a^b \left( \int_{\arctan 2}^{\arctan 4} d\theta \right) \rho^3 \, d\rho$$

$$= (\arctan 4 - \arctan 2) \int_a^b \rho^3 \, d\rho$$

$$= (\arctan 4 - \arctan 2) \left. \frac{\rho^4}{4} \right|_a^b$$

$$= (\arctan 4 - \arctan 2) \cdot \frac{b^4 - a^4}{4}.$$

3) Using the change of variables in polar coordinates from (8.27), in our case one obtains:

$$\left. \begin{array}{c} \rho^2 \leq 2 \\ \rho \geq 0 \end{array} \right\} \Longrightarrow \rho \in \left[0, \sqrt{2}\right],$$

$$x \leq y \Longrightarrow 1 \leq \tan\theta \Longrightarrow \theta \in \left[\frac{\pi}{4}, \frac{5\pi}{4}\right].$$

The domain $D$ will become $D'$, from the next figure:

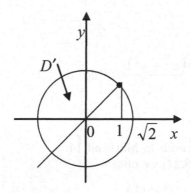

namely

$$D' = \left\{0 \leq \rho \leq \sqrt{2}, \ \frac{\pi}{4} \leq \theta \leq \frac{5\pi}{4}\right\}.$$

Such that,

$$\iint_D x\,\mathrm{d}x\mathrm{d}y = \int_0^{\sqrt{2}} \int_{\frac{\pi}{4}}^{\frac{5\pi}{4}} \rho^2 \cos\theta \;\mathrm{d}\rho \;\mathrm{d}\theta = \int_0^{\sqrt{2}} \left( \int_{\frac{\pi}{4}}^{\frac{5\pi}{4}} \cos\theta\mathrm{d}\theta \right) \rho^2 \;\mathrm{d}\rho$$

$$= \int_0^{\sqrt{2}} \left( \sin\theta\big|_{\frac{\pi}{4}}^{\frac{5\pi}{4}} \right) \rho^2 \;\mathrm{d}\rho = \left( \sin\frac{5\pi}{4} - \sin\frac{\pi}{4} \right) \frac{\rho^3}{3} \bigg|_0^{\sqrt{2}}$$

$$= \left( -\frac{\sqrt{2}}{2} - \frac{\sqrt{2}}{2} \right) \cdot \frac{2\sqrt{2}}{3} = -\frac{4}{3}.$$

We can also obtain this result in Matlab 7.9:
```
>> syms rho th
>> x=rho*cos(th);
>> y=rho*sin(th);
>> int(int(x*rho,rho,0,sqrt(2)),th,pi/4,5*pi/4)
     ans =
-4/3
```
and with Mathematica 8:

In[1]:= x := ρ * Cos[θ]

In[2]:= Integrate[Integrate[x * ρ, {ρ, 0, Sqrt[2]}], {θ, Pi / 4, 5 * Pi / 4}]

Out[2]= $-\dfrac{4}{3}$

and in Maple 15:

$$x := \rho \cdot \cos(\theta):$$

$$\int_{\frac{\pi}{4}}^{\frac{5\cdot\pi}{4}} \left( \int_0^{\sqrt{2}} x \cdot \rho \mathrm{d}\rho \right) \mathrm{d}\theta$$

$$-\frac{4}{3}$$

We can't compute this result in Mathcad 14.

## 8.5.2  Change of Variables in Generalized Polar Coordinates

Using the change of variables in generalized polar coordinates

$$\begin{cases} x = a\rho\cos\theta \\ y = b\rho\sin\theta \end{cases}, \ \rho \geq 0, \ \theta \in [0, 2\pi] \tag{8.29}$$

we shall have

$$\iint_D f(x,y)\,dxdy = \iint_{D'} f(a\rho\cos\theta, b\rho\sin\theta) \cdot |J| \ d\rho d\theta, \tag{8.30}$$

where

$$|J| = \frac{D(x,y)}{D(\rho,\theta)} = ab\rho$$

is the functional determinant (the Jacobian) of the functions $x$ and $y$.

**Example 8.27.** Use the change of variables in generalized polar coordinates to compute the double integral:

$$\iint_D \sqrt{2 - \frac{x^2}{a^2} - \frac{y^2}{b^2}}\,dxdy,$$

where

$$\frac{x^2}{a^2} + \frac{y^2}{b^2} \leq 1, \ x \geq 0, \ y \geq 0.$$

**Solution.**

After we apply the change of variables in the generalized polar coordinates (8.29), the domain $D$ will become $D'$:

$$D' = \left\{ 0 \leq \rho \leq 1, \ 0 \leq \theta \leq \frac{\pi}{2} \right\}$$

and

$$\iint_D \sqrt{2 - \frac{x^2}{a^2} - \frac{y^2}{b^2}}\,dxdy = \int_0^1 \int_0^{\pi/2} \sqrt{2 - \rho^2}\,ab\rho\ d\rho d\theta$$

$$= ab \int_0^1 \left( \int_0^{\frac{\pi}{2}} d\theta \right) \rho\sqrt{2 - \rho^2}\ d\rho$$

$$= \frac{ab\pi}{2} \int_0^1 \rho\sqrt{2 - \rho^2}\ d\rho =$$

$$= -\frac{ab\pi}{6} \int_0^1 \left( (2 - \rho^2)^{3/2} \right)'\ d\rho$$

$$= -\frac{ab\pi}{6} \left. (2 - \rho^2)^{3/2} \right|_0^1$$

$$= -\frac{ab\pi}{6} \left( 1 - 2\sqrt{2} \right) = \frac{ab\pi \left( 2\sqrt{2} - 1 \right)}{6}.$$

We shall have in Matlab 7.9:
>>syms rho th a b
>> I=simplify(int(int(a*b*rho*sqrt(2-rho^2),rho,0,1),
th,0,pi/2))
I =
1/6*a*b*(-1+2*2^(1/2))*pi
and in Mathcad 14:

$$\int_0^1 \int_0^{\frac{\pi}{2}} \sqrt{2 - \rho^2} \cdot a \cdot b \cdot \rho \, d\theta \, d\rho \ \ simplify \ \ \rightarrow \ \ \frac{\pi \cdot a \cdot b \cdot (2 \cdot \sqrt{2} - 1)}{6}$$

and with Mathematica 8:

In[5]:= **Integrate[Integrate[Sqrt[2 - ρ^2] \* a \* b \* ρ, {ρ, 0, 1}],**
　　　　**{θ, 0, Pi / 2}]**

Out[5]= $\dfrac{1}{6} \left( -1 + 2 \sqrt{2} \right) a b \pi$

and using Maple 15:

$$simplify \left( \int_0^1 \int_0^{\frac{\pi}{2}} \sqrt{2 - \rho^2} \cdot a \cdot b \cdot \rho \, d\theta d\rho \right)$$

$$\frac{1}{6} a b \pi \left( -1 + 2\sqrt{2} \right)$$

## 8.6   Riemann-Green Formula

The Riemann- Green formula is (see [48], p. 448):

$$\oint_C P \mathrm{d}x + Q \mathrm{d}y = \int \int_D \left( \frac{\partial Q}{\partial x} - \frac{\partial P}{\partial y} \right) \mathrm{d}x \mathrm{d}y, \qquad (8.31)$$

where:

- the direction of covering for the curve $C$ is the straight direction,
- $D$ is a closed domain, bounded by the curve $C$.

**Example 8.28.** Use the Riemann- Green formula to compute the following line integral and then check the obtained result by directly computing the line integral:

$$1) \ I = \oint_C e^{x^2+y^2} \left(-y dx + x dy\right), \text{where } (C) : x^2 + y^2 = 1;$$

$$2) \ J = \oint_C (x+y) \, dx - (x-y) \, dy,$$

where $(C)$ : $\dfrac{x^2}{a^2} + \dfrac{y^2}{b^2} = 1$ is an ellipse, covered counter- clockwise;

$$3) \ K = \oint_C 2 \left(x^2 + y^2\right) dx + (x+y)^2 \, dy, \text{where } C \text{ is the contour of the triangle,}$$

having the vertices: $A\,(1,1)$, $B\,(2,2)$, $C\,(1,3)$, covered in the straight direction.

**Solutions.**
1) For this integral:

$$\begin{cases} P\,(x,y) = -y e^{x^2+y^2} \\ Q\,(x,y) = x e^{x^2+y^2} \end{cases}$$

and

$$\begin{cases} \dfrac{\partial Q}{\partial x} = e^{x^2+y^2} + 2x^2 e^{x^2+y^2} \\ \dfrac{\partial P}{\partial y} = -e^{x^2+y^2} - 2y^2 e^{x^2+y^2}. \end{cases}$$

Therefore, we shall have:

$$I = \int\int_D \left(2 e^{x^2+y^2} + 2x^2 e^{x^2+y^2} + 2y^2 e^{x^2+y^2}\right) dx dy,$$

where

$$(D): \ x^2 + y^2 \le 1.$$

We choose polar coordinates since the disk is easily described in polar coordinates. Using the change of variables in polar coordinates:

$$\begin{cases} x = \rho \cos \theta \\ y = \rho \sin \theta \end{cases}, \ \rho \in [0,1], \ \theta \in [0, 2\pi]$$

one obtains

$$I = \int_0^1 \int_0^{2\pi} \left( 2e^{\rho^2} + 2e^{\rho^2}\rho^2 \right) \rho\,d\rho\,d\theta = 2\int_0^1 \int_0^{2\pi} e^{\rho^2} \left( 1 + \rho^2 \right) \rho\,d\rho\,d\theta,$$

$$(8.32)$$

namely

$$I = 2\int_0^1 \left( \int_0^{2\pi} d\theta \right) e^{\rho^2} \left( 1 + \rho^2 \right) \rho\,d\rho = 4\pi \left[ \int_0^1 e^{\rho^2} \rho\,d\rho + \int_0^1 e^{\rho^2}\rho^3\,d\rho \right]$$

$$= 4\pi \left[ \frac{1}{2} \int_0^1 \left( e^{\rho^2} \right)' d\rho + \frac{1}{2} \int_0^1 \left( e^{\rho^2} \right)' \rho^2\,d\rho \right]$$

$$= 2\pi \left[ e^{\rho^2} \Big|_0^1 + \rho^2 e^{\rho^2} \Big|_0^1 - 2\int_0^1 \rho \left( e^{\rho^2} \right)' d\rho \right]$$

$$= 2\pi \left( e - 1 + e - \int_0^1 \left( e^{\rho^2} \right)' d\rho \right) = 2\pi \left( e - 1 + e - e^{\rho^2} \Big|_0^1 \right)$$

$$= 2\pi \left( 2e - 1 - e + 1 \right) = 2\pi e.$$

We shall compute the integral from (8.32) in Matlab 7.9:
>> **syms rho th**
>> **I=2\*int(int(rho\*exp(rho^2)\*(1+rho^2),rho,0,1),th,0,2\*pi)**
**I=**

　　　**2\*exp(1)\*pi**
and with Mathcad 14:

$$2 \cdot \int_0^1 \int_0^{2\cdot\pi} e^{\rho^2} \cdot \left( 1 + \rho^2 \right) \rho\,d\theta\,d\rho \to 2 \cdot \pi \cdot e$$

and using Mathematica 8:

In[6]:= **2 \* Integrate[Integrate[Exp[ρ^2] \* (1 + ρ^2) \*ρ, {ρ, 0, 1}],**
　　　　　**{θ, 0, 2 \* Pi}]**

Out[6]= **2 e π**

and in Maple 15:

$$2 \cdot \int_0^1 \int_0^{2\cdot\pi} e^{\rho^2} \cdot \left( 1 + \rho^2 \right) \cdot \rho\,d\theta\,d\rho$$

$$2\,e\,\pi$$

We shall also compute directly the line integral, taking into account the parametrical equations of the circle

$$(C) : x^2 + y^2 = 1,$$

which are:

$$\begin{cases} x = \cos\theta \\ y = \sin\theta \end{cases}, \quad \theta \in [0, 2\pi].$$

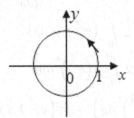

It will result that

$$I = \int_0^{2\pi} e\left[(-\sin\theta)(-\sin\theta) + \cos\theta \cdot \cos\theta\right] d\theta = e \int_0^{2\pi} d\theta = 2\pi e.$$

2) In the case of this integral:

$$\begin{cases} P(x,y) = x + y \\ Q(x,y) = -(x - y) \end{cases}$$

and

$$\begin{cases} \frac{\partial Q}{\partial x} = -1 \\ \frac{\partial P}{\partial y} = 1. \end{cases}$$

Therefore, one obtains:

$$J = \int\int_D (-1 - 1)\, dxdy = -2\int\int_D dxdy,$$

where

$$(D) : \frac{x^2}{a^2} + \frac{y^2}{b^2} \le 1.$$

Using the change of variables in generalized polar coordinates:

$$\begin{cases} x = a\rho\cos\theta \\ y = b\rho\sin\theta \end{cases}, \quad \rho \in [0, 1],\ \theta \in [0, 2\pi]$$

it will result:

$$J = 2 \int_0^1 \left( \int_0^{2\pi} d\theta \right) ab\rho d\rho = -2ab \cdot 2\pi \left. \frac{\rho^2}{2} \right|_0^1 = -2\pi ab.$$

We can also compute directly this line integral, taking into account the parametrical equations of the ellipse

$$(C) : \frac{x^2}{a^2} + \frac{y^2}{b^2} = 1,$$

which are

$$\begin{cases} x = a\cos\theta \\ y = b\sin\theta \end{cases}, \ \theta \in [0, 2\pi].$$

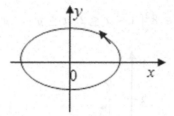

It will result that

$$J = \int_0^{2\pi} [(a\cos\theta + b\sin\theta)(-a\sin\theta) - (a\cos\theta - b\sin\theta) \cdot b\cos\theta] \, d\theta =$$

$$= \int_0^{2\pi} (-a^2 \sin\theta\cos\theta - ab\sin^2\theta - ab\cos^2\theta + b^2\sin\theta\cos\theta) \, d\theta$$

$$= (b^2 - a^2) \underbrace{\int_0^{2\pi} \sin\theta\cos\theta d\theta}_{I_1} - ab \int_0^{2\pi} d\theta = (b^2 - a^2) I_1 - ab \cdot 2\pi,$$

where

$$I_1 = \int_0^{2\pi} \sin\theta\cos\theta d\theta = \int_0^{2\pi} \sin\theta \cdot (\sin\theta)' \, d\theta$$

$$= \left. \frac{\sin^2\theta}{2} \right|_0^{2\pi} - \int_0^{2\pi} \sin\theta\cos\theta d\theta = 0 - 2I_1 \implies I_1 = 0.$$

Finally,

$$J = (b^2 - a^2) \cdot 0 - 2\pi ab = -2\pi ab.$$

3) For this integral:

$$\begin{cases} P(x,y) = 2\left(x^2 + y^2\right) \\ Q(x,y) = (x+y)^2 \end{cases}$$

and

$$\begin{cases} \frac{\partial Q}{\partial x} = 2(x+y) \\ \frac{\partial P}{\partial y} = 4y. \end{cases}$$

Therefore, one obtains:

$$K = \int\int_D (2x + 2y - 4y)\,\mathrm{d}x\mathrm{d}y = 2\int\int_D (x-y)\,\mathrm{d}x\mathrm{d}y,$$

where from the Figure 8.4 one can notice that:

$$D = \left\{(x,y) \in \mathbb{R}^2 \mid 1 \le x \le 2,\ x \le y \le -x + 4\right\}.$$

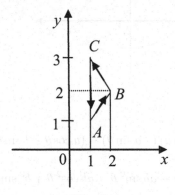

**Fig. 8.4**

Taking into account the equations of the straight lines $AB$ and $CB$:

$$AB : \frac{x-1}{2-1} = \frac{y-1}{2-1} \Longleftrightarrow$$
$$AB : x = y;$$

$$CB : \frac{x-1}{2-1} = \frac{y-3}{2-3} \Longleftrightarrow$$
$$CB : x - 1 = -y + 3 \Longleftrightarrow$$
$$CB : y = -x + 4$$

we shall have:

$$K = 2 \int_1^2 \left( \int_x^{4-x} (x-y)\, dy \right) dx = 2 \int_1^2 \left[ x(4-2x) - \frac{y^2}{2} \Big|_x^{4-x} \right] dx$$

$$= 2 \int_1^2 \left[ 4x - 2x^2 - \frac{1}{2}(16 - 8x + x^2 - x^2) \right] dx$$

$$= 2 \int_1^2 (4x - 2x^2 - 8 + 4x)\, dx = 4 \int_1^2 (-x^2 + 4x - 4)\, dx$$

$$= 4 \left( -\frac{x^3}{3} \Big|_1^2 + 4\frac{x^2}{2} \Big|_1^2 - 4x\Big|_1^2 \right) = -\frac{4}{3}.$$

One can check this result by computing directly the line integral:

$$K = \oint_C 2(x^2 + y^2)\, dx + (x+y)^2\, dy = K = \underbrace{\oint_{AB} 2(x^2 + y^2)\, dx + (x+y)^2\, dy}_{K_1}$$

$$+ \underbrace{\oint_{BC} 2(x^2 + y^2)\, dx + (x+y)^2\, dy}_{K_2} + \underbrace{\oint_{CA} 2(x^2 + y^2)\, dx + (x+y)^2\, dy}_{K_3}$$

$$= K_1 + K_2 + K_3.$$

Taking into account the parametrical equations of the straight lines $AB$, $BC$ and $CA$:

$$(AB) : \begin{cases} x = t \\ y = t \end{cases}, \ t \in [1, 2]$$

$$(BC) : \begin{cases} x = t \\ y = -t + 4 \end{cases}, \ t \in [2, 1]$$

$$(CA) : \begin{cases} x = 1 \\ y = t \end{cases}, \ t \in [3, 1]$$

we shall compute:

$$K_1 = \oint_{AB} 2\left(x^2 + y^2\right) dx + (x+y)^2 dy = \int_1^2 \left(2 \cdot 2t^2 \cdot 1 + 4t^2 \cdot 1\right) dt$$

$$= \int_1^2 8t^2 dt = 8 \left.\frac{t^3}{3}\right|_1^2 = 8 \cdot \frac{7}{3},$$

$$K_2 = \oint_{BC} 2\left(x^2 + y^2\right) dx + (x+y)^2 dy = -\int_1^2 \left[2\left(t^2 + t^2 - 8t + 16\right) - 16\right] dt$$

$$= -\int_1^2 \left(4t^2 - 16t + 16\right) dt = -4\int_1^2 \left(t^2 - 4t + 4\right) dt$$

$$= -\left(\left.\frac{t^3}{3}\right|_1^2 - 4 \left.\frac{t^2}{2}\right|_1^2 + 4\,t\big|_1^2\right) = -\frac{4}{3},$$

$$K_3 = \oint_{CA} 2\left(x^2 + y^2\right) dx + (x+y)^2 dy = -\int_1^3 \left[2\left(1 + t^2\right) \cdot 0 + \left(1 + t^2\right)^2 \cdot 1\right] dt$$

$$= -\int_1^3 \left(t^2 + 2t + 1\right) dt = -\left(\left.\frac{t^3}{3}\right|_1^3 + 2\left.\frac{t^2}{2}\right|_1^3 + t\big|_1^3\right) = -\frac{56}{3}$$

which make that

$$K = K_1 + K_2 + K_3 = \frac{56}{3} - \frac{4}{3} - \frac{56}{3} = -\frac{4}{3}.$$

## 8.7 Problems

1. Noting that the expression under the integral is a total differential calculate the line integral

$$\int_C x dx + y dy,$$

along the parabola $y = x^2$ between the points $A\left(0, 0\right)$ and $B\left(2, 4\right)$.

2. Compute the line integral:

$$\int_{AB} xy ds,$$

where $AB$ is the straight line segment which links the points $A\left(1, 2\right)$ and $B\left(2, 4\right)$.

3. Compute the line integral

$$I = \int_C \frac{y dx}{1 + y^2},$$

where $C$ is the arc: $y^2 = x$, with $y \geq 0$.

4. Compute the line integral:

$$\oint_\Gamma xy dx + z dy - x dz,$$

where $\Gamma$ is the closed curve from the following figure:

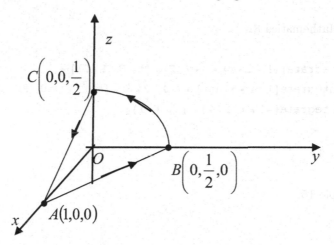

where

$$(AB) : \begin{cases} x = 1 - 2t \\ y = t \\ z = 0 \end{cases}, \quad t \in \left[0, \frac{1}{2}\right]$$

$$(BC) : \begin{cases} x = 0 \\ y = \frac{1}{2}\cos t \\ z = \frac{1}{2}\sin t \end{cases}, \quad t \in \left[0, \frac{\pi}{2}\right]$$

$$(CA) : \begin{cases} x = t \\ y = 0 \\ z = \frac{1-t}{2} \end{cases}, \quad t \in [0, 1].$$

**Computer solution.**
We shall compute this integral using Matlab 7.9:

>> syms t
>> int((1-2*t)*t*(-2),t,0,1/2)+int(1/2*sin(t)*(-1/2)*sin(t),
t,0,pi/2)+int(-t*(-1/2),t,0,1)
ans =
1/6 - pi/16

and in Mathcad 14:

$$\int_0^{\frac{1}{2}} (1-2{\cdot}t){\cdot}t{\cdot}(-2)\ dt + \int_0^{\frac{\pi}{2}} \frac{1}{2}{\cdot}\sin(t){\cdot}\left(-\frac{1}{2}{\cdot}\sin(t)\right)\ dt + \int_0^1 (-t){\cdot}\left(\frac{-1}{2}\right)\ dt \rightarrow \frac{1}{6} - \frac{\pi}{16}$$

and with Mathematica 8:

```
In[4]:= Integrate[(1 - 2 * t) * t * (-2), {t, 0, 1/2}] +
        Integrate[1/2 * Sin[t] * (-1/2 * Sin[t]), {t, 0, Pi/2}] +
        Integrate[-t * (-1/2), {t, 0, 1}]

Out[4]= 1/6 - π/16
```

and in Maple 15:

$$\int_0^{\frac{1}{2}} (1-2{\cdot}t){\cdot}t{\cdot}(-2)\ dt + \int_0^{\frac{\pi}{2}} \frac{1}{2}{\cdot}\sin(t){\cdot}\left(-\frac{1}{2}{\cdot}\sin(t)\right)\ dt + \int_0^1 (-t){\cdot}\left(-\frac{1}{2}\right)\ dt$$

$$\frac{1}{6} - \frac{1}{16}\pi \tag{1}$$

5. Compute the area of the cardioid:

$$(C): \begin{cases} x = 2a\cos t - a\cos 2t \\ y = 2a\sin t - a\sin 2t \end{cases}, t \in [0, 2\pi].$$

**Computer solution.**
We shall have in Matlab 7.9:
Step I. Select *File->New->M-file* and then write:
syms a t
x=2*a*cos(t)-a*cos(2*t);

```
y=2*a*sin(t)-a*sin(2*t);
yt=diff(y,t);
int(x*yt,0,2*pi)
```
Step II. Save the file area.m and then write in the command line:
```
>>area
ans =
6*a^2*pi
```
and using Mathcad 14:

$$x(t,a) := 2 \cdot a \cdot \cos(t) - a \cdot \cos(2 \cdot t) \qquad y(t,a) := 2 \cdot a \cdot \sin(t) - a \cdot \sin(2 \cdot t)$$

$$\int_0^{2 \cdot \pi} x(t,a) \cdot \frac{d}{dt} y(t,a) \, dt \to 6 \cdot \pi \cdot a^2$$

and with Mathematica 8:

In[2]:= x := 2 * a * Cos[t] - a * Cos[2 * t]

In[3]:= y := 2 * a * Sin[t] - a * Sin[2 * t]

In[4]:= yt = D[y, t];

In[5]:= A = Integrate[x * yt, {t, 0, 2 * Pi}]

Out[5]= 6 a² π

and in Maple 15:

$$x := 2 \cdot a \cdot \cos(t) - a \cdot \cos(2 \cdot t) :$$
$$y := 2 \cdot a \cdot \sin(t) - a \cdot \sin(2 \cdot t) :$$
$$\int_0^{2 \cdot \pi} x \cdot \left( \frac{\partial}{\partial t} y \right) dt$$

$$6 \pi a^2$$

6. Use the Riemann- Green formula to compute the following line integral:

$$\oint_C \sqrt{x^2 + y^2} dx + y \left[ xy + \ln \left( x + \sqrt{x^2 + y^2} \right) \right] dy,$$ where $C$ is the contour of the circle

$$(C) : (x - 1)^2 + (y - 1)^2 = 1.$$

**Computer solution.**

Using the Green's formula we have to compute

$$I = \int\int_D \left[ y \left( y + \frac{1 + \frac{2x}{2\sqrt{x^2+y^2}}}{x + \sqrt{x^2 + y^2}} \right) - \frac{y}{\sqrt{x^2 + y^2}} \right] dxdy = \int\int_D y^2 dxdy,$$

where

$$(D) : (x - 1)^2 + (y - 1)^2 \le 1.$$

We shall achieve in Matlab 7.9:

`>> syms rho th`
`>> int(int((rho*sin(th)+1)^2*rho,th,0,2*pi),rho,0,1)`
`ans =`
`(5*pi)/4`

and with Mathcad 14:

$$\int_0^{2 \cdot \pi} \left[ \int_0^1 (\rho \cdot \sin(\theta) + 1)^2 \cdot \rho \, d\rho \right] d\theta \rightarrow \frac{5 \cdot \pi}{4}$$

and using Mathematica 8:

```
In[1]:= Integrate[Integrate[(ρ*Sin[θ] + 1)^2*ρ, {ρ, 0, 1}],
        {θ, 0, 2*Pi}]

Out[1]= 5π
        ──
        4
```

and with Maple 15:

$$simplify \left( \int_0^1 \int_0^{2 \cdot \pi} (\rho \cdot \sin(\theta) + 1)^2 \cdot \rho \, d\theta d\rho \right)$$

$$\frac{5}{4} \pi$$

7. Applying Green's formula, evaluate:

$$\oint_C -x^2 y dx + xy^2 dy, \text{ where } C \text{ is the contour of the circle}$$

$$(C): x^2 + y^2 = r^2.$$

8. Find that the expression from under the integral sign is a total differential to compute the line integral:

$$I = \int_{AB} \left[ \frac{x - 2y}{(y - x)^2} + x \right] dx + \left[ \frac{y}{(y - x)^2} - y^2 \right] dy$$

where only the ends of the integration curve have specified.

9. Compute the double integral:

$$\int \int \arcsin \sqrt{x + y} dx dy,$$

$D$ being bounded by:

$$\begin{cases} x + y = 0 \\ x + y = 1 \\ y = -1 \\ y = 1. \end{cases}$$

**Computer solution.**
We shall compute this integral using Matlab 7.9:
>>syms x y
>> int(int(asin(sqrt(x+y)),x,-y,1-y),y,-1,1)
ans =
pi/2
and in Mathematica 8:

In[2]:= Integrate[Integrate[ArcSin[(x + y)^(1/2)], {x, -y, 1 - y}], {y, -1, 1}]

Out[2]= $\dfrac{\pi}{2}$

and with Maple 15:

$$\int_{-1}^{1} \left( \int_{-y}^{1-y} \arcsin\left( \sqrt{x + y} \right) dx \right) dy$$

$$\frac{1}{2} \pi$$

We can't compute this integral in Mathcad 14.

10.Evaluate the double integral:

$$\int\int_D \ln (x + y)\, dxdy,$$

where
$$D = \{(x, y) \in \mathbb{R}^2 \mid x \in [0, 1],\ y \in [1, 2]\}.$$

**Computer solution.**
We shall give a computer solution using Matllab 7.9:
```
>>syms x y
>> int(int(log(x+y),y,1,2),x,0,1)
   ans =
log((81*3^(1/2))/16) - 3/2
```
and in Mathcad 14:

$$\int_1^2 \left( \int_0^1 \ln(x + y)\, dx \right) dy \rightarrow \frac{9 \cdot \ln(3)}{2} - 4 \cdot \ln(2) - \frac{3}{2}$$

and with Mathematica 8:

In[17]:= **Integrate[Log[x + y], {y, 1, 2}, {x, 0, 1}]**

Out[17]= $-\dfrac{3}{2} + \dfrac{\text{Log}[3]}{2} + \text{Log}\left[\dfrac{81}{16}\right]$

and in Maple 15:

$$\int_0^1 \left( \int_1^2 \ln(x + y)\, dy \right) dx$$

$$-\frac{3}{2} - 4\ln(2) + \frac{9}{2}\ln(3)$$

11.Evaluate the integral

$$\int\int_D \sqrt{xy - y^2}\, dxdy,$$

where $D$ is bounded by the triangle with the vertices:

$$O\,(0,0)\,,\ A\,(4,1)\ \text{and}\ B\,(1,1)\,.$$

**Computer solution.**
We shall give a computer solution in Matlab 7.9:
```
>>x=0:0.1:4;
>>y1=x/4;
>> xx=0:0.1:1;
>> y=xx;
>>plot(0,0,'sb',4,1,'sb',1,1,'sb',xx,y,'r',x,y1,'r' ,'
    MarkerFaceColor','b','LineWidth',2)
>>a=[4 1];b=[0 0];
>>line(a,b)
```

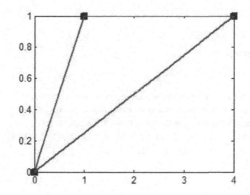

```
>>syms x y
>> int(int(sqrt(x*y-y^2),x,y,4*y),y,0,1)
ans =
(2*3^(1/2))/3
```
and with Mathcad 14:

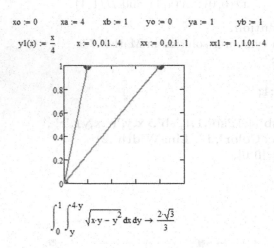

$$xo := 0 \qquad xa := 4 \qquad xb := 1 \qquad yo := 0 \qquad ya := 1 \qquad yb := 1$$

$$y1(x) := \frac{x}{4} \qquad x := 0, 0.1 .. 4 \qquad xx := 0, 0.1 .. 1 \qquad xx1 := 1, 1.01 .. 4$$

$$\int_0^1 \int_y^{4 \cdot y} \sqrt{x \cdot y - y^2} \, dx \, dy \to \frac{2 \cdot \sqrt{3}}{3}$$

and with Mathematica 8:

```
In[12]:= Show[Graphics[{PointSize[Large], Pink, Point[{0, 0}]}],
         Graphics[{PointSize[Large], Pink, Point[{1, 1}]}],
         Graphics[{PointSize[Large], Pink, Point[{4, 1}]}],
         Plot[x / 4, {x, 0, 4}], Plot[x, {x, 0, 1}], Plot[1, {x, 1, 4}]]
```

Out[12]=

```
In[13]:= Integrate[Sqrt[x * y - y^2], {y, 0, 1}, {x, y, 4 * y}]
```

Out[13]= $\dfrac{2}{\sqrt{3}}$

and in Maple 15:

*with*( *plottools*) :
*with*( *plots*) :
*display*( *line*( [ 0, 0], [ 1, 1], *color* = *blue*), *line*( [ 0, 0], [ 4, 1], *color*
   = *blue*), *line*( [ 4, 1], [ 1, 1], *color* = *blue*), *plot*( *Vector*( [ 0, 4, 1]),
   *Vector*( [ 0, 1, 1]), *style* = *point*))

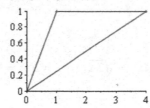

$$\int_0^1 \int_y^{4 \cdot y} \sqrt{x \cdot y - y^2} \; dx \, dy$$

$$\frac{2}{3}\sqrt{3}$$

12.Compute the moment of inertia of a triangle bounded by the straight lines $x + y = 2$, $x = 2$, $y = 2$ relative to the $Ox$- axis.

13.Compute the moment of inertia relative to the $Ox$- axis of the plane plate, by the density $\mu(x, y) = |x - y|$ , bounded by $y = \sqrt{2x}$, $y = 0$, with $0 \le x \le 2$.

**Computer solution.**
We shall give a computer solution using Matlab 7.9:
```
>> x=0:0.1:2;
>> y1=@(x) sqrt(2*x);
>> y2=@(x) 0*x;
>> y3=@(x) x;
>> plot(x,y1(x),'m',x,y2(x),'b',x,y3(x),'r','LineWidth',3)
```

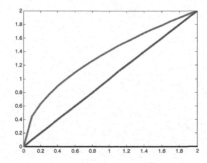

>> **syms x y**
>> **int(int(y^2\*(y-x),y,x,sqrt(2\*x)),x,0,2)+int(int(y^2\*(x-y),**
**y,0,x),x,0,2)**
  ans =
  **24/35**
and in Mathcad 14:

$$x := 0, 0.1 .. 2 \qquad y1(x) := \sqrt{2 \cdot x} \qquad y2(x) := 0 \quad y3(x) := x$$

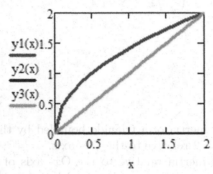

$$Ix := \int_0^2 \int_x^{\sqrt{2 \cdot x}} y^2 \cdot (y - x) \, dy \, dx + \int_0^2 \int_0^x y^2 \cdot (x - y) \, dy \, dx$$

$$Ix \to \frac{24}{35}$$

and with Mathematica 8:

In[6]:= **y1[x_] := Sqrt[2*x]**

In[7]:= **y2[x_] := 0**

In[8]:= **y3[x_] := x**

In[9]:= **Plot[{y1[x], y2[x], y3[x]}, {x, 0, 2}]**

Out[10]=
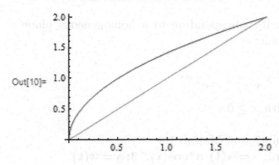

In[16]:= **Ix = Integrate[y^2 * (y - x), {x, 0, 2}, {y, x, Sqrt[2*x]}] +
Integrate[y^2 * (x - y), {x, 0, 2}, {y, 0, x}]**

Out[16]= $\dfrac{24}{35}$

and in Maple 15:

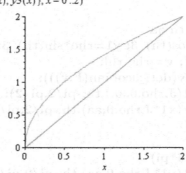

$y1 := x \to \sqrt{2 \cdot x}$ :
$y2 := x \to 0$ :
$y3 := x \to x$
$plot(\{y1(x), y2(x), y3(x)\}, x = 0 ..2)$

$$\int_0^2 \int_x^{\sqrt{2 \cdot x}} y^2 \cdot (y - x)\, dy\, dx + \int_0^2 \int_0^x y^2 \cdot (x - y)\, dy\, dx$$

$$\frac{24}{35}$$

14.Compute the centre of gravity coresponding to a homogeneous plane
   plate, bounded by the cycloid:

$$\begin{cases} x = a\,(t - \sin t) \\ y = a\,(1 - \cos t) \end{cases}, t \in [0, 2\pi]$$

and the straight line $y = 0$.

15.Compute the centre of gravity coresponding to a homogeneous plane
   plate, bounded by the astroide:

$$x^{2/3} + y^{2/3} = a^{2/3}$$

and the straight line $x = 0$, with $x \geq 0$.
   **Computer solution.**
   We shall give a computer solution in Matlab 7.9:
   >> a=1; t=0:0.01*pi:2*pi; x=@(t) a*cos(t).^3; y=@(t)
a*sin(t).^3;
   >> xx=-1:0.1:1; yy=@(xx) 0*xx;
   >> plot(yy(xx),xx,'b',xx,yy(xx),'b',x(t),y(t),'r','LineWidth',4)

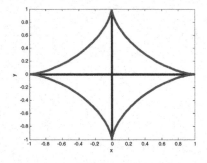

   >> syms rho th aa
   >> x1=rho*cos(th)^3; y1=rho*sin(th)^3;
   >> F=[x1 y1]; v=[rho th];
   >> J=simplify(det(jacobian(F,v)));
   >> I1=int(int(J,rho,0,aa),th,-pi/2,pi/2);
   >> I2=int(int(x1*J,rho,0,aa),th,-pi/2,pi/2);
   >> xG=I2/I1
   xG =
    (256*aa)/(315*pi)
   >> I3=int(int(y1*J,rho,0,aa),th,-pi/2,pi/2);
   >> yG=I3/I1
   yG =
   0
and in Mathcad 14:

$a := 1$    $t := 0, 0.01 \cdot \pi .. 2 \cdot \pi$    $xx := -1, -0.99 .. 1$

$x(t) := a \cdot \cos(t)^3$    $y(t) := a \cdot \sin(t)^3$    $yy(xx) := 0$

$x(t), xx, yy(xx)$

$$F(u) := \begin{pmatrix} u_0 \cdot \cos(u_1)^3 \\ u_0 \cdot \sin(u_1)^3 \end{pmatrix}$$

$x1(\rho, \theta) := F(u)_0$ substitute, $u_0 = \rho, u_1 = \theta \rightarrow \rho \cdot \cos(\theta)^3$ simplify $\rightarrow \rho \cdot \cos(\theta)^3$

$y1(\rho, \theta) := F(u)_1$ substitute, $u_0 = \rho, u_1 = \theta \rightarrow -\dfrac{\rho \cdot (\sin(3 \cdot \theta) - 3 \cdot \sin(\theta))}{4}$ simplify $\rightarrow \rho \cdot \sin(\theta)^3$

$J(u) := |\text{Jacob}(F(u), u)|$ simplify $\rightarrow \dfrac{3 \cdot \sin(2 \cdot u_1)^2 \cdot u_0}{4}$

$H(\rho, \theta) := J(u)$ substitute, $u_0 = \rho \rightarrow -\dfrac{3 \cdot \rho \cdot (\cos(4 \cdot u_1) - 1)}{8}$ substitute, $u_1 = \theta \rightarrow -\dfrac{3 \cdot \rho \cdot (\cos(4 \cdot \theta) - 1)}{8}$

$I1(aa) := \displaystyle\int_0^{aa} \int_{-\frac{\pi}{2}}^{\frac{\pi}{2}} H(\rho, \theta) \, d\theta \, d\rho \rightarrow \dfrac{3 \cdot \pi \cdot aa^2}{16}$    $I2(aa) := \displaystyle\int_0^{aa} \int_{-\frac{\pi}{2}}^{\frac{\pi}{2}} x1(\rho, \theta) \cdot H(\rho, \theta) \, d\theta \, d\rho \rightarrow \dfrac{16 \cdot aa^3}{105}$

$I3(aa) := \displaystyle\int_0^{aa} \int_{-\frac{\pi}{2}}^{\frac{\pi}{2}} y1(\rho, \theta) \cdot H(\rho, \theta) \, d\theta \, d\rho \rightarrow 0$

$xg(aa) := \dfrac{I2(aa)}{I1(aa)} \rightarrow \dfrac{256 \cdot aa}{315 \cdot \pi}$    $yg(aa) := \dfrac{I3(aa)}{I1(aa)} \rightarrow 0$

and with Mathematica 8:

In[1]:= **a := 1;**

In[2]:= **x[a_, t_] := a∗Cos[t]^3**

In[3]:= **y[a_, t_] := a∗Sin[t]^3**

In[4]:= **ParametricPlot[{x[a, t], y[a, t]}, {t, 0, 2∗Pi}]**

Out[4]=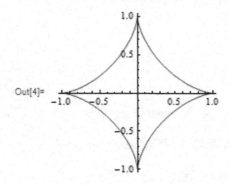

In[5]:= **J := Simplify[Det[D[{x[ρ, θ], y[ρ, θ]}, {{ρ, θ}}]]]**

In[6]:= **I1 := Integrate[J, {θ, -Pi/2, Pi/2}, {ρ, 0, aa}]**

In[7]:= **I2 := Integrate[x[ρ, θ]∗J, {θ, -Pi/2, Pi/2}, {ρ, 0, aa}]**

In[8]:= **I3 := Integrate[y[ρ, θ]∗J, {θ, -Pi/2, Pi/2}, {ρ, 0, aa}]**

In[9]:= **xG = I2/I1**

Out[9]= $\dfrac{256\,aa}{315\,\pi}$

In[10]:= **yG = I3/I1**

Out[10]= 0

and using Maple 15:

$$plot\left(\left[a \cdot \cos(t)^3\bigg|_{a=1}, a \cdot \sin(t)^3\bigg|_{a=1}, t=0 .. 2 \cdot \pi\right]\right)$$

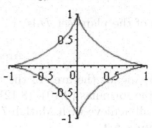

$x := (a, t) \rightarrow a \cdot \cos(t)^3$ :
$y := (a, t) \rightarrow a \cdot \sin(t)^3$ :
with( VectorCalculus ) :
with( LinearAlgebra ) :
$J := simplify\left(Determinant\left(Jacobian\left(\left[x(\rho, \theta), y(\rho, \theta)\right], [\rho, \theta]\right)\right)\right)$ :

$$I1 := \int_0^a \int_{-\frac{\pi}{2}}^{\frac{\pi}{2}} J\, d\theta\, d\rho :$$

$$I2 := \int_0^a \int_{-\frac{\pi}{2}}^{\frac{\pi}{2}} x(\rho, \theta) \cdot J\, d\theta\, d\rho :$$

$$I3 := \int_0^a \int_{-\frac{\pi}{2}}^{\frac{\pi}{2}} y(\rho, \theta) \cdot J\, d\theta\, d\rho :$$

$$xg = \frac{I2}{I1}; yg = \frac{I3}{I1}$$

$$xg = \frac{256}{315} \frac{a}{\pi}$$

$$yg = 0$$

16. Compute the area of the plane set $D$ bounded by the Lemniscate of Bernoulli:

$$\left(x^2 + y^2\right)^2 = 2a^2 \left(x^2 - y^2\right).$$

**Computer solution.**

Taking into account the polar equation of the Lemniscate of Bernoulli:

$$C : \rho^2 = 2a^2 \cos 2\theta, \quad -\frac{\pi}{4} \leq \theta \leq \frac{\pi}{4} \quad \text{(the first loop)}$$

the area of the plane set $D$ is

$$\text{Area} = 2 \cdot A_1,$$

where $A_1$ means the area of the first loop and it will be computed using each of the formulas (8.10)- (8.12).

We shall achieve with Matlab 7.9:

```
>> syms a t
>> x=a*sqrt(2*cos(2*t))*cos(t); y=a*sqrt(2*cos(2*t))*sin(t);
>> xt=diff(x,t); yt=diff(y,t);
>> A1=int(x*yt,t,-pi/4,pi/4); A2=-int(y*xt,t,-pi/4,pi/4);
A3=1/2*int(x*yt-y*xt,t,-pi/4,pi/4);
>> A=2*A1
A =
2*a^2
>> t=-pi/4:0.01*pi/4:pi/4;
>> x=a*sqrt(2*cos(2*t)).*cos(t); y=a*sqrt(2*cos(2*t)).*sin(t);
>>plot(x,y,'b',-x,y,'b','LineWidth',2)
```

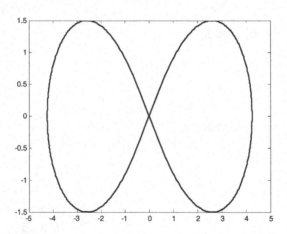

and in Mathcad 14:

$$x(a,t) := a \cdot \sqrt{2 \cdot \cos(2 \cdot t)} \cdot \cos(t) \quad y(a,t) := a \cdot \sqrt{2 \cdot \cos(2 \cdot t)} \cdot \sin(t) \quad yt(a,t) := \frac{d}{dt} y(a,t) \quad xt(a,t) := \frac{d}{dt} x(a,t)$$

$$A1(a) := \int_{-\frac{\pi}{4}}^{\frac{\pi}{4}} x(a,t) \cdot yt(a,t) \, dt \qquad A2(a) := \int_{-\frac{\pi}{4}}^{\frac{\pi}{4}} -y(a,t) \cdot xt(a,t) \, dt$$

$$A3(a) := \frac{1}{2} \int_{-\frac{\pi}{4}}^{\frac{\pi}{4}} x(a,t) \cdot yt(a,t) - y(a,t) \cdot xt(a,t) \, dt$$

$$A1(a) \to a^2 \qquad A2(a) \to a^2 \qquad A3(a) \to a^2 \qquad A(a) := 2 \cdot A1(a) \to 2 \cdot a^2$$

$$a := 3 \qquad t := -\frac{\pi}{4}, -\frac{\pi}{4} + 0.01 \cdot \frac{\pi}{4} \, .. \, \frac{\pi}{4}$$

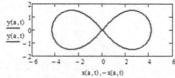

and in Mathematica 8:

```
In[22]:= x[a_, t_] := a * Sqrt[2 * Cos[2 * t]] * Cos[t];

In[23]:= y[a_, t_] := a * Sqrt[2 * Cos[2 * t]] * Sin[t];

In[24]:= xt = D[x[a, t], t]; yt = D[y[a, t], t];

In[25]:= A1 = Integrate[x[a, t] * yt, {t, -Pi / 4, Pi / 4}]
Out[25]= a²

In[26]:= A2 = -Integrate[y[a, t] * xt, {t, -Pi / 4, Pi / 4}]
Out[26]= a²

In[27]:= A3 = 1 / 2 * Integrate[x[a, t] * yt - y[a, t] * xt, {t, -Pi / 4, Pi / 4}]
Out[27]= a²

In[28]:= A = 2 * A1
Out[28]= 2 a²

In[30]:= ParametricPlot[{{x[3, t], y[3, t]}, {-x[3, t], y[3, t]}},
         {t, -Pi / 4, Pi / 4}, PlotStyle → Red]
```

Out[30]=

and using Maple 15:

$x := a \cdot \mathrm{sqrt}(2 \cdot \cos(2 \cdot t)) \cdot \cos(t) :$

$y := a \cdot \mathrm{sqrt}(2 \cdot \cos(2 \cdot t)) \cdot \sin(t) :$

$$A1 := \int_{-\frac{\pi}{4}}^{\frac{\pi}{4}} x \cdot \left( \frac{\partial}{\partial t} y \right) dt :$$

$$A2 := -\int_{-\frac{\pi}{4}}^{\frac{\pi}{4}} y \cdot \left( \frac{\partial}{\partial t} x \right) dt :$$

$$A3 := \frac{1}{2} \int_{-\frac{\pi}{4}}^{\frac{\pi}{4}} x \cdot \left( \frac{\partial}{\partial t} y \right) - y \cdot \left( \frac{\partial}{\partial t} x \right) dt :$$

$A = 2 \cdot A1$

$A = 2 a^2$

$$plot\left( \left[ \left[ x \Big|_{a=3} , y \Big|_{a=3} , t = -\frac{\pi}{4} .. \frac{\pi}{4} \right], \left[ -x \Big|_{a=3} , y \Big|_{a=3} , t = -\frac{\pi}{4} .. \frac{\pi}{4} \right] \right] \right)$$

17.Compute the double integral:

$$\iint_D (x + y)^4 (x - y)^2 \, dxdy,$$

where $D$ is the square bounded by the straight lines:

$$\begin{cases} x + y = 1 \\ x + y = -1 \\ x - y = -1 \\ x - y = -3. \end{cases}$$

18.Passing to polar coordinates, find the area bounded by:

$$\begin{cases} x^2 + y^2 = 2x \\ x^2 + y^2 = 4x \\ y = x \\ y = 0. \end{cases}$$

19.Determine the volume of the ellipsoid:

$$\frac{x^2}{a^2} + \frac{y^2}{b^2} + \frac{z^2}{c^2} = 1.$$

**Computer solution.**

We shall achieve the volume of the ellipsoid (see Fig. 8.4)

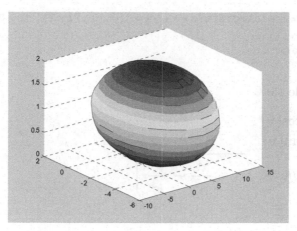

Fig. 8.4

with Matlab 7.9:

```
>> syms a b c rho th
>> x=a*rho*cos(th);
>>y=b*rho*sin(th);
>> z=simplify(c*sqrt(1-x^2/(a^2)-y^2/(b^2)));
>> Vol=8*int(int(z*a*b*rho,th,0,pi/2),rho,0,1)
Vol =
(4*pi*a*b*c)/3
```

and in Mathcad 14:

$$x(a,\rho,\theta) := a \cdot \rho \cdot \cos(\theta) \qquad y(b,\rho,\theta) := b \cdot \rho \cdot \sin(\theta)$$

$$z(a,b,c,\rho,\theta) := c \cdot \sqrt{1 - \frac{x(a,\rho,\theta)^2}{a^2} - \frac{y(b,\rho,\theta)^2}{b^2}} \quad \text{simplify} \rightarrow c \cdot \sqrt{1 - \rho^2}$$

$$\text{Vol}(a,b,c) := 8 \cdot \int_0^1 \int_0^{\frac{\pi}{2}} z(a,b,c,\rho,\theta) \cdot a \cdot b \cdot \rho \, d\theta \, d\rho \rightarrow \frac{4 \cdot \pi \cdot a \cdot b \cdot c}{3}$$

and using Mathematica 8:

In[1]:= $x := a * \rho * \text{Cos}[\theta]$

In[2]:= $y := b * \rho * \text{Sin}[\theta]$

In[3]:= $z := \text{Simplify}[c * \text{Sqrt}[1 - x^2 / (a^2) - y^2 / (b^2)]];$

In[4]:= $\text{Vol} = 8 * \text{Integrate}[z * a * b * \rho, \{\theta, 0, \text{Pi}/2\}, \{\rho, 0, 1\}]$

Out[4]= $\dfrac{4}{3} a b c \pi$

and with Maple 15:

$$x := a \cdot \rho \cdot \cos(\theta):$$
$$y := b \cdot \rho \cdot \sin(\theta):$$
$$z := simplify\left(c \cdot \sqrt{1 - \frac{x^2}{a^2} - \frac{y^2}{b^2}}\right):$$

$$Vol = 8 \cdot \int_0^1 \int_0^{\frac{\pi}{2}} z \cdot a \cdot b \cdot \rho d\theta \, d\rho$$

$$Vol = \frac{4}{3} c a b \pi$$

20.Find the volume of a solid bounded by the surfaces:

$$\begin{cases} x^2 + y^2 + z^2 = 4 \\ x^2 + y^2 = 2x \end{cases}$$

and the plane $(xOy)$, with $z \geq 0$.

# 9

# Triple and Surface Integral Calculus

## 9.1  Calculus Way of the Triple Integrals

The evaluation of a triple integral one reduces to the successive computation
of the three ordinary integrals or to the computation of one double and one
single integral (see [48], p. 491).

*Case 1.* Let $f(x, y, z)$ be a continuous function over a solid $V$ defined by

$$\begin{cases} a \leq x \leq b \\ \varphi_1(x) \leq y \leq \varphi_2(x) \\ \phi_1(x, y) \leq z \leq \phi_2(x, y). \end{cases}$$

Then, the triple integral is equal to the triple iterated integral:

$$\iiint_V f(x, y, z)\, \mathrm{d}x\mathrm{d}y\mathrm{d}z = \int_a^b \int_{\varphi_1(x)}^{\varphi_2(x)} \int_{\phi_1(x,y)}^{\phi_2(x,y)} f(x, y, z)\, \mathrm{d}z\mathrm{d}y\mathrm{d}x. \quad (9.1)$$

*Case 2.* In this case we define the domain $V$ as follows:

$$V = \{(x, y, z) \mid (x, y) \in D,\ \phi_1(x, y) \leq z \leq \phi_2(x, y)\},$$

where $(x, y) \in D$ is the notation that means that the point $(x, y)$ lies in the
region $D$ from the $xOy$ plane, i.e. $D =\mathrm{pr}_V xOy$.

G.A. Anastassiou and I.F. Iatan: Intelligent Routines, ISRL 39, pp. 475–572.
springerlink.com                    © Springer-Verlag Berlin Heidelberg 2013

In this case we shall evaluate the triple integral as follows:

$$\iiint_V f\left(x,y,z\right)\mathrm{d}x\mathrm{d}y\mathrm{d}z = \iint_D \left(\int_{\phi_1(x,y)}^{\phi_2(x,y)} f\left(x,y,z\right)\mathrm{d}z\right)\mathrm{d}y\mathrm{d}x. \quad (9.2)$$

**Example 9.1.** Compute the following triple integral:

$$I = \int_0^1 \int_0^1 \int_0^1 \frac{1}{\sqrt{x+y+z+1}}\mathrm{d}x\mathrm{d}y\mathrm{d}z.$$

**Solution.**
We shall have:

$$I = \int_0^1 \int_0^1 \int_0^1 \frac{1}{\sqrt{x+y+z+1}}\mathrm{d}x\mathrm{d}y\mathrm{d}z = \int_0^1 \mathrm{d}x \int_0^1 \mathrm{d}y \int_0^1 \frac{\mathrm{d}z}{\sqrt{x+y+z+1}}$$

$$= 2\int_0^1 \mathrm{d}x \int_0^1 \left[(x+y+z+1)^{1/2}\Big|_0^1\right]\mathrm{d}y$$

$$= 2\int_0^1 \mathrm{d}x \cdot \int_0^1 \left[(x+y+2)^{1/2} - (x+y+1)^{1/2}\right]\mathrm{d}y$$

$$= 2\cdot\frac{2}{3}\int_0^1 \left[(x+3)^{3/2} - (x+2)^{3/2} - (x+2)^{3/2} + (x+1)^{3/2}\right]\mathrm{d}x$$

$$= \frac{4}{3}\cdot\frac{2}{5}\left[4^{5/2} - 3^{5/2} - 2\cdot\left(3^{5/2} - 2^{5/2}\right) + 2^{5/2} - 1\right]$$

$$= \frac{8}{15}\left[32 - 27\sqrt{3} + 12\sqrt{2} - 1\right],$$

i.e.

$$I = 0.6428.$$

We can also solve this integral in Matlab 7.9:
```
>>syms x y z
>> I=eval(int(int(int(1/sqrt(x+y+z+1),z,0,1),y,0,1),x,0,1))
I =
  0.6428
```
or
```
>>syms x y z
>> f=@(x,y,z) 1./sqrt(x+y+z+1) ;
>> I=triplequad(f,0,1,0,1,0,1)
I =
  0.6428
```
and in Mathcad 14:

$$\int_0^1 \int_0^1 \int_0^1 \frac{1}{\sqrt{x+y+z+1}}\, dx\, dy\, dz = 0.6428$$

and with Mathematica 8:

In[4]:= `Integrate[1/Sqrt[x + y + z + 1], {x, 0, 1}, {y, 0, 1}, {z, 0, 1}];`

In[8]:= `SetPrecision[%, 4]`

Out[8]= `0.6428`

and using Maple 15:

*Digits* := 6

                                                                              6

$$evalf\left( \int_0^1 \int_0^1 \int_0^1 \frac{1}{\sqrt{x+y+z+1}}\, dx\, dy\, dz \right)$$

                                                                        0.6427

# 9.2   Change of Variables in Triple Integrals

## 9.2.1   *Change of Variables in Spherical Coordinates*

The spherical coordinates are a generalization of the polar coordinates.

The spherical coordinates are mostly used for the integrals over the solid that is bounded by a single sphere or more than one sphere.

Spherical coordinates $(\rho, \theta, \varphi)$ are represented in Fig. 9.1 (see [44], p. 667).

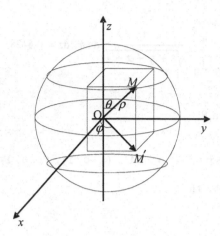

**Fig. 9.1** The spherical coordinates of a point from space

The relations between the Cartesian coordinates $(x, y, z)$ of a point from space and its spherical coordinates $(\rho, \theta, \varphi)$ are:

$$\begin{cases} x = \rho \sin\theta \cos\varphi \\ y = \rho \sin\theta \sin\varphi \\ z = \rho \cos\theta \end{cases}, \; \rho \geq 0, \; \theta \in [0, \pi], \; \varphi \in [0, 2\pi], \qquad (9.3)$$

where:

- $\rho$ represents the distance of the point $M$ from the origin,

- $\theta$ is the angle between the vector $\overrightarrow{OM}$ and the vector $\overrightarrow{k}$ (positive $Oz$-axis),

- $\varphi$ means the angle between the projection of vector $\overrightarrow{OM}$ on the $xOy$-plane and the vector $\overrightarrow{i}$ (positive $Ox$-axis).

Note that these equations satisfy

$$x^2 + y^2 + z^2 = \rho^2.$$

The functional determinant (Jacobian ) for the spherical coordinates is (see [44], p. 578):

$$|J| = \frac{D(x, y, z)}{D(\rho, \theta, \varphi)} = \begin{vmatrix} \frac{\partial x}{\partial \rho} & \frac{\partial x}{\partial \theta} & \frac{\partial x}{\partial \varphi} \\ \frac{\partial y}{\partial \rho} & \frac{\partial y}{\partial \theta} & \frac{\partial y}{\partial \varphi} \\ \frac{\partial z}{\partial \rho} & \frac{\partial z}{\partial \theta} & \frac{\partial z}{\partial \varphi} \end{vmatrix} = \rho^2 \sin\theta. \qquad (9.4)$$

So, we get that

$$\iiint_V f(x,y,z)\,dxdydz = \iiint_{V'} f(\rho\sin\theta\cos\varphi, \rho\sin\theta\sin\varphi, \rho\cos\theta)\cdot|J|\,d\rho d\theta d\varphi,$$

$$(9.5)$$

where

$$V' : \{\rho \geq 0,\ 0 \leq \theta \leq \pi,\ 0 \leq \varphi \leq 2\pi\}.$$

**Example 9.2.** Evaluate

$$I = \iiint_V \sqrt{1 + (x^2 + y^2 + z^2)^{3/2}}\,dxdydz,$$

where

$$V : x^2 + y^2 + z^2 \leq 1.$$

**Solution.**

Making the change of variable in spherical coordinates

$$\begin{cases} x = \rho\sin\theta\cos\varphi \\ y = \rho\sin\theta\sin\varphi \ ,\ \rho \in [0,1],\ \theta \in [0,\pi],\ \varphi \in [0,2\pi], \\ z = \rho\cos\theta \end{cases}$$

one deduces that

$$I = \int_0^1 \int_0^\pi \int_0^{2\pi} \sqrt{1 + (\rho^2)^{3/2}}\,\rho^2 \sin\theta\,d\rho d\theta d\varphi$$

$$- \int_0^{2\pi} \left[\int_0^\pi \left(\int_0^1 \rho^2\sqrt{1 + \rho^3}d\rho\right)\sin\theta d\theta\right]d\varphi$$

$$= \frac{2}{9}\int_0^{2\pi}\left[\int_0^\pi\left(\int_0^1\left((1+\rho^3)^{3/2}\right)'d\rho\right)\sin\theta d\theta\right]d\varphi$$

$$= \frac{2}{9}\cdot\left(\int_0^{2\pi}d\varphi\right)\left[\int_0^\pi\left((1+\rho^3)^{3/2}\Big|_0^1\right)\sin\theta d\theta\right]$$

$$= \frac{2}{9}\cdot 2\pi\left(2^{3/2}-1\right)\int_0^\pi\sin\theta d\theta = \frac{2}{9}\cdot 2\pi\left(2^{3/2}-1\right)\cdot(-\cos\theta|_0^\pi)$$

$$= \frac{4\pi}{9}\left(2\sqrt{2}-1\right)\cdot 2,$$

i.e.

$$I = \frac{8\pi}{9}\left(2\sqrt{2}-1\right) = 5.0159.$$

We have to build the following function in order to compute this integral in Matlab 7.9:

```
function yt=w(ro,th,phi)
x=ro*sin(th)*cos(phi);
y=ro*sin(th)*sin(phi);
z=ro*cos(th);
f=sqrt(1+(x.^2+y.^2+z.^2).^(3/2));
yt=f.*ro.^2.*sin(th);
end
```

In the command line one writes:

\>> I=triplequad(@w,0,1,0,pi,0,2*pi)

I =

  5.1059

We can also obtain this result using Mathcad 14:

$$\int_0^1 \int_0^\pi \int_0^{2\cdot\pi} \sqrt{1 + \left(\rho^2\right)^{\frac{3}{2}}} \cdot \rho^2 \cdot \sin(\theta) \, d\varphi \, d\theta \, d\rho = 5.1059$$

and with Mathematica 8:

In[13]:= **Integrate[Sqrt[1 + ($\rho$^2) ^ (3 / 2)] * $\rho$^2 * Sin[$\theta$], {$\varphi$, 0, 2 * Pi},**
        **{$\theta$, 0, Pi}, {$\rho$, 0, 1}];**

In[14]:= **SetPrecision[%, 6]**

Out[14]= 5.1059

and in Maple 15:

*Digits* := 5

$$5$$

$$evalf\left(\int_0^1 \int_0^\pi \int_0^{2\cdot\pi} \sqrt{1 + \left(\rho^2\right)^{\frac{3}{2}}} \cdot \rho^2 \cdot \sin(\theta) \, d\varphi \, d\theta \, d\rho\right)$$

$$5.1061$$

## 9.2.2  Change of Variables in Cylindrical Coordinates

Cylindrical coordinates are a generalization of two-dimensional polar coordinates to three dimensions by superposing a height $Oz$ axis.

If a point is described in cylindrical coordinates as in Figure 9.2 (see [44], p. 668), the equations of transformation between cylindrical coordinates $(\rho, \varphi, z)$ and Cartesian coordinates $(x, y, z)$ are as follows:

$$\begin{cases} x = \rho \cos \varphi \\ y = \rho \sin \varphi \quad, \rho \geq 0, \quad \varphi \in [0, 2\pi], \quad z \in \mathbb{R}, \\ z = z \end{cases}$$

where:

- $\rho$ represents the distance between the projection of the point $M$ on the $xOy$- plane and the origin O;

- $\varphi$ means the angle between the projection of the vector $\overrightarrow{OM}$ on the $xOy$-plane and the vector $\overrightarrow{i}$ (positive $Ox$ -axis);

- $z$ is the distance between the point $M$ and its projection on the $xOy$ -plane.

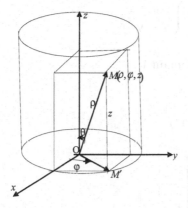

**Fig. 9.2** The cylindrical coordinates of a point from space

One can note that the relation between $x$, $y$, and $r$ is given by

$$r = \sqrt{x^2 + y^2}.$$

The Jacobian determinan for the cylindrical coordinates is (see [44], p. 578):

$$|J| = \frac{D(x, y, z)}{D(\rho, \varphi, z)} = \begin{vmatrix} \frac{\partial x}{\partial \rho} & \frac{\partial x}{\partial \varphi} & \frac{\partial x}{\partial z} \\ \frac{\partial y}{\partial \rho} & \frac{\partial y}{\partial \varphi} & \frac{\partial y}{\partial z} \\ \frac{\partial z}{\partial \rho} & \frac{\partial z}{\partial \varphi} & \frac{\partial z}{\partial z} \end{vmatrix} = \rho. \tag{9.6}$$

So, we get that

$$\iiint_V f(x,y,z)\,dxdydz = \iiint_{V'} f(\rho\cos\varphi, \rho\sin\varphi, z) \cdot |J|\,d\rho d\theta d\varphi,$$

(9.7)

where

$$V' : \{\rho \ge 0,\ 0 \le \varphi \le \pi,\ z \in \mathbb{R}\}.$$

**Example 9.4.** Compute the triple integral:

$$I = \iiint_V xy\,dxdydz,$$

where $V$ is bounded by the cylinder $x^2 + y^2 = R^2$ and by the planes

$$\begin{cases} z = 0 \\ z = 1 \\ y = x \\ y = \sqrt{3}x. \end{cases}$$

**Solution.**
The next figure shows the domain $V$.

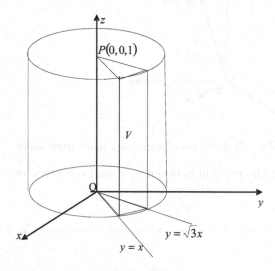

**Fig. 9.3**

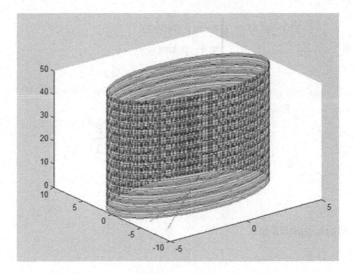

**Fig. 9.4**

In cylindrical coordinates, the region of integration $V$ is described by the domain

$$V' : \left\{ 0 \le \rho \le R, \ \frac{\pi}{4} \le \varphi \le \frac{\pi}{3}, \ 0 \le z \le 1 \right\}.$$

Writing the integral in cylindrical coordinates, we have:

$$I = \int\int\int_{V'} \rho^3 \cos\varphi \sin\varphi d\rho d\varphi dz = \int_0^R \int_{\pi/4}^{\pi/3} \int_0^1 \rho^3 \cos\varphi \sin\varphi d\rho d\varphi dz$$

$$= \int_0^R \rho^3 d\rho \int_{\pi/4}^{\pi/3} \cos\varphi \sin\varphi d\varphi \int_0^1 dz = \left( \frac{\rho^4}{4} \Big|_0^R \right) \cdot \left( \frac{\sin^2 \varphi}{2} \Big|_{\pi/4}^{\pi/3} \right) \cdot \left( z \Big|_0^1 \right) = \frac{R^4}{32}.$$

We can solve this problem using Matlab 7.9.

We have to select File->New->M-file and we have to write the following instructions:

**function yt=w(r,phi,z)**
**x=r*cos(phi); y=r*sin(phi); f=x*y; yt=f*r;**
**end**

The file will be saved with w.m and then one writes in the command line:

**>> syms R r phi z**
**>> I=int(int(int(w(r,phi,z),r,0,R),phi,pi/4,pi/3),z,0,1)**
**I =**
**1/32*R^4**

We shall also achieve this result in Mathcad 14:

$$yt(\rho, \varphi, z) := \begin{vmatrix} x \leftarrow \rho \cdot \cos(\varphi) \\ y \leftarrow \rho \cdot \sin(\varphi) \\ x \cdot y \cdot \rho \end{vmatrix}$$

$$\int_0^R \int_{\frac{\pi}{4}}^{\frac{\pi}{3}} \int_0^1 yt(\rho, \varphi, z) \, dz \, d\varphi \, d\rho \rightarrow \frac{R^4}{32}$$

and using Mathematica 8:

In[22]:= $x := \rho * Cos[\theta];$

In[23]:= $y := \rho * Sin[\theta];$

In[24]:= $yt[\rho\_, \theta\_, z\_] = x * y * \rho;$

In[25]:= $Integrate[yt[\rho, \theta, z], \{\rho, 0, R\}, \{\theta, Pi/4, Pi/3\}, \{z, 0, 1\}]$

Out[25]= $\dfrac{R^4}{32}$

and with Maple 15:

$x := \rho \cdot \cos(\varphi) :$
$y := \rho \cdot \sin(\varphi) :$
$yt := (\rho, \varphi, z) \rightarrow x \cdot y \cdot \rho$

$$(\rho, \varphi, z) \rightarrow xy\rho$$

$$\int_0^R \int_{\frac{\pi}{4}}^{\frac{\pi}{3}} \int_0^1 yt(\rho, \varphi, z) \, dz \, d\varphi \, d\rho$$

$$\frac{1}{32} R^4$$

## 9.3   Applications of the Triple Integrals

### 9.3.1   Mass of a Solid

The mass of a solid occupying the region $V$, which has the density $\delta\,(x,y,z)$ in the point $(x,y,z)$ is (see [42], p. 136):

$$\text{Mass} = \int\int\int_V \delta\,(x,y,z)\,\mathrm{d}x\mathrm{d}y\mathrm{d}z. \tag{9.8}$$

**Example 9.5.** Compute the mass of a solid body, by the form of the tetrahedron from the first octant, bounded by the planes: $x+y+z = 1$, $x = 0$, $y = 0$, $z = 0$, knowing that its density is $\delta\,(x,y,z) = xy$.
   **Solution.**
   We can notice that

$$V = \left\{(x,y,z) \in \mathbb{R}^3\,|\ 0 \le z \le 1-x-y,\ (x,y) \in D\right\},$$

where

$$D = \mathrm{pr}_V xOy = \left\{(x,y) \in \mathbb{R}^2\,|\ 0 \le x \le 1,\ 0 \le y \le 1-x\right\} = \Delta AOB,$$

see Fig. 9.5:

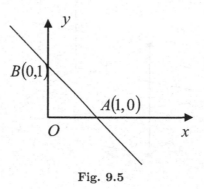

**Fig. 9.5**

Therefore,

$$\mathrm{M} = \int\int_D \left(\int_0^{1-x-y} xy\mathrm{d}z\right)\mathrm{d}x\mathrm{d}y = \int_0^1 \left(\int_0^{1-x}\left(\int_0^{1-x-y} xy\mathrm{d}z\right)\mathrm{d}y\right)\mathrm{d}x, \tag{9.9}$$

i.e.

$$M = \int_0^1 \left( \int_0^{1-x} xy\,(1-x-y)\,dy \right) dx = \int_0^1 \left( \left.\frac{xy^2}{2}\right|_0^{1-x} - \left.\frac{x^2y^2}{2}\right|_0^{1-x} - \left.\frac{xy^3}{3}\right|_0^{1-x} \right) dx$$

$$= \int_0^1 \left\{ \frac{x}{2} \left[ (1-x)^2 - x\,(1-x)^2 \right] - \frac{x}{3}\,(1-x)^3 \right\} dx$$

$$= \int_0^1 x\,(1-x)^2 \left[ \frac{1}{2} - \frac{1}{2}x - \frac{1}{3}\,(1-x) \right] dx$$

$$= \int_0^1 x\,(1-x)^2 \left[ \frac{1}{2} - \frac{1}{2}x - \frac{1}{3} + \frac{1}{3}x \right] dx = \int_0^1 x\,(1-x)^2 \left( \frac{1}{6} - \frac{1}{6}x \right) dx$$

$$= \frac{1}{6} \int_0^1 x\,(1-x)^3\,dx = \frac{1}{6} \int_0^1 x\,(1 - 3x + 3x^2 - x^3)\,dx$$

$$= \frac{1}{6} \left( \left.\frac{x^2}{2}\right|_0^1 - \left.\frac{3x^3}{3}\right|_0^1 + \left.\frac{3x^4}{4}\right|_0^1 - \left.\frac{x^5}{5}\right|_0^1 \right) = \frac{1}{120}.$$

A computer solution can be given in Matlab 7.9:

*Step I.* The following Matlab sequence allow us to represent the body.

```
>> x=[1 0 0 0 1 0]; y=[0 0 1 0 0 1]; z=[0 0 0 1 0 0];
>> plot3(x,y,z,1,0,0,'ob',0,1,0,'ob',0,0,1,'ob',0,0,0,'ob')
```

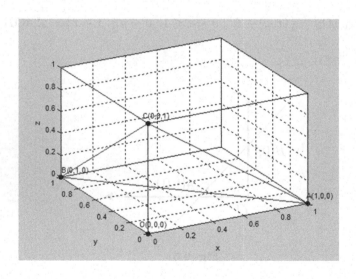

*Step II.* We compute the mass from (9.9) :

```
>> syms x y z
>> mass=int(int(int(x*y,z,0,1-x-y),y,0,1-x),x,0,1)
mass =
1/120
```

We shall also give a computer solution in Mathcad 14:

$$M := \int_0^1 \int_0^{1-x} x \cdot y \cdot (1 - x - y) \, dy \, dx \rightarrow \frac{1}{120}$$

and with Mathematica 8:

In[1]:= **M = Integrate[x * y * (1 - x - y), {x, 0, 1}, {y, 0, 1 - x}]**

Out[1]= $\dfrac{1}{120}$

and in Maple 15:

$$M = \int_0^1 \int_0^{1-x} x \cdot y \cdot (1 - x - y) \, dy \, dx$$

$$M = \frac{1}{120}$$

**Example 9.6.** Compute the mass of the body by the form

$$V = \left\{ (x, y, z) \in \mathbb{R}^3 \,\middle|\, \frac{x^2 + y^2}{a} \le z \le a, \ 0 \le y \le x \right\},$$

having the density $\delta(x, y, z) = xyz$.

**Solution.**

The next figure pictures the body.

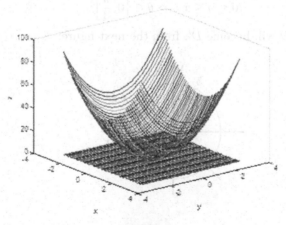

**Fig. 9.6**

One deduce that:

$$V = \left\{ (x, y, z) \in \mathbb{R}^3 \Big| \frac{x^2 + y^2}{a} \leq z \leq a, \ (x, y) \in D \right\},$$

where

$$D = \mathrm{pr}_V \ xOy = \left\{ (x, y) \in \mathbb{R}^2 \big| x^2 + y^2 \leq a^2, \ 0 \leq y \leq x \right\}.$$

We shall have:

$$M = \int \int \int_V xyz\,dxdydz = \int \int_D \left( \int_{(x^2+y^2)/a}^a z\,dz \right) xy\,dxdy, \qquad (9.10)$$

namely

$$M = \int \int_D \left( \frac{z^2}{2} \Big|_{\frac{x^2+y^2}{a}}^a \right) xy\,dxdy = \frac{1}{2} \int \int_D xy \left( a^2 - \frac{(x^2 + y^2)^2}{a^2} \right) dxdy.$$

Using the change of variables in polar coordinates:

$$\begin{cases} x = \rho\cos\theta \\ y = \rho\sin\theta \end{cases}, \ \rho \geq 0, \ \theta \in [0, 2\pi],$$

in our case, one obtains:

$$\begin{rcases} \rho^2 \leq a^2 \\ \rho \geq 0 \end{rcases} \Longrightarrow \rho \in [0, a],$$

$$0 \leq y \leq x \Longrightarrow \theta \in \left[0, \frac{\pi}{4}\right].$$

The domain $D$ will become $D'$, from the next figure:

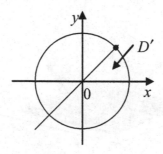

i.e.

$$D' : \left\{ 0 \leq \rho \leq a, \ 0 \leq \theta \leq \frac{\pi}{4} \right\}.$$

Such that,

$$M = \frac{1}{2} \int_0^a \int_0^{\pi/4} \rho^2 \sin\theta \cos\theta \left(a^2 - \frac{\rho^4}{a^2}\right) \rho \, d\rho \, d\theta$$

$$= \frac{1}{2} \int_0^a \left(\int_0^{\pi/4} \sin\theta \cos\theta \, d\theta\right) \cdot \rho^3 \left(a^2 - \frac{\rho^4}{a^2}\right) d\rho$$

$$= \frac{1}{2} \int_0^a \left(\left.\frac{\sin^2\theta}{2}\right|_0^{\pi/4}\right) \rho^3 \left(a^2 - \frac{\rho^4}{a^2}\right) d\rho$$

$$= \frac{1}{8} \left(\left.\frac{a^2\rho^4}{4}\right|_0^a - \left.\frac{\rho^8}{8a^2}\right|_0^a\right) = \frac{a^6}{64}.$$

We can also obtain this result in Matlab 7.9:
*Step 1.* One compute $\int_{(x^2+y^2)/a}^a z \, dz$.
`>> syms a x y z`
`>> I1=int(z,z,(x^2+y^2)/a,a)`
`I1 =`
`1/2*a^2-1/2*(x^2+y^2)^2/a^2`
*Step 2.* One computes the mass of the body:
`>> syms rho th`
`>> h=subs(subs(I1*x*y,x,rho*cos(th)),y,rho*sin(th));`
`>> Mass=int(int(h*rho,rho,0,a),th,0,pi/4)`
`Mass =`
`1/64*a^6`
The same result can be obtained with Mathcad 14:

$$h(a,\rho,\theta) := \int_{\sqrt{x^2+y^2}}^a z \, dz \text{ substitute}, x = \rho\cdot\cos(\theta), y = \rho\cdot\sin(\theta) \rightarrow \frac{a^4 - \rho^4}{2\cdot a^2}$$

$$x(\rho,\theta) := \rho\cdot\cos(\theta) \qquad y(\rho,\theta) := \rho\cdot\sin(\theta)$$

$$\int_0^a \int_0^{\frac{\pi}{4}} h(a,\rho,\theta)\cdot x(\rho,\theta)\cdot y(\rho,\theta)\cdot\rho \, d\theta \, d\rho \rightarrow \frac{a^6}{64}$$

and with Mathematica 8:

In[2]:= **x := ρ*Cos[θ]**

In[3]:= **y := ρ*Sin[θ]**

In[9]:= **I1 := Integrate[z, {z, (x^2 + y^2) / a, a}]**

In[10]:= **M = Integrate[I1*x*y*ρ, {ρ, 0, a}, {θ, 0, Pi/4}]**

Out[10]= $\dfrac{a^6}{64}$

and in Maple 15:

$$x := \rho \cdot \cos(\theta) :$$
$$y := \rho \cdot \sin(\theta) :$$

$$I1 := simplify \left( \int_{\frac{(x^2 + y^2)}{a}}^{a} z\, dz \right) :$$

$$M = \int_0^a \int_0^{\frac{\pi}{4}} I1 \cdot x \cdot y \cdot \rho \, d\theta \, d\rho$$

$$M = \frac{1}{64} a^6$$

**Example 9.7.** Use the change of variables in spherical coordinates to compute the mass of the body which has the form of the sphere

$$x^2 + y^2 + z^2 = R^2$$

and the density

$$\delta(x, y, z) = (x - y)^2 + z^2.$$

**Solution.**

In our case, the domain $V$ from the computing formula corresponding to the mass of the spherical body (see Fig. 9.8) is

$$V : \left\{ x^2 + y^2 + z^2 \leq R^2 \right\}.$$

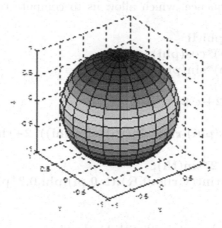

Fig. 9.8

Using the change of variables in spherical coordinates (9.3), the domain $V$ will become $V'$:

$$V' : \{0 \le \rho \le R,\ 0 \le \theta \le \pi,\ 0 \le \varphi \le 2\pi\}$$

and the mass of the solid will be:

$$M = \int_0^R \int_0^\pi \int_0^{2\pi} \delta\left(\rho \sin\theta \cos\varphi, \rho \sin\theta \sin\varphi, \rho \cos\theta\right) \cdot \rho^2 \sin\theta\ d\rho d\theta d\varphi.$$

As

$$\begin{aligned}
\delta\left(\rho\sin\theta\cos\varphi, \rho\sin\theta\sin\varphi, \rho\cos\theta\right) &= (\rho\sin\theta\cos\varphi - \rho\sin\theta\sin\varphi)^2 + \rho^2\cos^2\theta \\
&= \rho^2\sin^2\theta\,(\cos\varphi - \sin\varphi)^2 + \rho^2\cos^2\theta \\
&= \rho^2\sin^2\theta\,(1 - 2\sin\varphi\cos\varphi) + \rho^2\cos^2\theta \\
&= \rho^2\sin^2\theta - 2\rho^2\sin^2\theta\sin\varphi\cos\varphi + \rho^2\cos^2\theta \\
&= \rho^2 - 2\rho^2\sin^2\theta\sin\varphi\cos\varphi
\end{aligned}$$

it will result that:

$$\begin{aligned}
M &= \int_0^R \int_0^\pi \int_0^{2\pi} \left(\rho^2 - 2\rho^2\sin^2\theta\sin\varphi\cos\varphi\right) \cdot \rho^2\sin\theta\ d\rho d\theta d\varphi \\
&= \int_0^R \left[\int_0^\pi \left(2\pi\rho^4\sin\theta - 2\rho^4\sin^3\theta \cdot \left.\frac{\sin^2\varphi}{2}\right|_0^{2\pi}\right) d\theta\right] d\rho \\
&= \left.\frac{4\pi\rho^5}{5}\right|_0^R = \frac{4\pi R^5}{5}.
\end{aligned}$$

The Matlab 7.9 sequence, which allow us to compute the mass of the solid is:

```
>> syms rho th phi R
>> x=rho*sin(th)*cos(phi);
>> y=rho*sin(th)*sin(phi);
>> z=rho*cos(th);
>> delta=(x-y)^2+z^2
 delta =
 (rho*sin(th)*cos(phi)-rho*sin(th)*sin(phi))^2+rho^
2*cos(th)^2
>> v=delta*rho^2*sin(th);
>> masa=int(int(int(v,rho,0,R),th,0,pi),phi,0,2*pi)
 masa =
 4/5*R^5*pi
```

This result can also be obtained with Mathcad 14:

$$x(\rho,\theta,\varphi) := \rho \cdot \sin(\theta) \cdot \cos(\varphi) \quad y(\rho,\theta,\varphi) := \rho \cdot \sin(\theta) \cdot \sin(\varphi) \quad z(\rho,\theta,\varphi) := \rho \cdot \cos(\theta)$$

$$\delta(\rho,\theta,\varphi) := (x(\rho,\theta,\varphi) - y(\rho,\theta,\varphi))^2 + z(\rho,\theta,\varphi)^2$$

$$\int_0^R \int_0^\pi \int_0^{2\cdot\pi} \delta(\rho,\theta,\varphi) \cdot \rho^2 \cdot \sin(\theta) \, d\varphi \, d\theta \, d\rho \rightarrow \frac{4 \cdot \pi \cdot R^5}{5}$$

and in Mathematica 8:

```
In[7]:= x := ρ * Sin[θ] * Cos[φ]

In[8]:= y := ρ * Sin[θ] * Sin[φ]

In[9]:= z := ρ * Cos[θ]

In[10]:= δ[ρ, θ, φ] := (x - y)^2 + z^2

In[11]:= Integrate[δ[ρ, θ, φ] * ρ^2 * Sin[θ], {φ, 0, 2*Pi}, {θ, 0, Pi}, {ρ, 0, R}]

Out[11]= 4 π R^5
         ─────
           5
```

and using Maple 15:

$x := \rho \cdot \sin(\theta) \cdot \cos(\varphi) :$
$y := \rho \cdot \sin(\theta) \cdot \sin(\varphi) :$
$z := \rho \cdot \cos(\theta) :$
$\delta := (x - y)^2 + z^2 :$
$$\int_0^R \int_0^\pi \int_0^{2\cdot\pi} \delta \cdot \rho^2 \cdot \sin(\theta) \, d\varphi \, d\theta \, d\rho$$

$$\frac{4}{5} \pi R^5$$

### 9.3.2  Volume of a Solid

The volume of a region $V$ of space is (see [42], p. 136):

$$\mathrm{Vol} = \int \int \int_V dx dy dz. \qquad (9.11)$$

**Example 9.8.** Use a triple integral to compute the common volume for the sphere

$$x^2 + y^2 + z^2 = R^2$$

and the cylinder

$$x^2 + y^2 = r^2, \ r < R$$

see Fig. 9.9.

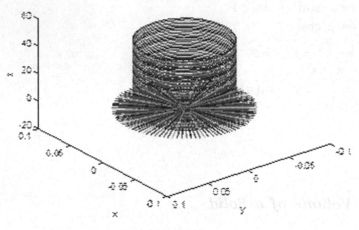

**Fig. 9.9**

**Solution.**
We shall have:

$$V = \left\{ (x, y, z) \in \mathbb{R}^3 \mid -\sqrt{R^2 - x^2 - y^2} \leq z \leq \sqrt{R^2 - x^2 - y^2}, \ (x, y) \in D \right\},$$

where

$$D = \left\{ (x, y) \in \mathbb{R}^2 \mid x^2 + y^2 \leq r^2 \right\}.$$

Therefore,

$$\mathrm{Vol} = \int\int_D \left( \int_{-\sqrt{R^2-x^2-y^2}}^{\sqrt{R^2-x^2-y^2}} \mathrm{d}z \right) \mathrm{d}x\mathrm{d}y = 2\int\int_D \sqrt{R^2 - x^2 - y^2}\,\mathrm{d}x\mathrm{d}y.$$

$$(9.12)$$

Making the chance of variables in polar coordinates:

$$\begin{cases} x = \rho\cos\theta \\ y = \rho\sin\theta \end{cases}, \ \rho \in [0, r], \ \theta \in [0, 2\pi]$$

gives

$$\mathrm{Vol} = 2\int_0^r \int_0^{2\pi} \rho\sqrt{R^2 - \rho^2}\,\mathrm{d}\rho\mathrm{d}\theta = 4\pi \underbrace{\int_0^r \rho\sqrt{R^2 - \rho^2}\,\mathrm{d}\rho}_{I_1} = 4\pi I_1,$$

where

$$I_1 = \int_0^r \rho\sqrt{R^2 - \rho^2}\,d\rho = \int_0^r \frac{\rho R^2}{\sqrt{R^2 - \rho^2}}\,d\rho - \int_0^r \frac{\rho^3}{\sqrt{R^2 - \rho^2}}\,d\rho$$

$$= -R^2 \int_0^r \left(\sqrt{R^2 - \rho^2}\right)'\,d\rho + \int_0^r \rho^2 \left(\sqrt{R^2 - \rho^2}\right)'\,d\rho$$

$$= -R^2 \left.\sqrt{R^2 - \rho^2}\right|_0^r + \rho^2 \left.\sqrt{R^2 - \rho^2}\right|_0^r - 2\int_0^r \rho\sqrt{R^2 - \rho^2}\,d\rho$$

$$= -R^2\sqrt{R^2 - r^2} + R^3 + r^2\sqrt{R^2 - r^2} - 2I_1$$

$$= R^3 - \left(R^2 - r^2\right)\sqrt{R^2 - r^2} - 2I_1;$$

therefore

$$3I_1 = R^3 - \left(R^2 - r^2\right)^{\frac{3}{2}} \implies$$

$$I_1 = \frac{1}{3}\left[R^3 - \left(R^2 - r^2\right)^{\frac{3}{2}}\right].$$

Finally,

$$\text{Vol} = \frac{4\pi}{3}\left[R^3 - \left(R^2 - r^2\right)^{\frac{3}{2}}\right].$$

We shall compute in Matlab 7.9 the volume from (9.12):
>> **syms rho th r R**
>>**x=rho\*cos(th);**
>>**y=rho\*sin(th);**
>> **u=simplify(sqrt(R^2-x^2-y^2));**
>> **Vol=2\*simple(int(int(u\*rho,rho,0,r),th,0,2\*pi))**
**Vol =**
**-4/3\*(R^2-r^2)^(3/2)\*pi+4/3\*(R^2)^(3/2)\*pi**
and using Mathcad 14:

$$\text{I1}(R,\rho,\theta) := \int_{-\sqrt{R^2-x^2-y^2}}^{\sqrt{R^2-x^2-y^2}} 1\,dz \quad \text{substitute}, x = \rho\cdot\cos(\theta), y = \rho\cdot\sin(\theta) \;\rightarrow\; 2\cdot\sqrt{R^2 - \rho^2}$$

$$\text{Vol}(R,r) := \int_0^r \int_0^{2\cdot\pi} \text{I1}(R,\rho,\theta)\cdot\rho\,d\theta\,d\rho \;\rightarrow\; \frac{4\cdot\pi\cdot\left[\left(R^2 - r^2\right)^{\frac{3}{2}} - \left(R^2\right)^{\frac{3}{2}}\right]}{3}$$

and in Mathematica 8:

```
In[46]:= x := ρ * Cos[θ]

In[47]:= y := ρ * Sin[θ]

In[48]:= I1 := Integrate[1, {z, -Sqrt[R^2 - x^2 - y^2], Sqrt[R^2 - x^2 - y^2]}]

In[49]:= Vol = Integrate[I1 * ρ, {θ, 0, 2 * Pi}, {ρ, 0, r}]
```

$$Out[49]= \frac{4}{3} \pi \sqrt{R^2} \left[ R^2 + \sqrt{1 - \frac{r^2}{R^2}} \ (r - R) \ (r + R) \right]$$

and with Maple 15:

$$x := \rho \cdot \cos(\theta) :$$
$$y := \rho \cdot \sin(\theta) :$$
$$I1 := simplify\left( \int_{-\sqrt{R^2 - x^2 - y^2}}^{\sqrt{R^2 - x^2 - y^2}} 1 \, dz \right) :$$
$$Vol = simplify\left( \int_0^r \left( \int_0^{2 \cdot \pi} I1 \cdot \rho \, d\theta \right) d\rho \right)$$
$$Vol = \frac{4}{3} \pi \left( R^3 \operatorname{csgn}(R) - \sqrt{R^2 - r^2} \, R^2 + \sqrt{R^2 - r^2} \, r^2 \right)$$

**Example 9.9.** Use a triple integral to find the volume of solid that lies between the paraboloid

$$x^2 + y^2 = az,$$

the cylinder

$$x^2 + y^2 = 2ax$$

and the plane $z = 0$, where $a =$ constant$> 0$.
**Solution.**
One can notice (see Fig. 9.10) that:

$$V = \left\{ (x, y, z) \in \mathbb{R}^3 | 0 \leq z \leq \frac{x^2 + y^2}{a}, \ (x, y) \in D \right\},$$

where

$$D = \left\{(x,y) \in \mathbb{R}^2 \mid x^2 + y^2 \le 2ax\right\} \Longleftrightarrow$$
$$D = \left\{(x,y) \in \mathbb{R}^2 \mid (x-a)^2 + y^2 \le a^2\right\}.$$

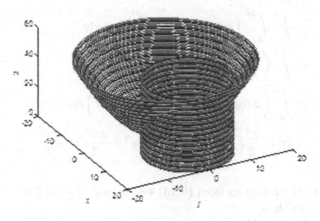

**Fig. 9.10**

We shall have

$$\text{Vol} = \int \int \int_V \mathrm{d}x\mathrm{d}y\mathrm{d}z = \int \int_D \left(\int_0^{(x^2+y^2)/a} \mathrm{d}z\right) \mathrm{d}x\mathrm{d}y = \int \int_D \frac{x^2+y^2}{a}\mathrm{d}x\mathrm{d}y. \quad (9.13)$$

We shall make the change of variable in polar coordinates:

$$\begin{cases} x = a + \rho\cos\theta \\ y = \rho\sin\theta \end{cases}, \ \rho \in [0,a], \ \theta \in [0, 2\pi].$$

Making this substitution gives

$$\text{Vol} = \int_0^a \int_0^{2\pi} \frac{(a + \rho \cos \theta)^2 + \rho^2 \sin^2 \theta}{a} \rho \, d\rho d\theta$$

$$= \int_0^a \int_0^{2\pi} \frac{a^2 + 2a\rho + \rho^2 \cos^2 \theta + \rho^2 \sin^2 \theta}{a} \rho \, d\rho d\theta$$

$$= \int_0^a \int_0^{2\pi} \left( a + 2\rho \cos \theta + \frac{\rho^2}{a} \right) \rho \, d\rho d\theta$$

$$= \int_0^a \left( \int_0^{2\pi} \rho \left( a + 2\rho \cos \theta + \frac{\rho^2}{a} \right) d\theta \right) d\rho$$

$$= \int_0^a \left( 2\pi a\rho + 2\rho^2 \sin \theta \big|_0^{2\pi} + \frac{\rho^3}{a} \cdot 2\pi \right) d\rho$$

$$= 2\pi a \cdot \frac{\rho^2}{2} \Big|_0^a + 2\pi \cdot \frac{\rho^4}{4a} \Big|_0^a = \pi a^3 + \frac{\pi a^3}{2} = \frac{3\pi a^3}{2}.$$

Tto evaluate the integrals from (9.13) we can use Matlab 7.9:
>> **syms rho th a**
>> **x=a+rho*cos(th);**
>>**y=rho*sin(th);**
>> **Vol=int(int(((x^2+y^2)/a)*rho,rho,0,a),th,0,2*pi)**
 Vol =
**3/2*a^3*pi**
and in Mathcad 14:

$$\text{I1}(a,\rho,\theta) := \int_0^{\frac{x^2+y^2}{a}} 1 \, dz \text{ substitute}, x = a + \rho \cdot \cos(\theta), y = \rho \cdot \sin(\theta) \rightarrow \frac{a^2 + 2 \cdot \cos(\theta) \cdot a \cdot \rho + \rho^2}{a}$$

$$\text{Vol}(a) := \int_0^a \int_0^{2 \cdot \pi} \text{I1}(a,\rho,\theta) \cdot \rho \, d\theta \, d\rho \rightarrow \frac{3 \cdot \pi \cdot a^3}{2}$$

and with Mathematica 8:

In[50]:= **x := a + ρ ∗ Cos [θ]**

In[51]:= **y := ρ ∗ Sin[θ]**

In[52]:= **I1 := Integrate[1, {z, 0, (x^2 + y^2) / a}]**

In[53]:= **Vol = Integrate[I1 ∗ ρ, {θ, 0, 2 ∗ Pi}, {ρ, 0, a}]**

Out[53]= $\dfrac{3\,a^3\,\pi}{2}$

and using Maple 15:

$$x := a + \rho \cdot \cos(\theta) :$$
$$y := \rho \cdot \sin(\theta) :$$
$$I1 := simplify\left(\int_0^{\left(\frac{x^2+y^2}{a}\right)} 1\,dz\right) :$$
$$Vol = simplify\left(\int_0^a\left(\int_0^{2\cdot\pi} I1\cdot\rho\,d\theta\right)d\rho\right)$$

$$Vol = \frac{3}{2}\,\pi a^3$$

## 9.3.3 Centre of Gravity

The coordinates of the centre of gravity corresponding to a solid by the form of a domain $V$ is the point $G\,(x_G, y_G, z_G)$, where (see [42], p. 136):

$$\begin{cases} x_G = \frac{\int\int\int_V x\delta(x,y,z)dxdydz}{\int\int\int_V \delta(x,y,z)dxdydz} \\ y_G = \frac{\int\int\int_V y\delta(x,y,z)dxdydz}{\int\int\int_V \delta(x,y,z)dxdydz} \\ z_G = \frac{\int\int\int_V z\delta(x,y,z)dxdydz}{\int\int\int_V \delta(x,y,z)dxdydz} \end{cases} \qquad (9.14)$$

If the solid is homogeneous, we can set $\delta\,(x, y, z) = 1$ in the formulae for the coordinates of the centre of gravity.

**Example 9.10.** Find the centre of gravity of a homogeneous body, bounded by the surfaces

$$y^2 + 2z^2 = 4x$$

and

$$x = 2.$$

**Solution.**

The next figure pictures the homogeneous body.

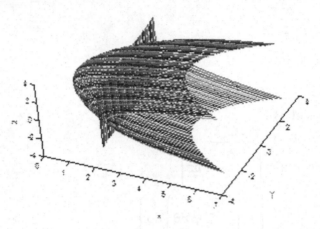

**Fig. 9.11**

We can notice that

$$V = \left\{ (x, y, z) \in \mathbb{R}^3 \mid \frac{y^2 + 2z^2}{4} \leq x \leq 2, \ (x, y) \in D \right\},$$

where

$$D = \left\{ (y, z) \in \mathbb{R}^2 \mid y^2 + 2z^2 \leq 8 \right\} \Longleftrightarrow$$

$$D = \left\{ (y, z) \in \mathbb{R}^2 \mid \frac{y^2}{8} + \frac{z^2}{4} \leq 1 \right\}.$$

We shall compute

$$I_1 = \int\int\int_V \mathrm{d}x \mathrm{d}y \mathrm{d}z = \int\int_D \left( \int_{(y^2+2z^2)/a}^2 \mathrm{d}x \right) \mathrm{d}y \mathrm{d}z = \int\int_D \left( 2 - \frac{y^2 + 2z^2}{4} \right) \mathrm{d}y \mathrm{d}z.$$

(9.15)

Using the change of variables in the generalized polar coordinates

$$\begin{cases} y = 2\sqrt{2}\rho \cos\theta \\ z = 2\rho \sin\theta \end{cases}, \ \rho \in [0, 1], \ \theta \in [0, 2\pi]$$

we shall get

$$I_1 = \int_0^1 \int_0^{2\pi} 2 \cdot 2\sqrt{2} \left(2 - 2\rho^2\right) \rho d\rho d\theta = 8\sqrt{2} \int_0^1 \left(\int_0^{2\pi} d\theta\right) \cdot \rho \left(1 - \rho^2\right) d\rho$$

$$= 16\pi\sqrt{2} \int_0^1 \rho \left(1 - \rho^2\right) d\rho = 16\pi\sqrt{2} \left(\frac{\rho^2}{2}\bigg|_0^1 - \frac{\rho^4}{4}\bigg|_0^1\right)$$

$$= 16\pi\sqrt{2} \cdot \frac{1}{4} = 4\pi\sqrt{2}.$$

Similarly, it will result:

$$I_2 = \int\int\int_V x dx dy dz = \int\int_D \left(\int_{(y^2+2z^2)/4}^2 x dx\right) dy dz = \frac{1}{2} \int\int_D \left[4 - \left(\frac{y^2 + 2z^2}{4}\right)^2\right] dy dz,$$

(9.16)

i.e.

$$I_2 = \frac{1}{2} \int_0^1 \left(\int_0^{2\pi} d\theta\right) \cdot 4\sqrt{2}\rho \left(4 - 4\rho^4\right) d\rho$$

$$= 16\pi\sqrt{2} \int_0^1 \rho \left(1 - \rho^4\right) d\rho = 16\pi\sqrt{2} \left(\frac{\rho^2}{2}\bigg|_0^1 - \frac{\rho^6}{6}\bigg|_0^1\right)$$

$$= 16\pi\sqrt{2} \cdot \left(\frac{1}{2} - \frac{1}{6}\right) = \frac{16\pi\sqrt{2}}{3}.$$

We have also to compute

$$I_3 = \int\int\int_V y dx dy dz = \int\int_D \left(\int_{(y^2+2z^2)/4}^2 dx\right) y dy dz = \int\int_D \left(2 - \frac{y^2 + 2z^2}{4}\right)^2 y dy dz,$$

(9.17)

such that

$$I_3 = \left(2\sqrt{2}\right)^2 \cdot 2 \int_0^1 \int_0^{2\pi} \cos\theta \cdot \rho^2 \left(2 - 2\rho^2\right) d\rho d\theta$$

$$= 32\sqrt{2} \int_0^1 \left(\int_0^{2\pi} \cos\theta d\theta\right) \cdot \rho^2 \left(1 - \rho^2\right) d\rho = 0.$$

Analogously,

$$I_4 = \int\int\int_V z dx dy dz = \int\int_D \left(\int_{(y^2+2z^2)/4}^2 dx\right) z dy dz = \int\int_D \left(2 - \frac{y^2 + 2z^2}{4}\right)^2 z dy dz,$$

(9.18)

namely

$$I_4 = 8\sqrt{2} \int_0^1 \int_0^{2\pi} \rho^2 \left(2 - 2\rho^2\right) \sin\theta d\rho d\theta$$

$$= 16\sqrt{2} \int_0^1 \left(\int_0^{2\pi} \sin\theta d\theta\right) \cdot \rho^2 \left(1 - \rho^2\right) d\rho = 0.$$

Hence,

$$\begin{cases} x_G = \frac{I_2}{I_1} = \frac{4}{3} \\ y_G = \frac{I_3}{I_1} = 0 \\ z_G = \frac{I_4}{I_1} = 0 \end{cases}$$

and the gravity center will be the point $G\left(\frac{4}{3}, 0, 0\right)$.

**Remark. 9.11.** As the homogeneous body is a body of rotation, with the rotation axis $Ox$ it results that $y_G = z_G = 0$.

We can also find this point in Matlab 7.9:

*Step 1.* Compute the integral from (9.15).

```
>> syms rho th
>> y=2*sqrt(2)*rho*cos(th);
>> z=2*rho*sin(th);
>> I1=int(int((2-(y^2+2*z^2)/4)*2*2*sqrt(2)*rho,rho,0,1),
th,0,2*pi)
I1 =
4*pi*2^(1/2)
```

*Step 2.* Compute the integral from (9.16).

```
>>I2= (1/2)*int(int((4-((y^2+2*z^2)/4)^2)*2*2*sqrt(2)*rho,
rho,0,1),th,0,2*pi)
I2 =
16/3*pi*2^(1/2)
```

*Step 3.* Find $x_G$.

```
>> xg=I2/I1
xg =
4/3
```

*Step 4.* Compute the integral from (9.17).

```
>> I3=int(int((2-(y^2+2*z^2)/4)*y*2*2*sqrt(2)*rho,rho,0,1),
th,0,2*pi)
I3 =
0
```

*Step 5.* Find $y_G$.

```
>> yg=I3/I1
yg =
0
```

*Step 6.* Compute the integral from (9.18).

```
>> I4=int(int((2-(y^2+2*z^2)/4)*z*2*2*sqrt(2)*rho,rho,0,1),
th,0,2*pi)
I4 =
0
```

*Step 7.* Find $z_G$.
`>> zg=I4/I1`
  zg =
      0
and using Mathcad 14:

$$y(\rho,\theta) := 2\cdot\sqrt{2}\cdot\rho\cdot\cos(\theta) \qquad z(\rho,\theta) := 2\cdot\rho\cdot\sin(\theta)$$

$$xg := \frac{\displaystyle\int_0^1 \int_0^{2\cdot\pi} \left(\int_{\frac{y(\rho,\theta)^2+2\cdot z(\rho,\theta)^2}{4}}^2 x\, dx\right)\cdot\rho\, d\theta\, d\rho}{\displaystyle\int_0^1 \int_0^{2\cdot\pi} \left(\int_{\frac{y(\rho,\theta)^2+2\cdot z(\rho,\theta)^2}{4}}^2 1\, dx\right)\cdot\rho\, d\theta\, d\rho} \qquad\qquad xg \to \frac{4}{3}$$

$$yg := \frac{\displaystyle\int_0^1 \int_0^{2\cdot\pi} \left(\int_{\frac{y(\rho,\theta)^2+2\cdot z(\rho,\theta)^2}{4}}^2 1\, dx\right)\cdot y(\rho,\theta)\cdot\rho\, d\theta\, d\rho}{\displaystyle\int_0^1 \int_0^{2\cdot\pi} \left(\int_{\frac{y(\rho,\theta)^2+2\cdot z(\rho,\theta)^2}{4}}^2 1\, dx\right)\cdot\rho\, d\theta\, d\rho} \qquad\qquad yg \to 0$$

$$zg := \frac{\displaystyle\int_0^1 \int_0^{2\cdot\pi} \left(\int_{\frac{y(\rho,\theta)^2+2\cdot z(\rho,\theta)^2}{4}}^2 1\, dx\right)\cdot z(\rho,\theta)\cdot\rho\, d\theta\, d\rho}{\displaystyle\int_0^1 \int_0^{2\cdot\pi} \left(\int_{\frac{y(\rho,\theta)^2+2\cdot z(\rho,\theta)^2}{4}}^2 1\, dx\right)\cdot\rho\, d\theta\, d\rho} \qquad\qquad zg \to 0$$

and with Mathematica 8:

```
In[42]:= y := 2 * Sqrt[2] * ρ * Cos[θ]

In[43]:= z := 2 * ρ * Sin[θ]

In[44]:= I1 := Integrate[Integrate[1, {x, (y^2 + 2 * z^2) / 4, 2}] * ρ, {θ, 0, 2 * Pi}, {ρ, 0, 1}]

In[45]:= I2 := Integrate[Integrate[x, {x, (y^2 + 2 * z^2) / 4, 2}] * ρ, {θ, 0, 2 * Pi}, {ρ, 0, 1}]

In[46]:= I3 := Integrate[Integrate[1, {x, (y^2 + 2 * z^2) / 4, 2}] * y * ρ, {θ, 0, 2 * Pi}, {ρ, 0, 1}]

In[47]:= I4 := Integrate[Integrate[1, {x, (y^2 + 2 * z^2) / 4, 2}] * z * ρ, {θ, 0, 2 * Pi}, {ρ, 0, 1}]

In[48]:= xg = I2 / I1

Out[48]=  4
          -
          3

In[49]:= yg = I3 / I1

Out[49]= 0

In[50]:= zg = I3 / I1

Out[50]= 0
```

and in Maple 15:

$y := 2 \cdot \sqrt{2} \cdot \rho \cdot \cos(\theta):$

$z := 2 \cdot \rho \cdot \sin(\theta):$

$I1 := simplify \left( \int_0^1 \left( \int_0^{2 \cdot \pi} \left( \int_{\frac{y^2 + 2 \cdot z^2}{4}}^2 1 \, dx \right) \cdot \rho \, d\theta \right) d\rho \right):$

$I2 := simplify \left( \int_0^1 \left( \int_0^{2 \cdot \pi} \left( \int_{\frac{y^2 + 2 \cdot z^2}{4}}^2 x \, dx \right) \cdot \rho \, d\theta \right) d\rho \right):$

$I3 := simplify \left( \int_0^1 \left( \int_0^{2 \cdot \pi} \left( \int_{\frac{y^2 + 2 \cdot z^2}{4}}^2 1 \, dx \right) \cdot y \cdot \rho \, d\theta \right) d\rho \right):$

$I4 := simplify \left( \int_0^1 \left( \int_0^{2 \cdot \pi} \left( \int_{\frac{y^2 + 2 \cdot z^2}{4}}^2 1 \, dx \right) \cdot z \cdot \rho \, d\theta \right) d\rho \right):$

$xg := \dfrac{I2}{I1}$

$$\dfrac{4}{3}$$

$yg := \dfrac{I3}{I1}$

$$0$$

$zg := \dfrac{I4}{I1}$

$$0$$

**Example 9.12.** Find the centre of gravity of the homogeneous hemisphere

$$V : \{x^2 + y^2 + z^2 = R^2, \; z \geq 0\}$$

using the change of variables in spherical coordinates.

**Solution.**

In our case, we shall have $\delta(x, y, z) = 1$.

In spherical coordinates, the integral over the homogeneous hemisphere is the integral over the region

$$V' : \left\{0 \leq \rho \leq R, \; 0 \leq \theta \leq \frac{\pi}{2}, \; 0 \leq \varphi \leq 2\pi\right\}.$$

We shall have:

$$I_1 = \int\int\int_V \delta(x,y,z)\,dxdydz = \int_0^R \int_0^{\frac{\pi}{2}} \int_0^{2\pi} \rho^2 \sin\theta d\rho d\theta d\varphi$$

$$= \int_0^R \left( \int_0^{\frac{\pi}{2}} \left( \int_0^{2\pi} d\varphi \right) \sin\theta d\theta \right) \rho^2 d\rho$$

$$= -2\pi \int_0^R \left( \cos\theta \big|_0^{\frac{\pi}{2}} \right) \rho^2 d\rho = 2\pi \int_0^R \rho^2 d\rho = 2\pi \left. \frac{\rho^3}{3} \right|_0^R = \frac{2\pi R^3}{3},$$

$$I_2 = \int\int\int_V x\delta(x,y,z)\,dxdydz = \int_0^R \int_0^{\frac{\pi}{2}} \int_0^{2\pi} \rho^3 \sin^2\theta \cos\varphi d\rho d\theta d\varphi = 0,$$

$$I_3 = \int\int\int_V y\delta(x,y,z)\,dxdydz = \int_0^R \int_0^{\frac{\pi}{2}} \int_0^{2\pi} \rho^3 \sin^2\theta \sin\varphi d\rho d\theta d\varphi = 0,$$

$$I_4 = \int\int\int_V z\delta(x,y,z)\,dxdydz = \int_0^R \int_0^{\frac{\pi}{2}} \int_0^{2\pi} \rho^3 \sin\theta \cos\theta d\rho d\theta d\varphi$$

$$= \int_0^R \left( \int_0^{\frac{\pi}{2}} \left( \int_0^{2\pi} d\varphi \right) \sin\theta \cos\theta d\theta \right) \rho^3 d\rho$$

$$= 2\pi \int_0^R \left( \int_0^{\frac{\pi}{2}} \sin\theta \cos\theta d\theta \right) \rho^3 d\rho = \pi \int_0^R \left( \sin^2\theta \big|_0^{\frac{\pi}{2}} \right) \rho^3 d\rho$$

$$= \pi \int_0^R \rho^3 d\rho = \left. \frac{\pi \rho^4}{4} \right|_0^R = \frac{\pi R^4}{4}.$$

Finally, one finds that the gravity center of the solid is the point $G\left(0,0,\frac{3}{8}R\right)$.

We need the following steps to find the gravity center in Matlab 7.9 :

*Step 1.* One computes $I_1 = \int\int\int_V \delta(x,y,z)dxdydz$.

```
>> syms rho th phi R
>> x=rho*sin(th)*cos(phi);
>> y=rho*sin(th)*sin(phi);
>> z=rho*cos(th);
>> delta=1;
>> v=delta*rho^2*sin(th);
>> I1=int(int(int(v,rho,0,R),th,0,pi/2),phi,0,2*pi);
```

*Step 2.* One computes $I_2 = \int\int\int_V x\delta(x,y,z)dxdydz$.

```
>> I2=int(int(int(x*v,rho,0,R),th,0,pi/2),phi,0,2*pi);
```

*Step 3.* One calculates $x_G$.

```
>> xg=I2/I1
```

**xg =**
**0**
*Step 4.* One computes $I_3 = \int\int\int_V y\delta\,(x,y,z)\mathrm{d}x\mathrm{d}y\mathrm{d}z$.
$>>$ **I3=int(int(int(y\*v,rho,0,R),th,0,pi/2),phi,0,2\*pi);**
*Step 5.* One calculates $y_G$.
$>>$ **yg=I3/I1**
      **yg =**
**0**
*Step 6.* One computes $I_4 = \int\int\int_V y\delta\,(x,y,z)\mathrm{d}x\mathrm{d}y\mathrm{d}z$.
$>>$ **I4=int(int(int(z\*v,rho,0,R),th,0,pi/2),phi,0,2\*pi);**
*Step 7.* One calculates $z_G$.
$>>$ **zg=I4/I1**
**zg =**
**3/8\*R**
We can also achieve this result using Mathcad 14:

$$x(\rho,\theta,\varphi) := \rho\cdot\sin(\theta)\cdot\cos(\varphi) \quad y(\rho,\theta,\varphi) := \rho\cdot\sin(\theta)\cdot\sin(\varphi) \quad z(\rho,\theta) := \rho\cdot\cos(\theta)$$

$$xg := \frac{\displaystyle\int_0^R \int_0^{\frac{\pi}{2}} \int_0^{2\cdot\pi} x(\rho,\theta,\varphi)\cdot\rho^2\cdot\sin(\theta)\, \mathrm{d}\varphi\, \mathrm{d}\theta\, \mathrm{d}\rho}{\displaystyle\int_0^R \int_0^{\frac{\pi}{2}} \int_0^{2\cdot\pi} \rho^2\cdot\sin(\theta)\, \mathrm{d}\varphi\, \mathrm{d}\theta\, \mathrm{d}\rho} \qquad xg \to 0$$

$$yg := \frac{\displaystyle\int_0^R \int_0^{\frac{\pi}{2}} \int_0^{2\cdot\pi} y(\rho,\theta,\varphi)\cdot\rho^2\cdot\sin(\theta)\, \mathrm{d}\varphi\, \mathrm{d}\theta\, \mathrm{d}\rho}{\displaystyle\int_0^R \int_0^{\frac{\pi}{2}} \int_0^{2\cdot\pi} \rho^2\cdot\sin(\theta)\, \mathrm{d}\varphi\, \mathrm{d}\theta\, \mathrm{d}\rho} \qquad yg \to 0$$

$$zg := \frac{\displaystyle\int_0^R \int_0^{\frac{\pi}{2}} \left(\int_0^{2\cdot\pi} 1\,\mathrm{d}\varphi\right)\cdot z(\rho,\theta)\cdot\rho^2\cdot\sin(\theta)\, \mathrm{d}\theta\, \mathrm{d}\rho}{\displaystyle\int_0^R \int_0^{\frac{\pi}{2}} \int_0^{2\cdot\pi} \rho^2\cdot\sin(\theta)\, \mathrm{d}\varphi\, \mathrm{d}\theta\, \mathrm{d}\rho} \qquad zg \to \frac{3\cdot R}{8}$$

and with Mathematica 8:

```
In[1]:= x := ρ * Sin[θ] * Cos[φ]

In[2]:= y := ρ * Sin[θ] * Sin[φ]

In[3]:= z := ρ * Cos[θ]

In[4]:= δ[ρ, θ, φ] := 1

In[8]:= I1 := Integrate[δ[ρ, θ, φ] * ρ^2 * Sin[θ], {φ, 0, 2 * Pi}, {θ, 0, Pi / 2}, {ρ, 0, R}]

In[9]:= I2 := Integrate[δ[ρ, θ, φ] * x * ρ^2 * Sin[θ], {φ, 0, 2 * Pi}, {θ, 0, Pi / 2}, {ρ, 0, R}]

In[10]:= I3 := Integrate[δ[ρ, θ, φ] * y * ρ^2 * Sin[θ], {φ, 0, 2 * Pi}, {θ, 0, Pi / 2}, {ρ, 0, R}]

In[11]:= I4 := Integrate[δ[ρ, θ, φ] * z * ρ^2 * Sin[θ], {φ, 0, 2 * Pi}, {θ, 0, Pi / 2}, {ρ, 0, R}]

In[12]:= xg = I2 / I1

Out[12]= 0

In[13]:= yg = I3 / I1

Out[13]= 0

In[14]:= zg = I4 / I1

Out[14]= 3 R
        ───
         8
```

and in Maple 15:

$x := \rho \cdot \sin(\theta) \cdot \cos(\varphi)$ :
$y := \rho \cdot \sin(\theta) \cdot \sin(\varphi)$ :
$z := \rho \cdot \cos(\theta)$ :

$$xg := \frac{\displaystyle\int_0^R \int_0^{\frac{\pi}{2}} \int_0^{2\cdot\pi} x \cdot \rho^2 \cdot \sin(\theta)\, d\varphi\, d\theta\, d\rho}{\displaystyle\int_0^R \int_0^{\frac{\pi}{2}} \int_0^{2\cdot\pi} \rho^2 \cdot \sin(\theta)\, d\varphi\, d\theta\, d\rho}$$

$$0$$

$$yg := \frac{\displaystyle\int_0^R \int_0^{\frac{\pi}{2}} \int_0^{2\cdot\pi} y \cdot \rho^2 \cdot \sin(\theta)\, d\varphi\, d\theta\, d\rho}{\displaystyle\int_0^R \int_0^{\frac{\pi}{2}} \int_0^{2\cdot\pi} \rho^2 \cdot \sin(\theta)\, d\varphi\, d\theta\, d\rho}$$

$$0$$

$$zg := \frac{\displaystyle\int_0^R \int_0^{\frac{\pi}{2}} \left( \int_0^{2\cdot\pi} 1\, d\varphi \right) \cdot z \cdot \rho^2 \cdot \sin(\theta)\, d\theta\, d\rho}{\displaystyle\int_0^R \int_0^{\frac{\pi}{2}} \int_0^{2\cdot\pi} \rho^2 \cdot \sin(\theta)\, d\varphi\, d\theta\, d\rho}$$

$$\frac{3}{8} R$$

## 9.3.4  Moments of Inertia

The moment of inertia of a solid body by the form of a domain $V$, with the density $\delta(x, y, z)$ relative to the coordinate axes are (see [42], p. 136):

$$I_{Ox} = \int\int\int_V \left(y^2 + z^2\right) \delta\left(x, y, z\right) \mathrm{d}x\mathrm{d}y\mathrm{d}z \qquad (9.19)$$

$$I_{Oy} = \int\int\int_V \left(z^2 + x^2\right) \delta\left(x, y, z\right) \mathrm{d}x\mathrm{d}y\mathrm{d}z \qquad (9.20)$$

$$I_{Oz} = \int\int\int_V \left(x^2 + y^2\right) \delta\left(x, y, z\right) \mathrm{d}x\mathrm{d}y\mathrm{d}z. \qquad (9.21)$$

**Example 9.13.** Use the change of variables in cylindrical coordinates to compute the moments of inertia for a homogeneous cylinder, with the density $\delta_0$, right circular, by the height $h$ and the radius of the base circle $R$, with respect to its axis of symmetry.

**Solution.**

The Figure 9.12 pictures the cylinder.

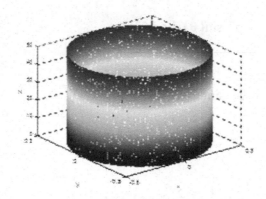

**Fig. 9.12**

In our case, we shall have

$$\begin{cases} \delta\left(x, y, z\right) = \delta_0 \\ V : \left\{x^2 + y^2 \le R^2,\ 0 \le z \le h\right\} \\ V' : \left\{0 \le \rho \le R,\ 0 \le \varphi \le 2\pi, 0 \le z \le h\right\} \end{cases}$$

therefore

$$I_{Oz} = \rho_0 \int\int\int_V \left(x^2 + y^2\right) \mathrm{d}x\mathrm{d}y\mathrm{d}z = \rho_0 \int_0^R \int_0^{2\pi} \int_0^h \rho^2 \cdot \rho\mathrm{d}\rho\mathrm{d}\varphi\mathrm{d}z$$

$$= \rho_0 \int_0^R \left(\int_0^{2\pi} \left(\int_0^h \mathrm{d}z\right) \mathrm{d}\varphi\right) \rho^3 \mathrm{d}\rho$$

$$= h\rho_0 \int_0^R \left(\int_0^{2\pi} \mathrm{d}\varphi\right) \rho^3 \mathrm{d}\rho = 2\pi h\rho_0 \int_0^R \rho^3 \mathrm{d}\rho = 2\pi h\rho_0 \cdot \left.\frac{\rho^4}{4}\right|_0^R = \pi h\rho_0 \cdot \frac{R^4}{2}.$$

We need the following sequence in order to compute the inertia moment of our body with respect to the axis, in Matlab 7.9:

```
>> syms rho phi z delta0 R h
>> x=rho*cos(phi);
>> y=rho*sin(phi);
>> Ioz=int(int(int((x^2+y^2)*rho,rho,0,R),phi,0,2*pi),z,0,h)
Ioz =
1/2*pi*R^4*h
```

and with Mathcad 14:

$$x(\rho,\theta) := \rho\cdot\cos(\theta) \qquad y(\rho,\theta) := \rho\cdot\sin(\theta)$$

$$Ioz(R,h) := \int_0^R \int_0^{2\cdot\pi} \int_0^h \left(x(\rho,\theta)^2 + y(\rho,\theta)^2\right)\cdot\rho \; dz \, d\theta \, d\rho$$

$$Ioz(R,h) \rightarrow \frac{\pi\cdot R^4\cdot h}{2}$$

and using Mathematica 8:

```
In[4]:= x := ρ * Cos[θ]

In[5]:= y := ρ * Sin[θ]

In[6]:= Ioz = Integrate[(x^2 + y^2) * ρ, {z, 0, h}, {θ, 0, 2 * Pi}, {ρ, 0, R}]

Out[6]= 1/2 h π R^4
```

and in Maple 15:

$$x := \rho\cdot\cos(\theta) :$$
$$y := \rho\cdot\sin(\theta) :$$

$$Ioz = \int_0^R \int_0^{2\cdot\pi} \int_0^h \left(x^2 + y^2\right)\cdot\rho \; dz \, d\theta \, d\rho$$

$$Ioz = \frac{1}{2}\,\pi h R^4$$

## 9.4   Surface Integral of the First Type

A surface integral is an integral over a surface, in three-dimensional space.

As the line integral generalizes the simple definite integral, the surface integral is an analogue generalization of the double integral in some plane domains.

The value of the surface integral of the first type does not depend on the choice of the side of the surface $\sum$ over which the integration is performed.

*Case 1.* Let $\sum$ be a surface in the three-dimensional Euclidean space $\mathbb{R}^3$, specified by the parametrical representation:

$$\left(\sum\right): \begin{cases} x = x\left(u, v\right) \\ y = y\left(u, v\right) \\ z = z\left(u, v\right) \end{cases}, \quad (u, v) \in D \subseteq \mathbb{R}^2, \tag{9.22}$$

then the following (see [2], p. 62) defines the surface integral of the first type (or the integral over the surface area):

$$\int\int_{\sum} F\left(x, y, z\right) d\sigma = \int\int_{D} F\left(x\left(u, v\right), y\left(u, v\right), z\left(u, v\right)\right) \sqrt{A^2 + B^2 + C^2} du dv, \tag{9.23}$$

where

$$\begin{cases} A = \frac{D(y,z)}{D(u,v)} \\ B = \frac{D(z,x)}{D(u,v)} \\ C = \frac{D(x,y)}{D(u,v)} \end{cases} \tag{9.24}$$

are functional determinants.

*Case 2.* Let $\sum$ be a surface in the three-dimensional Euclidean space $\mathbb{R}^3$, specified by its explicitly form representation:

$$z = f\left(x, y\right), (x, y) \in D \subseteq \mathbb{R}^2, \tag{9.25}$$

then the following (see [15], p. 258) defines the surface integral of the first type:

$$\int\int_{\sum} F\left(x, y, z\right) d\sigma = \int\int_{D} F\left(x, y, f\left(x, y\right)\right) \sqrt{1 + p^2 + q^2} dx dy, \tag{9.26}$$

where

$$\begin{cases} p = \frac{\partial f}{\partial x} \\ q = \frac{\partial f}{\partial y}. \end{cases} \tag{9.27}$$

**Remark 9.14** (see [48], p. 463). For $F = 1$, the integral from (9.23) and, respectively from (9.26) expresses the area of the surface $\sum$.

**Example 9.15.** Compute the surface integral of the first type:

$$I = \int \int_{\Sigma} (x^2 + y^2)\, d\sigma,$$

where ($\sum$) is the sphere

$$x^2 + y^2 + z^2 = a^2.$$

**Solution.**

The parametrical representation of the sphere is:

$$\begin{cases} x = a \sin\theta \cos\varphi \\ y = a \sin\theta \sin\varphi \\ z = a \cos\theta \end{cases}, \ a > 0, \ \theta \in [0, \pi], \ \varphi \in [0, 2\pi].$$

We shall deduce that:

$$A = \frac{D(y, z)}{D(\theta, \varphi)} = \begin{vmatrix} a\cos\theta\sin\varphi & a\sin\theta\cos\varphi \\ -a\sin\theta & 0 \end{vmatrix} = a^2 \sin^2\theta\cos\varphi,$$

$$B = \frac{D(z, x)}{D(\theta, \varphi)} = \begin{vmatrix} -a\sin\theta & 0 \\ a\cos\theta\cos\varphi & -a\sin\theta\sin\varphi \end{vmatrix} = a^2 \sin^2\theta\sin\varphi,$$

$$C = \frac{D(x, y)}{D(\theta, \varphi)} = \begin{vmatrix} a\cos\theta\cos\varphi & -a\sin\theta\sin\varphi \\ a\cos\theta\sin\varphi & a\sin\theta\cos\varphi \end{vmatrix} = a^2 \sin\theta\cos\theta;$$

hence

$$A^2 + B^2 + C^2 = a^4 \sin^4\theta\cos^2\varphi + a^4 \sin^4\theta\sin^2\varphi + a^4 \sin^2\theta\cos^2\theta$$
$$= a^4 \sin^4\theta + a^4 \sin^2\theta\cos^2\theta$$
$$= a^4 \sin^2\theta \left(\sin^2\theta + \cos^2\theta\right) = a^4 \sin^2\theta$$

and

$$I = \int \int_D \left(a^2 \sin^2\theta\cos^2\varphi + a^2\sin^2\theta\sin^2\varphi\right) \sqrt{a^4\sin^2\theta}\,d\theta d\varphi$$
$$= \int \int_D a^2 \sin^2\theta \cdot a^2 \sin\theta d\theta d\varphi = a^4 \int \int_D \sin^3\theta d\theta d\varphi,$$

where

$$D : \{0 \leq \theta \leq \pi, 0 \leq \varphi \leq 2\pi\}.$$

It results:

$$I = a^4 \int_0^\pi \int_0^{2\pi} \sin^3 \theta d\theta d\varphi = a^4 \int_0^{2\pi} d\varphi \underbrace{\int_0^\pi \sin^3 \theta d\theta}_{I_1} = 2\pi a^4 I_1,$$

where

$$I_1 = \int_0^\pi \sin^3 \theta d\theta = -\int_0^\pi \sin^2 \theta \, (\cos \theta)' \, d\theta = -\sin^2 \theta \cos \theta \big|_0^\pi + 2 \int_0^\pi \sin \theta \cos^2 \theta d\theta$$

$$= 2 \int_0^\pi \sin \theta \, (1 - \sin^2 \theta) \, d\theta = 2 \int_0^\pi \sin \theta d\theta - 2 \int_0^\pi \sin^3 \theta d\theta$$

$$= -2 \cos \theta \big|_0^\pi - 2I_1 = 4 - 2I_1,$$

i.e.

$$3I_1 = 4 \Longrightarrow I_1 = \frac{4}{3}.$$

Finally, one achieves that

$$I = 2\pi a^4 \cdot \frac{4}{3} = \frac{8\pi a^4}{3}.$$

We can also compute this integral in Matlab 7.9:

```
>> syms a th phi
>> x=a*sin(th)*cos(phi);
>> y=a*sin(th)*sin(phi);
>> z=a*cos(th);
>> A=det(jacobian([y,z],[th phi]))
    A =
 a^2*sin(th)^2*cos(phi)
>> B=det(jacobian([z x],[th phi]))
    B =
 a^2*sin(th)^2*sin(phi)
>> C=det(jacobian([x y],[th phi]))
C =
 a^2*cos(th)*cos(phi)^2*sin(th)+a^2*sin(th)*
 sin(phi)^2*cos(th)
>> I=simple(int(int((x^2+y^2)*sqrt(A^2+B^2+C^2),
    th,0,pi),phi,0,2*pi));
>> I
I =
8/3*a^4*pi
```

and with Mathcad 14:

$$f1(a,x) := \begin{pmatrix} a \cdot \sin(x_0) \cdot \cos(x_1) \\ a \cdot \sin(x_0) \cdot \sin(x_1) \end{pmatrix} \quad f2(a,x) := \begin{pmatrix} a \cdot \cos(x_0) \\ a \cdot \sin(x_0) \cdot \cos(x_1) \end{pmatrix} \quad f3(a,x) := \begin{pmatrix} a \cdot \sin(x_0) \cdot \sin(x_1) \\ a \cdot \cos(x_0) \end{pmatrix}$$

$$A(a,x) := |Jacob(f3(a,x),x)| \quad B(a,x) := |Jacob(f2(a,x),x)| \quad C(a,x) := |Jacob(f1(a,x),x)|$$

$$u(a,x) := A(a,x)^2 + B(a,x)^2 + C(a,x)^2$$

$$m(a,\theta,\varphi) := f1(a,x)_0 \text{ substitute}, x_0 = \theta \;\rightarrow\; a \cdot \cos(x_1) \cdot \sin(\theta) \text{ substitute}, x_1 = \varphi \;\rightarrow\; a \cdot \cos(\varphi) \cdot \sin(\theta)$$

$$n(a,\theta,\varphi) := f1(a,x)_1 \text{ substitute}, x_0 = \theta, x_1 = \varphi \;\rightarrow\; \frac{a \cdot (\cos(\varphi - \theta) - \cos(\varphi + \theta))}{2} \text{ simplify} \;\rightarrow\; a \cdot \sin(\varphi) \cdot \sin(\theta)$$

$$w(a,\theta) := m(a,\theta,\varphi)^2 + n(a,\theta,\varphi)^2 \text{ simplify} \;\rightarrow\; a^2 \cdot \sin(\theta)^2$$

$$uu(a,\theta) := u(a,x) \text{ substitute}, x_0 = \theta, x_1 = v \;\rightarrow\; -\frac{a^4 \cdot (\cos(2 \cdot \theta) - 1)}{2} \text{ simplify} \;\rightarrow\; a^4 \cdot \sin(\theta)^2$$

$$\int_0^\pi \left( \int_0^{2 \cdot \pi} 1 \, d\varphi \right) \cdot a^2 \cdot \sin(\theta) \cdot \left( a^2 \cdot \sin(\theta)^2 \right) d\theta \;\rightarrow\; \frac{8 \cdot \pi \cdot a^4}{3}$$

and with Mathematica 8:

In[33]:= **F[m_, n_] := m^2 + n^2**

In[34]:= **x := a * Sin[θ] * Cos[φ]**

In[35]:= **y := a * Sin[θ] * Sin[φ]**

In[36]:= **z := a * Cos[θ]**

In[37]:= **Aa := Det[D[{y, z}, {{θ, φ}}]]**

In[38]:= **Bb := Det[D[{z, x}, {{θ, φ}}]]**

In[39]:= **Cc := Det[D[{x, y}, {{θ, φ}}]]**

In[40]:= **u = Simplify[Aa^2 + Bb^2 + Cc^2]**

Out[40]= $a^4 \sin[\theta]^2$

In[42]:= **Integrate[F[x, y] * Sqrt[u], {θ, 0, Pi}, {φ, 0, 2 * Pi}]**

Out[42]= $\dfrac{8}{3} a^2 \sqrt{a^4} \, \pi$

and using Maple 15:

$$F := (m, n) \rightarrow m^2 + n^2 :$$
$$x := a \cdot \sin(\theta) \cdot \cos(\varphi) :$$

$$y := a \cdot \sin(\theta) \cdot \sin(\varphi) :$$
$$z := a \cdot \cos(\theta) :$$
$with(VectorCalculus) :$
$M1, A := Jacobian([y, z], [\theta, \varphi], 'determinant') :$
$M2, B := Jacobian([z, x], [\theta, \varphi], 'determinant') :$
$M3, C := Jacobian([x, y], [\theta, \varphi], 'determinant') :$
$u := simplify(A^2 + B^2 + C^2)$

$$a^4 \sin(\theta)^2$$

$$\int_0^\pi \int_0^{2 \cdot \pi} F(x, y) \cdot \sqrt{u} \, d\theta \, d\varphi$$

$$\frac{8}{3} a^4 \operatorname{csgn}(a^2) \pi$$

**Example 9.16.** Evaluate

$$I = \int \int_\Sigma \sqrt{x^2 + y^2} d\sigma,$$

where $(\Sigma)$ is is the lateral surface of the cone:

$$\frac{x^2}{a^2} + \frac{y^2}{a^2} - \frac{z^2}{b^2} = 0, \ 0 \le z \le b.$$

**Solution.**
For our surface we shall have:

$$\begin{cases} p = \dfrac{\partial f}{\partial x} = \dfrac{b}{a} \cdot \dfrac{2x}{2\sqrt{x^2+y^2}} = \dfrac{b}{a} \cdot \dfrac{x}{\sqrt{x^2+y^2}} \\[2mm] q = \dfrac{\partial f}{\partial y} = \dfrac{b}{a} \cdot \dfrac{y}{\sqrt{x^2+y^2}}; \end{cases}$$

the area element for the surface is

$$d\sigma = \sqrt{1 + p^2 + q^2} dx dy \Longleftrightarrow$$

$$d\sigma = \sqrt{1 + \frac{b}{a}^2 \cdot \frac{x^2}{x^2 + y^2} + \frac{b}{a}^2 \cdot \frac{y^2}{x^2 + y^2}} dx dy,$$

i.e.

$$d\sigma = \frac{\sqrt{a^2 + b^2}}{a} dx dy.$$

Hence

$$I = \int\int_{\Sigma} \sqrt{x^2 + y^2} d\sigma = \int\int_{D} \sqrt{x^2 + y^2} \cdot \frac{\sqrt{a^2 + b^2}}{a} dx dy$$

$$= \frac{\sqrt{a^2 + b^2}}{a} \int\int_{D} \sqrt{x^2 + y^2} dx dy,$$

where $D$ is that region in the $xOy$- plane such that

$$D = \left\{ (x, y) \in \mathbb{R}^2 \mid x^2 + y^2 \le a^2 \right\};$$

this results since

$$\left. \begin{array}{c} z = \frac{b}{a}\sqrt{x^2 + y^2} \\ 0 \le z \le b \end{array} \right\} \implies x^2 + y^2 \le a^2.$$

The region $D$ can be parameterized by:

$$\begin{cases} x = \rho \cos \theta \\ y = \rho \sin \theta \end{cases}$$

for $(\rho, \theta)$ in the region $D'$, defined by:

$$D' = \left\{ (\rho, \theta) \in \mathbb{R}^2 \mid \rho \in [0, a], \ \theta \in [0, 2\pi] \right\}.$$

Therefore, the region of integration is simpler to describe using polar coordinates and the double integral can be written:

$$I = \int_0^a \int_0^{2\pi} \sqrt{\rho^2} \cdot \frac{\sqrt{a^2 + b^2}}{a} \cdot \rho d\rho d\theta = \frac{\sqrt{a^2 + b^2}}{a} \cdot 2\pi \int_0^a \int_0^{2\pi} \rho^2 d\rho, \quad (9.28)$$

i.e.

$$I = 2\pi \cdot \frac{\sqrt{a^2 + b^2}}{a} \cdot \frac{\rho^3}{3} \Big|_0^a = 2\pi \cdot \frac{\sqrt{a^2 + b^2}}{a} \cdot \frac{a^3}{3} = \frac{2\pi a^2}{3} \sqrt{a^2 + b^2}.$$

We can also evaluate on the computer our surface integral, using Matlab 7.9:

*Step 1.* One computes $\sqrt{1 + p^2 + q^2}$.

```
>> syms a b x y
>> z=(b/a)*sqrt(x^2+y^2);
>> p=diff(z,x);
```

```
>> q=diff(z,y);
 >> w=simple(sqrt(1+p^2+q^2))
w =
(a^2+b^2)^(1/2)/a
```

*Step 2.* One computes the double integral from (9.28):

```
>> syms rho th
>> x=rho*cos(th);
>> y=rho*sin(th);
>> I=simple(int(int(w*sqrt(x^2+y^2)*rho,rho,0,a),th,0,2*pi))
I =
2/3*pi*a^2*(a^2+b^2)^(1/2)
```

and in Mathcad 14:

$$z(a,b,x,y) := \frac{b}{a} \cdot \sqrt{x^2 + y^2}$$

$$p(a,b,x,y) := \frac{d}{dx} z(a,b,x,y) \qquad q(a,b,x,y) := \frac{d}{dy} z(a,b,x,y)$$

$$\sqrt{1 + p(a,b,x,y)^2 + q(a,b,x,y)^2} \;\; \text{simplify} \;\rightarrow\; \sqrt{\frac{b^2}{a^2} + 1}$$

$$x(\rho,\theta) := \rho \cdot \cos(\theta) \qquad y(\rho,\theta) := \rho \cdot \sin(\theta)$$

$$I(a,b) := \int_0^a \int_0^{2 \cdot \pi} \sqrt{x(\rho,\theta)^2 + y(\rho,\theta)^2} \cdot \rho \cdot \sqrt{\frac{b^2}{a^2} + 1} \; d\theta \, d\rho$$

$$I(a,b) \;\; \text{simplify} \;\rightarrow\; \frac{2 \cdot \pi \cdot a^3 \cdot \sqrt{\frac{b^2}{a^2} + 1}}{3}$$

and with Mathematica 8:

In[1]:= **z := (b / a) ∗ Sqrt [x^2 + y^2]**

In[2]:= **p := D[z, x]**

In[3]:= **q := D[z, y]**

In[4]:= **w = Simplify [Sqrt [1 + p^2 + q^2]]**

Out[4]= $\sqrt{1 + \dfrac{b^2}{a^2}}$

In[5]:= **x := ρ ∗ Cos [θ]**

In[6]:= **y := ρ ∗ Sin [θ]**

In[7]:= **Integrate [w ∗ Sqrt [x^2 + y^2] ∗ ρ, {θ, 0, 2 ∗ Pi}, {ρ, 0, a}]**

Out[7]= $\dfrac{2}{3}\, a^3 \sqrt{1 + \dfrac{b^2}{a^2}}\ \pi\,(-1 + 2\,\text{HeavisideTheta}[a])$

and using Maple 15:

$$z := \frac{b}{a}\sqrt{x^2 + y^2}\ :$$

$$p := \frac{\partial}{\partial x}z\ :$$

$$q := \frac{\partial}{\partial y}z\ :$$

$$w := simplify\!\left(\sqrt{1 + p^2 + q^2}\right)\ :$$

$$x := \rho \cdot \cos(\theta)\ :$$

$$y := \rho \cdot \sin(\theta)\ :$$

$$\int_0^a \int_0^{2\cdot\pi} w \cdot \sqrt{x^2 + y^2}\ \cdot \rho\, d\theta\, d\rho$$

$$\frac{2}{3}\,\operatorname{signum}(\Re(a))\, a^3\, \pi\, \sqrt{\frac{a^2 + b^2}{a^2}}$$

## 9.5   Surface Integral of the Second Type

The surface integral of the second type one defines by analogy with the line integral of the second type.

The value of the surface integral of the second type depends on the choice of the side of the surface $\sum$ over which the integration is performed.

Any surface has two side: the upper side, which corresponds to the directed upwards normal (that makes an acute angle with the positive direction of the $Oz$ axis) and the lower side, which corresponds to the directed downwards normal, i.e. the normal makes an obtuse angle with the positive direction of the $Oz$ axis.

**Definition 9.17** (see [15], p. 258). If a smooth surface $\sum$ is the boundary of a bounded domain, $\sum_+$ denotes its orientation by means of the outward normal (hence $\sum_-$ is determined by the inward normal) and $P(x, y, z)$, $Q(x, y, z)$, $R(x, y, z)$ are continuous functions then the corresponding **surface integral of the second type** is:

$$\iint_{\sum_+} P(x, y, z)\, \mathrm{d}y\mathrm{d}z + Q(x, y, z)\, \mathrm{d}z\mathrm{d}x + R(x, y, z)\, \mathrm{d}x\mathrm{d}y \quad (9.29)$$

$$= \iint_{\sum} [P(x, y, z) \cos\alpha + Q(x, y, z) \cos\beta + R(x, y, z) \cos\gamma]\, \mathrm{d}\sigma,$$

where the unit normal vector which specify the surface orientation is:

$$\overrightarrow{n} = \cos\alpha\ \overrightarrow{i} + \cos\beta\ \overrightarrow{j} + \cos\gamma\ \overrightarrow{k}, \quad (9.30)$$

$\cos\alpha, \cos\beta, \cos\gamma$ being the direction cosines of the normal to the surface $\sum$.

**Remark 9.18.**

1) When we go to the other side $\sum_-$ of the surface, this integral reverses its sign.
2) We call the side of the surface with the normal vector, the positive side of the surface. The outside is the positive side of the surface $\sum$, namely the face for which the cosine angle between the $Oz$ axis and the unit normal vector is positive, namely $\cos\gamma > 0$.
3) The surface integrals of the second kind over the oriented surface $\sum_+$ are called surface integrals with respect to the outside of the surface, while those over $\sum_-$ are called surface integrals with respect to the inside.

**Proposition 9.19** (see [2], p. 64). If the surface $\sum$ is represented by the parametrical representation (9.22), then the direction cosines of the normal to this surface are given by

$$\begin{cases} \cos\alpha = \dfrac{A}{\pm\sqrt{A^2+B^2+C^2}} \\ \cos\beta = \dfrac{B}{\pm\sqrt{A^2+B^2+C^2}} \\ \cos\gamma = \dfrac{C}{\pm\sqrt{A^2+B^2+C^2}} \end{cases} \qquad (9.31)$$

where $A, B, C$ are defined in (9.24) and the choice of sign in front of the root should be brought into agreement with the side of the surface $\sum$.
**Proposition 9.20** (see [42], p. 148). In the case when the surface $\sum$ is represented explicitly as (9.25), then the direction cosines of the normal to this surface are given by

$$\begin{cases} \cos\alpha = \dfrac{-p}{\pm\sqrt{1+p^2+q^2}} \\ \cos\beta = \dfrac{-q}{\pm\sqrt{1+p^2+q^2}} \\ \cos\gamma = \dfrac{1}{\pm\sqrt{1+p^2+q^2}} \end{cases} \qquad (9.32)$$

where $p, q$ are defined in (9.27) and the choice of sign in front of the root should be brought into agreement with the side of the surface $\sum$.
**Example 9.21.** Compute

$$I = \int\!\!\int_{\Sigma_+} x\,dydz + y\,dzdx + z\,dxdy,$$

$\sum_+$ being the external side of the surface of a tetrahedron bounded by the planes:

$$\begin{cases} x = 0 \\ y = 0 \\ z = 0 \\ x + y + z = a. \end{cases}$$

**Solution.**
Using (9.29), we shall have

$$I = \int\!\!\int_{\Sigma} [x\cos\alpha + y\cos\beta + z\cos\gamma]\,d\sigma.$$

The triangle (see Fig. 9.13) bounds the surface $\sum$.

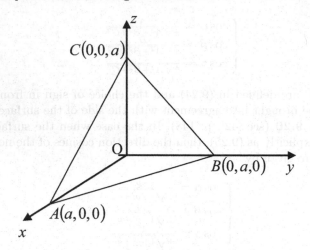

**Fig. 9.13**

Denoting by $D$, the projection of the surface $\sum$ in the $xOy$- plane, we can write:

$$z = a - x - y, \quad (x,y) \in D,$$

where

$$D : \{0 \leq x \leq a, \ 0 \leq y \leq a - x\}.$$

We compute

$$\begin{cases} p = \frac{\partial z}{\partial x} = -1 \\[2mm] q = \frac{\partial z}{\partial y} = -1 \end{cases}$$

and

$$\begin{cases} \cos \alpha = \frac{1}{\sqrt{3}} \\ \cos \beta = \frac{1}{\sqrt{3}} \\ \cos \gamma = \frac{1}{\sqrt{3}}; \end{cases}$$

hence, using (9.26) it will result:

$$I = \int \int_D \frac{1}{\sqrt{3}} \cdot (x + y + a - x - y) \cdot \sqrt{3}\,dxdy = a \int \int_D dxdy,$$

i.e.

$$I = a \int_0^a \left( \int_0^{a-x} dy \right) dx = a \int_0^a \left( y\big|_0^{a-x} \right) dx = a \int_0^a (a - x)\, dx$$

$$= a \left( ax\big|_0^a - \frac{x^2}{2}\bigg|_0^a \right) = a \left( a^2 - \frac{a^2}{2} \right) = \frac{a^3}{2}.$$

We shall achieve this result using Matlab 7.9:

```
>> syms a x y
>> z=a-x-y;
>> p=diff(z,x);
>> q=diff(z,y);
>> u=sqrt(1+p^2+q^2);
>> cosg=1/u;
>> cosa=-p/u;
>> cosb=-q/u;
>> int(int((x*cosa+y*cosb+z*cosg)*u,y,0,a-x),x,0,a)
 ans =
 a^3/2
```

and in Mathcad 14:

$$z(a,x,y) := a - x - y \qquad p(a,x,y) := \frac{d}{dx} z(a,x,y) \qquad q(a,x,y) := \frac{d}{dy} z(a,x,y)$$

$$u(a,x,y) := \sqrt{1 + p(a,x,y)^2 + q(a,x,y)^2}$$

$$\cos\gamma(a,x,y) := \frac{1}{u(a,x,y)} \qquad \cos\alpha(a,x,y) := -\frac{p(a,x,y)}{u(a,x,y)} \qquad \cos\beta(a,x,y) := -\frac{q(a,x,y)}{u(a,x,y)}$$

$$\int_0^a \int_0^{a-x} (x\cdot\cos\alpha(a,x,y) + y\cdot\cos\beta(a,x,y) + z(a,x,y)\cdot\cos\gamma(a,x,y))\cdot u(a,x,y)\, dy\, dx \text{ simplify } \rightarrow \frac{a^3}{2}$$

and with Mathematica 8:

```
In[1]:= z := a - x - y;

In[2]:= p := D[z, x];

In[3]:= q := D[z, y];

In[4]:= u := Sqrt[1 + p^2 + q^2];

In[5]:= cosγ := 1 / u;

In[6]:= cosα := - p / u;

In[7]:= cosβ := - q / u;

In[8]:= Integrate[(x * cosα + y * cosβ + z * cosγ) * u, {x, 0, a}, {y, 0, a - x}]

Out[8]= a³
        ──
        2
```

and using Maple 15:

$$z := a - x - y:$$

$$p := \frac{d}{dx} z:$$

$$q := \frac{d}{dy} z:$$

$$u := \sqrt{1 + p^2 + q^2} :$$

$$\cos\gamma := \frac{1}{u} :$$

$$\cos\alpha := -\frac{p}{u} :$$

$$\cos\beta := -\frac{q}{u} :$$

$$\int_0^a \int_0^{a-x} (x \cdot \cos\alpha + y \cdot \cos\beta + z \cdot \cos\gamma) \cdot u \, dy \, dx$$

$$\frac{1}{2} a^3$$

## 9.5.1  Flux of a Vector Field

**Example 9.22** (see [42], p. 169). The **flux** of a vector field

$$\overrightarrow{F}(x, y, z) = P(x, y, z)\,\overrightarrow{i} + P(x, y, z)\,\overrightarrow{j} + R(x, y, z)\,\overrightarrow{k}$$

through a surface in a direction defined by the unit vector of the normal $\overrightarrow{n}$ from (9.30) to the surface $\sum$ is the integral

$$\text{Flux} = \iint_{\sum} \overrightarrow{F} \cdot \overrightarrow{n}\,d\sigma, \tag{9.33}$$

where:

- $\overrightarrow{F} \cdot \overrightarrow{n}$ means the dot product between $\overrightarrow{F}$ and $\overrightarrow{n}$,
- $d\sigma$ is the differential surface element.

**Example 9.23.** Find the flux of the vector field

$$\overrightarrow{F}(x, y, z) = x^2\,\overrightarrow{i} + y^2\,\overrightarrow{j} + z^2\,\overrightarrow{k}$$

across the lateral surface of the cone

$$\left(\sum\right) : x^2 + y^2 = z^2,\ 0 \le z \le h.$$

**Solution.**
We shall have

$$z = \sqrt{x^2 + y^2},\ 0 \le z \le h,$$

$$\begin{cases} p = \dfrac{\partial z}{\partial x} = \dfrac{2x}{2\sqrt{x^2+y^2}} = \dfrac{x}{\sqrt{x^2+y^2}} \\ q = \dfrac{\partial z}{\partial y} = \dfrac{y}{\sqrt{x^2+y^2}}; \end{cases}$$

therefore

$$\sqrt{1 + p^2 + q^2}\,dxdy = \sqrt{1 + \frac{x^2}{x^2 + y^2} + \frac{y^2}{x^2 + y^2}} = \sqrt{2}$$

and

$$\overrightarrow{n} = \frac{-x}{\sqrt{2}\cdot\sqrt{x^2 + y^2}}\,\overrightarrow{i} - \frac{-y}{\sqrt{2}\cdot\sqrt{x^2 + y^2}}\,\overrightarrow{j} + \frac{1}{\sqrt{2}}\,\overrightarrow{k}.$$

It will result

$$\overrightarrow{F} \cdot \overrightarrow{n} = -\frac{x^3}{\sqrt{2}\cdot\sqrt{x^2 + y^2}} - \frac{y^3}{\sqrt{2}\cdot\sqrt{x^2 + y^2}} + \frac{z^2}{\sqrt{2}}.$$

Hence,

$$\text{Flux} = \int\int_D \left( -\frac{x^3}{\sqrt{2} \cdot \sqrt{x^2 + y^2}} - \frac{y^3}{\sqrt{2} \cdot \sqrt{x^2 + y^2}} + \frac{x^2 + y^2}{\sqrt{2}} \right) \cdot \sqrt{2} dx dy,$$

namely

$$\text{Flux} = \int\int_D \left( -\frac{x^3}{\sqrt{x^2 + y^2}} - \frac{y^3}{\sqrt{x^2 + y^2}} + x^2 + y^2 \right) dx dy,$$

where

$$D = \left\{ (x, y) \in \mathbb{R}^2 \mid x^2 + y^2 \le h^2 \right\}.$$

Using the change of variable in polar coordinates

$$\begin{cases} x = \rho \cos \theta \\ y = \rho \sin \theta \end{cases}, \quad \rho \in [0, h], \ \theta \in [0, 2\pi]$$

we get that

$$\text{Flux} = \int_0^h \int_0^{2\pi} \left( -\frac{\rho^3 \cos^3 \theta}{\rho} - \frac{\rho^3 \sin^3 \theta}{\rho} + \rho^2 \right) \rho d\rho d\theta$$

$$= \int_0^h \left( \int_0^{2\pi} \left( -\cos^3 \theta - \sin^3 \theta + 1 \right) d\theta \right) \rho^3 d\rho.$$

We have to compute

$$I_1 = \int_0^{2\pi} \cos^3 \theta d\theta = \int_0^{2\pi} \cos^2 \theta \cdot (\sin \theta)' \, d\theta$$

$$= \sin \theta \cos^2 \theta \Big|_0^{2\pi} + 2 \int_0^{2\pi} \cos \theta \cdot \sin^2 \theta d\theta$$

$$= 2 \int_0^{2\pi} \cos \theta \cdot (1 - \cos^2 \theta) \, d\theta = 2 \int_0^{2\pi} \cos \theta d\theta - 2 \int_0^{2\pi} \cos^3 \theta d\theta$$

$$= 2 \sin \theta \Big|_0^{2\pi} - 2I_1 \implies I_1 = 0;$$

$$I_2 = \int_0^{2\pi} \sin^3 \theta d\theta = -\int_0^{2\pi} \sin^2 \theta \cdot (\cos \theta)' \, d\theta$$

$$= -\sin^2 \theta \cos \theta \Big|_0^{2\pi} + 2 \int_0^{2\pi} \sin \theta \cos^2 \theta d\theta$$

$$= 2 \int_0^{2\pi} \sin \theta \left( 1 - \sin^2 \theta \right) d\theta = 2 \int_0^{2\pi} \sin \theta d\theta - 2 \int_0^{2\pi} \sin^3 \theta d\theta$$

$$= -2 \cos \theta \Big|_0^{2\pi} - 2I_2 \implies I_2 = 0.$$

Finally,

$$\text{Flux} = \int_0^h 2\pi\rho^3 d\rho = 2\pi \left.\frac{\rho^4}{4}\right|_0^h = \frac{\pi h^4}{2}.$$

The following Matlab 7.9 sequence allows us to find the flux of our vector field:

```
>> syms x y
>> z=sqrt(x^2+y^2);
>> p=diff(z,x);
>> q=diff(z,y);
>> w=sqrt(1+p^2+q^2);
>> n=[-p/w -q/w 1/w];
>> F=[x^2 y^2 z^2];
>> v=dot(F,n);
>> syms rho th h
>> k=subs(v*w,{x,y},{rho*cos(th),y,rho*sin(th)});
>> flux=int(int(k*rho,rho,0,h),th,0,2*pi)
flux =
(pi*h^4)/2
```

We can also obtain this result with Mathcad 14:

$$z(x,y) := \sqrt{x^2 + y^2} \qquad p(x,y) := \frac{d}{dx}z(x,y) \qquad q(x,y) := \frac{d}{dy}z(x,y)$$

$$u(x,y) := \sqrt{1 + p(x,y)^2 + q(x,y)^2}$$

$$v(x,y,z) := -\frac{p(x,y)}{u(x,y)}\cdot x^2 - \frac{p(x,y)}{u(x,y)}\cdot y^2 + \frac{1}{u(x,y)}\cdot z(x,y)^2$$

$$xx(\rho,\theta) := \rho\cdot\cos(\theta) \qquad yy(\rho,\theta) := \rho\cdot\sin(\theta)$$

$$w(\rho,\theta) := v(x,y,z)\cdot u(x,y) \text{ substitute}, x = xx(\rho,\theta), y = yy(\rho,\theta) \rightarrow -\frac{\rho^3\cdot\cos(\theta) - \left(\rho^2\right)^{\frac{3}{2}}}{\sqrt{\rho^2}}$$

$$\int_0^h \int_0^{2\cdot\pi} w(\rho,\theta)\cdot\rho \, d\theta \, d\rho \rightarrow \frac{\pi\cdot h^4}{2}$$

and using Mathematica 8:

In[119]:= `z := Sqrt[x^2 + y^2];`

In[120]:= `p := D[z, x];`

In[121]:= `q := D[z, y];`

In[122]:= `u := Simplify[Sqrt[1 + p^2 + q^2]];`

In[123]:= `xx[ρ_, θ_] := ρ * Cos[θ];`

In[124]:= `yy[ρ_, θ_] := ρ * Sin[θ];`

In[125]:= `({x^2, y^2, z^2} . {-p/u, -q/u, 1/u}) * u /. x → xx[ρ, θ];`

In[126]:= `m = % /. y → yy[ρ, θ];`

In[127]:= `Integrate[m * ρ, {θ, 0, 2 * Pi}, {ρ, 0, h}]`

Out[127]= $\dfrac{h^4 \pi}{2}$

and in Maple 15:

$z := \sqrt{x^2 + y^2} :$

$p := \dfrac{d}{dx} z :$

$q := \dfrac{d}{dy} z :$

$u := simplify\left( \sqrt{1 + p^2 + q^2} \right) :$

$with(LinearAlgebra) :$

$F := \langle x^2, y^2, z^2 \rangle :$

$n := \left\langle -\dfrac{p}{u}, -\dfrac{q}{u}, \dfrac{1}{u} \right\rangle :$

$w := simplify\left( F \cdot n \cdot u \Big|_{x = \rho \cdot \cos(\theta),\, y = \rho \cdot \sin(\theta)} \right) :$

$\displaystyle\int_0^h \int_0^{2 \cdot \pi} w \cdot \rho \, d\theta \, d\rho$

$\dfrac{1}{2} \pi \operatorname{csgn}(h)^2 h^4$

## 9.5.2 Gauss-Ostrogradski Formula

The Gauss-Ostrogradski formula establishes a relationship between a triple integral over a three-dimensional bounded domain and the surface integrals over its boundary.

**Theorem 9.24 (Gauss-Ostrogradski formula,** see [8] and [42], p. 158 and [2], p. 76). If $\sum$ is a closed smooth surface bounding the volume $V$ and $P(x, y, z)$, $Q(x, y, z)$, $R(x, y, z)$ are functions which are continuous together with their first partial derivatives continuous in the closed domain $V$, then we have the *Ostrogradsky-Gauss formula:*

$$\int\int_{\Sigma_+} P(x, y, z)\, dydz + Q(x, y, z)\, dzdx + R(x, y, z)\, dxdy \quad (9.34)$$

$$= \int\int\int_V \left[ \frac{\partial P}{\partial x}(x, y, z) + \frac{\partial Q}{\partial y}(x, y, z) + \frac{\partial R}{\partial z}(x, y, z) \right] dxdydz.$$

**Example 9.25.** Evaluate the following integral, by transforming it into a triple integral:

$$I = \int\int_{\Sigma_+} xyz\,(xdydz + ydzdx + zdxdy),$$

where $\sum$ is the closed surface bounding the field of space given by the inequalities:

$$x^2 + y^2 + z^2 \le R^2, \ x \ge 0, \ y \ge 0, \ z \ge 0.$$

**Solution.**
One can notice that

$$\begin{cases} P(x, y, z) = x^2yz \\ Q(x, y, z) = xy^2z \\ R(x, y, z) = xyz^2; \end{cases}$$

hence

$$I = \int\int\int_V (2xyz + 2xyz + 2xyz)\, dxdydz = 6 \int\int\int_V xyz\,dxdydz,$$

where

$$V : \left\{ x^2 + y^2 + z^2 \le R^2, \ x \ge 0, \ y \ge 0, \ z \ge 0 \right\}.$$

Passing to the spherical coordinates:

$$\begin{cases} x = \rho \sin\theta \cos\varphi \\ y = \rho \sin\theta \sin\varphi \ , \ \rho \in [0, R], \ \theta \in \left[0, \frac{\pi}{2}\right], \ \varphi \in \left[0, \frac{\pi}{2}\right] \\ z = \rho \cos\theta \end{cases} \quad (9.35)$$

one deduces that:

$$I = 6 \int_0^R \int_0^{\frac{\pi}{2}} \int_0^{\frac{\pi}{2}} \rho^3 \sin^2 \theta \cos \theta \cdot \sin \varphi \cos \varphi \cdot \rho^2 \sin \theta d\rho d\theta d\varphi$$

$$= 6 \int_0^R \left[ \int_0^{\frac{\pi}{2}} \left( \int_0^{\frac{\pi}{2}} \sin \varphi \cos \varphi d\varphi \right) \sin^3 \theta \cos \theta \right] \rho^5 d\rho$$

$$= 6 \int_0^R \left( \int_0^{\frac{\pi}{2}} \frac{\sin^2 \varphi}{2} \Big|_0^{\frac{\pi}{2}} \cdot \sin^3 \theta \cos \theta d\theta \right) \rho^5 d\rho$$

$$= 3 \int_0^R \left( \int_0^{\frac{\pi}{2}} \sin^3 \theta \cdot (\sin \theta)' d\theta \right) \rho^5 d\rho = 3 \int_0^R \frac{\sin^4 \theta}{4} \Big|_0^{\frac{\pi}{2}} \cdot \rho^5 d\rho$$

$$= \frac{3}{4} \int_0^R \rho^5 d\rho = \frac{R^6}{8}.$$

We need the following steps in order to compute this integral in Matlab 7.9:

*Step 1.* Compute $w(x, y, z) = \frac{\partial P}{\partial x} + \frac{\partial Q}{\partial y} + \frac{\partial R}{\partial z}$.

>> **syms x y z**
>> **u=x\*y\*z;**
>> **P=u\*x**
  P =
  x^2\*y\*z
>> **Q=u\*y**
  Q =
  x\*y^2\*z
>> **R=u\*z**
  R =
  x\*y\*z^2
>> **w=diff(P,x)+diff(Q,y)+diff(R,z)**
  w =
  6\*x\*y\*z

*Step 2.* Compute $I = \int \int \int_V w(x, y, z) dx dy dz$, where

$$V : \{ x^2 + y^2 + z^2 \leq r^2, \ x \geq 0, \ y \geq 0, \ z \geq 0 \}.$$

using the change of variable in spherical coordinates from (9.35).

>> **syms rho th phi r**
>> **w1=subs(w,{x,y,z},{rho\*sin(th)\*cos(phi),rho\*sin(th)**
   **\*sin(phi),rho\*cos(th)});**
>> **int(int(int(w1\*rho^2\*sin(th),rho,0,r),th,0,pi/2),phi,0,pi/2)**
  ans =
  1/8\*r^6
and in Mathcad 14:

$$u(x,y,z) := x \cdot y \cdot z \quad P(x,y,z) := u(x,y,z) \cdot x \quad Q(x,y,z) := u(x,y,z) \cdot y \quad R(x,y,z) := u(x,y,z) \cdot z$$

$$w(x,y,z) := \frac{d}{dx} P(x,y,z) + \frac{d}{dy} Q(x,y,z) + \frac{d}{dz} R(x,y,z)$$

$$x(\rho,\theta,\varphi) := \rho \cdot \sin(\theta) \cdot \cos(\varphi) \quad y(\rho,\theta,\varphi) := \rho \cdot \sin(\theta) \cdot \sin(\varphi) \quad z(\rho,\theta,\varphi) := \rho \cdot \cos(\theta)$$

$$\int_0^r \int_0^{\frac{\pi}{2}} \int_0^{\frac{\pi}{2}} w(x(\rho,\theta,\varphi),y(\rho,\theta,\varphi),z(\rho,\theta,\varphi)) \cdot \rho^2 \cdot \sin(\theta) \, d\varphi \, d\theta \, d\rho \rightarrow \frac{r^6}{8}$$

and using Mathematica 8:

```
In[58]:= u := x * y * z

In[59]:= P := u * x

In[60]:= Q := u * y

In[61]:= R := u * z

In[62]:= w := D[P, x] + D[Q, y] + D[R, z]

In[63]:= xx[ρ_, θ_, φ_] := ρ * Sin[θ] * Cos[φ]

In[64]:= yy[ρ_, θ_, φ_] := ρ * Sin[θ] * Sin[φ]

In[65]:= zz[ρ_, θ_, φ_] := ρ * Cos[θ]

In[66]:= w1 = w /. {x -> xx[ρ, θ, φ], y → yy[ρ, θ, φ], z → zz[ρ, θ, φ]};

In[67]:= Integrate[w1 * ρ^2 * Sin[θ], {φ, 0, Pi/2}, {θ, 0, Pi/2}, {ρ, 0, r}]

Out[67]= r^6
         ──
         8
```

and in Maple 15:

$u := x \cdot y \cdot z :$
$P := u \cdot x :$
$Q := u \cdot y :$
$R := u \cdot z :$

$$w := \frac{d}{dx} P + \frac{d}{dy} Q + \frac{d}{dz} R :$$

$wl := w\Big|$

$\Big|x = \rho \cdot \sin(\theta) \cdot \cos(\varphi), \, y = \rho \cdot \sin(\theta) \cdot \sin(\varphi), \, z = \rho \cdot \cos(\theta)$

$$\int_0^r \int_0^{\frac{\pi}{2}} \int_0^{\frac{\pi}{2}} wl \cdot \rho^2 \cdot \sin(\theta) \, d\varphi \, d\theta \, d\rho$$

$$\frac{1}{8} r^6$$

**Example 9.26.** Compute the surface integral of the second type from the *Example 9.21* using the Gauss-Ostrogradski formula.

**Solution.**

One can notice that

$$\begin{cases} P(x,y,z) = x \\ Q(x,y,z) = y \\ R(x,y,z) = z; \end{cases}$$

hence

$$I = \int \int \int_V (1 + 1 + 1) \, dx dy dz = 3 \int \int \int_V dx dy dz,$$

where

$$V : \{ x \in [0, a], \; y \in [0, a - x], \; z \in [0, a - x - y] \}.$$

Therefore,

$$I = 3 \int_0^a \int_0^{a-x} \int_0^{a-x-y} dx dy dz = 3 \int_0^a \left( \int_0^{a-x} \left( \int_0^{a-x-y} dz \right) dy \right) dx$$

$$= 3 \int_0^a \left( \int_0^{a-x} (a - x - y) \, dy \right) dx = 3 \int_0^a \left( ay \Big|_0^{a-x} - xy \Big|_0^{a-x} - \frac{y^2}{2} \Big|_0^{a-x} \right) dx$$

$$= 3 \int_0^a \left( a(a - x) - x(a - x) - \frac{(a - x)^2}{2} \right) dx$$

$$= 3 \int_0^a \left( -ax + \frac{a^2}{2} + \frac{x^2}{2} \right) dx = 3 \left( -a \frac{x^2}{2} \Big|_0^a + \frac{a^2}{2} x \Big|_0^a + \frac{x^3}{6} \Big|_0^a \right),$$

i.e.

$$I = 3 \left( -\frac{a^3}{2} + \frac{a^3}{2} + \frac{a^3}{6} \right) = \frac{a^3}{2}.$$

We shall give a computer solution using Matlab 7.9:

```
>> syms a x y z
>> P=x;Q=y;R=z;
>> w=diff(P,x)+diff(Q,y)+diff(R,z);
>> int(int(int(w,z,0,a-x-y),y,0,a-x),x,0,a)
ans =
a^3/2
```

and in Mathcad 14:

$$P(x,y,z) := x \qquad\qquad Q(x,y,z) := y \qquad\qquad R(x,y,z) := z$$

$$w(x,y,z) := \frac{d}{dx}P(x,y,z) + \frac{d}{dy}Q(x,y,z) + \frac{d}{dz}R(x,y,z)$$

$$\int_0^a \int_0^{a-x} \int_0^{a-x-y} w(x,y,z)\, dz\, dy\, dx \rightarrow \frac{a^3}{2}$$

and with Mathematica 8:

In[6]:= P := x

In[7]:= Q := y

In[8]:= R := z

In[9]:= w := D[P, x] + D[Q, y] + D[R, z]

In[10]:= Integrate[w, {x, 0, a}, {y, 0, a - x}, {z, 0, a - x - y}]

Out[10]= $\frac{a^3}{2}$

and in Maple 15:

$P := x:$

$Q := y:$

$R := z:$

$w := \dfrac{d}{dx}P + \dfrac{d}{dy}Q + \dfrac{d}{dz}R :$

$\displaystyle\int_0^a \int_0^{a-x} \int_0^{a-x-y} w \, dz \, dy \, dx$

$\dfrac{1}{2}a^3$

**Example 9.27.** Compute the following surface integral of the second type in two ways: both as a surface integral and using the Gauss-Ostrogradski formula:

$$I = \iint_{\Sigma_-} x^3 dydz + y^3 dzdx + z^3 dxdy,$$

where $(\Sigma)$ is the sphere

$$x^2 + y^2 + z^2 = R^2.$$

**Solution.**

The next figure shows the surface $\Sigma$.

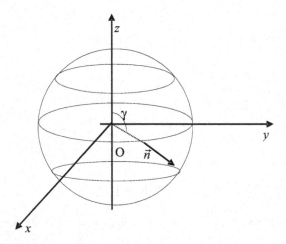

**Fig. 9.14**

Using the formula (9.29) we shall have

$$I = -\iint_{\Sigma} \left[ x^3 \cos\alpha + y^3 \cos\beta + z^3 \cos\gamma \right] d\sigma$$

which can be computed with (9.23).

Using the parametrical representation of the sphere $\sum$:

$$\begin{cases} x = R\sin\theta\cos\varphi \\ y = R\sin\theta\sin\varphi \ , \theta \in [0,\pi]\,, \varphi \in [0,2\pi] \\ z = R\cos\theta \end{cases}$$

we shall get

$$A = \frac{D\,(y,z)}{D\,(\theta,\varphi)} = \begin{vmatrix} R\cos\theta\sin\varphi & R\sin\theta\cos\varphi \\ -R\sin\theta & 0 \end{vmatrix} = R^2\sin^2\theta\cos\varphi,$$

$$B = \frac{D\,(z,x)}{D\,(\theta,\varphi)} = \begin{vmatrix} -R\sin\theta & 0 \\ R\cos\theta\cos\varphi & -R\sin\theta\sin\varphi \end{vmatrix} = R^2\sin^2\theta\sin\varphi,$$

$$C = \frac{D\,(x,y)}{D\,(\theta,\varphi)} = \begin{vmatrix} R\cos\theta\cos\varphi & -R\sin\theta\sin\varphi \\ R\cos\theta\sin\varphi & R\sin\theta\cos\varphi \end{vmatrix} = R^2\sin\theta\cos\theta$$

and

$$A^2 + B^2 + C^2 = R^4\sin^2\theta.$$

One obtains that

$$\cos\gamma = \frac{R^2\sin\theta\cos\theta}{\pm R^2\sin\theta} = \pm\cos\theta.$$

As in our case $\cos\gamma < 0$ it will result

$$\cos\gamma = \begin{cases} -\cos\theta, & \theta \in \left[0,\frac{\pi}{2}\right] \\ \cos\theta, & \theta \in \left(\frac{\pi}{2},\pi\right], \end{cases}$$

$$\cos\alpha = \frac{R^2\sin^2\theta\cos\varphi}{\pm R^2\sin\theta} = \pm\sin\theta\cos\varphi = \begin{cases} -\sin\theta\cos\varphi, & \theta \in \left[0,\frac{\pi}{2}\right] \\ \sin\theta\cos\varphi, & \theta \in \left(\frac{\pi}{2},\pi\right] \end{cases}$$

$$\cos\beta = \frac{R^2\sin^2\theta\sin\varphi}{\pm R^2\sin\theta} = \pm\sin\theta\sin\varphi = \begin{cases} -\sin\theta\sin\varphi, & \theta \in \left[0,\frac{\pi}{2}\right] \\ \sin\theta\sin\varphi, & \theta \in \left(\frac{\pi}{2},\pi\right]. \end{cases}$$

Hence

$$\begin{aligned} I &= -\int_0^{\pi/2}\int_0^{2\pi} \left(-R^3\sin^4\theta\cos^4\varphi - R^3\sin^4\theta\sin^4\varphi - R^3\cos^4\theta\right)\left(-R^2\sin\theta\right)d\theta d\varphi \\ &\quad - \int_{\pi/2}^{\pi}\int_0^{2\pi} \left(R^3\sin^4\theta\cos^4\varphi + R^3\sin^4\theta\sin^4\varphi + R^3\cos^4\theta\right)R^2\sin\theta d\theta d\varphi \\ &= -\int_0^{\pi}\int_0^{2\pi} \left(R^3\sin^4\theta\cos^4\varphi + R^3\sin^4\theta\sin^4\varphi + R^3\cos^4\theta\right)R^2\sin\theta d\theta d\varphi \\ &= -R^5\int_0^{\pi}\int_0^{2\pi} \left(\sin^5\theta\cos^4\varphi + \sin^5\theta\sin^4\varphi + \cos^4\theta\sin\theta\right)d\theta d\varphi, \end{aligned}$$

i.e.

$$I = -R^5 \left[ \int_0^\pi \left( \int_0^{2\pi} \cos^4 \varphi d\varphi \right) \sin^5 \theta d\theta + \int_0^\pi \left( \int_0^{2\pi} \sin^4 \varphi d\varphi \right) \sin^5 \theta d\theta \right.$$
$$\left. + \int_0^\pi \left( \int_0^{2\pi} d\varphi \right) \cos^4 \theta \sin \theta d\theta \right].$$

We shall compute:

$$I_1 = \int_0^{2\pi} \cos^4 \varphi d\varphi = \int_0^{2\pi} \cos^3 \varphi \cdot (\sin \varphi)' d\varphi$$
$$= \sin \varphi \cos^3 \varphi \Big|_0^{2\pi} + 3 \int_0^{2\pi} \cos^2 \varphi \sin^2 \varphi d\varphi$$
$$= 3 \int_0^{2\pi} \cos^2 \varphi \left( 1 - \cos^2 \varphi \right) d\varphi = 3 \int_0^{2\pi} \cos^2 \varphi d\varphi - 3 \int_0^{2\pi} \cos^4 \varphi d\varphi,$$

i.e.

$$4 I_1 = 3\pi \Longrightarrow I_1 = \frac{3\pi}{4};$$

$$I_2 = \int_0^{2\pi} \sin^4 \varphi d\varphi = - \int_0^{2\pi} \sin^3 \varphi \cdot (\cos \varphi)' d\varphi$$
$$= - \sin^3 \varphi \cos \varphi \Big|_0^{2\pi} + 3 \int_0^{2\pi} \sin^2 \varphi \cos^2 \varphi d\varphi$$
$$= 3 \int_0^{2\pi} \sin^2 \varphi \left( 1 - \sin^2 \varphi \right) d\varphi = 3 \int_0^{2\pi} \sin^2 \varphi d\varphi - 3 \int_0^{2\pi} \sin^4 \varphi d\varphi,$$

i.e.

$$4 I_2 = 3\pi \Longrightarrow I_2 = \frac{3\pi}{4}.$$

Therefore,

$$I = -R^5 \left[ 2 \cdot \frac{3\pi}{4} \underbrace{\int_0^\pi \sin^5 \theta d\theta}_{I_3} + 2\pi \underbrace{\int_0^\pi \cos^4 \theta \sin \theta d\theta}_{I_4} \right]$$
$$= -R^5 \left( \frac{3\pi}{2} I_3 + 2\pi I_4 \right),$$

where

$$I_3 = \int_0^\pi \sin^5 \theta d\theta = -\int_0^\pi \sin^4 \theta \cdot (\cos \theta)' \, d\theta = -\sin^4 \theta \cos \theta \Big|_0^\pi + 4 \int_0^\pi \sin^3 \theta \cos^2 \theta d\theta$$

$$= 4 \int_0^\pi \sin^3 \theta \left(1 - \sin^2 \theta\right) d\theta = 4 \int_0^\pi \sin^3 \theta d\theta - 4 \int_0^\pi \sin^5 \theta d\theta,$$

i.e.

$$5I_3 = 4 \cdot \frac{4}{3} \implies I_3 = \frac{16}{15}$$

and

$$I_4 = \int_0^\pi \cos^4 \theta \sin \theta d\theta = -\frac{1}{5} \int_0^\pi \left(\cos^5 \theta\right)' d\theta = -\frac{\cos^5 \theta}{5} \Big|_0^\pi = \frac{2}{5}.$$

Finally, it will result

$$I = -R^5 \left[ \frac{3\pi}{2} \cdot \frac{16}{15} + 2\pi \cdot \frac{2}{5} \right] = -\frac{12\pi R^5}{5}.$$

The next function, which is defined in Matlab 7.9 allows the calculation of $\cos \alpha$, $\cos \beta$, $\cos \gamma$.

```
function [al,be,ga]=f(t,s,A,B,C,a,b,c)
if a<=t<=b
 u=-s;
elseif b<t<=c
 u=s;
end
ga=C/u;
be=B/u;
al=A/u;
end
```

We shall save the next file, having the following content and the name g.m:

```
syms R th phi
x=R*sin(th)*cos(phi);y=R*sin(th)*sin(phi);
z=R*cos(th);
A=det(jacobian([y z],[th phi]));
B=det(jacobian([z x],[th phi]));
C=det(jacobian([x y],[th phi]));
s=sqrt(A^2+B^2+C^2);
a=0;
b=pi/2;
c=pi;
```

```
t=a:b;
t1=b:c;
[al,be,ga]=f(t,s,A,B,C,a,b,c);
 r=(x^3*al+y^3*be+z^3*ga)*s;
[al1,be1,ga1]=f(t1,s,A,B,C,a,b,c);
 r1=(x^3*al1+y^3*be1+z^3*ga1)*s;
 int(int(r,th,a,b),phi,a,2*c)+int(int(r1,th,a,b),phi,a,2*c)
```

Then, we have to write in the command line:

>> g

ans =

-12/5*R^5*pi

In the case when we shall apply the Gauss-Ostrogradski formula we achieve:

$$I = \int\int_{\Sigma_-} P(x,y,z)\,dydz + Q(x,y,z)\,dzdx + R(x,y,z)\,dxdy$$

$$= -\int\int\int_V \left[\frac{\partial P}{\partial x}(x,y,z) + \frac{\partial Q}{\partial y}(x,y,z) + \frac{\partial R}{\partial z}(x,y,z)\right]\,dxdydz,$$

where

$$V : \{x^2 + y^2 + z^2 \le R^2\}$$

and

$$\begin{cases} P(x,y,z) = x^3 \\ Q(x,y,z) = y^3 \\ R(x,y,z) = z^3. \end{cases}$$

As

$$\begin{cases} \frac{\partial P}{\partial x} = 3x^2 \\ \frac{\partial Q}{\partial y} = 3y^2 \\ \frac{\partial R}{\partial z} = 3z^2 \end{cases}$$

we deduce that:

$$I = -3 \int\int\int_V (x^2 + y^2 + z^2)\,dxdydz,$$

which can be computed with the change of variables in spherical coordinates:

$$\begin{cases} x = \rho\sin\theta\cos\varphi \\ y = \rho\sin\theta\sin\varphi \\ z = \rho\cos\theta \end{cases}, \rho \in [0, R],\ \theta \in [0, \pi],\ \varphi \in [0, 2\pi];$$

using (9.3) we shall have:

$$I = (-3) \int_0^R \int_0^\pi \int_0^{2\pi} \rho^2 \cdot \rho^2 \sin\theta \rho d\rho d\theta d\varphi = (-3) \int_0^R \left[ \int_0^\pi \left( \int_0^{2\pi} d\varphi \right) \sin\theta d\theta \right] \rho^4 d\rho$$

$$= -6\pi \int_0^R \left( \int_0^\pi \sin\theta d\theta \right) \rho^4 d\rho = -6\pi \cdot (-\cos\theta)|_0^\pi \cdot \int_0^R \rho^4 d\rho$$

$$= -\frac{12\pi\rho^5}{5}\bigg|_0^R = -\frac{12\pi R^5}{5}.$$

We shall obtain this result with Maple 7.9:
```
>> syms x y z
>> P=x^3;Q=y^3;R=z^3;
>> w=diff(P,x)+diff(Q,y)+diff(R,z);
>> syms r rho th phi
>> k=simplify(subs(w,{x,y,z},{rho*sin(th)*cos(phi),
    rho*sin(th)*sin(phi),rho*cos(th)}));
>> -int(int(int(k*rho^2*sin(th),phi,0,2*pi),th,0,pi),rho,0,r)
ans =
-(12*pi*r^5)/5
```
and in Mathcad 14:

$$P(x,y,z) := x^3 \qquad Q(x,y,z) := y^3 \qquad \underset{\sim}{R}(x,y,z) := z^3$$

$$w(x,y,z) := \frac{d}{dx}P(x,y,z) + \frac{d}{dy}Q(x,y,z) + \frac{d}{dz}R(x,y,z) \rightarrow 3 \cdot x^2 + 3 \cdot y^2 + 3 \cdot z^2$$

$$x(\rho,\theta,\varphi) := \rho \cdot \sin(\theta) \cdot \cos(\varphi) \qquad y(\rho,\theta,\varphi) := \rho \cdot \sin(\theta) \cdot \sin(\varphi) \qquad z(\rho,\theta,\varphi) := \rho \cdot \cos(\theta)$$

$$u(\rho,\theta,\varphi) := w(x(\rho,\theta,\varphi),y(\rho,\theta,\varphi),z(\rho,\theta,\varphi)) \text{ simplify } \rightarrow 3 \cdot \rho^2$$

$$-\int_0^r \int_0^\pi \int_0^{2\cdot\pi} w(x(\rho,\theta,\varphi),y(\rho,\theta,\varphi),z(\rho,\theta,\varphi)) \cdot \rho^2 \cdot \sin(\theta) \, d\varphi \, d\theta \, d\rho \rightarrow -\frac{12 \cdot \pi \cdot r^5}{5}$$

and with Mathematica 8:

In[10]:= **P := x^3**

In[11]:= **Q := y^3**

In[12]:= **R := z^3**

In[13]:= **w := D[P, x] + D[Q, y] + D[R, z]**

In[14]:= **xx[$\rho$_, $\theta$_, $\phi$_] := $\rho$ \* Sin[$\theta$] \* Cos[$\phi$]**

In[15]:= **yy[$\rho$_, $\theta$_, $\phi$_] := $\rho$ \* Sin[$\theta$] \* Sin[$\phi$]**

In[16]:= **zz[$\rho$_, $\theta$_, $\phi$_] := $\rho$ \* Cos[$\theta$]**

In[17]:= **w1 = w /. {x -> xx[$\rho$, $\theta$, $\phi$], y → yy[$\rho$, $\theta$, $\phi$], z → zz[$\rho$, $\theta$, $\phi$]};**

In[18]:= **-Integrate[w1 \* $\rho$^2 \* Sin[$\theta$], {$\phi$, 0, 2 \* Pi}, {$\theta$, 0, Pi}, {$\rho$, 0, r}]**

Out[18]= $-\dfrac{12\,\pi\,r^5}{5}$

and using Maple 15:

$$P := x^3 :$$
$$Q := y^3 :$$
$$R := z^3 :$$
$$w := \frac{d}{dx}P + \frac{d}{dy}Q + \frac{d}{dz}R :$$
$$w1 := simplify\left( w\Big|_{x = \rho\,\sin(\theta)\,\cos(\varphi),\, y = \rho\,\sin(\theta)\,\sin(\varphi),\, z = \rho\,\cos(\theta)} \right) :$$
$$\int_0^r \int_0^\pi \int_0^{2\cdot\pi} w1 \cdot \rho^2 \cdot \sin(\theta)\, d\varphi\, d\theta\, d\rho$$

$$\frac{12}{5}\pi r^5$$

## 9.5.3   Stokes Formula

The Stokes formula is the analogue from $\mathbb{R}^3$ of the Green's formula from $\mathbb{R}^2$ namely, Green's formula is simply Stokes' formula in the plane. Green's formula deals with 2-dimensional regions, and Stokes' formula deals with 3-dimensional regions.

**Theorem 9.28. (Stokes formula, see** [48], p. 472 and [42], p. 159 and [8] and [2], p. 79)**.** If the functions $P(x, y, z)$, $Q(x, y, z)$, $R(x, y, z)$ are continuously differentiable and $C$ is a closed contour bounding a two-sided surface $\sum$, we have the *Stokes' formula*:

$$I = \oint_C P(x, y, z) \, dx + Q(x, y, z) \, dy + R(x, y, z) \, dz \tag{9.36}$$

$$= \int\int_{\sum_+} \left( \frac{\partial R}{\partial y} - \frac{\partial Q}{\partial z} \right) dydz + \left( \frac{\partial P}{\partial z} - \frac{\partial R}{\partial x} \right) dzdx + \left( \frac{\partial Q}{\partial x} - \frac{\partial P}{\partial y} \right) dxdy$$

$$= \int\int_{\sum} \left[ \left( \frac{\partial R}{\partial y} - \frac{\partial Q}{\partial z} \right) \cos\alpha + \left( \frac{\partial P}{\partial z} - \frac{\partial R}{\partial x} \right) \cos\beta + \left( \frac{\partial Q}{\partial x} - \frac{\partial P}{\partial y} \right) \cos\gamma \right] d\sigma,$$

where: $\cos\alpha, \cos\beta, \cos\gamma$ are the direction cosines of the normal to the surface $\sum$ and the direction of the normal is defined so that on the side of the normal, the contour of $C$ is traced counter- clockwise.

The Stokes' formula can be also written in the form (see [42], p. 173 and [8]):

$$I = \oint_\Gamma \vec{F} \cdot d\vec{r} = \int\int_{\sum} \text{curl} \, \vec{F} \cdot \vec{n} \, d\sigma, \tag{9.37}$$

where:

- $\vec{F}$ is a vector field,

- $\vec{n}$ constitutes the unit normal vector at the considered face of the surface $\sum$,

- the curve $\Gamma$ means the boundary of the surface $\sum$.

The Stokes formula says the surface integral of curl $\vec{F} \cdot \vec{n}$ over a surface $\sum$ is the circulation of the vector field $\vec{F}$ around the contour of $\Gamma$.

The curve $\Gamma$ and the surface $\sum$ are oriented such that: the meaning to go on the curve $\Gamma$ is associated with the corresponding face of the surface $\sum$. Another way of thinking about the proper orientation is the following: for someone which is walking near the edge of the surface $\sum$, in the direction corresponding to the orientation of $\Gamma$, the surface must be to your left and the edge $\Gamma$ must be to your right.

**Example 9.29.** Compute the line integral

$$I = \int_\Gamma (z - y) \, dx + (x - z) \, dy + (y - x) \, dz,$$

where the closed curve $\Gamma$ is the contour of the triangle with the vertices $A(a, 0, 0)$, $B(0, b, 0)$ and $C(0, 0, c)$ using the following two methods:

1) directly as a line integral,
2) with the Stokes formula.
**Solutions.**
1) The next figure the closed curve $\Gamma$.

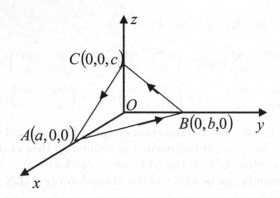

**Fig. 9.15**

From the Cartesian equations of the straight lines $AB$, $BC$, $CA$ we shall obtain their parametrical equations:

$$(AB): \begin{cases} \frac{x-a}{-a} = \frac{y-b}{b-0} \\ z = 0 \end{cases} \implies (AB): \begin{cases} x = t \\ y = -\frac{b}{a}(t-a), \ t \in [a,0]; \\ z = 0 \end{cases}$$

$$(BC): \begin{cases} x = 0 \\ \frac{y-b}{-b} = \frac{z-0}{c} \end{cases} \implies (BC): \begin{cases} x = 0 \\ y = t \\ z = -\frac{c}{b}(t-b) \end{cases}, \ t \in [b,0];$$

$$(CA): \begin{cases} y = 0 \\ \frac{x-a}{-a} = \frac{z-0}{c} \end{cases} \implies (CA): \begin{cases} y = 0 \\ x = t \\ z = -\frac{c}{a}(t-a) \end{cases}, \ t \in [0,a].$$

Hence,

$$I = \int_\Gamma (z-y)\,\mathrm{d}x + (x-z)\,\mathrm{d}y + (y-x)\,\mathrm{d}z = \underbrace{\int_{AB} (z-y)\,\mathrm{d}x + (x-z)\,\mathrm{d}y + (y-x)\,\mathrm{d}z}_{I_1}$$

$$+ \underbrace{\int_{BC} (z-y)\,\mathrm{d}x + (x-z)\,\mathrm{d}y + (y-x)\,\mathrm{d}z}_{I_2} + \underbrace{\int_{CA} (z-y)\,\mathrm{d}x + (x-z)\,\mathrm{d}y + (y-x)\,\mathrm{d}z}_{I_3}$$

$$= I_1 + I_2 + I_3.$$

We shall evaluate:

$$I_1 = -\int_0^a \left\{ \left[ 0 + \frac{b}{a}(t-a) \right] \cdot 1 + t \cdot \left( -\frac{b}{a} \right) + \left[ -\frac{b}{a}(t-a) - t \right] \cdot 0 \right\} dt$$

$$= -\int_0^a \left( \frac{b}{a}t - b - \frac{b}{a}t \right) dt = bt|_0^a = ba,$$

$$I_2 = -\int_0^b \left\{ \left[ -\frac{c}{b}(t-b) - t \right] \cdot 0 + \frac{c}{b}(t-b) \cdot 1 + t \cdot \left( -\frac{c}{b} \right) \right\} dt$$

$$= -\int_0^b \left( \frac{c}{b}t - c - \frac{c}{b}t \right) dt = ct|_0^b = cb,$$

$$I_3 = \int_0^a \left\{ \left( -\frac{c}{a} \right) \cdot (t-a) \cdot 1 + \left[ t + \frac{c}{a}(t-a) \right] \cdot 0 + (0-t) \cdot \left( -\frac{c}{a} \right) \right\} dt$$

$$= \int_0^a \left( -\frac{c}{a}t + c + \frac{c}{a}t \right) dt = ct|_0^a = ca$$

and then we shall obtain:

$$I = ab + bc + ca.$$

2) For our integral we shall have

$$\begin{cases} P(x,y,z) = z - y \\ Q(x,y,z) = x - z \\ R(x,y,z) = y - x; \end{cases}$$

therefore

$$\begin{cases} \frac{\partial R}{\partial y} = 1 \\ \frac{\partial Q}{\partial z} = -1 \\ \frac{\partial P}{\partial z} = 1 \\ \frac{\partial R}{\partial x} = -1 \\ \frac{\partial Q}{\partial x} = 1 \\ \frac{\partial P}{\partial y} = -1. \end{cases}$$

One gets:

$$I = 2 \int\int_\Sigma (\cos\alpha + \cos\beta + \cos\gamma) \, d\sigma.$$

Taking into account that the equation of the plane defined by three non-collinear points $A, B, C$ is:

$$
\begin{vmatrix}
x - x_A & y - y_A & z - z_A \\
x_B - x_A & y_B - y_A & z_B - z_A \\
x_C - x_A & y_C - y_A & z_C - z_A
\end{vmatrix} = 0
$$

it results that the equation of the plane defined by the points $A(a,0,0)$, $B(0,b,0)$, $C(0,0,c)$ is:

$$
\frac{x}{a} + \frac{y}{b} + \frac{z}{c} = 1;
$$

therefore

$$
z = c\left(1 - \frac{x}{a} - \frac{y}{b}\right).
$$

Using the relation (9.27) we shall have:

$$
\begin{cases}
p = \dfrac{\partial z}{\partial x} = -\dfrac{c}{a} \\
q = \dfrac{\partial z}{\partial y} = -\dfrac{c}{b} \\
\sqrt{1 + p^2 + q^2} = \sqrt{1 + \dfrac{c^2}{a^2} + \dfrac{c^2}{b^2}} = \dfrac{\sqrt{a^2b^2 + b^2c^2 + a^2c^2}}{ab}.
\end{cases}
$$

From the condition $\cos\gamma > 0$ and and the relation (9.32) one deduces:

$$
\begin{cases}
\cos\alpha = \dfrac{c}{a} \cdot \dfrac{ab}{\sqrt{a^2b^2 + b^2c^2 + a^2c^2}} = \dfrac{cb}{\sqrt{a^2b^2 + b^2c^2 + a^2c^2}} \\
\cos\beta = \dfrac{ca}{\sqrt{a^2b^2 + b^2c^2 + a^2c^2}} \\
\cos\gamma = \dfrac{ab}{\sqrt{a^2b^2 + b^2c^2 + a^2c^2}}.
\end{cases}
$$

We shall achieve:

$$
\begin{aligned}
I &= 2 \int\int_D \frac{cb + ca + ab}{\sqrt{a^2b^2 + b^2c^2 + a^2c^2}} \cdot \frac{\sqrt{a^2b^2 + b^2c^2 + a^2c^2}}{ab} \, dxdy \\
&= \frac{2}{ab}(cb + ca + ab) \int\int_D dxdy,
\end{aligned}
$$

where

$$
D = \mathrm{pr}_{\Sigma} xOy = \left\{(x,y) \in \mathbb{R}^2 \mid 0 \le x \le a,\ 0 \le y \le -\frac{b}{a}x + b\right\}.
$$

Finally,

$$
\begin{aligned}
I &= \frac{2}{ab}(cb + ca + ab) \int_0^a \left(\int_0^{-\frac{b}{a}x + b} dy\right) dx = \frac{2}{ab}(cb + ca + ab) \int_0^a \left(-\frac{b}{a}x + b\right) dx \\
&= \frac{2}{ab}(cb + ca + ab)\left(-\frac{b}{a} \cdot \frac{x^2}{2}\Big|_0^a + +bx|_0^a\right) = \frac{2}{ab}(cb + ca + ab)\left(-\frac{ab}{2} + ab\right) \\
&= cb + ca + ab.
\end{aligned}
$$

**Example 9.30.** Compute the circulation of the vector field

$$\overrightarrow{F}(x, y, z) = x^2 y^3 \overrightarrow{i} + \overrightarrow{j} + z\overrightarrow{k}$$

along the circumference

$$(\Gamma) : \begin{cases} x^2 + y^2 = R^2 \\ z = 0 \end{cases}$$

using the following two methods:
1) directly, as a line integral;
2) with the Stokes' formula, taking as a surface

$$\left(\sum\right) : z = \sqrt{R^2 - x^2 - y^2}.$$

**Solutions.**
1) We shall have

$$\oint_\Gamma \overrightarrow{F} \cdot d\overrightarrow{r} = \oint_\Gamma \left( x^2 y^3 \overrightarrow{i} + \overrightarrow{j} + z\overrightarrow{k} \right) \cdot \left( \overrightarrow{i}\, dx + \overrightarrow{j}\, dy + \overrightarrow{k}\, dz \right)$$

$$= \oint_\Gamma \left( x^2 y^3 dx + dy + z dz \right),$$

where

$$(\Gamma) : \begin{cases} x = R\cos\varphi \\ y = R\sin\varphi \ , \ \varphi \in [0, 2\pi]. \\ z = 0 \end{cases} \tag{9.38}$$

Using the parametric representation of the curve from (9.38) one gets:

$$I = \int_0^{2\pi} \left[ R^2 \cos^2\varphi \cdot R^3 \sin^3\varphi \cdot (-R\sin\varphi) + R\cos\varphi + 0 \right] d\varphi$$

$$= \int_0^{2\pi} \left( -R^6 \cos^2\varphi \sin^4\varphi + R\cos\varphi \right) d\varphi$$

$$= -R^6 \underbrace{\int_0^{2\pi} \cos^2\varphi \sin^4\varphi\, d\varphi}_{I_1} + R\sin\varphi\big|_0^{2\pi} = -R^6 I_1,$$

where

$$I_1 = \int_0^{2\pi} (1 - \sin^2\varphi) \sin^4\varphi\, d\varphi = \underbrace{\int_0^{2\pi} \sin^4\varphi\, d\varphi}_{I_2} - \underbrace{\int_0^{2\pi} \sin^6\varphi\, d\varphi}_{I_3}$$

$$= I_2 - I_3$$

and

$$I_2 = \int_0^{2\pi} \sin^4 \varphi d\varphi = -\int_0^{2\pi} (\cos \varphi)' \sin^3 \varphi d\varphi = -\sin^3 \varphi \cos \varphi \Big|_0^{2\pi} + 3 \int_0^{2\pi} \sin^2 \varphi \cos^2 \varphi d\varphi$$

$$= 3 \int_0^{2\pi} \sin^2 \varphi \left(1 - \sin^2 \varphi\right) d\varphi = 3 \int_0^{2\pi} \sin^2 \varphi d\varphi - 3 \int_0^{2\pi} \sin^4 \varphi d\varphi$$

$$\Longrightarrow 4I_2 = 3\pi \Longrightarrow I_2 = \frac{3\pi}{4},$$

respectively

$$I_3 = \int_0^{2\pi} \sin^6 \varphi d\varphi = -\int_0^{2\pi} (\cos \varphi)' \sin^5 \varphi d\varphi = -\sin^5 \varphi \cos \varphi \Big|_0^{2\pi} + 5 \int_0^{2\pi} \sin^4 \varphi \cos^2 \varphi d\varphi$$

$$= 5 \int_0^{2\pi} \sin^4 \varphi \left(1 - \sin^2 \varphi\right) d\varphi = 5 \int_0^{2\pi} \sin^4 \varphi d\varphi - 5 \int_0^{2\pi} \sin^6 \varphi d\varphi$$

$$\Longrightarrow 6I_3 = 5 \cdot \frac{3\pi}{4} \Longrightarrow I_3 = \frac{5\pi}{8}.$$

Finally,

$$I = -R^6 \left(\frac{3\pi}{4} - \frac{5\pi}{8}\right) = -\frac{R^6 \pi}{8}.$$

2) The components of the vector field $\overrightarrow{F}$ are:

$$\begin{cases} P(x,y,z) = x^2 y^3 \\ Q(x,y,z) = 1 \\ R(x,y,z) = z; \end{cases}$$

therefore

$$\operatorname{curl} \overrightarrow{F} = \left(\frac{\partial R}{\partial y} - \frac{\partial Q}{\partial z}\right) \overrightarrow{i} + \left(\frac{\partial P}{\partial z} - \frac{\partial R}{\partial x}\right) \overrightarrow{j} + \left(\frac{\partial Q}{\partial x} - \frac{\partial P}{\partial y}\right) \overrightarrow{k} = -3x^2 y^3 \overrightarrow{k}. \qquad (9.39)$$

Since $\cos \gamma > 0$, from (9.32) it results that

$$\cos \gamma = \frac{1}{\sqrt{1 + p^2 + q^2}},$$

where

$$\begin{cases} p = \dfrac{\partial z}{\partial x} = -\dfrac{2x}{2\sqrt{R^2 - x^2 - y^2}} = -\dfrac{x}{\sqrt{R^2 - x^2 - y^2}} \\ q = \dfrac{\partial z}{\partial y} = -\dfrac{y}{\sqrt{R^2 - x^2 - y^2}}; \end{cases}$$

therefore

$$\sqrt{1+p^2+q^2} = \sqrt{1 + \frac{x^2}{R^2-x^2-y^2} + \frac{y^2}{R^2-x^2-y^2}} = \frac{R}{\sqrt{R^2-x^2-y^2}}$$

and

$$\begin{cases} \cos\gamma = \frac{\sqrt{R^2-x^2-y^2}}{R} \\ \cos\alpha = -\frac{p}{\sqrt{1+p^2+q^2}} = \frac{x}{R} \\ \cos\beta = -\frac{q}{\sqrt{1+p^2+q^2}} = \frac{y}{R}. \end{cases}$$

Hence, the unit normal vector from (9.30) will be

$$\vec{n} = \frac{x}{R}\vec{i} + \frac{y}{R}\vec{j} + \frac{\sqrt{R^2-x^2-y^2}}{R}\vec{k}. \tag{9.40}$$

Taking into account the relations (9.39) and (9.40) we obtain

$$\operatorname{curl}\vec{F}\cdot\vec{n} = -3x^2y^2\frac{\sqrt{R^2-x^2-y^2}}{R}.$$

Let's determine now

$$J = \iint_\Sigma \operatorname{curl}\vec{F}\cdot\vec{n}\,d\sigma = -\frac{3}{R}\iint_D x^2y^2\sqrt{R^2-x^2-y^2}\cdot\frac{R}{\sqrt{R^2-x^2-y^2}}dxdy$$

$$= -3\iint_D x^2y^2dxdy,$$

where

$$D = \left\{ (x,y)\in\mathbb{R}^2 \mid x^2+y^2 \le R^2 \right\}.$$

The region of integration is simpler to describe using polar coordinates. Using the change of variables in polar coordinates

$$\begin{cases} x = \rho\cos\theta \\ y = \rho\sin\theta \end{cases}, \ \rho\in[0,R],\ \theta\in[0,2\pi]$$

one obtains

$$J = -3\int_0^R\int_0^{2\pi}\rho^4\sin^2\theta\cos^2\theta\cdot\rho d\rho d\theta = -3\int_0^R\underbrace{\left(\int_0^{2\pi}\sin^2\theta\cos^2\theta d\theta\right)}_{J_1}\rho^5 d\rho$$

$$= -3J_1\int_0^R\rho^5 d\rho = -J_1\cdot\frac{R^6}{2},$$

where

$$J_1 = \int_0^{2\pi} \sin^2 \theta \left(1 - \sin^2 \theta\right) d\theta = \int_0^{2\pi} \sin^2 \theta d\theta - \int_0^{2\pi} \sin^4 \theta d\theta$$

$$= \pi - \frac{3\pi}{4} = \frac{\pi}{4}.$$

Finally,

$$J = -\frac{\pi R^6}{8}.$$

We shall have in Matlab 7.9:

```
>> syms x y z R
>> F=[x^2*y^3 1 z];
>> rot=[diff(F(3),y)-diff(F(2),z) diff(F(2),z)-diff(F(3),x)
diff(F(2),x)-diff(F(1),y)];
>> z=sqrt(R^2-x^2-y^2);
>> p=diff(z,x);
>> q=diff(z,y);
>> w=sqrt(1+p^2+q^2)
>> n=[-p/w -q/w 1/w];
>> v=dot(rot,n)*w;
>> syms rho th
>> k=subs(v,{x,y},{rho*cos(th),rho*sin(th)});
>> c=int(int(k*rho,rho,0,R),th,0,2*pi)
c =
-1/8*pi*R^6
```

and using Mathcad 14:

$$F(x,y,z) := \begin{pmatrix} x^2 \cdot y^3 \\ 1 \\ z \end{pmatrix}$$

$$z(r,x,y) := \sqrt{r^2 - x^2 - y^2} \qquad p(r,x,y) := \frac{d}{dx} z(r,x,y) \qquad q(r,x,y) := \frac{d}{dy} z(r,x,y)$$

$$w(r,x,y) := \sqrt{1 + p(r,x,y)^2 + q(r,x,y)^2}$$

$$n(r,x,y) := \begin{pmatrix} \dfrac{-p(r,x,y)}{w(r,x,y)} \\ \dfrac{-q(r,x,y)}{w(r,x,y)} \\ \dfrac{1}{w(r,x,y)} \end{pmatrix} \qquad rot(x,y,z) := \begin{pmatrix} \dfrac{d}{dy} F(x,y,z)_2 - \dfrac{d}{dz} F(x,y,z)_1 \\ \dfrac{d}{dz} F(x,y,z)_1 - \dfrac{d}{dx} F(x,y,z)_2 \\ \dfrac{d}{dx} F(x,y,z)_1 - \dfrac{d}{dy} F(x,y,z)_0 \end{pmatrix}$$

$$v(r,x,y,z) := \left( n(r,x,y)^T \cdot rot(x,y,z) \right) \cdot w(r,x,y)$$

$$k(r,\rho,\theta) := v(r,x,y,z) \text{ substitute}, x = \rho \cdot \cos(\theta), y = \rho \cdot \sin(\theta) \rightarrow 3 \cdot \rho^4 \cdot \cos(\theta)^2 \cdot \left( \frac{\cos(2 \cdot \theta)}{2} - \frac{1}{2} \right)$$

$$\int_0^r \int_0^{2 \cdot \pi} k(r,\rho,\theta) \cdot \rho \, d\theta \, d\rho \rightarrow -\frac{\pi \cdot r^6}{8}$$

and with Mathematica 8:

In[1]:= F := {x^2 * y^3, 1, z}

In[2]:= rot := {D[F[[3]], y] - D[F[[2]], z], D[F[[2]], z] - D[F[[3]], x],
        D[F[[2]], z] - D[F[[1]], y]}

In[3]:= z[x_, y_] := Sqrt[Rr^2 - x^2 - y^2]

In[4]:= p := D[z[x, y], x]

In[5]:= q := D[z[x, y], y]

In[6]:= w := Sqrt[1 + p^2 + q^2]

In[7]:= n := {-p / w, -q / w, 1 / w}

In[8]:= v :=
        ({D[F[[3]], y] - D[F[[2]], z], D[F[[2]], z] - D[F[[3]], x],
            D[F[[2]], x] - D[F[[1]], y]} . {-p / w, -q / w, 1 / w}) * w;

In[9]:= k := v /. {x → ρ * Cos[θ], y → ρ * Sin[θ]}

In[10]:= Integrate[k * ρ, {ρ, 0, r}, {θ, 0, 2 * Pi}]

Out[10]= $-\dfrac{\pi\, r^6}{8}$

and in Maple 15:

with( *VectorCalculus* ) :

SetCoordinates( *'cartesian'*$_{x,y,z}$ ) :

$F := VectorField(\langle x^2 \cdot y^3, 1, z\rangle)$ :

*rot* := *Curl*(F) :

$z := \sqrt{r^2 - x^2 - y^2}$ :

$p := \dfrac{d}{dx} z$ :

$q := \dfrac{d}{dy} z$ :

$w := \sqrt{1 + p^2 + q^2}$ :

$n := VectorField\left(\left\langle -\dfrac{p}{w}, -\dfrac{q}{w}, \dfrac{1}{w}\right\rangle\right)$ :

$v := (rot.n) \cdot w$ :

$k := v\Big|_{x = \rho \cdot \cos(\theta), y = \rho \cdot \sin(\theta)}$ :

$\displaystyle \int_0^r \int_0^{2\cdot\pi} k \cdot \rho \, d\theta \, d\rho$

$$-\frac{1}{8}\,\pi\, r^6$$

## 9.6   Problems

1. Compute the triple integral:

$$I = \int\int\int_V \frac{xyz}{(x^2+y^2+z^2+1)^4}\,dx\,dy\,dz,$$

with $V = [0,1] \times [0,1] \times [0,1]$.

**Computer solution.**

We shall achieve using Matlab 7.9:

```
>> syms x y z
f=@(x,y,z)(x.*y.*z)./((x.^2+y.^2+z.^2+1).^4);
>> triplequad(f,0,1,0,1,0,1)
ans =
 0.0052
```

and with Mathcad 14:

$$\int_0^1 \left[ \int_0^1 \left[ \int_0^1 \frac{x \cdot y \cdot z}{\left(x^2 + y^2 + z^2 + 1\right)^4} \, dx \right] dy \right] dz = 0.00521$$

and in Mathematica 8:

In[1]:= **Integrate[ (x \* y \* z) / ((x^2 + y^2 + z^2 + 1) ^4) , {x, 0, 1}, {y, 0, 1}, {z, 0, 1}]**

Out[1]= $\dfrac{1}{192}$

We can not compute this triple integral using Maple 15.

2. Compute:

$$I = \int\int\int_V \frac{dx\,dy\,dz}{(a^2 + x^2 + y^2 - z)^{3/2}},$$

with $V$ defined by:

$$\begin{cases} x^2 + y^2 \geq az \\ x^2 + y^2 \leq a^2 \\ z \geq 0. \end{cases}$$

3. Find the volume of the homogeneous body, bounded by the sphere

$$x^2 + y^2 + z^2 = 4$$

and the paraboloid

$$x^2 + y^2 = 3z.$$

**Computer solution.**
We shall compute the volume of the body from the Figure 9.16:

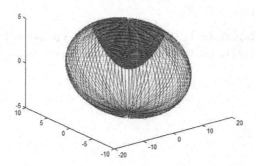

**Fig. 9.16**

using Matlab 7.9:

```
>> syms z rho th
>> x=rho*cos(th);
>> y=rho*sin(th);
>> int(int(int(1,z,(x^2+y^2)/3,sqrt(4-x^2-y^2))*rho,rho,0,
sqrt(3)),th,0,2*pi)
   ans =
   (19*pi)/6
```

and in Mathcad 14:

$$I1(x,y) := \int_{\frac{x^2+y^2}{3}}^{\sqrt{4-x^2-y^2}} 1 \, dz$$

$$x(\rho,\theta) := \rho\cdot\cos(\theta) \qquad y(\rho,\theta) := \rho\cdot\sin(\theta)$$

$$\int_0^{\sqrt{3}} \int_0^{2\cdot\pi} I1(x(\rho,\theta),y(\rho,\theta))\cdot\rho \, d\theta \, d\rho \rightarrow \frac{19\cdot\pi}{6}$$

and with Mathematica 8:

In[4]:= **x** := ρ * Cos [θ]

In[5]:= **y** := ρ * Sin [θ]

In[6]:= **Integrate [Integrate [1, {z, (x^2 + y^2) /3, Sqrt [4 - x^2 - y^2]}] * ρ,**
       **{ρ, 0, Sqrt [3]}, {θ, 0, 2 * Pi}]**

Out[6]= $\dfrac{19\,\pi}{6}$

and using Maple 15:

$$x := \rho \cdot \cos(\theta) :$$
$$y := \rho \cdot \sin(\theta) :$$

$$I1 := simplify\left(\int_{\frac{x^2+y^2}{3}}^{\sqrt{4-x^2-y^2}} 1 dz\right) :$$

$$\int_0^{2\cdot\pi}\int_0^{\sqrt{3}} I1 \cdot \rho d\rho\, d\theta$$

$$\frac{19}{6}\pi$$

4. Find the triple integral

$$I = \int\int\int_V \frac{z\,dx\,dy\,dz}{\sqrt{1+x^2+y^2+z^2}},$$

where $V$ is bounded by the sphere:

$$x^2 + y^2 + z^2 = z.$$

**Computer solution.**
We shall achieve in Matlab 7.9:
```
>> rho phi th k
>> x=rho*cos(th)*sin(phi);
>> y=rho*sin(th)*sin(phi);
>> z=rho*cos(phi);
>> v=[rho th phi];
>> F=[x y z];
```

>> d=abs(simplify(det(jacobian(F,v))));
>> int(int(int(z*d/sqrt(1+x^2+y^2+z^2),rho,0,cos(phi)),th,
0,2*pi),phi,0,pi/2)
ans =
-(2*pi*(2*2^(1/2) - 3))/5

and with Mathcad 14:

$$f(x) := \begin{pmatrix} x_0 \cdot \cos(x_1) \cdot \sin(x_2) \\ x_0 \cdot \sin(x_1) \cdot \sin(x_2) \\ x_0 \cdot \cos(x_2) \end{pmatrix}$$

$$d(\rho, \theta, \varphi) := |Jacob(f(x), x)| \text{ substitute}, x_0 = \rho, x_1 = \theta, x_2 = \varphi \rightarrow -\rho^2 \cdot \sin(\varphi)$$

$$x(\rho, \theta, \varphi) := \rho \cdot \cos(\theta) \cdot \sin(\varphi) \qquad y(\rho, \theta, \varphi) := \rho \cdot \sin(\theta) \cdot \sin(\varphi) \qquad z(\rho, \varphi) := \rho \cdot \cos(\varphi)$$

$$u(\rho, \theta, \varphi) := x(\rho, \theta, \varphi)^2 + y(\rho, \theta, \varphi)^2 + z(\rho, \varphi)^2 \text{ simplify} \rightarrow \rho^2$$

$$Ioz := \int_0^{\frac{\pi}{2}} \left[ \int_0^{2 \cdot \pi} \left( \int_0^{\cos(\varphi)} \frac{\rho}{\sqrt{1+\rho^2}} \cdot \rho^2 \, d\rho \right) d\theta \right] \cdot \cos(\varphi) \cdot \sin(\varphi) \, d\varphi \rightarrow -\frac{2 \cdot \pi \cdot (2 \cdot \sqrt{2} - 3)}{5}$$

and using Mathematica 8:

```
x := ρ * Cos[θ] * Sin[φ]

y := ρ * Sin[θ] * Sin[φ]

z := ρ * Cos[φ]

d := Abs[Simplify[Det[D[{x, y, z}, {{ρ, θ, φ}}]]]]

Ii = Integrate[z / Sqrt[1 + x^2 + y^2 + z^2] * d, {φ, 0, Pi / 2},
    {θ, 0, 2 * Pi}, {ρ, 0, Cos[φ]}]
```

$$-\frac{2}{5} \left( -3 + 2\sqrt{2} \right) \pi$$

and with Maple 15:

$x := \rho \cdot \cos(\theta) \cdot \sin(\varphi)$ :
$y := \rho \cdot \sin(\theta) \cdot \sin(\varphi)$ :
$z := \rho \cdot \cos(\varphi)$ :
$with(VectorCalculus)$ :
$M, d := Jacobian([x, y, z], [\rho, \theta, \varphi], 'determinant')$ :

$dl := simplify(|d|)$

$$\left|\sin(\varphi)\,\rho^2\right|$$

$$factor\left(\int_0^{\frac{\pi}{2}} \int_0^{2\cdot\pi} \int_0^{\cos(\varphi)} \frac{z}{\sqrt{1+x^2+y^2+z^2}} \cdot dl\, d\rho\, d\theta\, d\varphi\right)$$

$$-\frac{2}{5}\pi\left(-3+2\sqrt{2}\right)$$

5. Compute

$$I = \int\int_{\Sigma} xyz\, d\sigma,$$

where $\sum$ is the surface of the paraboloid

$$x^2 + y^2 = z$$

between the planes $z = 0$, $z = 1$, with $x \geq 0$, $y \geq 0$.

6. Find the moment of inertia relative to the $Oz$- axis of the solid body, having its density

$$\delta(x, y, z) = k,$$

bounded by the ellipsoid

$$\frac{x^2}{a^2} + \frac{y^2}{b^2} + \frac{z^2}{c^2} = 1.$$

**Computer solution.**
We shall give a computer solution in Matlab 7.9:
>> **syms a b c rho phi th k**

```
>> x=a*rho*cos(th)*sin(phi);
>> y=b*rho*sin(th)*sin(phi);
>> z=c*rho*cos(phi);
>> v=[rho th phi];
>> F=[x y z];
>> d=simplify(det(jacobian(F,v)));
>> int(int(int((x^2+y^2+z^2)*k*d,phi,0,pi),th,0,2*pi),
rho,0,1)
   ans =
   -(4*pi*a*b*c*k*(a^2 + b^2 + c^2))/15
```

and with Mathcad 14:

$$f(a,b,c,x) := \begin{pmatrix} a \cdot x_0 \cdot \cos(x_1) \cdot \sin(x_2) \\ b \cdot x_0 \cdot \sin(x_1) \cdot \sin(x_2) \\ c \cdot x_0 \cdot \cos(x_2) \end{pmatrix}$$

$$d(a,b,c,\rho,\theta,\varphi) := |Jacob(f(a,b,c,x),x)| \; substitute, x_0 = \rho, x_1 = \theta, x_2 = \varphi \; \to \; -a \cdot \rho^2 \cdot b \cdot c \cdot \sin(\varphi)$$

$$x(a,\rho,\theta,\varphi) := a \cdot \rho \cdot \cos(\theta) \cdot \sin(\varphi) \quad y(b,\rho,\theta,\varphi) := b \cdot \rho \cdot \sin(\theta) \cdot \sin(\varphi) \quad z(c,\rho,\theta,\varphi) := c \cdot \rho \cdot \cos(\varphi)$$

$$Ioz(a,b,c,k) := \int_0^1 \int_0^{2 \cdot \pi} \int_0^\pi k \cdot \left( x(a,\rho,\theta,\varphi)^2 + y(b,\rho,\theta,\varphi)^2 + z(c,\rho,\theta,\varphi)^2 \right) \cdot d(a,b,c,\rho,\theta,\varphi) \, d\varphi \, d\theta \, d\rho$$

$$Ioz(a,b,c,k) \to -\frac{4 \cdot \pi \cdot a \cdot b \cdot c \cdot k \cdot \left( a^2 + b^2 + c^2 \right)}{15}$$

and using Mathematica 8:

```
In[31]:= x := a * ρ * Cos[θ] * Sin[φ]

In[32]:= y := b * ρ * Sin[θ] * Sin[φ]

In[33]:= z := c * ρ * Cos[φ]

In[34]:= d = Simplify[Det[D[{x, y, z}, {{ρ, θ, φ}}]]];

In[35]:= Ioz = Integrate[k * (x^2 + y^2 + z^2) * d, {φ, 0, Pi}, {θ, 0, 2 * Pi}, {ρ, 0, 1}]

Out[35]= -4/15 a b c (a^2 + b^2 + c^2) k π
```

and in Maple 15:

$x := a \cdot \rho \cdot \cos(\theta) \cdot \sin(\varphi)$ :
$y := b \cdot \rho \cdot \sin(\theta) \cdot \sin(\varphi)$ :
$z := c \cdot \rho \cdot \cos(\varphi)$ :
$with(\ VectorCalculus)$ :
$M,\ d := Jacobian(\ [x,\ y,\ z],\ [\rho,\ \theta,\ \varphi],\ 'determinant')$ :

$dl := simplify(d)$

$$-a \sin(\varphi)\ b \rho^2 c$$

$$factor\left( \int_0^1 \int_0^{2\cdot\pi} \int_0^{\pi} k \cdot (x^2 + y^2 + z^2) \cdot dl\ d\varphi d\theta\ d\rho \right)$$

$$-\frac{4}{15} \pi abck (b^2 + a^2 + c^2)$$

7. Find the moment of inertia relative to the $Oz$- axis of the homogeneous body

$$\frac{x^2}{a^2} + \frac{y^2}{b^2} \le \frac{z^2}{c^2},\ 0 \le z \le h.$$

8. Determine the coordinates of the centre of gravity of a homogeneous parabolic envelope

$$az = x^2 + y^2,\ 0 \le z \le a.$$

9. Compute the mass of the surface of the cube

$$\begin{cases} 0 \le x \le 1 \\ 0 \le y \le 1 \\ 0 \le z \le 1 \end{cases}$$

if the surface density at the point $M\ (x, y, z)$ is

$$\delta\ (x, y, z) = xyz.$$

10. Compute the surface integral of the first type:

$$I = \int \int_{\Sigma} \left( 2x + \frac{4}{3} y + z \right) d\sigma,$$

where $(\sum)$ is

$$\frac{x}{2} + \frac{y}{2} + \frac{z}{4} = 1,$$

with $x \geq 0$, $y \geq 0$, $z \geq 0$.

**Computer solution.**

We shall give a computer solution in Matlab 7.9:

```
>> syms x y
>> z=4*(1-x/2-y/3);
>> p=diff(z,x);
>> q=diff(z,y);
>> w=simple(sqrt(1+p^2+q^2)) ;
>> int(int(F*w,y,0,(-3/2)*(x-2)),x,0,2)
ans =
4*61^(1/2)
```

and with Mathcad 14:

$$z(x,y) := 4 \cdot \left( 1 - \frac{x}{2} - \frac{y}{3} \right) \qquad p(x,y) := \frac{d}{dx} z(x,y) \qquad q(x,y) := \frac{d}{dy} z(x,y)$$

$$F(x,y) := 2 \cdot x + \frac{4}{3} \cdot y + z(x,y)$$

$$I := \int_0^2 \int_0^{\left(-\frac{3}{2}\right) \cdot (x-2)} F(x,y) \cdot \sqrt{1 + p(x,y)^2 + q(x,y)^2} \; dy \, dx \to 4 \cdot \sqrt{61}$$

and using Mathematica 8:

```
In[1]:= z := 4 * (1 - x / 2 - y / 3)

In[2]:= F := 2 * x + 4 / 3 * y + z

In[3]:= p := D[z, x]

In[4]:= q := D[z, y]

In[5]:= w := Simplify[Sqrt[1 + p^2 + q^2]]

In[6]:= Integrate[w * F, {x, 0, 2}, {y, 0, (-3/2) * (x - 2)}]

Out[6]= 4 √61
```

and in Maple 15:

$$z := 4 \cdot \left( 1 - \frac{x}{2} - \frac{y}{3} \right):$$

$$p := \frac{\partial}{\partial x} z:$$

$$q := \frac{\partial}{\partial y} z:$$

$$w := simplify\left( \sqrt{1 + p^2 + q^2} \right):$$

$$F := 2 \cdot x + \frac{4}{3} \cdot y + z:$$

$$\int_0^2 \int_0^{\left( -\frac{3}{2} \right) \cdot (x-2)} w \cdot F \, dy \, dx$$

$$4\sqrt{61}$$

11.Compute the surface integral of the first type:

$$I = \int\int_{\Sigma} xyz \, d\sigma,$$

where $(\Sigma)$ is the surface of the paraboloid $x^2 + y^2 = z$, situated between the planes $z = 0$, $z = 1$, with $x \geq 0$, $y \geq 0$.

**Computer solution.**

We shall compute this surface integral of the first type, for the surface $(\Sigma)$ showed in Fig. 9.17

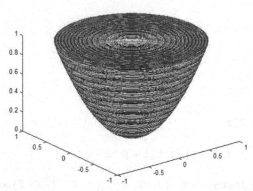

**Fig. 9.17**

using Matlab 7.9:

```
>> syms x y
>> z=x^2+y^2;
>> p=diff(z,x);
>> q=diff(z,y);
>> w=simple(sqrt(1+p^2+q^2));
>> F=x*y*z;
>> syms rho th
>> h=simple(subs(F*w,{x,y},{rho*cos(th),rho*sin(th)}));
>> int(int(h,th,0,pi/2)*rho,rho,0,1)
ans =
(25*5^(1/2))/336 - 1/1680
```

and with Mathcad 14:

$$z(x,y) := x^2 + y^2 \qquad p(x,y) := \frac{d}{dx}z(x,y) \qquad q(x,y) := \frac{d}{dy}z(x,y)$$

$$\underset{\wedge\wedge\wedge}{F}(x,y) := x \cdot y \cdot z(x,y)$$

$$w(x,y) := \sqrt{1 + p(x,y)^2 + q(x,y)^2} \text{ simplify } \rightarrow \sqrt{4 \cdot x^2 + 4 \cdot y^2 + 1}$$

$$x(\rho,\theta) := \rho \cdot \cos(\theta) \qquad y(\rho,\theta) := \rho \cdot \sin(\theta)$$

$$I := \int_0^1 \int_0^{\frac{\pi}{2}} F(x(\rho,\theta),y(\rho,\theta)) \cdot w(x(\rho,\theta),y(\rho,\theta)) \cdot \rho \; d\theta \; d\rho \rightarrow \frac{25 \cdot \sqrt{5}}{336} - \frac{1}{1680}$$

and using Mathematica 8:

In[1]:= z := x^2 + y^2

In[2]:= p := D[z, x]

In[3]:= q := D[z, y]

In[4]:= w = Simplify[Sqrt[1 + p^2 + q^2]]

Out[4]= $\sqrt{1 + 4x^2 + 4y^2}$

In[5]:= x := ρ * Cos[θ]

In[6]:= y := ρ * Sin[θ]

In[7]:= Integrate[w * x * y * z * ρ, {θ, 0, Pi/2}, {ρ, 0, 1}]

Out[7]= $\dfrac{-1 + 125\sqrt{5}}{1680}$

and in Maple 15:

$$z := x^2 + y^2 :$$
$$p := \frac{\partial}{\partial x} z :$$
$$q := \frac{\partial}{\partial y} z :$$
$$w := simplify\left(\sqrt{1 + p^2 + q^2}\right) :$$
$$x := \rho \cdot \cos(\theta) :$$
$$y := \rho \cdot \sin(\theta) :$$
$$\int_0^1 \int_0^{\frac{\pi}{2}} w \cdot x \cdot y \cdot z \cdot \rho \, d\theta \, d\rho$$

$$\frac{25}{336}\sqrt{5} - \frac{1}{1680}$$

12. Use the Stokes' formula to find the given integrals and then verify the
results by direct calculations:

$$\oint_\Gamma (x + y)^2 \, dx - (x - y)^2 \, dy,$$

where the curve $\Gamma$ is closed by the parabola $y = x^2$ and by the straight line $y = x$, passed in the directly sense.

13.Applying the Stokes' formula, find the given integrals and verify the results by direct calculations:

$$\oint_{ABCA} y^2 dx + z^2 dy + x^2 dz,$$

where $ABCA$ is the contour of the triangle $ABC$ with the vertices $A\,(a, 0, 0)$, $B\,(0, a, 0)$ and $C\,(0, 0, a)$.

14.Compute the line integral

$$I = \oint_C (y - z) \, dx + (z - x) \, dy + (x - y) \, dz,$$

where $C$ is the ellipse achieved by the intersection of the cylinder $x^2 + y^2 = 1$ with the plane $x + z = 1$, crossed so that the projection of curve $C$ on the plane $(xOy)$ to be positively oriented, using the following two methods:
    1) directly as a line integral,
    2) with the Stokes formula.
    **Computer solution.**
    We shall compute this line integral with the Stokes formula using Matlab 7.9:

```
>> syms x y z
>> F=[y-z z-x x-y];
>> rot=[diff(F(3),y)-diff(F(2),z) diff(F(2),z)-diff(F(3),x)
diff(F(2),x)-diff(F(1),y)]
   rot =
   [ -2, 0, -2]
>> z=1-x;
>> p=diff(z,x);
>> q=diff(z,y);
>> w=sqrt(1+p^2+q^2);
>> n=[-p/w -q/w 1/w];
>> v=dot(rot,n)*w;
>> syms rho th
>> k=subs(v,{x,y},{rho*cos(th),rho*sin(th)});
>> c=int(int(k*rho,rho,0,1),th,0,2*pi)
```

c =
(-4)*pi

and in Mathcad 14:

$$F(x,y,z) := \begin{pmatrix} y - z \\ z - x \\ x - y \end{pmatrix}$$

$$z(x,y) := 1 - x \qquad\qquad p(x,y) := \frac{d}{dx}z(x,y) \qquad q(x,y) := \frac{d}{dy}z(x,y)$$

$$w(x,y) := \sqrt{1 + p(x,y)^2 + q(x,y)^2}$$

$$n(x,y) := \begin{pmatrix} \dfrac{-p(x,y)}{w(x,y)} \\ \dfrac{-q(x,y)}{w(x,y)} \\ \dfrac{1}{w(x,y)} \end{pmatrix} \qquad rot(x,y,z) := \begin{pmatrix} \dfrac{d}{dy}F(x,y,z)_2 - \dfrac{d}{dz}F(x,y,z)_1 \\ \dfrac{d}{dz}F(x,y,z)_1 - \dfrac{d}{dx}F(x,y,z)_2 \\ \dfrac{d}{dx}F(x,y,z)_1 - \dfrac{d}{dy}F(x,y,z)_0 \end{pmatrix}$$

$$v(x,y,z) := \left(n(x,y)^T \cdot rot(x,y,z)\right) \cdot w(x,y)$$

$$k(\rho,\theta) := v(x,y,z) \ substitute, x = \rho \cdot \cos(\theta), y = \rho \cdot \sin(\theta) \ \to -4$$

$$\int_0^1 \int_0^{2 \cdot \pi} k(\rho,\theta) \cdot \rho \ d\theta \ d\rho \to -4 \cdot \pi$$

and in Mathematica 8:

```
In[24]:= F := {y - z, z - x, x - y}

In[25]:= rot := {D[F[[3]], y] - D[F[[2]], z], D[F[[1]], z] - D[F[[3]], x],
            D[F[[2]], z] - D[F[[1]], y]}

In[26]:= zz[x_, y_] := 1 - x

In[27]:= p := D[zz[x, y], x]

In[28]:= q := D[zz[x, y], y]

In[29]:= w := Sqrt[1 + p^2 + q^2]

In[30]:= n := {-p/w, -q/w, 1/w}

In[31]:= v =
        ({D[F[[3]], y] - D[F[[2]], z], D[F[[2]], z] - D[F[[3]], x],
            D[F[[2]], x] - D[F[[1]], y]} . {-p/w, -q/w, 1/w}) *w

Out[31]= -4

In[32]:= k := v /. {x → ρ*Cos[θ], y → ρ*Sin[θ]}

In[33]:= Integrate[k*ρ, {ρ, 0, 1}, {θ, 0, 2*Pi}]

Out[33]= -4 π
```

and in Maple 15:

*with*(*VectorCalculus*) :

*SetCoordinates*('*cartesian*'$_{x, y, z}$) :

$F := VectorField(\langle y - z, z - x, x - y \rangle)$ :

*rot* := *Curl*(*F*) :

$z := 1 - x$ :

$p := \dfrac{d}{dx} z$ :

$q := \dfrac{d}{dy} z$ :

$w := \sqrt{1 + p^2 + q^2}$ :

$n := VectorField\left(\left\langle -\dfrac{p}{w}, -\dfrac{q}{w}, \dfrac{1}{w} \right\rangle\right)$ :

$v := (rot.n) \cdot w$ :

$k := v\Big|_{x = \rho \cdot \cos(\theta), \, y = \rho \cdot \sin(\theta)}$ :

$k := \displaystyle\int_0^1 \int_0^{2\cdot\pi} k \cdot \rho \, d\theta \, d\rho$

$$-4\,\pi$$

15. Use the Gauss- Ostrogradski formula to compute the following surface integral:

$$I = \int\!\!\int_{\Sigma_+} x^2 dydz + y^2 dzdx + z^2 dxdy,$$

where ($\sum$) is the cube

$$0 \le x \le a, \; 0 \le y \le a, \; 0 \le z \le a.$$

**Computer solution.**

We shall have in Matlab 7.9:

>> **syms a x y z**
>> **P=x^2;Q=y^2;R=z^2;**
>> **w=diff(P,x)+diff(Q,y)+diff(R,z);**
>> **int(int(int(w,z,0,a),y,0,a),x,0,a)**

ans =

3*a^4

and using Mathcad 14:

$$P(x,y,z) := x^2 \qquad Q(x,y,z) := y^2 \qquad R(x,y,z) := z^2$$

$$w(x,y,z) := \frac{d}{dx}P(x,y,z) + \frac{d}{dy}Q(x,y,z) + \frac{d}{dz}R(x,y,z)$$

$$\int_0^a \int_0^a \int_0^a w(x,y,z)\, dz\, dy\, dx \to 3 \cdot a^4$$

and with Mathematica 8:

In[6]:= **P := x^2**

In[7]:= **Q := y^2**

In[8]:= **R := z^2**

In[9]:= **w := D[P, x] + D[Q, y] + D[R, z]**

In[10]:= **Integrate[w, {z, 0, a}, {y, 0, a}, {x, 0, a}]**

Out[10]= 3 a⁴ → $3\,a^4$

and using Maple 15:

$$P := x^2 :$$
$$Q := y^2 :$$
$$R := z^2 :$$
$$w := \frac{d}{dx}P + \frac{d}{dy}Q + \frac{d}{dz}R :$$
$$\int_0^a \int_0^a \int_0^a w\, dz\, dy\, dx$$

$$3\,a^4$$

16.Apply the Gauss- Ostrogradski formula to compute the surface integral:

$$I = \int\!\!\int_{\Sigma_+} x\,dy\,dz + y\,dz\,dx + z\,dx\,dy,$$

where $(\Sigma)$ is the cylinder $x^2 + y^2 = a^2$, $z \in [-H, H]$.

17.Evaluate the following surface integral of the second type:

$$\int\int_{\Sigma} ydzdx,$$

$\Sigma$ being the external side of the ellipsoid:

$$\frac{x^2}{a^2} + \frac{y^2}{b^2} + \frac{z^2}{c^2} - 1 = 0.$$

**Computer solution.**

We shall give a computer solution in Matlab 7.9:

```
>> syms a b c rho phi
>> x=a*sin(th)*cos(phi);
>> y=b*sin(th)*sin(phi); z=c*cos(th);
>> A=det(jacobian([y z],[th phi]));
>>B=det(jacobian([z x],[th phi]));
>> C=det(jacobian([x y],[th phi])); s=simplify
(sqrt(A^2+B^2+C^2));
>> c2=B/s;
>> int(int(y*c2*s,th,0,pi),phi,0,2*pi)
ans =
(4*pi*a*b*c)/3
```

and using Mathematica 8:

```
In[11]:= x := a * Sin[θ] * Cos[φ]

In[12]:= y := b * Sin[θ] * Sin[φ]

In[13]:= z := c * Cos[θ]

In[14]:= Aa := Simplify[Det[D[{y, z}, {{θ, φ}}]]]

In[15]:= Bb := Simplify[Det[D[{z, x}, {{θ, φ}}]]]

In[16]:= Cc := Simplify[Det[D[{x, y}, {{θ, φ}}]]]

In[17]:= s := Simplify[Sqrt[Aa^2 + Bb^2 + Cc^2]]

In[20]:= c2 := Bb / s

In[19]:= Integrate[y * c2 * s, {θ, 0, Pi}, {φ, 0, 2 * Pi}]

Out[19]=  4/3 a b c π
```

and in Maple 15:

$x := a \cdot \sin(\theta) \cdot \cos(\varphi)$ :
$y := b \cdot \sin(\theta) \cdot \sin(\varphi)$ :
$z := c \cdot \cos(\theta)$ :
$with( VectorCalculus)$ :
$M1, A := Jacobian( [y, z], [\theta, \varphi], 'determinant')$ :
$M2, B := Jacobian( [z, x], [\theta, \varphi], 'determinant')$ :
$M3, C := Jacobian( [x, y], [\theta, \varphi], 'determinant')$ :
$u := simplify( sqrt(A^2 + B^2 + C^2))$ :
$c2 := \dfrac{B}{u}$ :

$$\int_0^{\pi} \int_0^{2 \cdot \pi} y \cdot c2 \cdot u \, d\varphi \, d\theta$$

$$\dfrac{4}{3} \pi b c a$$

18.Find the flux of the vector field

$$\vec{F}(x, y, z) = 2x \, \vec{i} + y \, \vec{j} - z \, \vec{k}$$

across the surface

$$(\Sigma) : y^2 + z^2 = ax, \ 0 \le x \le a$$

according to the normal $\vec{n}$ on the surface, which makes an acute angle with the negative $Oz$ semi-axis.

19.Find the centre of gravity of the body $\Omega$, by the density $\mu = k$, bounded by the cone $x^2 + y^2 = z^2$ and by the planes $z = 0$ and $z = h$.

**Computer solution.**
We shall find the centre of gravity of the body $\Omega$ (see Fig. 9.18)

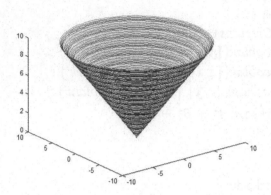

Fig. 9.18

using Matlab 7.9:
>> **syms rho th h k**
>> **x=rho*cos(th);**
>> **y=rho*sin(th);**
>> **I1=int(int(int(k,z,sqrt(x^2+y^2),h)*rho,th,0,2*pi),**
rho,0,h);
>> **I2=int(int(int(k,z,sqrt(x^2+y^2),h)*x*rho,th,0,2*pi),**
rho,0,h);
>> **I3=int(int(int(k,z,sqrt(x^2+y^2),h)*y*rho,th,0,2*pi),**
rho,0,h);
>> **I4=int(int(int(k*z,z,sqrt(x^2+y^2),h)*rho,th,0,2*pi),**
rho,0,h);
>> **xg=I2/I1**
xg =
0
>> **yg=I3/I1**
yg =
0
>> **zg=I4/I1**
zg =
(3*h)/4
and in Mathcad 14:

$$x(\rho,\theta) := \rho \cdot \cos(\theta) \qquad y(\rho,\theta) := \rho \cdot \sin(\theta)$$

$$xg(h,k) := \frac{\displaystyle\int_0^h \int_0^{2\cdot\pi} \left(\int_{\sqrt{x(\rho,\theta)^2+y(\rho,\theta)^2}}^h k\, dz\right)\cdot \rho \cdot x(\rho,\theta)\, d\theta\, d\rho}{\displaystyle\int_0^h \int_0^{2\cdot\pi} \left(\int_{\sqrt{x(\rho,\theta)^2+y(\rho,\theta)^2}}^h k\, dz\right)\cdot \rho\, d\theta\, d\rho} \qquad xg(h,k) \to 0$$

$$yg(h,k) := \frac{\displaystyle\int_0^h \int_0^{2\cdot\pi} \left(\int_{\sqrt{x(\rho,\theta)^2+y(\rho,\theta)^2}}^h k\, dz\right)\cdot \rho \cdot y(\rho,\theta)\, d\theta\, d\rho}{\displaystyle\int_0^h \int_0^{2\cdot\pi} \left(\int_{\sqrt{x(\rho,\theta)^2+y(\rho,\theta)^2}}^h k\, dz\right)\cdot \rho\, d\theta\, d\rho} \qquad yg(h,k) \to 0$$

$$zg(h,k) := \frac{\displaystyle\int_0^h \int_0^{2\cdot\pi} \int_{\sqrt{x(\rho,\theta)^2+y(\rho,\theta)^2}}^h k\cdot z\, dz \cdot \rho\, d\theta\, d\rho}{\displaystyle\int_0^h \int_0^{2\cdot\pi} \left(\int_{\sqrt{x(\rho,\theta)^2+y(\rho,\theta)^2}}^h k\, dz\right)\cdot \rho\, d\theta\, d\rho} \qquad zg(h,k) \to \frac{3\cdot h}{4}$$

and with Mathematica 8:

```
In[19]:= x := ρ * Cos[θ]

In[20]:= y := ρ * Sin[θ]

In[21]:= I1 := Integrate[Integrate[k, {z, Sqrt[x^2 + y^2], h}] * ρ, {θ, 0, 2 * Pi}, {ρ, 0, h}]

In[22]:= I2 := Integrate[Integrate[k, {z, Sqrt[x^2 + y^2], h}] * x * ρ, {θ, 0, 2 * Pi}, {ρ, 0, h}]

In[23]:= I3 := Integrate[Integrate[k, {z, Sqrt[x^2 + y^2], h}] * y * ρ, {θ, 0, 2 * Pi}, {ρ, 0, h}]

In[24]:= I4 := Integrate[Integrate[k * z, {z, Sqrt[x^2 + y^2], h}] * ρ, {θ, 0, 2 * Pi}, {ρ, 0, h}]

In[25]:= xg = I2 / I1

Out[25]= 0

In[26]:= yg = I3 / I1

Out[26]= 0

In[30]:= zg = I4 / I1

Out[30]=      3 h
        ────────────────────────────
        4 (5 - 4 HeavisideTheta[h])
```

and with Maple 15:

$x := \rho \cdot \cos(\theta) :$

$y := \rho \cdot \sin(\theta) :$

$I1 := simplify\left(\int_0^h \left(\int_0^{2 \cdot \pi} \left(\int_{\sqrt{x^2+y^2}}^h kdz\right) \cdot \rho \, d\theta\right) d\rho\right) :$

$I2 := simplify\left(\int_0^h \left(\int_0^{2 \cdot \pi} \left(\int_{\sqrt{x^2+y^2}}^h kdz\right) \cdot x \cdot \rho \, d\theta\right) d\rho\right) :$

$I3 := simplify\left(\int_0^h \left(\int_0^{2 \cdot \pi} \left(\int_{\sqrt{x^2+y^2}}^h kdz\right) \cdot y \cdot \rho \, d\theta\right) d\rho\right) :$

$I4 := simplify\left(\int_0^h \left(\int_0^{2 \cdot \pi} \left(\int_{\sqrt{x^2+y^2}}^h k \cdot z \, dz\right) \cdot \rho \, d\theta\right) d\rho\right) :$

$xg := \dfrac{I2}{I1}$

$\qquad\qquad\qquad\qquad\qquad\qquad\qquad 0$

$yg := \dfrac{I3}{I1}$

$\qquad\qquad\qquad\qquad\qquad\qquad\qquad 0$

$zg := \dfrac{I4}{I1}$

$\qquad\qquad\qquad\qquad\qquad -\dfrac{3}{4}\, \dfrac{h}{-3 + 2 \, \text{signum}(\Re(h))}$

20.If $\sum$ is the external side of the surface:

$$\frac{x^2}{a^2} + \frac{y^2}{b^2} - \frac{z^2}{c^2} = 0 .$$

between the planes $z = 0$ and $z = b$, compute the surface integral:

$$J = \int\int_{\Sigma} x^2 dydz + y^2 dxdz + z^2 dxdy.$$

# References

[1]   Apostol, T.M.: Linear algebra: a first course, with applications to differential equations. Wiley, New York (1997)
[2]   Armeanu, I., Petrehus, V.: Matematici avansate cu aplicatii. Sitech, Craiova (2009)
[3]   Avramescu, C.: Ecuatii diferentiale si sisteme dinamice, note de curs, Universitatea din Craiova, Facultatea de Matematica- Informatica (1994)
[4]   Avramescu, C., Vladimirescu, C.: Ecuatii Diferentiale si Integrale. Reprografia Universitatii din Craiova (2003)
[5]   Avramescu, C., Vladimirescu, C.: Curs de Calcul Stiintific. Reprografia Universității din Craiova (2002)
[6]   Axler, S.J.: Linear algebra done right. Springer, New York (1997)
[7]   Bapat, R.B.: Linear algebra and linear models. Springer, Heidelberg (2000)
[8]   Baranenkov, G., Demidovich, B., Efimenko, V., Kogan, S., Lunts, G., Porshneva, E., Sycheva, E., Frolov, S., Shostak, R., Yanpolsky, A.: 3193 Problems in Mathematical Analysis,
      http://mpec.sc.mahidol.ac.th/RADOK/physmath/mat12/start.htm
[9]   Beardon, A.F.: Algebra and geometry. Cambridge University Press, New York (2005)
[10]  Bărbulescu, I., Barbulescu, A.: Algebra si Analiza matematica. Rapsodia Romana, Craiova (1995)
[11]  Bourne, M.: Area Under a Curve,
      http://www.intmath.com/Applications-integration/2_
      Area-under-curve.php
[12]  Brannan, J.R., Boyce, W.E.: Differential equations: an introduction to modern methods and applications. John Wiley, Hoboken (2007)
[13]  Canuto, C., Tabacco, A.: Mathematical Analysis I. Springer, Heidelberg (2008)

[14] Chiriac, V., Chiriac, M.: Probleme de Algebra. Tehnica, Bucuresti (1977)

[15] Chirita, S.: Probleme de matematici superioare. Didactica si Pedagogica, Bucuresti (1989)

[16] Chitescu, I., Cristescu, R., Grigore, G., Gussi, G., Halanay, A., Jurchescu, M., Marcus, S.: Dictionar de Analiza matematica. Stiintifica si Enciclopedica, Bucuresti (1989)

[17] Fulton, S.R.: Integrals of Rational Functions, http://people.clarkson.edu/~sfulton/ma132/parfrac.pdf

[18] Emanouvilov, O.: Representation of Functions as Power Series, http://orion.math.iastate.edu/vika/cal3_files/Lec26.pdf

[19] Ghinea, M., Fireteanu, V.: Matlab: Calcul numeric- Grafică-Aplicatii. Teora, Bucuresti (1998)

[20] Gârban, V., Udrea, C.: Ecuaţii diferentiale si ecuatii cu derivate partiale de ordinul I. Academia Tehnica Militara, Bucuresti (2001)

[21] Giaquinta, M., Modica, G.: Mathematical Analysis. An Introduction to Functions of Several Variables. Springer, Heidelberg (2009)

[22] Hazewinkel, M.: Encyclopaedia of Mathematics Supplement I. Springer, Heidelberg (2011), http://eom.springer.de/S/s091350.htm

[23] Iatan, I.: Advances Lectures on Linear Algebra with Applications. Lambert Academic Publishing AG& Co. KG, Saarbrücken (2011)

[24] Iatan, I.: Îndrumător de laborator în Matlab 7.0. Conspress, Bucureşti (2009)

[25] Ivanovici, M.: Ecuaţii diferenţiale. Reprografia Universităţii din Craiova (1993)

[26] Khovanskïi, A., Varchenko, A., Vassiliev, V.: Geometry of differential equations. American Mathematical Society (1998)

[27] King, A.C., Billingham, J., Otto, S.R.: Differential equations: linear, nonlinear, ordinary, partial. Cambridge University Press, New York (2003)

[28] Kokoska, S.: Applications of the Definite Integral, Department of Mathematics, Computer Science and Statistics. Bloomsburg University, Bloomsburg, Pennsylvania 17815, http://facstaff.bloomu.edu/skokoska/apps.pdf

[29] Krantz, S.G., Chen, G., Aron, R.M.: Journal of Mathematical Analysis and Applications 381(1) (2011)

[30] Larson, R., Hostetler, R.P., Edwards, B.H.: Calculus with Analytic Geometry. Houghton Mifflin Co., Boston (2006)

[31] Lay, D.C.: Linear Algebra and Its Applications. Addison- Wesley Publishing Company, USA (1994)

[32] McMahon, D.: Linear algebra demystified. McGraw-Hill (2006)

[33] Newns, W.F.: Functional Dependence. The American Mathematical Monthly 74, 911–920 (1967)

[34] Nicholson, W.K.: Linear Algebra and with Applications. PWS Publishing Company, Boston (1995)

[35] Nicolescu, C.P.: Analiza matematica. Albatros, Bucuresti (1987)

[36] Olariu, V., Olteanu, O.: Analiza matematica. Semne, Bucuresti (1999)

[37] Paltineanu, G.: Analiza matematica. Calcul Diferenţial. Agir, Bucuresti (2002)

[38] Paltineanu, G.: Analiza matematica. Calcul Integral. Agir, Bucuresti (2004)

[39] Paltineanu, G., Matei, P.: Ecuatii diferentiale si ecuatii cu derivate partiale cu aplicatii. Matrix Rom, Bucureşti (2007)

[40] Polyanin, A.D., Manzhirov, A.V.: Handbook of Mathematics for Engineers and Scientists. Chapman & Hall/CRC Press (2006)

[41] Popa, I.: Analiza matematica. Calcul diferential. MatrixRom, Bucuresti (2000)

[42] Popa, I.: Analiza matematica. Calcul integral. MatrixRom, Bucuresti (2001)

[43] Popescu, A.: Mathematical Analysis. Conspress, Buc. (2009)

[44] Postelnicu, V., Coatu, S.: Mica enciclopedie matematica. Tehnica, Bucuresti (1980)

[45] Postolache, M.: Metode numerice. Fair Parteners, Bucuresti (2003)

[46] Repka, J.: Calculus with analytic geometry. Wm. C. Brown Publishers, Iowa (1994)

[47] Richter, W.D.: Generalized spherical and simplicial coordinates. Journal of Mathematical Analysis and Applications 336(2), 1187–1202 (2007)

[48] Roşculeţ, M.: Analiză matematica. Didactica si Pedagogica, Bucuresti (1984)

[49] Rosculet, M., Toma, G., Masgras, V., Stanciu, V., Braileanu, E., Dimcevici, N.: Probleme de analiza matematica. Tehnica, Bucuresti (1993)

[50] Schenck, H.: Computational algebraic geometry. Cambridge University Press, New York (2003)

[51] Schoen, R., Yau, S.-T.: Lectures on differential geometry. International Press, Cambridge (1994)

[52] Siretchi, G.: Calcul diferential si integral, vol. 2. Stiintifica si Enciclopedica, Bucuresti (1985)

[53] Stanasila, O.: Analiza Matematica. Didactica si Pedagogica, Bucuresti (1981)

[54] Stein, S., Barcellos, A.: Calculus and analytic geometry. McGraw-Hill, New York (1992)

[55] Strang, G.: Linear Algebra and Its Applications. Thomson Brooks/Cole, USA (2006)

[56] Toma, I., Soare, M.V., Teodorescu, P.P.: Ecuaţ ii diferenţiale cu aplicaţii în mecanica construcţiilor. Tehnică, Bucureşti (1999)

[57] Toma, I., Iatan, I.: Analiză numerică. Curs, aplicaţii, algoritmi în pseudocod şi programe de calcul. MatrixRom, Bucureşti (2005)

[58] Toma, I.: Analiză matematică. Conspress, Bucureşti (2010)

[59] Tuduce, R.A.: Signal Theory. Bren, Bucharest (1998)

[60] Trench, W.F.: Introduction to Real Analysis. Prentice Hall (2002)

[61] Ving, P.K.: Calculus of One Real Variable. A Tutorial, http://www.phengkimving.com/calc_of_one_real_var/ contents.htm#chap_11

[62] Zorich, V.A., Cooke, R.: Mathematical Analysis II. Springer, Heidelberg (2004)

# List of Symbols

$\sum_{n\geq 1} a_n$, 3

$\sum_{n\geq 1} (-1)^{n+1} a_n$, 14

$\sum_{n\geq 1} a_n x^n$, 41

$\sum_{n\geq 1} \frac{f^{(n)}(a)}{n!} (x-a)^n$, 49

$\sum_{n\geq 1} \frac{f^{(n)}(0)}{n!} x^n$, 50

$\lim_{n\to\infty} \sum_{k=0}^{n} a_k x^k$, 60

$f'_x$, 72

$f''_x$, 72

$\mathrm{d}f$, 83

$\mathrm{d}^2 f$, 83

$\mathrm{d}f(x_0, y_0)(h, k)$, 91

$\mathrm{d}f(x_0, y_0, z_0)(h, k, r)$, 91

$J_f(a)$, 93

$\frac{D(f_1, f_2, f_3)}{D(x,y,z)}$, 97

$f(x_0 + h, y_0 + k)$, 127

$\frac{\partial z}{\partial \vec{s}}$, 157

$\operatorname{grad} \varphi$, 163

$\operatorname{div} \vec{v}$, 163

$\operatorname{rot} \vec{v}$, 164

$\int \frac{P(x)}{Q(x)} \mathrm{d}x$, 245

$\int R(\sin x, \cos x) \mathrm{d}x$, 251

$\int x^m \left(ax^n + b\right)^p \mathrm{d}x$, 252

$x_G$, 277, 433

$y_G$, 277, 433

$P\left(\lambda\right)$, 349

$y_o\left(x\right)$, 357

$y_p\left(x\right)$, 357

$Y^o\left(x\right)$, 378

$Y^p\left(x\right)$, 378

$\int_{AB} f\left(x, y, z\right)\mathrm{d}s$, 395

$\rho\left(x, y, z\right)$, 396

$\int_C P\left(x, y\right)\mathrm{d}x + Q\left(x, y\right)\mathrm{d}y$, 406

$\overrightarrow{F}\left(x, y, z\right)$, 407

$\oint_C x\mathrm{d}y$, 407

$\int\int_D f\left(x, y\right)\mathrm{d}x\mathrm{d}y$, 421

$I_x$, 436, 510

$\int\int\int_V f\left(x, y, z\right)\mathrm{d}x\mathrm{d}y\mathrm{d}z$, 475

$\int_\Sigma f\left(x, y, z\right)\mathrm{d}\sigma$, 512

$\sum_+$, 520

$\sum_-$, 520

$\overrightarrow{n}$, 520

$\int\int_{\Sigma_+} P\left(x, y, z\right)\mathrm{d}y\mathrm{d}z + Q\left(x, y, z\right)\mathrm{d}z\mathrm{d}x + R\left(x, y, z\right)\mathrm{d}x\mathrm{d}y$, 520

$\cos\alpha$, $\cos\beta$, $\cos\gamma$, 520, 521

# Index

Cauchy sequence, 1
partial sums, 3
sum of the series, 3
alternating series, 14
Leibnitz's test, 14
comparison test I, 16
root test, 18
ratio test, 21
Raabe's and Duhamel's test, 26
comparison test II, 29
comparison test III, 30
power series, 41

radius of convergence, 41
interval of convergence, 41
set of convergence, 42
Taylor series, 49
Mac Laurin series, 50
sum of power series, 60
partial derivatives, 71
Schwarz's criterion, 72
differentiable, 83
approximate calculations, 90

Jacobian matrix, 93
functional determinant, 97
homogeneous function, 99
composite functions, 102
change of variables, 118
Taylor's formula, 126
Mac Laurin's formula, 127
derivative in a given direction, 157
level surfaces, 158
level straight lines, 158
scalar field, 158
vector field, 162
gradient, 163
divergence, 163
rotation (curl), 163
potential, 164
derivative of implicit
functions, 187
differentiation of implicit
functions, 193
system of implicit functions, 203
functionally dependent, 209
extreme, 213

## 580    Index

positive definite matrix, 214
negative definite matrix, 214
conditional extremum, 214
Lagrange function, 214
integrals of rational functions, 245
integrating trigonometric
functions, 251
integrating irrational
functions, 252
binomial integral, 252
area, 260, 265, 430
arc length of a curve, 269
volumes of solids, 276, 438, 493
centre of gravity, 277, 433, 499
improper integrals, 280
comparison criterion, 293
parameter integrals, 296
Cauchy problem, 317
succesive approximation
method, 317
family of curves, 321
general solution, 321
particular solution, 321
singular solution, 321
envelope, 321
equation with separable
variables, 322
homogeneous equation, 324
nonhomogeneous equation, 330
total differential equation, 332
exact differential equation, 332
Bernoulli's equation, 339
Riccati's equation, 340
Lagrange's equation, 342
Clairaut's equation, 346
higher order differential
equation, 349
characteristic polynomial, 349
nonhomogeneous linear differential equation, 357

method of variation of
constants, 357
method of the undetermined
coefficients, 360
Euler's equation, 367
system of first order differential
equations, 369
characteristic equation of the
system, 371
method of characteristic
equation, 371
elimination method, 373
nonhomogeneous system, 377
smooth curve, 395
element of arc, 395
line integral of the first type, 395
linear density, 396
integral of the second type, 406
variable force, 407
mechanical work, 407
counter-clockwise, 407
double integral, 421
mass, 431, 485
moments of inertia, 436, 509
polar coordinates, 441
generalized polar coordinates, 445
Riemann- Green formula, 447
triple integral, 475
sferical coordinates, 477
cylindrical coordinates, 480
surface integral of the first
type, 512
surface integral of the second
type, 520
unit normal vector, 520
direction cosines, 521
flux, 524
Gauss-Ostrogradski formula, 529
Stokes formula, 540
circulation, 545